SCIENTIFIC, HEALTH AND SOCIAL ASPECTS OF THE FOOD INDUSTRY

Edited by **Benjamin Valdez,
Michael Schorr** and **Roumen Zlatev**

INTECHOPEN.COM

Scientific, Health and Social Aspects of the Food Industry

Edited by Benjamin Valdez, Michael Schorr and Roumen Zlatev

CBS, Edition **2017**

Published by InTech

Janeza Trdine 9, 51000 Rijeka, Croatia

Notice
Statements and opinions expressed in the chapters are these of the individual contributors and not necessarily those of the editors or publisher. No responsibility is accepted for the accuracy of information contained in the published chapters. The publisher assumes no responsibility for any damage or injury to persons or property arising out of the use of any materials, instructions, methods or ideas contained in the book.

Publishing Process Manager Masa Vidovic

Technical Editor Teodora Smiljanic

Cover Designer InTech Design Team

Additional hard copies can be obtained from orders@intechopen.com

Scientific, Health and Social Aspects of the Food Industry,
Edited by Benjamin Valdez, Michael Schorr, Roumen Zlatev
p. cm.
ISBN 978-953-307-916-5

Contents

Contents

Contents

Preface

In a prehistoric era, human beings were moving through inhospitable grounds, obtaining their daily sustenance by hunting and gathering fruits, seeds and roots. Massive food production and supply started with the agricultural revolution, paralleled with the development in the fertile valley of the rivers Tigris-Euphrates in Mesopotamia and the Nile in pharaonic Egypt. Joseph, the son of Jacob, organized the food stores and saved the Egyptian people during famine.

This book presents the wisdom, knowledge and expertise of the food industry that ensures the supply of food to maintain the health, comfort, and wellbeing of humankind. The global food industry has the largest market: the world population of seven billion people.

This book pioneers life-saving innovations and assists in the combat against world hunger and food shortages that threaten human essentials, such as water and energy supply. Floods, droughts, fires, storms, climate change, global warming and greenhouse gas emissions can be devastating, altering the environment and, ultimately, the production of foods.

This volume is well-organized into four parts:

- Scientific and technological aspects
- Social and economical issues
- Health aspects
- Quality control

It comprises 23 chapters, arranged in a hierarchical sequence, starting with the seminal basics of food science and technology developments, methods for food processing and other food industry related activities. It continues with the influential role food plays in the global society and economy, emphasizing the relevance of food to human health. Lastly, quality control of food in all its stages of production: manufacturing, packaging and marketing. Special articles deal with foods as medicine, particularly, the benefits of wine to maintain health of the body and soul. Experts from industry and academia, as well as food producers, designers of food processing equipment, and corrosion practitioners have written special chapters for this rich compendium, based on their encyclopedic knowledge and practical experience. This is a multi-authored book. The writers, who come from diverse areas of food science and technology, enrich this volume by presenting different approaches and orientations.

The book will be useful to chemists, biologists, chemical engineers, and economists who design, build and operate plants for food production. Also maintenance personnel, materials, and mechanical engineers should benefit from this compendium. University professors, researchers, and their undergraduate and postgraduate students will use this varied book for the preparation of their thesis on food-related projects.

Nowadays, people follow the international normative for environmental protection, and the food industry must take care to use environmentally friendly methods in the processing steps.

This volume is illustrated with hundreds of compact tables, well-drawn diagrams, and clear graphs and photographs. Also valuable are the thousands of scientific and technical references in the book from the international literature displaying the latest advances in food science and technology.

Finally, it is our pleasant duty to express our sincere thanks to the authors for providing their important contributions to the food industry and to this book.

Benjamin Valdez, Michael Schorr and Roumen Zlatev
Institute of Engineering
Universidad Autónoma Baja California, Mexicali,
México

Part 1

Scientific and Technological Aspects

A Discussion Paper on Challenges and Proposals for Advanced Treatments for Potabilization of Wastewater in the Food Industry

D. B. Luiz[1,2], H. J. José[1] and R. F. P. M. Moreira[1]
[1]*Federal University of Santa Catarina*
[2]*Embrapa Fisheries and Aquaculture*
Brazil

1. Introduction

The World Commission on Environment and Development (WCED, 1987, apud Burkhard et al., 2000) defines sustainable development as "development that meets the needs of the present without compromising the ability of future generations to meet their own needs". Sustainable production is a realistic dilemma in food industries, especially in slaughterhouses and the meat processing industry due to the many types of processes involved (Oliver et al., 2008; Casani et al., 2005).

Unfortunately, in many cases the implementation of environmental actions for industrial purposes depends on large-scale economic incentives (Cornel et al., 2011). Concomitantly, the increased costs of fresh water, wastewater treatment and waste disposal are important economic incentives for wastewater reuse and biomass-to-energy actions in industries. Hence, environmental, financial and social sustainability must be achieved together. Since pollution is predominantly due to human behavior, social sustainability requires the commitment of managers, workers and consumers (Burkhard et al., 2000).

Water scarcity is a reality in many world regions. Water pollution and overexploitation, climate change, urbanization, industrialization and increases in the world population and, consequently, food consumption and production are the main factors aggravating the global fresh water crisis (EPA, 2004; Clevelario et al., 2005; Khan et al., 2009; Luiz et al., 2009, 2011). Hence, sustainability in water and wastewater management is a current requirement of industries in order to promote the minimization of fresh water consumption and reduction of wastewater production preserving high-quality groundwaters (Cornel et al., 2011).

Concerning the water consumption in food industries, this input is mostly used for cleaning and disinfection, cooling and heating (Oliver et al., 2008), and also for processing food for human consumption and for sanitary uses. International organizations (e.g. Codex Alimentarius, 2001, 2007) have recognized and stimulated the implementation of direct and indirect wastewater reuse and indirect potable reuse with techniques that take into account hygienic concerns (Casani et al., 2005) to avoid risking adverse effects on the product integrity, the environment and the health of workers and customers. However, these procedures will be universal in the near future.

Currently, conventional wastewater techniques are not satisfactory to produce reuse wastewater with drinking water quality, further procedures being necessary. Advanced oxidation processes (AOPs) are the most suitable tertiary treatment. AOPs are based on chemical oxidation and can degraded any kind of organic matter with highly reactive and non-selective radicals, mainly hydroxyl radicals ($\bullet OH$) (Oller et al., 2010; Luiz et al., 2010). AOPs can involve a combination of strong oxidants (e.g. O_3 and H_2O_2), ultraviolet (UV), semiconductor catalysts (e.g. TiO_2 and ZnO) and ultrasound (15 kHz to 1 MHz), and the most common examples are: H_2O_2/UV, H_2O_2/O_3, $H_2O_2/O_3/UV$, O_3/UV, TiO_2/UV, H_2O_2/catalyst, Fenton (Matilainen and Sillanpää, 2010).

There are many studies and reviews on advanced processes to treat secondary wastewater from industries which also address the challenges and limitations to water reuse and the hygienic concerns especially in food industries. However, the challenges associated with treating a secondary wastewater to provide drinking water quality become clear when real wastewater is treated in batch and in pilot scale. The limitations are several, particularly in food industries. Hence, this review aims to present an overview and a discussion regarding the challenges and proposals related to the advanced treatment of food industry wastewater to provide drinking water quality, elucidating some experimental and theoretical questions surrounding such processes. Valuable experimental advice which is not usually found in research papers will be provided, for instance, which oxidation treatment should be chosen, which initial experiments should be carried, and which methodology should be followed to evaluate the kinetics constants in different situations will be provided, along with methodology issues. Some previous papers by the authors will be discussed and also some unpublished data on experiences in slaughterhouses and the meat processing industry. The oxidation treatments discussed are ozonation and many AOPs; ultraviolet treatment will also be addressed.

Due to the wide variety of processes and food products, food-processing wastewater can be a complex mixture, and this must be taken into account when considering the recycling, reuse, reconditioning for recycling or reuse, treatment or disposal of water. It is known that the segregation of wastewater streams in at an industrial plant – separating the effluent of each process – or at least the combining of the most similar streams in terms of physical-chemical and microbiological characteristics, enables an optimal treatment for each type of wastewater, energy savings, greater efficiency and lower cost of disposal and reuse. Nevertheless, in most plants the wastewater from all process – including from the toilets – are collected together, making the treatment difficult and costly. Hence, ascertaining the basic nature of food industry wastewater and its variability is the first challenge. This information is important to selecting the best combination of processes through which to treat the effluent.

Besides the need to remove minerals (by filtration) and organic matter, it is also necessary eliminate nitrogen compounds, especially in the dairy and meat industries. Advanced oxidation processes can remove simultaneously organic matter and nitrate; however, there are many parameters that should be taken into account, e.g., concentration of organic compounds and oxygen-free media. Some industrial plants also need to remove a specific recalcitrant organic compound which can be oxidized by advanced treatments; however, the effectiveness of the treatment is mainly dependent on whether the oxidant is selective or non-selective, the presence of oxidant scavengers and the dosage of oxidant.

Given that the growing demand for water, limited access to water in some regions and increased concern regarding the environmental impact of industrial activities on the environment are aspects driving the research on and implementation of water reuse in industrial plants, the theme discussed in this review will be of interest to several fields of chemistry, as well as food, chemical and environmental engineering.

2. UV radiation

UV radiation can degrade organic compounds via two routes: direct photolysis and photo-oxidation via radical generation. In the first route, direct photolysis or photodegradation, the efficiency is directly related to the ability of the target organic compound to absorb UV radiation at the wavelength used (λ) (Beltran et al., 1993). The UV absorption leads to direct excitation and the breakdown of organic pollutants (Rincón; Pulgarin, 2006). Lau et al. (2007) leading to the formation of excited radicals (R•) by UV excitation of organic compounds (RH). These radicals can be converted into stable molecules (dimers) via a dimerization process (Equations 1-6) which is favored by the presence of O_2 in water. If the free radical chain is interfered with (termination reactions, Equations 4-6), UV treatment may be less effective (Lau et al., 2007).

Initiation:

$$RH \xrightarrow{\ hv\ } R^{\bullet} + H^{\bullet} \tag{1}$$

Propagation:

$$R^{\bullet} + O_2 \longrightarrow ROO^{\bullet} \tag{2}$$

$$RH + ROO^{\bullet} \longrightarrow ROOH + R^{\bullet} \tag{3}$$

Termination:

$$ROO^{\bullet} + ROO^{\bullet} \longrightarrow ROOR + O_2 \tag{4}$$

$$ROO^{\bullet} + R^{\bullet} \longrightarrow ROOR \tag{5}$$

$$R^{\bullet} + R^{\bullet} \xrightarrow{\ hv\ } RR \tag{6}$$

According to Beltran-Heredia et al. (2001), for pH <7.0, the rate of photodecomposition increases with increasing pH, which may be due to an increase in the generation of free radicals (R•) (Equations 1-3).

The second route is photo-oxidation via oxidative processes by radical generation. When natural water matrices are used, the presence of nitrate, iron (III) and/or organic matter can provide •OH due to photo-oxidation of these compounds by UV or other AOP in combination with UV, such as UV/H_2O_2, UV/O_3, $UV/H_2O_2/O_3$.

Nitrate ions absorb UV radiation by acting as an internal filter to UV light and, in parallel, can form •OH through the mechanism detailed below (Equations 7-11) (Neamtu and Frimmel, 2006).

$$NO_3^- \xrightarrow{\;hv\;} NO_2^- + O \tag{7}$$

$$NO_3^- \xrightarrow{\;hv\;} NO_2^{\bullet} + O^{- \bullet} \tag{8}$$

$$2NO_2^{\bullet} + H_2O \longrightarrow NO_2^- + NO_3^- + 2H^+ \tag{9}$$

$$O + H_2O \longrightarrow 2 \bullet OH \tag{10}$$

$$O^{- \bullet} + H_2O \longrightarrow \bullet OH + HO^- \tag{11}$$

For wavelengths longer than 300 nm, complex aqueous ferric iron $Fe(OH)_2^+$, $Fe(OH)_4^-$, $Fe(OH)^{2+}$, Fe^{3+}, $Fe_2(OH)_2^{4+}$) can generate ferrous ions and •OH radicals by photolysis through the internal transfer of electrons (Equation 12) (Espinoza et al., 2007).

$$Fe(OH)^{2+} \xrightarrow{\;hv\;} Fe^{2+} + HO^{\bullet} \tag{12}$$

Dissolved organic matter absorbs UV light and can generate reactive radicals such as singlet oxygen, superoxide anions ($\bullet O_2^-$), •OH radicals and peroxyl radicals (ROO•). These reactive transients can degrade organic pollutants following different paths, and consequently they are not degraded only by direct photolysis (Chin et al. 2004; Neamtu and Frimmel, 2006).

The UV radiation is also used for disinfection, especially of reuse water. Ultraviolet radiation is bactericidal because it causes damage to the nucleic acids (DNA and RNA) of microorganisms (bacteria and viruses), inactivating them. Consequently, UV radiation prevents micro-organisms from multiplying because the photoproducts formed from nucleic acids (e.g., pyrimidine dimers) inhibit replication and transcription (Achilleos et al., 2005). The peak absorbance of nucleic acid is around 260 nm, but below 230 nm absorbance is also high. UV lamps, most of which emitted light at 254 nm (low pressure mercury lamps) are commonly used in studies on the disinfection of aqueous matrices, but polychromatic UV lamps can also be effective in inactivating certain microorganisms (Hijnen et al., 2006).

The use of UV disinfection is increasing mainly due to the fact that it is less expensive than chlorine disinfection, safer than chlorine gas, does not form organochlorine and is effective against *Cryptosporidium* and *Giardia*, while chlorine is not. Additionly, UV disinfection does not form by-products or residual toxicity, and has a low regrowth of bacteria. However, UV radiation requires water with low turbidity to allow the radiation to be effective (Achilleos et al., 2005).

As solar radiation is composed of approximately 4% of UVA/UVB (295-400 nm), besides visible light (400-800 nm) and infrared radiation, this natural radiation has been used for disinfection, color reduction and dissolved organic matter removal in surface waters (Kulovaara et al., 1995, Martín-Domínguez et al., 2005, Davies et al., 2009).

3. Ozonation

Ozone decomposes to oxygen after generation, so it cannot be stored and must be generated on-site. As this oxidant is a gas with low solubility in water (1.05 g L^{-1}), the efficiency of the gas–liquid transfer of ozone is the main variable affecting the effectiveness of organic compound oxidation by ozone in water matrices (Sievers, 2011). The main decomposition reactions of ozone in pure water at ambient temperature (around 20°C) are Equations 13-23.

Initiation:

$$O_3 + OH^- \longrightarrow HO_2^- + O_2 \qquad k = 70 \ M^{-1}s^{-1} \qquad \text{(Staehling; Hoigne, 1982)} \qquad (13)$$
$$k = 48 \ M^{-1}s^{-1} \qquad \text{(Forni et al., 1982)}$$

Propagation:

$$O_3 + HO_2^- \longrightarrow \cdot OH + O_2^- \cdot + O_2 \qquad k = 2,8 \times 10^6 \ M^{-1} \ s^{-1} \qquad \text{(Staehling; Hoigne, 1982)} \qquad (14)$$

$$O_3 + O_2^- \cdot \longrightarrow O_3^- \cdot + O_2 \qquad k = 1,6 \times 10^9 \ M^{-1} \ s^{-1} \qquad \text{(Bühler et al., 1984)} \qquad (15)$$

$$(\text{pH} \geq 8) \ O_3 + O_2^- \cdot \rightleftharpoons O_3^- \cdot + O_2 \qquad k_+ = 1,9 \times 10^3 \ s^{-1} \qquad \text{(Elliot; McCracken, 1989)} \qquad (16)$$
$$k_- = 3,5 \times 10^9 \ M^{-1} \ s^{-1} \qquad \text{(Elliot; McCracken, 1989)}$$

$$O^- \cdot + H_2O \longrightarrow \cdot OH + OH^- \qquad k = 10^8 \ s^{-1} \qquad \text{(von Gunten, 2003a)} \qquad (17)$$

$$(\text{pH} \leq 8) \ \ O_3^- \cdot + H^+ \rightleftharpoons HO_3 \cdot \qquad k_+ = 5,2 \times 10^{10} \ M^{-1} \ s^{-1} \qquad \text{(Bühler et al., 1984)} \qquad (18)$$
$$k_- = 3,3 \times 10^2 \ s^{-1} \qquad \text{(Bühler et al., 1984)}$$

$$HO_3 \cdot \longrightarrow \cdot OH + O_2 \qquad k = 1,1 \times 10^5 \ s^{-1} \qquad \text{(Bühler et al., 1984)} \qquad (19)$$

$$O_3 + \cdot OH \longrightarrow HO_2 \cdot + O_2 \qquad k = 3,0 \times 10^9 \ M^{-1} \ s^{-1} \qquad \text{(Bahnemann; Hart, 1982)} \qquad (20)$$

$$HO_2^- + H^+ \rightleftharpoons H_2O_2 \qquad pK_a = 11,7 \pm 0,2 \qquad \text{(Behar et al., 1970)} \qquad (21)$$

$$HO_2 \cdot \rightleftharpoons H^+ + O_2^- \cdot \qquad pK_a = 4,8 \qquad \text{(Behar et al., 1970)} \qquad (22)$$
$$K_{equilíbrio} = 1,3 \times 10^{-5} \ M^{-1} \quad \text{(Behar et al., 1970)}$$

$$\cdot OH \rightleftharpoons O^- \cdot + H^+ \qquad pK_a = 11,84 \qquad \text{(Elliot; McCracken, 1989)} \qquad (23)$$

Where pK_a is the acid dissociation constant. The termination reactions are any reactions between $\bullet O_2^-$, $\bullet O_3^-$, HO_2^-, $HO_2 \bullet$ and $\bullet OH$ (Bühler et al. 1984; Nadezhdin, 1988).

Analyzing the mechanism described above (Equations 13-23), it appears that the decomposition of ozone can be accelerated by increasing the pH, since hydroxyl ions (OH^-) (Equation 13) start this process and the HO_2^- concentration is pH-dependent (Equation 21) (Andreozzi et al., 1999). As the decomposition of ozone leads to the formation of H_2O_2 (Equation 21), the addition of hydrogen peroxide will promote the decomposition of ozone and increase the formation of $\bullet OH$. Therefore, the combination $O_3 + H_2O_2$ provides the advanced oxidation process (AOP) O_3/H_2O_2.

In the decomposition reactions (organic matter oxidation reactions), there are two mechanisms: direct attack of ozone and free radical attack. The direct attack of ozone is selective and usually occurs in the case of unsaturated compounds (carbon-carbon double or triple bonds - π bonds), aromatic rings, amines and sulfides (von Gunten, 2003a). The primary reaction of ozone with a certain compound (C) occurs through the generation of an electrophilic additional intermediate compound (C-O_3) which decomposes via consecutive reactions until the formation of stable compounds (von Gunten, 2003a).

The direct attack of ozone on tertiary amine occurs by hydrolysis and demethylation, generating the corresponding aldehyde and secondary amine, which is the most stable compound containing the nitrogen atom. Similarly, the same reactions occur with the secondary and primary amine until the formation of the most stable organic compound (Muñoz and von Sonntag, 2000, Lange et al., 2006, Luiz et al., 2010). The oxidation of ammonia through direct attack by ozone is very slow, but it generates the same product as the radical attack: NO_3^-. The replacement of hydrogen atoms by alkyl groups (C_nH_{2n+1}) in the amine nitrogen increases the rate constant of oxidation via ozone (k). The k value for oxidation via ozonation of ammonia (NH_3) is 20 M^{-1} s^{-1}, of diethylamine is 9.1 x 10^5 M^{-1} s^{-1} and of triethylamine is 4.1 x 10^6 M^{-1} s^{-1} (von Gunten, 2003a).

In the radical attack, radical ions are formed during the oxidative degradation of ozone in water, the main ones being: oxygen free radical (O-$\bullet$), superoxide radical ($\bullet O_2$-) and hydroxyl radical ($\bullet OH$) (Hoigné; Bader, 1983, Luiz et al., 2010).

The free radical attack is usually faster than the direct attack of ozone. However, in an ozonation process, the concentration of ozone is higher than that of its degradation radicals and hence the oxidation is mainly due to molecular attack. There are molecules that are slowly degraded by ozone treatments (because the molecular attack is selective) and rapidly degraded in advanced oxidation processes in which the attack by $\bullet OH$ radicals predominates (non-selective, higher removal efficiency and the mineralization of organic compounds) (Lau et al., 2007). Therefore, in tertiary wastewater treatment, it is necessary to know which organic contaminants are present in the wastewater in order to verify whether they can be oxidized by simple ozonation or if an AOP would be more effective.

Ozone is a powerful disinfectant, causing cell inactivation through direct damage to the membrane and cell wall, disruption of enzymatic reactions and DNA damage, including in the case of highly resistant pathogens such as protozoan *Cryptosporidium parvum* (EPA, 2004; von Gunten, 2003b). Compared with chlorination, disinfection with ozone is a better option for water bodies which are not organochlorine receivers and its excess can be easily dissipated (Achilleos et al. 2005; EPA, 2004). However, since the concentration required to inactivate the most resistant organisms is very high, there may be the formation of non-desirable disinfection by-products, especially bromates which have carcinogenic potential (von Gunten, 2003b).

4. Advanced oxidation processes (AOP)

Advanced oxidation processes (AOPs) degrade organic contaminants into carbon dioxide, water and inorganic anions through the action of transient oxidizing species, especially hydroxyl radicals ($\bullet OH$). AOPs are clean processes because they do not generally require post-treatment or the final disposal of potential waste.

The •OH radicals are not selective, thus they have the ability to degrade all organic substances present in a liquid or gas to be treated. The abstraction of H for the reaction of •OH and organic compounds in the presence of saturated oxygen is the main route for the shortening of the chains of ketones, aldehydes and carboxylic acids (Mellouk et al., 2003).

4.1 UV/H₂O₂

Ultraviolet (UV) treatment promotes photolysis (or photo-oxidation) of organic compounds. To achieve a greater degradation, there are processes that combine UV radiation with other chemical agents such as O_3 and H_2O_2 (Parkinson et al., 2001). Among the variety of AOPs available and their combinations, the reagent H_2O_2 is frequently used due to its stability during transport and storage, almost infinite solubility in water, and low installation and operation costs compared to other processes, such as the application of O_3 (Alfano et al., 2001). In UV/H₂O₂ treatment, radiation with a wavelength shorter than 300 nm (UV-C irradiation) breaks H_2O_2 into •OH radicals (Equation 24) which oxide the pollutant compounds. However, for the organic compound degradation to be effective, and mineralization of the compounds to occur (into CO_2 and H_2O), the H_2O_2 concentration must be above the stoichiometric demand and the reaction time under UV radiation must be sufficient, as previously determined experimentally (Alfano et al., 2001).

The photolysis of water yields dissolved oxygen (Equation 24) which can favour the photodegradation of organic compounds (Equation 2). Nevertheless the photolysis of hydrogen peroxide by UV radiation promotes the formation of stronger oxidants: hydroxyl radicals (•OH) (Equation 25). The mechanism of AOP UV/H₂O₂ treatment includes initiation, propagation and termination reactions, and degradation reactions of the organic compounds (RH) (Equations 25-32) (Alfano *et al.*, 2001).

$$2H_2O \xrightarrow{h\nu} O_2 + 4H^+ + 4e^- \tag{24}$$

Initiation:

$$H_2O_2 \xrightarrow{h\nu} 2\,{}^{\bullet}OH \tag{25}$$

Propagation:

$$H_2O_2 + {}^{\bullet}OH \xrightarrow{k_2} HO_2{}^{\bullet} + H_2O \tag{26}$$

$$H_2O_2 + HO_2{}^{\bullet} \xrightarrow{k_3} {}^{\bullet}OH + H_2O + O_2 \tag{27}$$

Termination:

$$2\,{}^{\bullet}OH \xrightarrow{k_4} H_2O_2 \tag{28}$$

$$2HO_2{}^{\bullet} \xrightarrow{k_5} H_2O_2 + O_2 \tag{29}$$

$$\bullet OH + HO_2\bullet \xrightarrow{k_6} H_2O + O_2 \tag{30}$$

Decomposition:

$$RH + \bullet OH \xrightarrow{k_7} products \tag{31}$$

$$RH + HO_2\bullet \xrightarrow{k_8} products \tag{32}$$

The propagation (Equations 26-27) and termination (Equations 28-30) reactions indicate that when there is H_2O_2 in excess and, subsequently, an excess of $\bullet OH$ radicals, the recombination of these radicals is favored (Equation 28) reversing the initiation reaction (Equation 25) and inhibiting the oxidation reactions of organic compounds (Equations 31-32).

Yang et al. (2005) found that there is no significant effect of pH on the degradation efficiency if the treatment is only via photodegradation by UV radiation. However, in a UV/H_2O_2 reaction the generation of hydroxyl radicals ($\bullet OH$) may be affected by the presence of high concentrations of H^+ and OH^-. Many authors have observed higher rates of decomposition of organic contaminants at low pH (Song et al., 2008). However, Yang et al. (2005) found the efficiency of mineralization of the complex sodium ethylenediamine tetraacetic acid (Na-EDTA) decreased from 98% (in ultra-pure water and without pH correction) to around 70% with initial pH values of 2.3 and 11.6, respectively.

The commercial and industrial applications of UV/H_2O_2 processes require the determination of optimal oxidant and irradiation dosages. Therefore, the development of a mathematical model for these processes can help the determination of such variables (Crittenden et al., 1999). Usually, the kinetic model used for common pollutants in aqueous solutions is a pseudo-first-order model by photodegradation (UV radiation) and also by the AOP H_2O_2/UV (Chen et al. 2007; Song et al. 2008; Luiz et al., 2009).

Dosages below 15 mg L^{-1} of hydrogen peroxide do not have a significant toxic effect on bacteria when the system is not exposed to light (in darkness) (Rincón; Pulgarin, 2006; Sciacca et al., 2010). However, hydrogen peroxide in the extracellular region weakens the cell walls, making the bacteria more sensitive to oxidative stress and this may explain the observed inactivation at lower concentrations. Therefore, although H_2O_2 is not directly toxic to the bacteria, the cleavage products, for instance, the hydroxyl ($\bullet OH$) and superoxide ($HO_2\bullet$ and $\bullet O_2^-$) radicals generated by the addition of another agent in the system, such as UV, O_3, ferric or ferrous ions (Fenton reaction) (Sciacca et al., 2010) can be toxic. Such species can induce reactions in lipids, proteins and DNA, inactivating the cells.

4.2 O_3/H_2O_2, O_3/UV and $O_3/H_2O_2/UV$

The AOPs O_3/H_2O_2, O_3/UV and $O_3/H_2O_2/UV$ have been developed to treat wastewater highly contaminated by both organic matter and microorganisms. The effectiveness of these treatments is due to synergistic interaction of the positive effects of three reactions: direct ozonation (molecular attack), photolysis and oxidation by hydroxyl radicals ($\bullet OH$) (Lau et al., 2007, Luiz et al., 2010). Some important factors, such as quantifying the speed of photons absorbed in the heterogeneous medium (ozone gas in water matrices), and assessing the degree of ozone reactivity with unsaturated organic compounds, still need to be better understood (Legrini et al., 1993).

The mechanism of the AOPs O_3/H_2O_2, O_3/UV and $O_3/H_2O_2/UV$ are very similar to each other, and all have the hydroxyl radical ($\bullet OH$) as the most important intermediate compound (Beltran et al. 1995; Legrini et al., 1993). In the AOP O_3/H_2O_2, besides the initiation reaction of ozone decomposition, another initiation reaction is the decomposition of hydrogen peroxide by ozone into hydroxyl radicals (Equation 33). The decomposition of hydrogen peroxide by ozone (Equation 33) is very slow (Legrini et al., 1993), but it enhances the other ozone decomposition reactions which generate more $\bullet OH$ radicals (Equations 13-23). Therefore, the reaction rate can be increased with increasing the pH, because the ionic form of H_2O_2 (HO_2^-) reacts with O_3 (Staehelin; Hoigné, 1982) (Equation 14) followed by other reactions of propagation, termination and decomposition.

$$O_3 + H_2O_2 \rightarrow O_2 + HO_2^{\bullet} + {}^{\bullet}OH \tag{33}$$

The absorbance spectrum of ozone indicates that it has a greater absorption at 254 nm than H_2O_2, 3300 and 18.6 M^{-1} cm^{-1}, respectively, and the influence of the internal effects of the system is lower in an O_3/UV system, due to the effect of suspended solids and aromatic compounds which can act as a UV filters. Nevertheless, the quantum yield of $\bullet OH$ radicals in the AOP O_3/UV (around 0.1) is lower than in H_2O_2/UV (Sievers, 2011). The quantum yield of a photochemical reaction is the number of events divided by the number of absorbed photons of a specific wavelength during the same period time (Khun et al., 2004). The absorption of UV light by O_3 in water leads, firstly, to the generation of H_2O_2 (Equation 34) (Legrini et al., 1993). Therefore, the other reactions of propagation, termination and decomposition are all cited for O_3 treatments, and for O_3/H_2O_2 H_2O_2/UV (Equations 14-33).

$$O_3 + H_2O \xrightarrow{h\nu} H_2O_2 + O_2 \tag{34}$$

The AOP $O_3/H_2O_2/UV$ should be used to enhance the generation of hydroxyl radicals, as its initiation reactions are the same as those of ozonation, H_2O_2/UV, O_3/H_2O_2 and O_3/UV (Equations 13, 25, 33 and 34). All other reactions are of the types propagation, termination and decomposition (Equations 14-32).

AOPs promote a higher degree of mineralization compared to other tertiary oxidative treatments (such as simple ozonation), ensuring decreased levels or the absence (if the mineralization is total) of oxidized intermediate products that, usually, reduce the initial toxicity (Rosal et al., 2008).

4.3 Homogeneous fenton and fenton-like pocess

A widespread AOP is the homogeneous Fenton process involving the reaction of hydrogen peroxide (H_2O_2) with the Fenton catalyst (ferrous ion - Fe(II) - dissolved in acid aqueous media) generating hydroxyl radicals ($\bullet OH$) and ferric ions (Fe(III)). Fe(III) ions undergo reduction to Fe(II) mainly through the action of the oxidant H_2O_2 ensuring the continuation of the catalytic reaction. The main reactions of the Fenton process are represented by Equations 35 to 41 (Garrido-Ramirez et al. 2010; Herney-Ramirez et al. 2010; Navalon et al., 2010, Umar et al., 2010).

$$Fe^{2+} + H_2O_2 \rightarrow Fe^{3+} + OH + OH^- \tag{35}$$

$$Fe^{3+} + H_2O_2 \rightarrow Fe^{2+} + HO_2 + H^+ \tag{36}$$

$$OH + H_2O_2 \rightarrow HO_2 + H_2O \tag{37}$$

$$OH + Fe^{2+} \rightarrow Fe^{3+} + OH^- \tag{38}$$

$$Fe^{3+} + HO_2 \rightarrow Fe^{2+} + O_2 + H^+ \tag{39}$$

$$Fe^{2+} + HO_2 + H^+ \rightarrow Fe^{3+} + H_2O_2 \tag{40}$$

$$2HO_2 \rightarrow H_2O_2 + O_2 \tag{41}$$

There are many variations of the Fenton process and the main ones are:
- Modified Fenton or Fenton-like reaction: ferric ion or another transition metal (such as copper) is used instead of ferrous ion (Herney-Ramirez *et al.*, 2010);
- Heterogeneous Fenton: supported catalyst is used in a solid matrix (heterogeneous system);
- Photo-Fenton: the regeneration of Fe(II) is accelerated due to photoreduction of Fe(III), and the generation of hydroxyl radicals is increased due to the photolysis of hydrogen peroxide (Equation 25) (Umar et al., 2010);
- Electro-Fenton: continuous generation of H_2O_2 in acidic solution by reducing gaseous O_2 in carbon cathodes (Sirés et al., 2007);

Other variations: heterogeneous Fenton-like, heterogeneous photo-Fenton, Photo-Fenton-like, and heterogeneous Photo-Fenton-like processes.

Equation 42 describes the Fenton reaction and its variations should occur at acid pH through the occurrence of hydrogen peroxide decomposition in the presence of a ferrous ion catalyst (Umar et al., 2010).

$$2Fe^{2+} + H_2O_2 + 2H^+ \rightarrow 2Fe^{3+} + 2H_2O \tag{42}$$

The homogeneous Fenton process is carried out in four stages: (1) pH adjustment (optimal pH is around 2.8), (2) oxidation reaction through the action of oxidizing radicals (especially •OH), (3) neutralization followed by coagulation (formation of ferric hydroxide complexes), and (4) precipitation (Umar et al., 2010). Therefore, the organic substances are removed by oxidation and by coagulation. Thus, the homogeneous Fenton process has the great disadvantage of producing, at the end of the process, solid waste (sludge) with a high metal concentration, generating environmental risks and extra costs due to the post-treatment of sludge to recover part of the catalyst (Herney-Ramirez et al., 2010, Umar et al., 2010). Hence, the heterogeneous Fenton process eliminates the step of catalyst precipitation and removal at the end of the process (Garrido-Ramirez et al. 2010; Herney-Ramirez et al. 2010 ; Navalon et al., 2010).

Among the various factors that affect the efficiency of the Fenton process, pH and the $H_2O_2/Fe(II)$ ratio are the most important ones. In reactions with a low $H_2O_2/Fe(II)$ ratio, chemical coagulation is predominant in the removal of organic matter, whereas with high ratios the chemical oxidation is dominant (Umar et al., 2010).

4.4 Heterogeneous AOPs

The heterogeneous AOPs using a semiconductor as the catalyst and ultraviolet radiation have the advantage of not requiring the addition of a strong oxidant to produce hydroxyl radicals (photocatalysis process, e.g., the AOP TiO_2/UV). However, the addition of an oxidant can accelerate the production of oxidizing radicals and increase the speed of organic matter removal (e.g. the AOP $TiO_2/UV/H_2O_2$).

Photocatalysis is the catalysis of a photochemical reaction on the surface of a solid, usually a semiconductor, TiO_2 being the most used (Fujishima et al., 2008). In this reaction the oxidation of organic pollutants occurs in the presence of an oxidizing agent, followed by activation of a semiconductor with light energy (Ahmed et al., 2010) at wavelengths $\leq$ 400 nm (Malato et al., 2009).

The catalyst is activated when it absorbs photons with energies greater than the value of its "band gap" energy resulting in the excitation of electrons from the valence band to the conduction band forming a hole (h^+) in the valence band. This hole promotes the oxidation of pollutants (in the case of photo-catalytic oxidation of organic pollutants) or water, in the latter case producing hydroxyl radicals (OH •). The electron (e^-) of the conduction band reduces the oxygen or other oxidative agent in the absence of O_2 (Ahmed et al., 2010). The oxidizing agent binds to the conduction band, where the reduction occurs through the gaining of electrons (e^-), while the organic compound to be degraded binds to the valence band, where oxidation occurs, donating e^- and receiving h^+. Thus the organic compound is the reducing agent or electron donor (hole scavenger) (Malato et al., 2009).

Photocatalysis is an AOP because, during the photochemical reactions and the activation of the catalyst, hydroxyl radicals are formed (Malato et al., 2009, Hermann, 2010). Hence, these treatments may reduce the toxicity of wastewaters by total or partial mineralization of synthetic organic compounds transforming them into more biodegradable substances.

There are photocatalytic processes using solar energy, representing a Green Chemistry technology approach (Hermann, 2010), with the following benefits: 1) process carried out at ambient temperature and pressure, 2) the oxidizing agent can be oxygen obtained from the atmosphere, 3) many catalysts are low-cost, safe and reusable, and 4) use of solar energy as an energy source (Malato et al., 2009).

5. Water and wastewater management

Prior to the implementation of a tertiary process for wastewater treatment aiming at water reuse to reduce the consumption of fresh drinking water, it is essential to minimize and to optimize the water consumption in all processes of an industrial plant. Thus, it is essential to understand the industrial process in detail, and to develop a methodology for water management, presenting alternatives for minimizing water consumption and effluent generation, obeying the specific laws of the industry sector.

The authors of this review carried out a water and wastewater management (W2M) study in a slaughterhouse based on water consumption and effluent generation in the various sectors. The W2M proposed was aimed at minimizing water consumption and the evaluation of possibilities for water and indirect potable reuse in food industries.

5.1 Case study in a meat processing plant

The case study was carried out in a meat processing plant located in the west of Santa Catarina State (southern Brazil) where water pollution and overexploitation, the uneven distribution of rainfall throughout the seasons and long periods of drought especially in summer, have become a problem. The industrial activities include the slaughter and processing of poultry and swine. The industry has its own drinking water treatment plant (DWTP) and wastewater treatment plant (WWTP). The major water resource of this industry is a river. Its DWTP produces around 8,600 m³/day of drinking water and the WWTP treats around 7,900 m³/day of wastewater. The first step of the wastewater treatment is equalization, followed by coagulation, flocculation, and flotation. The last step is the secondary treatment: activated sludge.

5.1.1 Water minimization

To evaluate the possibilities for water minimization, it was first necessary to analyze the current Brazilian laws concerning slaughterhouses for poultry and swine. These guidelines provide the technical standards required for the facilities and equipment used in the slaughter of animals and meat industrialization, including water quality and minimum water consumption in each process. As a further step, a water balance was carried out for all processes. The processes or equipment which have the greatest water demand are usually the most important and the easiest points for which to verify the possibilities to avoid the misuse and overexploitation of fresh water (drinking water from plant's DWTP). Hence, four high consumption points were identified with the potential for reducing the fresh water consumption in-line with the current Brazilian legislation: (1) pre-cooling of giblets, (2) washing of poultry carcasses before pre-chiller, (3) transportation of giblets, poultry necks and feet, and (4) washing of swine carcasses after buckling. It was found that in the first two processes the water consumption could be reduced; in the third process the devices could be substituted by others which do not need water to function properly; and in the last process there was an unnecessary step, according to the legislation. The minimization of the water demand achieved in these four process steps alone was approximately 806 m³/d.

5.1.2 Direct wastewater reuse

After the minimization of water use, the most important action is to prioritize direct recycling and reuse of wastewater, without the need for advanced reconditioning or treatment. Hence, following the water balance, the wastewater with the possibility for direct or indirect recycling or reuse was evaluated physically, chemically and microbiologically to verify if and where it could be recycled and reused. Four wastewaters were identified as having a real possibility of reuse after carrying out all of the characterization analysis: (1)

defrosting of refrigerating and freezing chambers, (2) purging of condensers, (3) cooling of smoke fumigator chimneys, and (4) sealing and cooling of vacuum pumps. These four wastewaters totalized approximately 1,383 m³/day of wastewater with the possibility for reuse in processes without direct contact with food products, that is, in non-potable uses (e.g., as cooling water, for flushing toilets or as irrigation around the plant), thus saving fresh potable water. These wastewaters had similar water quality parameters indicating that they could be mixed before undergoing the same reconditioning treatment. Hence, the mixed wastewater was also characterized and it was shown that it could be simply treated to remove total suspended solids and chlorinated/disinfected before reuse as, for example, cooling water.

The successful results achieved from the proposed W2M (*i.e.*, financial saving of 28.2% and 25.4% in the fresh water requirement - Table 4) prompted the managers of the pilot industrial plant to suggest and to develop other research projects related to water and wastewater reuse in partnership with Brazilian government funded research centers, to develop indirect wastewater technologies. These results are consistent with the work of Tokos et al. (2009) who presented several mixed-integer nonlinear programming models for the optimization of a water network in a brewery plant. The authors presented a theoretical reduction of around 27% in the fresh water demand and associated costs.

Condition	Water flow [m³/dia]	Water saving [%]	Annual Costs[1] [$]
Production in 2007	8,616.0	-	1,539,353
Theoretical production after water minimization	7,810.0	9.4	1,366,000
Theoretical production after wastewater reuse	7,216.8	16.0	1,256,996
Theoretical production after water minimization and wastewater reuse	6,410.0	25.4	1,104,731

[1]Considering costs in 2007: $0.10 and $0.42 per m³ to treat water (DWTP) and wastewater (WWTP) respectively, in pilot industrial plant.

Table 1. Water and financial savings

5.1.3 Indirect wastewater reuse: indirect potable reuse

Additionally, to reduce even further the percentage of fresh water consumption, indirect wastewater reuse can be carried out after reconditioning of the secondary effluent (*i.e.*, after secondary activated-sludge treatment) by applying tertiary treatments such as advanced oxidation processes (AOPs). This tertiary treated water could be used in other processes without contact with food products, that is, non-potable uses (*e.g.* as cooling water, boiler feed water, toilet flushing water or for irrigation around the plant) (Cornel et al., 2011; Luiz et al., 2011).

Luiz and co-workers (2009, 2011) evaluated some AOPs in bench- and pilot-scale tests with the aim of providing reclaimed water with drinking water quality (according to Brazilian legislation) for industrial water reuse in processes without direct contact with food products. In the first study (Luiz et al., 2009) evaluated the kinetics of the photo-induced degradation of color and UV_{254} under UV radiation with and without the addition of H_2O_2 to treat secondary wastewater after ferric sulfate coagulation. The H_2O_2/UV treatment was 5.2 times faster than simple UV application in removing aromatic compounds.

In the second study (Luiz et al., 2011) four tertiary hybrid treatments using a pilot plant with a capacity of 500 L/h were evaluated. This pilot plant consisted of a pre-filtration system, an oxidation (H_2O_2) or second filtration system and a UV radiation device. The best combination was pre-filtration followed by H_2O_2 addition and UV radiation (AOP H_2O_2/UV). In addition to H_2O_2/UV treatment, the authors also carried out experiments in pilot scale with O_3, O_3/UV and O_3/H_2O_2. However, H_2O_2/UV was found to be a faster and more efficient option to treat slaughterhouse wastewater to the drinking water quality standard, with the exception of one quality parameter, that is, nitrate.

Hence, another treatment was evaluated by the authors to remove the nitrate content, allowing the reuse of the wastewater with drinking water quality in cases without direct contact with food products, reducing the fresh water consumption and conserving natural water resources. The p-n junction photocatalyst p-ZnO/n-TiO_2 was prepared by decomposition of zinc nitrate and photodeposition on TiO_2 and it was used to reduce nitrate ions in aqueous solution and in synthetic and real slaugherthouse wastewater.

Heterogeneous photocatalytic reduction of nitrate over semiconductor materials have also been developed as a promising method for controlling the concentration of nitrate in water. Metal doping (Pt, Pd, Rh, Pt-Cu, Cu, Fe, Bi) and addition of hole scavengers are essential for the reductive removal of nitrate (Rengaraj and Li, 2007; Sá et al., 2009; García-Serrano et al., 2009; Kim et al., 2008; Li et al., 2010). Hole scavengers are electron donors, such as methanol, benzene, oxalic acid, formic acid and humic acids which have been used to improve photocatalytic efficiency (Xu et al., 2010; Li and Wasgestian, 1998; Rengaraj and Li; 2007).

There were a number of studies related to the photocatalytic activity of TiO_2 or ZnO coupled with metals, and the results showed that these catalysts presented more efficient charge separation, and increased lifetime of the charge carriers, and an enhanced interfacial charge transfer to adsorb substrates (Shifu et al., 2008). Coupled photocatalyst ZnO/TiO_2 has showed higher photocatalytic activity than that of single one, since the p-n junction that is formed by the integration of p-type ZnO and n-type TiO_2 contributes to enhance the electrons/holes separation (Shifu et al., 2008).

The selectivity, conversion and activity data indicated ZnO-TiO_2 was a better catalyst than TiO_2 (Table 2). This indicates that Zn ion exerted its role as a doping ion, promoting the charge separation of the pairs of vacancy-electron photoproduced by acting as an electron sink, and consequently increase of electrons reaching the surface for reaction to take place (Sá et al., 2009; Ranjit and Viswanathan, 1997).

Catalyst	Selectivity [%]	Conversion [%]	Activity [$\mu mol_{NO_3^-}$ (min $g_{catalisador}$)$^{-1}$]
TiO_2	70.71	87.46	0.92
ZnO-TiO_2	95.45	91.67	14.24

Table 2. Selectivity (S_N) and conversion ($C_\%$) within reaction time = 120 min and activity within reaction time = 20 min.

Prior to the photocatalytic reactions, the real wastewater (RSW) was filtered 50 micron cartridge filter, activated carbon filter and 10 micron cartridge filter to remove suspended particles (RSW-F) simulating the microfiltration of the previous studies (Luiz et al., 2009, 2011). Due to low TOC concentration in RSW-F, formic acid was added as hole scavenger (Zhang et al., 2005; Rengaraj; Li, 2007; Sá et al., 2009; Malato et al., 2009; Wehbe et al., 2009). Formic acid was added in enough amount to satisfy the stoichiometric relationship nitrate:formic acid (Equation 43) without formation of intermediates (including nitrite and ammonia) (Zhang et al., 2005). Hence, the evaluated molar ratios were CHOOH:nitrate = 2.7, 1.6 and 1.0, respectively [HCOOH] = 1636, 1000 and 500 mg L^{-1} or 427, 261 and 130 mg C L^{-1}, and catalyst 1 g L^{-1}.

$$2NO_3^- + 5HCOO^- + 7H^+ \rightarrow N_2 + 5CO_2 + 6H_2O \tag{43}$$

Therefore, after the application of microfiltration for the removal of mainly suspended solids and the catalytic photoreduction of nitrate with the addition of a hole scavenger (carbon source), it was necessary to remove the excess of this carbon source. Hence, after nitrate was below the Brazilian legislation limit, (CHOOH$_{initial\ (0\ min)}$:H$_2$O$_{2,\ initial\ (120\ min)}$ = 1:1 (M:M)) was added into the reaction to promote another advanced oxidative process (AOP): H$_2$O$_2$/TiO$_2$/UV (Figure 1). Hence, after all these steps, the wastewater achieved drinking water quality, being the main critical parameters shown in Table 3.

	RSW[2]	RSW-F[2]	A[1]	B[1]	C[1]	BRL[3]
pH	6.8 ± 0.6	6.2 – 6.5	6.20	6.15	6.00	6.0 - 9.50
Turbidity, FTU	40 ± 33	22 – 25	2.00	2.00	1.00	5
Apparent color, Pt-Co unit	217 ± 177	11 – 13	3.00	3.00	10.00	15
Total coliforms, CFU/100 mL	> 10^5	> 10^5	0	0	0	0
Nitrate, mg N L^{-1}	45.9 ± 17.7	39.8 – 41.9	1.20	4.50	4.00	10
Nitrite, mg N L^{-1}	-	3.74 – 3.77	0.06	0.08	0.21	1
Ammonia, mg N L^{-1}	-	0.4 – 1.0	0.30	0.90	0.30	1.24

[1] Composite samples (RSW). Composite samples after filtration (RSW-F)
[2] A: [CHOOH]$_i$ = 1636 mg L^{-1} / H$_2$O$_2$ = 1209 mg L^{-1}; B: [CHOOH]$_i$ = 1000 mg L^{-1} / H$_2$O$_2$ = 739 mg L^{-1}; C: [CHOOH]$_i$ = 500 mg L^{-1} / H$_2$O$_2$ = 370 mg L^{-1}. Zn-TiO$_2$ catalyst = 1 g L^{-1}. Reaction time: 300 min.
[3] BRL: Brazilian regulatory legislation (Brazilian drinking water standards: Brazilian Ministry of Health Administrative Ruling 518/2004).

Table 3. Average quality of the secondary effluent of the slaughterhouse and quality parameters of the wastewater after the treatments: filtration (RSW-F, Table 1) + TiO$_2$/UV/Argon (reaction time 120 min) + H$_2$O$_2$/TiO$_2$/UV (reaction time 300 min).

Table 3 shows the average characterization of three composite samples collected to characterize the secondary effluent from the slaughterhouse (RSW). The real effluent slaughterhouse wastewater used in those studies was collected after treatment (coagulation+flocculation followed by aerobic biological treatment) from a Brazilian poultry slaughterhouse. The point to collect real treated effluent was the one located immediately outside the secondary sedimentation tank. A composite sample of 24h (1 L of sample collected per hour for 24 hours) was collected and characterized. All organic, inorganic and microbiological parameters of Brazilian drinking water standards (Brazilian Ministry of

Health Administrative Ruling 518/2004) were analyzed, and only nitrate, nitrite, color, turbidity and total coliform were above the limit. The proposal treatments were able to reduce those parameters and also remove micropollutants that are not removed by secondary treatment (e.g. activated sludge).

The influent of the slaughterhouse RSW contained the mixture of sewage from meat processing and sanitary sewage, the treated wastewater still contained persistent pollutants arising from animal and human use, e.g., drugs and personal care products. The target compounds nonyl- (NP) and octylphenol (OP), linear alkylbenzene sulfonates (LAS), triclosan and ibuprofen could be detected and confirmed applying a LTQ Orbitrap mass spectrometer (Thermo Electron, Bremen, Germany) to the SPE-extracts after HPLC-(HR)MS. Besides these pollutants octylphenol- (OPEO) and nonylphenol ethoxylates (NPEO), polyethylene glycols (PEG), erythromycin, sulfamethoxazole and sulfadimidine were also observed.

The AOPs evaluated were also very effective to remove the micropollutants present in the already treated slaughterhouse wastewater. Among all the compounds found by LC-MS in the samples only NP, NPEO and PEG were still observable in the samples after the tertiary treatments proposed, however, in lower amounts.

Figure 1 shows the suggested flowchart of the process suggested by the authors to treat secondary wastewater with high concentrations of nitrogen compounds and recalcitrant organic compounds as antibiotics, medicament and personal care products which are commonly found in sanitary, domestic and industrial wastewater.

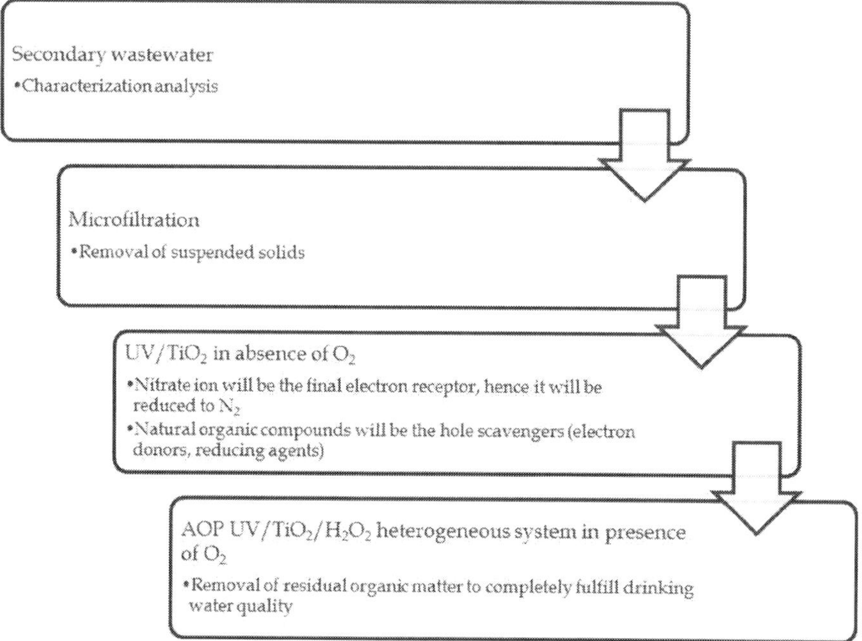

Fig. 1. Suggested process to treat secondary wastewaters with high concentration of nitrate and residual organic matter.

6. Selection of an AOP and alternatives for water reuse

As proposed in section 5.1 (Case Study in a Meat Processing Plant), experiments in bench and pilot scale should be carried out before the selection of an AOP for the tertiary treatment of the wastewater intended for reused (Sievers, 2011). This initial research would define: (i) the best treatment for each kind of industrial wastewater considering the reused water quality required and hence which improvements should be sought (*e.g.* color or turbidity removal, target recalcitrant organic compound removal, disinfection); (ii) the kind and quantity of radical scavengers present in the wastewater (*e.g.* humic substances); (iii) the dosage of oxidant (if used) and the impact of its residual concentration on the subsequent use or disposal of the tertiary treated wastewater, and (iv) the energy and investment costs (Sievers, 2011).

The tertiary treated water could be used in other processes without contact with food products (non-potable uses) where public health and product integrity would not be compromised. Therefore, "this major health concern makes it imperative for governments and the global community to implement proper reuse planning and practices, emphasizing public health and environmental protection" (EPA, 2004). Boiler feed water, cooling water, toilet flushing water or irrigation around the plant are some alternatives for the use of reclaimed water (Cornel et al., 2011; Luiz et al., 2011). Cooling water may be the most attractive option because it does not need to be of high quality, and usually a large amount is required (Cornel et al., 2011).

A sentence at EPA (2004) resume the past and the future of wastewater reuse (indirect potable reuse): "Notwithstanding the fact that some proposed, high profile, indirect potable reuse projects have been defeated in recent years due to public or political opposition to perceived health concerns, indirect potable reuse will likely increase in the future" (EPA, 2004).

7. References

Achilleos, A., Kythreotou, N. & Fatta, D. (2005). Development of tools and guidelines for the promotion of sustainable urban wastewater treatment and reuse in agricultural production in the Mediterranean countries: Task 5 - Technical Guidelines on Wastewater Utilisation. European Commission - Euro Mediterranean Partnership.

Ahmed, S., Rasul, M.G., Martens, W. N., Brown, R. & Hashib, M.A. (2010). Heterogeneous photocatalytic degradation of phenols in wastewater: A review on current status and developments, *Desalination*, 261 (1-2), 3-18.

Alfano, O. M., Brandi, R. J. & Cassano, A. E. (2001). Degradation kinetics of 2,4-D in water employing hydrogen peroxide and UV radiation, *Chem. Eng. J.*, 82, 209-218.

Andreozzi, R., Caprio, V., Insola, A. & Marotta, R. (1999) Advanced oxidation processes (AOP) for water purification and recovery, *Catal. Today*, 53, 51-59.

Bahnnemann, D. & Hart, E. J. (1982). Rate Constants of the Reaction of the Hydrated Electron and Hydroxyl Radical with Ozone In Aqueous Solution, *J. Phys. Chem.*, 86, 252-255.

Behar, C., Czapski, G. & Duchovny, I. (1970). Carbonate radical in flash photolysis and pulse radiolysis of aqueous carbonate solutions, *J. Phys. Chem.*, 74 (10), 2206-2210.

Beltrán, F. J., Ovejero, G., Acedo, B. (1993). Oxidation of atrazine in water by ultraviolet radiation combined with hydrogen peroxide, *Water Res.*, 27 (6), 1013-1021.

Beltrán, F. J., Ovejero, G., Garcia-Araya, J. F. & Rivas, J. (1995). Oxidation of Polynuclear Aromatic Hydrocarbons in Water. 2. UV Radiation and Ozonation in the Presence of UV Radiation, *Ind. Eng. Chem. Res.*, 34, 1607-1615.

Beltran-Heredia, J., Torregrosa, J., Dominguez, J. R. & Peres, J. A. (2001). Kinetics of the Oxidation of p-Hydroxybenzoic Acid by the H_2O_2/UV System, *Ind. Eng. Chem. Res.*, 40, 3104-3108.

Bühler, R. E., Staehlin, J. & Hoigné, J. (1984). Ozone decomposition in water studied by pulse radiolysis. 1. HO_2/O_2^- and HO_3/O_3^- as intermediates, *J. Phys. Chem.*, 88, 2560-2564.

Burkhard, R., Deletic, A. & Craig, A. (2000) Techniques for water and wastewater management: a review of techniques and their integration in planning, *Urban Water*, 2, 197-221.

Casani, S., Rouhany, M. & Knøchel, S. (2005). A discussion paper on challenges and limitations to water reuse and hygiene in the food industry. *Water Res.* 39, 1134–1146.

Chen, L., Zhou, H.-Y. & Deng, Q.-Y. (2007). Photolysis of nonylphenol ethoxylates: the determination of the degradation kinetics and the intermediate products. *Chemosphere*, 68 (2), 354-359.

Chin, Y.-P., Miller, P. L., Zeng, L., Cawley, K. & Weavers, L. K. (2004). Photosensitized degradation of bisphenol A by dissolved organic matter. *Envir. Sci. Tech.*, 38 (22), 5888-5894.

Clevelario, J., Neves, V., Oliveira, P. T. T. M., Costa, V. G., Amendola, P., Rocha, R. M. & Costa, J. J. G. (2005). Water Statistics in Brazil: An Overview. IWG-Env, International Work Session on Water Statistics, Vienna.

Codex Alimentarius (2001) Proposed Draft Guidelines for the Hygienic Reuse of Processing Water in Food Plants, CX/FH 01/9. Joint FAO/WHO Food Standards Programme, 34th Session, Bangkok, Thailand.

Codex Alimentarius (2007) Report of the Thirty-Eighth Session of the Codex Committee on Food Hygiene, Alinorm 07/30/13. Joint FAO/WHO Food Standards Programme, 38th Session, Rome, Italy.

Cornel P, Meda A and Bieker S (2011) Wastewater as a Source of Energy, Nutrients, and Service Water. In: Peter Wilderer (ed.) *Treatise on Water Science*, Oxford: Academic Press, vol. 4, 337–375.

Crittenden, J. C., Hu, S., Hand, D. W. & Green, S. A. (1999). A kinetic model for H_2O_2/UV process in a completely mixed batch reactor. *Water Res.*, 33 (10), 2315-2328.

Davies, C. M., Roser, D. J., Feitz, A. J. & Ashbolt, N. J. (2009). Solar radiation disinfection of drinking water at temperate latitudes: Inactivation rates for an optimised reactor configuration. *Water Res.*, 43 (3), 643-652.

Elliot, A. J. & McCracken, D. R. (1989). Effect of temperature on O^- reactions and equilibria: a pulse radiolysis study. *Radiat. Phys. Chem.*, 33 (1), 69-74, 1989.

EPA. *Guidelines for Water Reuse*. Washington, DC: Environmental Protection Agency, 2004.

Espinoza, L. A., Neamtu, M. & Frimmel, F. H. The effect of nitrate, Fe(III) and bicarbonate on the degradation of bisphenol A by simulated solar UV-irradiation, *Water Res.*, 41 (19), 4479-4487, 2007.

Forni, L., Bahnemann, D. & Hart, E.J. (1982). Mechanism of the Hydroxide Ion Initiated Decomposition of Ozone in Aqueous Solution, *J. Physical Chem.*, 86 (2), 255-259.

Fujishima, A., Zhang, X. & Tryk, D.A. (2008). TiO_2 photocatalysis and related surface phenomena, *Surf. Sci. Rep.*, 63, 515–582.

Garrido-Ramírez, E.G., Theng, B.K.G. & Mora, M.L. (2010). Clays and oxide minerals as catalysts and nanocatalysts in Fenton-like reactions — A review, *Appl. Clay Sci.*, 47, 182-192.

Herney-Ramirez, J., Vicente, M. A. & Madeira, L.M. (2010). Heterogeneous photoFenton oxidation with pillared claybased catalysts for wastewater treatment: A review, *Appl. Catal. B Environ.*, 98, 10–26.

Herrmann, J-M. (2010). Fundamentals and misconceptions in photocatalysis, *J. Photoch. Photobio. A.*, 216(2-3), 85-93.

Hijnen, W. A., Beerendonk, E. F. & Medema, G. J.(2006). Inactivation credit of UV radiation for viruses, bacteria and protozoan (oo)cysts in water: A review, *Water. Res.*, 40 (1), 3-22.

Hoigné, J. & Bader, H. (1983) Rate constants of reactions of ozone with organic and inorganic compounds in water – I: Non-dissociating organic compounds. *Water Res.*, 17, 173-183.

J. García-Serrano, E. Gómez-Hernández, M. Ocampo-Fernández & U. Pal. (2009). Effect of Ag doping on the crystallization and phase transition of TiO2 nanoparticles, Current Applied Physics, 9, 1097–1105.

Khan, S., Khan, M.A., Hanjra, M.A. & Mu, J. (2009). Pathways to reduce the environmental footprints of water and energy inputs in food production, *Food Policy*, 34, 141–149,

Khun, H. J., Braslavsky, S. E. & Schmidt, R. (2004). Chemical Actinometry (IUPAC Technical Report), *Pure Appl. Chem.*, 76 (12), 2105-2146.

Kim, Y., Lee, J., Jeong, H., Lee, Y., Um, M.-H., Jeong, K. M., Yeo, M.-K. & Kang, M. (2008). Methyl orange removal over Zn-incorporated TiO_2 photo-catalyst, *Journal of Industrial and Engineering Chemistry*, 14 (3), 396-400.

Kulovaara, M., Backlund, P. & Corin, N. (1995). Light-induced degradation of DDT in humic water, *Sci. Total Environ.*, 170, 185-191.

Lange, F., Cornelissen, S., Kubac, D., Sein, M. M., von Sonntag, J. & Hannich, C. B., et al. (2006). Degradation of macrolide antibiotics by ozone: A mechanistic case study with clarithromycin, *Chemosphere*, 65, 17-23.

Lau, T. K., Chu, W. & Graham, N. Reaction pathways and kinetics of butylated hydroxyanisole with UV, ozonation and UV/O_3 processes. Water Res., 41, 765-774, (2007).

Legrini, O., Oliveros, E. & Braun, A. M. (1993). Photochemical Processes for Water Treatment. *Chem. Rev.,.* 93, 671-678.

Li, L., Xu, Z., Liu, F., Shao, Y., Wang, J., Wan, H. & Zheng, S. (2010). Photocatalytic nitrate reduction over $Pt-Cu/TiO_2$ catalysts with benzene as hole scavenger, *Journal of Photochemistry and Photobiology A: Chemistry*, 212, 113-121.

Luiz, D. B., Genena, A. K., José, H. J., Moreira, R. F. & Schröder, H. F. (2009). Tertiary treatment of slaughterhouse effluent: degradation kinetics applying UV radiation or H_2O_2/UV, *Water Sci. Technol.*, 60 (7), 1869-1874.

Luiz, D. B., Genena, A. K., Virmond, E., José, H. J., Moreira, R. F., Gebhardt, W. & Schröder, H. F. (2010). Identification of degradation products of erythromycin A arising from ozone and AOP treatment, *Water Environ. Res.*, 82, 797-805.

Luiz, D.B., Silva, G. S., Vaz, E.A.C., José H.J. & Moreira, R.F.P.M. (2011). Evaluation of Hybrid Treatments to Produce High Quality Reuse Water, *Water Sci. Technol.*, 63(9), 46-51.

Malato, S., Fernández-Ibáñez, P., Maldonado, M.I., Blanco, J. & Gernjak, W. (2009). Decontamination and disinfection of water by solar photocatalysis: Recent overview and trends, *Catalysis Today*, 147, 1–59.

Malato, S., Fernández-Ibáñez, P., Maldonado, M.I., Blanco, J. & Gernjak, W. (2009). Decontamination and disinfection of water by solar photocatalysis: Recent overview and trends, *Catalysis Today*, 147, 1–59.

Martín-Domínguez, A., Alarcón-Herrera, M. T., Martín-Domínguez, I. R. & González-Herrera, A. (2005). Efficiency in the disinfection of water for human consumption in rural communities using solar radiation, *Sol. Energy*, 78 (1), 31-40.

Matilainen, A. & Sillanpää, M. (2010). Removal of natural organic matter from drinking water by advanced oxidation processes, *Chemosphere*, 80, 351–365

Mellouki, A., Le Bras, G. & Sidebottom, H. (2003). Kinetics and Mechanisms of the Oxidation of Oxygenated Organic Compounds in the Gas Phase, *Chem. Rev.*, 103, 5077–5096.

Muñoz, F. & von Sonntag, C. (2000). The reactions of ozone with tertiary amines including the complexing agents nitrilotriacetic acid (NTA) and ethylenediaminetetraacetic acid (EDTA) in aqueous solution. *J. Chem. Soc., Perkin Trans.*, 2, 2029-2033.

Nadezhdin, A. D. (1988). Mechanism of Ozone Decomposition in Water. The Role of Termination. *Ind. Eng. Chem. Res.*, 27 (4), 548-550.

Navalon, S., Alvaro, M. & Garcia, H. (2010). Heterogeneous Fenton catalysts based on clays, silicas and zeolites. *Applied Catalysis B: Environmental*, 99, 1–26.

Neamtu, M. & Frimmel, F. H. (2006). Photodegradation of endocrine disrupting chemical nonylphenol by simulated solar UV-irradiation. *Sci. Total Environ.* , 369 (1-3), 295-306.

Oliver, P., Rodríguez, R. & Udaquiola, S. (2008). Water use optimization in batch process industries. Part 1: design of the water network. J. *Cleaner Product.*, 16, 1275-1286.

Oller, I., Malato, S. & Sánchez-Pérez, J.A. (2010). Combination of Advanced Oxidation Processes and biological treatments for wastewater decontamination - A review. *Science of the Total Environmen*, 409 (20), 4141-4166.

Parkinson, A., Barry, M. J., Roddick, F. A. & Hobday, M. D. (2001). Preliminary toxicity assessment of water after treatment with UV-irradiation and UVC/H_2O_2. *Water Res.* , 35 (15), 3656-3664.

Ranjit, K. T.; Viswanathan, B. (1997). Photocatalytic reduction of nitrate and nitrite ions over doped TiO_2 catalysts. *Journal of Photochemistry and Photobiology A: Chemistry*, 107, 215-220.

Rengaraj, S. & Li, X.Z. Enhanced photocatalytic reduction reaction over Bi^{3+}-TiO_2 nanoparticles in presence of formic acid as a hole scavenger. *Chemosphere*, 66, 930–938, 2007.

Rincón, A. G. & Pulgarin, C. (2006). Comparative evaluation of Fe^{3+} and TiO_2 photoassisted processes in solar photocatalytic of water. *Appl. Catal. B-Environ.*, 63, 222-231.

Rosal, R., Rodríguez, A., Perdigón-Melón, J. A., Mezcua, M., Hernando, M. D. & Léton, P., et al. (2008). Removal of pharmaceuticals and kinetics of mineralization by O_3/H_2O_2 in a biotreated municipal wastewater. Water Res., 42, 3719-3728.

Sá, J., Agüera, C. A., Gross, S. & Anderson, J. A. (2009). Photocatalytic nitrate reduction over metal modified TiO_2. *Applied Catalysis B: Environmental*, 85, 192–200.

Sciacca, F., Rengifo-Herrera, J. A., Wéthé, J. & Pulgarin, C. (2010). Dramatic enhancement of solar disinfection (SODIS) of wild Salmonella sp. in PET bottles by H_2O_2 addition on natural water of Burkina Faso containing dissolved iron. *Chemosphere*, 78 (9), 1186-1191.

Shifu, C., Wei, Z. & Sujuan, Z. (2008). Preparation, characterization and activity evaluation of p-n junction photocatalyst p-ZnO/n-TiO_2. *Applied Surface Science*, 255, 2478-2484.

Sievers, M. (2011). Advanced Oxidation Processes. In: Peter Wilderer (ed.) *Treatise on Water Science*, Oxford: Academic Press, vol. 4, pp. 377–408.

Sirés, I., Oturan, N., Oturan, M.A., Rodríguez, R.M. & Garrido, J.A., Brillas, E. (2007). Electro-Fenton degradation of antimicrobials triclosan and triclocarban. *Electrochimica Acta*, 52, 5493–5503.

Song, W., Ravindran, V. & Pirbazari, M. (2008). Process optimization using a kinetics model for the ultraviolet radiation-hydrogen peroxide decomposition of natural and synthetic organic compounds in groundwater. *Chem. Eng. Sci.*, 63, 3249-3270.

Staehelin, J. & Hoigne, J. Decomposition of Ozone in Water: Rate of Initiation by Hydroxide Ions and Hydrogen Peroxide. *Environ. Sci. Technol.*, 16, 676–681, 1982.

Tokos,H. & Zorka, N.P. (2009). Synthesis of batch water network for a brewery plant. *Journal of Cleaner Production*, 17, 1465–1479

Umar, M., Aziz, H. A. & Yusoff, M. S. (2010). Trends in the use of Fenton, electro-Fenton and photo-Fenton for the treatment of landfill leachate. *Waste Management*, 30 (11), 2113-2121.

von Gunten, U. (2003a). Ozonation of drinking water: Part I. Oxidation kinetics and product formation. *Water Res.*, 37, 1443-1467.

von Gunten, U. (2003b). Ozonation of drinking water: Part II. Disinfection and by-product formation in presence of bromide, iodide or chlorine. *Water Res.*, 37, 1469-1487.

Wehbe, N., Jaafar, M., Guillard, C., Herrmann, J.-M., Miachon, S., Puzenat, E. & Guilhaume, N. (2009). Comparative study of photocatalytic and nonphotocatalytic reduction of nitrates in water. *Applied Catalysis A: General*, 368, 1-8.

Y. Li & F. Wasgestian. (1998). Photocatalytic reduction of nitrate ions on TiO2 by oxalic acid. *Journal of Photochemistry and Photobiology A: Chemistry*, 112, 255-259.

Yang, C., Xu, Y. R., Teo, K. C., Goh, N. K., Chia, L. S. & Xie, R. J. (2005). Destruction of organic pollutants in reusable wastewater using advanced oxidation technology. *Chemosphere*, 59 (3), 441-445.

Zhang, F., Jin, R., Chen, J., Shao, C., Gao, W., Li, L. & Guan, N. (2005). High photocatalytic activity and selectivity for nitrogen in nitrate reduction on Ag/TiO$_2$ catalyst with fine silver clusters. *Journal of Catalysis*, 232, 424–431.

2

The Application of Vacuum Impregnation Techniques in Food Industry

A. Derossi, T. De Pilli and C. Severini
Department of Food Science, University of Foggia
Italy

1. Introduction

The interest of food scientists in the filed of microstructure is recently exponentially increased. This interest has raised after the recognition of the importance that chemical reactions and physical phenomena occurring at microscopic scale have on safety and quality of foods. This concept was well resumed from Aguilera (2005) which stated that "*...the majority of elements that critically participate in transport properties, physical and rheological behavior, textural and sensorial traits of foods are below the 100 μm range*". As example, Torquato (2000) reported that through microscopic observation it is possible to observe that only a portion of cells of the crumb bread solid matrix are connected, even though at macroscopic level it may appear as completely connected; so, the three dimensional spatial distribution of cell crumbs greatly affects the sensorial quality of bread. Before the scientific recognition of the above consideration, food scientists which focused their efforts on the effects of traditional and innovative industrial processes on food quality, analyzed only macroscopic indexes such as color, texture, taste, concentration of several nutritional compounds, etc. without to consider that they essentially are the result of chemical and physical phenomena occurring at microscopic level. However, with the aim to be more precise in the use of the term "microscopic" we may generalize the classification that Mebatsion et al. (2008) reported for the study of fruit microstructure. The authors considered three different spatial scales: 1. The macroscale which refers the food as a whole or a continuum of biological tissues with homogeneous properties; 2. The mesoscale which refers the topology of biological tissues; 3. The microscale which addresses the difference of individual cells in terms of cell walls, cell membranes, internal organelles, etc. Here, we would like focus the attention on the importance to study the foods at mesoscopic scale by considering them as mesoscopic divided material (MDM). On this basis the majority of foods may be defined as "*biological systems where an internal surface partitions and fills the space in a very complex way*". At mesoscopic scale the three dimensional architecture of foods may be studied analyzing the relation between void and solid phases where the voids (capillaries, pores) may be partially or completely filled with liquids or gases. The relations between these two phases and their changes during processing is one of the most important factors affecting the safety and the sensorial and/or nutritional quality of foods. Until few years ago only the porosity fraction of foods was reported in literature as experimental index of the internal microstructure but it gives us only low level of information. Instead, a second level of structure characterization may be reached by analyzing pore dimension, shape, length, surface roughness, tortuosity, connectivity, etc. So, one of the most important future challenge will be the precise characterization of the three dimensional architecture of foods, its changes during food processing and its relation with safety and quality. However, some pioneering researchers have focused their attention on this research field and the first results are reported in literature (Datta, 2007a; Datta, 2007b; Halder et al., 2007). For instance, the term "food matrix engineering (FME)" has been used to define a branch of food engineering that applies the knowledge about food matrix composition, structure and properties with the aim to promote and control adequate changes that improve some sensorial and/or functional properties in the food products as well as their stability (Fito et al., 2003). The authors reported that food gels and emulsions, extruded, deep fried and puffed foods may be considered as current FME. Instead, among the non-traditional processing, one of the most important and newest techniques based on the properties of food microstructure is the vacuum impregnation (VI). With this term are classified some technologies based on the exploitation of void fraction of foods with the aim to introduce, in a controlled way, chemical/organic compounds into the capillaries of biological tissues. VI is based on the application of a partial vacuum pressure which allows to remove native liquids and gases entrapped into the capillaries and to impregnate them with a desired external solution after the restoration of atmospheric pressure. At first, a vacuum osmotic dehydration (VOD) treatment was proposed and studied as method to accelerate water loss and solid gain during the immersion of vegetables into hypertonic solution. More recently, VI treatments have been studied as method to enrich food with nutritional and functional compounds, to introduce some ingredients with the aim to obtain food with innovative sensorial properties as well as to introduce compounds able to inhibit the most important degradation reactions and the microbial growth. This chapter has the aim to analyze and to discuss the application of vacuum impregnation techniques in food industry. At first, the theoretical principles and the mathematical modeling of the phenomena involved during the application of VI will be discussed. Also, the response of biological tissues to vacuum impregnation will be reviewed. In a second section, focusing the attention of the reader on different potential applications in food industry, the effect of process variables on both impregnation level and quality of final products will be analyzed.

2. Theoretical background and structural changes

The main phenomenon on the basis of vacuum impregnation treatment is the hydrodynamic mechanism (HDM) which was well described from Fito & Pastor (1994), Fito (1994) and Chiralt & Fito (2003). The authors, discussing the results of the water diffusion coefficients (De) obtained with both microscopic and macroscopic analysis during osmotic dehydration (OD), reported that the values were in general 100 times greater when measured at microscopic scale; in particular, De values were about 10^{12} m^2/s and 10^{10} m^2/s when relieved respectively at microscopically and macroscopically. These observations suggested that a fast mass transfer mechanism, in addition with the traditional diffusion, is involved during OD treatments. This mechanism is the hydrodynamic mechanism which is based on pressure gradients generated from the changes of sample volume and/or externally imposed at non-compartmented section such as intercellular spaces, capillaries, pores, etc. These phenomena play a key role in the solid-liquid operation increasing the rate of several processes during which mass transfer occurs. The action of HDM was well explained from Chiralt & Fito (2003). When food pieces are immersed in an external solution the surface of samples is washed and the solution partially penetrates into the open pores. After, in line with a deformation of cell membranes due to the loss of native liquids and gases, a pressure gradient is generated and HDM is promoted.

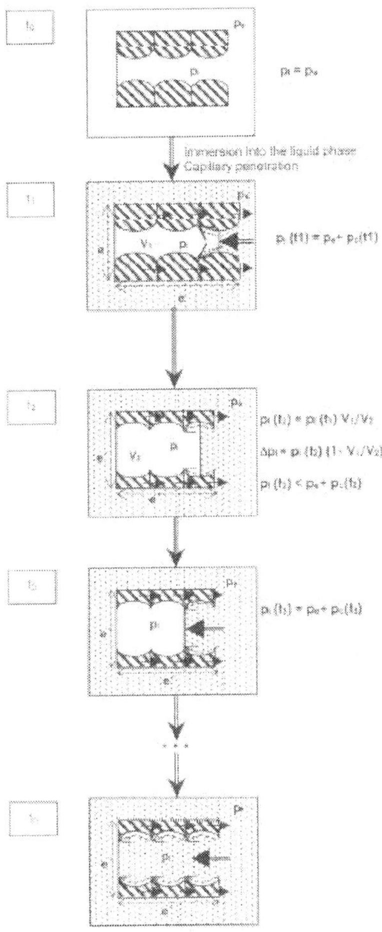

Fig. 1. Schematic representation of hydrodynamic mechanisms due to capillary action and pressure gradients as a consequence of internal volume changes (From Chiralt and Fito, 2003).

In figure 1 a schematic representation of HDM during osmotic dehydration is shown. Before the immersion of vegetable tissues into hypertonic solution the capillary pressure inside the pore is equal to the external (atmospheric) pressure. At t_1 capillary pressure promotes the initial gain of osmotic solution and the compression of gases inside the pores; so, internal pressure becomes greater than external one. In these conditions gases irreversibly tend to flow out and the cells in contact with hypertonic solution dehydrated due to osmotic pressure gradient. The water loss produces an increase of internal volume as a result of cells shrinkage during dehydration. Moreover, in line with the volume increase the pressure inside the capillary becomes lower to the external one promoting the suction of additional external solution. This process proceeds until the capillary is completely impregnated with the osmotic solution (Barat et al., 1998; Chiralt and Fito, 2003). All these phenomena becomes marked during vacuum impregnation treatments when a vacuum pressure gradient is externally imposed. The treatments are performed through two subsequent steps: 1. The immersion of foods into the solution and the application of a vacuum pressure (p) for a vacuum period (t_1) (also called vacuum time); 2. The restoration of atmospheric pressure maintaining the samples immersed into the solution for a relaxation period (t_2) (also called relaxation time). During VI in addition to HDM a deformation-relaxation phenomena (DRP) simultaneously occurs. HDM and DRP both affect the reaching of an equilibrium situation and their intensities are strictly related with the three dimensional food microstructure and mechanical properties of solid matrix.

Figure 2 schematically shows the phenomena involved during vacuum solid-liquid operations of an ideal pore. At time zero the samples are immersed into the external liquid and the internal pressure of the pore (pi) is equal to external (atmospheric) pressure (p_e). After, a vacuum pressure (p) is applied in the head space of the system for a time (t_1) promoting a situation in which p_i is greater than p_e. In this condition, the internal gases expand producing the deformation (enlargement) of capillary and the increase of internal volume. Moreover, native liquids and gases partially flow out on the basis of the pressure gradient (step 1-A). At this time hydrodynamic mechanism begins and external liquid partially flows inside the capillary as a consequence of the pressure gradient. These phenomena simultaneously occur until the equilibrium is reached (step 1-B). In the second step the atmospheric pressure in the head space of the system is restored and the samples are maintained into the solution for a relaxation time (t_2). During this period, the generated pressure gradient ($p_i < p_e$) promotes both HDM and solid matrix deformation (compression) which respectively produce capillary impregnation and the reduction of pore volume until a new equilibrium is reached (Fito & Chiralt, 1994; Fito et al., 1996).

2.1 Mathematical modeling of vacuum impregnation and related structural changes

Fito (1994) and Fito et al. (1996) were the first scientists which translate in mathematical language the phenomena involved at mesoscopic scale during VI treatments. The model is based on the analysis of the contributions of both liquid penetration and solid matrix deformation of an ideal pore of volume $Vg_0 = 1$ during each step of VI. From time t = 0 to the step 1B a situation expressed by the following equation is obtained:

$$Vg_{1b} = 1 + Xc_1 - Xv_1 \tag{1}$$

Where Vg_{1b} is the pore volume at the step 1-B, Xc_1 is the increment of pore volume due to the expansion of internal gases and Xv_1 is the partial reduction of pore volume due to the initial suction of external liquid as a consequence of HDM.

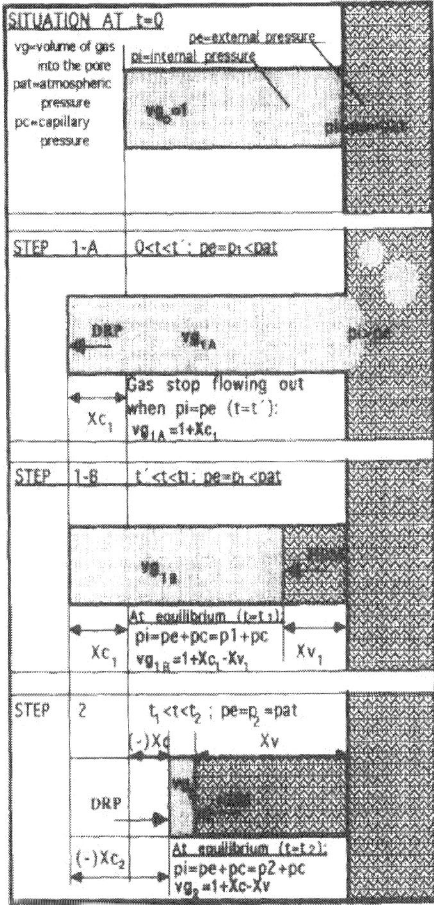

Fig. 2. Schematic representation of vacuum impregnation of an ideal pore. Deformation-relaxation phenomena and hydrodynamic mechanism occurring during vacuum period (t1) and relaxation time (t2) (from Fito et al., 1996).

At the end of step 2 the total liquid penetration and matrix deformation may be described respectively by the equations:

$$Xv = Xv_1 + Xv_2 \qquad (2)$$

$$Xc = Xc_1 + Xc_2 \qquad (3)$$

where Xv_1 and Xv_2 are the volume reduction due to liquid penetration respectively at the end of step 1 and 2; Xc_1 and Xc_2 are the volume pore changes as a result of solid matrix deformation (enlargement and compression) after the steps 1 and 2. Also, the total volume variation at the end of the process may be described as follow:

$$Vg_2 = 1 + Xc - Xv \tag{4}$$

Equations 2 and 3 may be used to calculate liquid penetration and solid matrix deformation of the total sample taking into account its porosity fraction value (ε_e):

$$X = e_e Xv \tag{5}$$

$$g = e_e Xc \tag{6}$$

As reported from Fito et al. (1996) when a pressure variation applied in a solid-liquid system and an equilibrium situation is reached, HDM assumes an isothermal compression of gas into the pores. So, the situation reached at the end of step 1-B may be mathematically express as:

$$\frac{V_{g1B}}{V_{g1A}} = \frac{1 + Xc - Xv}{1 + Xc} = \frac{1}{r} \tag{7}$$

Where r is the apparent compression rate ($\sim$ atmospheric pressure/vacuum pressure, dimensionless) (Zhao and Xie, 2004). From equation 7 may be obtained the following:

$$\frac{Xv1}{1 + Xc1} = 1 - \frac{1}{r} \tag{8}$$

Also, by using the equation 5 and 6 it is possible to obtain:

$$X_1 = (\varepsilon_e + \gamma_1)\left(1 - \frac{1}{r_1}\right) \tag{9}$$

On these basis the equilibrium situation at the end of step 1-B may be mathematically expressed as:

$$X_1 - Y_1 = \varepsilon_e\left(1 - \frac{1}{r_1}\right) - \frac{\gamma_1}{r_1} \tag{10}$$

Furthermore, the same considerations may be extended for the phenomena involved from step 1 and step 2. So, between $t = t_1$ and $t = t_2$ the equilibrium situation may be expressed as:

$$X - Y = (\varepsilon_e + \gamma)\left(1 - \frac{1}{r_2}\right) - \gamma_1 \tag{11}$$

Starting from the above equations it is possible to calculate the porosity value (ε_e) at the end of VI process from the value of X, γ and γ_1 by:

$$\varepsilon_e = \frac{(X - \gamma)r_2 + \gamma_1}{r_2 - 1} \tag{12}$$

Where X is the volume fraction of sample impregnation by the external liquid at the end of VI treatments (m³ of liquid/m³ of sample a t=0), ε_e is the effective porosity, γ_1 is the relative volume deformation at the end of vacuum period (t₁, m³ of sample deformation/m³ of sample at t=0); γ is the volume deformation at the end of process (m³ of sample deformation/m³ of sample at t=0). However, although the measure of porosity fraction value is easily obtained from the apparent (ρ_a) and real density (ρ_r) values of the sample (Lewis, 1993; Gras, Vidal-Brotons, Betoret, Chiralt & Fito, 2002), the experimental estimation of X, γ and r are not easy to measure and some modifications of the equipment used for the experiments are necessary (Salvatori et al., 1998). Briefly, the experimental methodology is performed precisely weighting the sample at different steps of the treatment even if it is under vacuum condition (Fito et al., 1996; Salvatori et al., 1998). The authors defined a parameter called magnitude (H):

$$H_t = X_t - \gamma_t = \frac{L_0 - L_t - M_w}{V_0 \rho} \tag{13}$$

Where L is the weight measured in the balance (kg), M_w is the liquid evaporated during the experiment, V is the sample volume (m³) and ρ is the liquid density (kg/m³). Instead the subscripts 0 and t refer respectively the time zero and any time t_i corresponding at each step of the process. Practically speaking the magnitude H is an overall index of the contribution of both liquid penetration (X) and solid matrix deformation (γ) on sample volume. A theoretical curve of H value as a function of time during vacuum impregnation is shown in figure 3.

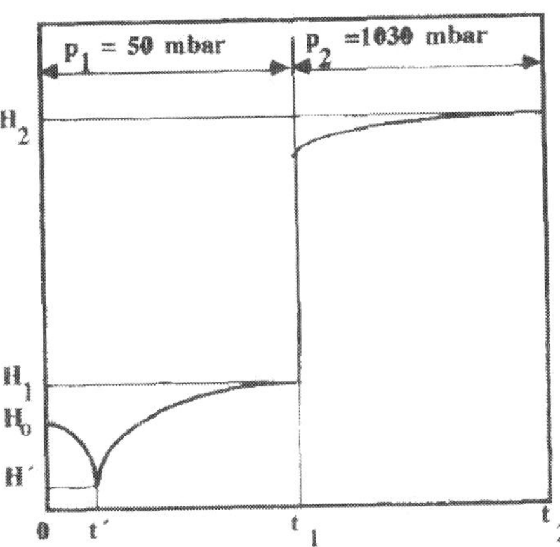

Fig. 3. Theoretical curve of H value as a function of time during vacuum impregnation treatments (From Fito et al., 1998).

From the figure it is possible to observe that as vacuum pressure is applied in the system, H value quickly decreases until H′ at t = t′. This is because sample volume increases under the action of the gas expansion due to the positive pressure gradient (pi > pe). During the vacuum period (from t = t′ to t = t₁) H value slowly increases because native liquids and gases flow out, solid matrix relaxes and external liquid begins to impregnate pores due to

HDM. H_1 is obtained when the equilibrium condition is reached at the end of step 1B. At time $t = t_1$ the restoration of atmospheric pressure (p = 1030 mbar) leads a sudden increase of H value as a consequence of liquid penetration. Also, during relaxation time (from $t = t_1$ to $t = t_2$) pore volume reduction due to capillary compression and HDM simultaneously occur. H_2 value is obtained when the new equilibrium situation will be reached. From the above consideration, the crucial importance of microscopic properties of foods such as dimension and shape of samples, their three dimensional architecture, the resistance of biological tissue to gas and liquid flow, solid matrix deformation, etc., is obvious. In particular, these factors affect the kinetics of all phenomena simultaneously involved during VI; so, the quality of foods will be a result of the rates by which each phenomena occur. For instance, if the solid matrix relaxation of food is slow, capillary impregnation could occur without significant deformation. On the other hand a fast deformation could significantly reduce capillary impregnation. Salvatori et al. (1998) studied the time evolution of H value of several vegetables submitted to VI at 50 mbar for different vacuum period (t_1) and relaxation time (t_2). As example, figure 4 shows the results obtained from apple samples (Granny Smith) submitted to VI for a $t_1 = t_2 = 15$ minutes.

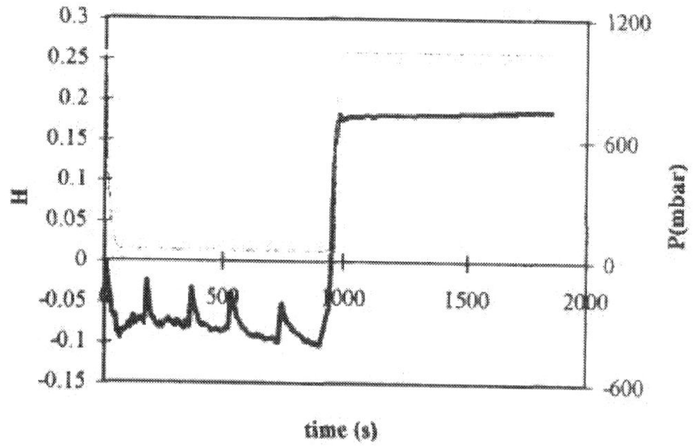

Fig. 4. Experimental curve of apple sample submitted to VI treatment at 50 mbar for a t1 = t2 = 15 minutes (From Salvatori et al., 1998).

As expected, H value decreased quickly after the application of vacuum pressure in line with pores expansion. However, in disagreement with the behavior of an ideal pore reported in figure 3, at the end of vacuum period (~ 1000 s), H value was negative (H1 < 0) stating that pore volume deformation still was greater than liquid penetration due to the enlargement of pores. In particular, the authors reported that at the end of vacuum period the X_1 and γ_1 values (an average of all experiments performed in different operative conditions) respectively were – 4.2% and 1.7%. The negative value of the liquid penetration at the end of vacuum period was explained on the basis of native liquid release from the pores under the action of the negative pressure gradient ($p_e < p_i$). The peaks values observed in figure 1 are in line with the variability of both tissue structure characteristics and the surrounding fluid properties (Salvatori et al., 1998). After, the restoring of atmospheric pressure led to a significant increase of H value due to the compression of pores ($\gamma < 0$) and the suction of external liquid (X > 0). Indeed at the end of the process the authors reported a value of X and γ of 15% and -0.6% respectively. However, foods may show significant different trends as a function of their unique microstructure properties which are very different from the behavior of an ideal pore. For instance, mango and peach samples as well as oranges showed a positive solid matrix deformation (enlargement of the pores) also after the restoration of atmospheric pressure (Salvatori et al., 1998; Fito et al., 2001). Gras et al. (2002) reported that carrot, diced zucchini and beetroot, submitted to VI treatment in a sucrose isotonic solution, showed a pore volume increase (γ > 0) also at the end of process coupled with significant impregnation values of 16%, 20% and 7% respectively. These behaviors were explained with an overall situation in which the rate of liquid impregnation being very fast in comparison with the solid matrix deformation allowed to keep a residual capillary expansion.

2.2 Gas, liquid and solid matrix volume changes during VI

With the aim to have a complete theoretical analysis of the phenomena involved during VI it is important to consider the gases, liquid and solid matrix volume changes occurring during vacuum and relaxation times. Barat et al. (2001) proposed an experimental approach to study the volume changes during VI assuming food constituted of three phases: solid matrix (SM), liquid phase (LP) and gas phase (GP). On this basis the authors reported that total volume changes which occur during vacuum impregnation treatment may be mathematically express as:

$$\Delta V = \Delta V^{LP} + \Delta V^{GP} + \Delta V^{SM} \tag{14}$$

Moreover, since the variation of solid matrix volume may be assumed equal to zero equation 14 may be reduced at:

$$\Delta V = \Delta V^{LP} + \Delta V^{GP} \tag{15}$$

Here, ΔV may be obtained by pycnometer method of fresh and impregnated food whereas the volume changes of liquid phase (ΔV^{LP}) may be estimated by the following equation (Barat et al., 2001: Atares, Chiralt & Gonzales-Martinez, 2008):

$$\Delta V^{L} = \left\{ \left[m^{t}\left(x_{w}^{t} - x_{s}^{t}\right)\right] / r_{LP}^{t} - \left[m^{0}\left(x_{w}^{0} - x_{s}^{0}\right)\right] / r_{LP}^{0} \right\} / V^{0} \tag{16}$$

Where m is the mass of sample (g), x_{w} and x_{s} are respectively the moisture and solid content of sample (water or solid, g/g of fresh vegetable), ρ_{LP} is the density of liquid phase (g/mL) and V is the volume of samples (mL). Moreover, the superscripts t and 0 refer respectively to each time treatment (t_i) and fresh vegetable.

The liquid phase density values may be estimated by the follow equation (Atares et al., 2008):

$$r_{LP} = \left(230z_{s}^{2} + 339z_{s} + 1000\right) / 1000 \tag{17}$$

where z_{s} is the solid content of sample liquid phase (g solid/g liquid phase). This method has been proved to give precise results during osmotic dehydration process (Barat et al., 2000; Barat et al., 2001). In particular, Barat et al. (2001) plotted the ΔV as a function of ΔV^{LP}

of apple (Granny Smith) samples submitted to traditional osmotic dehydration (OD) and pulsed vacuum osmotic dehydration (PVOD) performed applying a pressure of 180 mbar for a t_1 = 5 minutes.

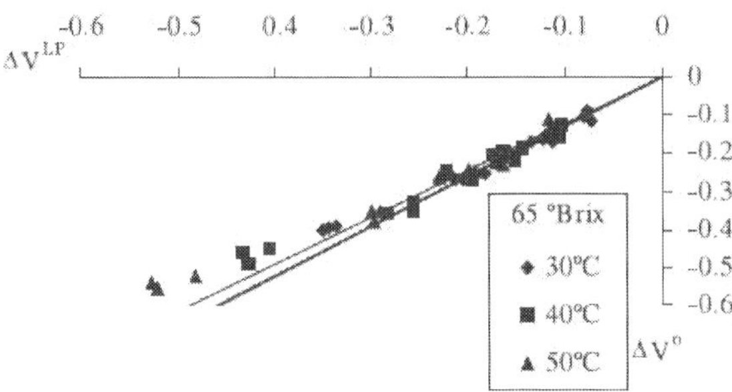

Fig. 5. Total volume changes as a function of liquid phase volume changes of apple samples submitted to OD and PVOD in different operative conditions.

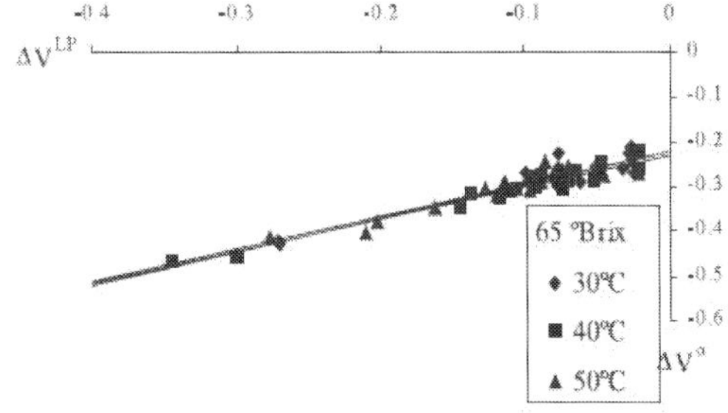

Fig. 6. Total volume changes as a function of liquid phase volume changes of apple samples submitted to OD and PVOD in different operative conditions.

Samples submitted to OD showed a linear trend with no significant intercept value stating that no significant total volume variation was observed before liquid penetration occurring under the action of HDM. By linear regression of experimental data it is possible to estimate the relative contribution of ΔV^{LP} and ΔV^{GP} by using the following equation:

$$\Delta V = s_1 \Delta V^{LP} \tag{18}$$

$$\frac{\Delta V^{GP}}{\Delta V^0} = 1 - \frac{1}{s_1} \tag{19}$$

In comparison with OD treatments, it is worth noting (Figure 6) that samples submitted to PVOD showed a significant intercept with an average value of 0.20. This may be considered as the result of the initial vacuum pulse which promotes the pores expansion coupled with the removal of native liquid and gases and the action of HDM. In this case the linear regression of experimental data shown in figure 6 assumes the following form:

$$\Delta V = i_2 + s_2 \Delta V^{LP} \tag{20}$$

Also, in the case of PVOD the general form of the equation 15 was modified with the following:

$$\Delta V = \Delta V^{LP} + \Delta V^{LP-VI} + \Delta V^{CR} \tag{21}$$

Where ΔV^{LP-VI} and ΔV^{CR} are the volume change due to liquid impregnation and volume changes due to compression-relaxation of solid matrix respectively. In equation 21 the two additional terms substituted ΔV^{GP}. Also, it is important to note that these two terms explain the intercept value of equation 20 and their sum is close to the porosity of food (for apples sample about the 0.22). In their conclusion the authors stated that osmotic dehydration in porous fruits may be explained in terms of LP and GP changes in line with the flow of gases and external liquid and the changes of pore volume as a consequence of the enlargement and compression of pores.

3. Process variables

As previous reported, vacuum impregnation is a technique that allows to introduce several chemical compounds and/or ingredients in the void phase of foods. The process is performed by applying a vacuum pressure in the head space of the system for a time t_1 and then by restoring atmospheric pressure for a relaxation time t_2. The numerous phenomena involved during the process are affected by several variables which may be classified as external and internal of foods. In the first class, vacuum pressure (p), time length of vacuum period (t_1), time length of relaxation time (t_2), viscosity of external solution, temperature, concentration of solution, product/solution mass ratio and size and shape of the samples may be introduced. Instead, the three dimensional architecture of food and the mechanical properties of the biological tissues may be considered as internal variables. However, it is worth nothing that the term "three dimensional architecture" refers to an ensemble of microscopic and mesoscopic characteristics such as porosity, size and shape of pores, connectivity, tortuosity, capillary curvatures, etc. which greatly affect the vacuum impregnation treatment. In this paragraph the effects of each variable above reported, considered both singularly and as synergistic effects will be discussed.

3.1 External variables

Among the external variables, vacuum pressure may be considered as the most important because it represents the force that produces the pressure gradient between the void phase of food and the atmosphere surrounding the external liquid. Briefly, vacuum pressure is the variable by which all phenomena previously reported may occur. In general, a vacuum pressure ranged between 50 and 600 mbar is reported in literature (Fito et al., 1996; Rastogi et al., 1996; Salvatori et al., 1998; Barat et al., 2001; Fito et al., 2001a; Fito et al., 2001b; Giraldo et al., 2003; Mujica-Paz et al., 2003; Zhao & Xie, 2004; Silva Paes et al., 2007; Corzo et al., 2007; Derossi et al., 2010; Derossi et al. 2011). Also, vacuum level is generally considered directly related to an increase of impregnation level (X) as a consequence of a higher release of native liquids and gases coupled with a greater HDM and DRP. Mujica-Paz et al. (2003a) studied the effect of vacuum pressure in a range of 135-674 mbar on the volume of pores impregnated from an isotonic solution of several fruits. The authors showed that for apple, peach, papaya and melon samples a greater impregnation was observed in line with an increase of vacuum

　　　　　　　　Scientific, Health and Social Aspects of the Food Industry

pressure; instead, for mango, papaya and namey X values increased with the increasing of vacuum level until a maximum after that impregnation slightly decreased. Mujica-Paz et al. (2003b) studied the effect of vacuum pressure (135-674 mbar) on the weight reduction (WR), water loss (WL) and solid gain (SG) of apple, melon and mango slices kept in a hypertonic solution for a $t_1 = t_2 = 10$ minutes. For all experiments, melon and mango samples showed a positive WR, stating that fruits lost a significant fraction of their weight; instead, apples showed a negative WR values, stating a weight gain. In particular, for melon samples the authors reported a direct correlation between vacuum pressure and WR probably because as lower the vacuum pressure as greater the capillary impregnation, which decreased the weight reduction of the samples caused by both the osmotic dehydration and the removal of native liquids from the pores. Indeed the negative values observed for apple samples were explained on the basis of their high porosity fraction (~27.3%). In this conditions, the high free volume of apples increased the impregnation level more than water loss leading to an increase of weight of samples (figure 7a and 7b). Also, similar results were observed for water loss which assumed positive values for mango and melon and negative values in certain operative condition for apple samples (figures 8a and 8b).

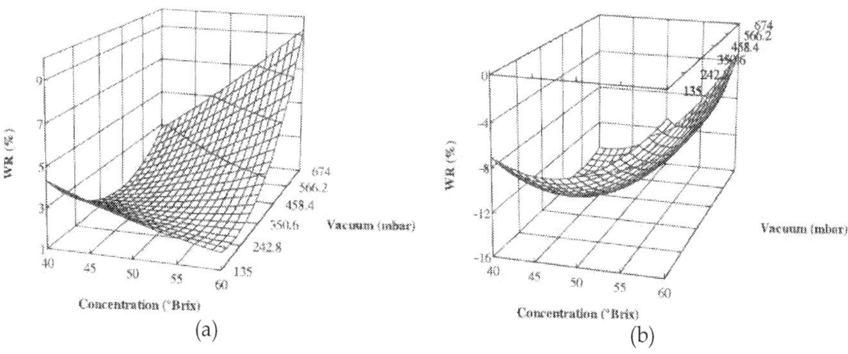

Fig. 7. Effect of vacuum pressure and osmotic solution concentration on weight reduction of melon and apple samples (from Mujica-Paz et al., 2003).

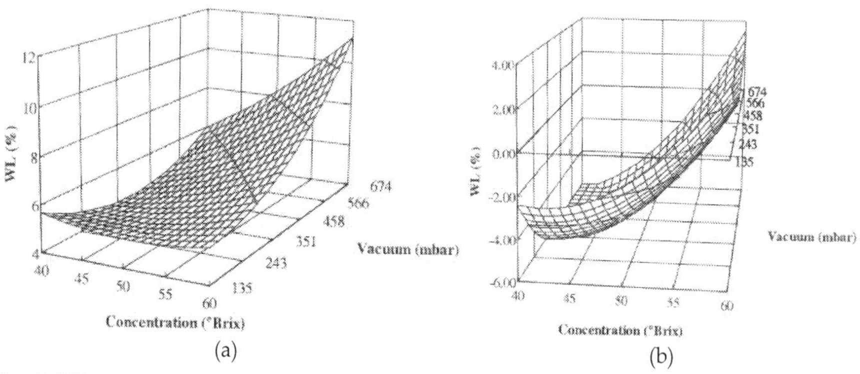

Fig. 8. Effects of vacuum pressure and osmotic solution concentration on water loss of mango and apples samples (from Mujica-Paz et al., 2003).

Derossi et al. (2010), studying the application of a vacuum acidification treatment on pepper slices, reported that pH ratio values (RpH) were lower when a pressure of 200 mbar was used in comparison with the results obtained applying a pressure of 400 mbar. Furthermore, some authors studied the effect of a decrease of vacuum pressure on the acidification of zucchini slices, showing that RpH values were directly correlated with vacuum pressure (Derossi et al., 2011). Nevertheless, the same authors did not observe a statistically significant variation of the porosity of the impregnated samples when vacuum level increased from 400 to 200 mbar. However, Derossi et al. (2010; 2011) showed that the use of a vacuum pressure significantly improved the rate of acidification in comparison with a traditional acidifying-dipping at atmospheric pressure. The authors attributed this result to the increase of acid-solution contact area due to the capillary impregnation. Hofmeister et al. (2005) studied the visual aspect of Mina cheese samples submitted to vacuum impregnation in different operative conditions. The authors reported a greater impregnation level when a pressure of 85.3 kPa was applied in comparison with the experiments performed at 80 kPa. In agreement with the theoretical principles this result was assumed as a consequence of a greater removal of air from the pores of cheese samples. Moreover, Andres et al. (2001) studied the effect of vacuum level on apple samples showing that impregnation level was affected by the applied vacuum pressure.

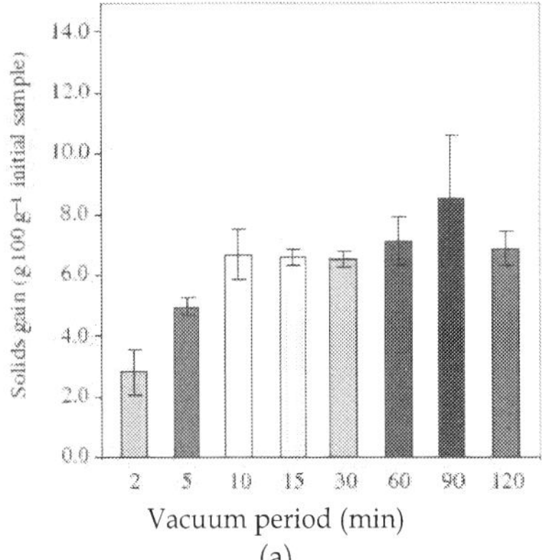

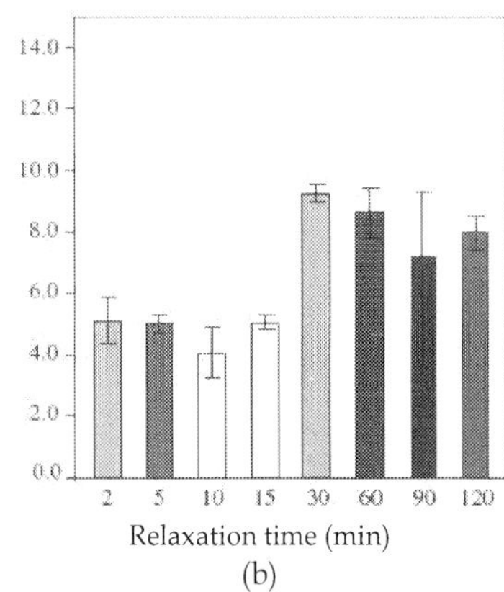

(a)   (b)

Fig. 9 Solid gain values of apple cylinders submitted to vacuum osmotic dehydration at 10 mbar as a function of vacuum and relaxation times (from Silva Paes et al (2007).

Vacuum period and relaxation times represent two important variables affecting the results of the application of VI. Vacuum period (t_1) refers to the time during which food microstructure tends to reach an equilibrium situation after the application of the vacuum pulse. As previously reported, several phenomena such as deformation (enlargement) of capillaries, the expulsion of gases and native liquids from the pores and the partial impregnation of pores, simultaneously occur during t_1. Instead, relaxation time (t_2) is the period during which food structure moves toward an equilibrium situation after the restoration of atmospheric pressure. During t_2, capillary impregnation (by HDM) and deformation (compression) of pores occur under the action of a negative pressure gradient.

In general an increase of impregnation level of foods as a function of the time length of both t_1 and t_2 should be expected. Nevertheless due to the complex equilibrium occurring among the several phenomena involved the interpretation of the results is very difficult. Fito et al. (1996) and Salvatori et al. (1998) studied the response of several fruits (apple, mango, strawberry, kiwi, peach, banana, apricot) on both the X and γ values when submitted to vacuum impregnation at 50 mbar for a t_1 ranged between 5 and 15 minutes and a relaxation time ranged between 5 and 20 minutes. In both cases the authors did not observe significant differences of volume impregnated and solid matrix deformation as a function of t_1 and t_2 variation. Derossi et al. (2010) did not show a significant difference of total mass variation of pepper slices submitted to vacuum acidification at 200 mbar for a vacuum time of 2 and 5 minutes. On the contrary the results reported from Silva Paes et al (2007) stated a great influence of vacuum period and relaxation time on water gain (WG), solid gain (SG) and weight reduction (WR) of apple cylinders submitted to vacuum osmotic dehydration. As example figures 9a and 9b respectively show the solid gain of apple samples as a function of vacuum period and relaxation time.

It is worth nothing that SG values were directly correlated with an increase of t_1 in the range of 2 and 10 minutes. Instead, when a greater vacuum time was applied, no significant differences were observed. The authors suggested that when a short t_1 was applied, the removal of gases from the pores was not complete; so, their residues could have hindered the osmotic solution penetration. When vacuum period increased from 10 to 120 minutes, the complete removal of gases made the increase of t_1 unable to influence solid gain. In figure 9b it is possible to observe an increase of solid gain until a relaxation time of 30 minutes after that SG values are approximately constant. The authors hypothesized that 30 minutes is the time necessary to reach the equilibrium situation after which no differences were observed. Mujica-Paz et al. (2003a), studying the application of vacuum treatment to mango, papaya, peach and melon samples, showed that vacuum time directly affected the pore volume impregnated from external solution in a range of 3-25 minutes. Instead, the authors reported that vacuum time did not influence the X values for banana and apple samples. In the same way Salvatori (1997) reported that t_1 in a range of 5 and 15 minutes did not affect the impregnation level for mango and peach samples. Guillemin et al. (2008) studied the effect of sodium chloride concentration, sodium alginate concentration and vacuum time on weight reduction, chloride ions concentrations and pectinmethylesterase activity (PMEa) of apple cubes. The results showed that the effect of vacuum period was less significant and the more difficult to explain. Hofmeister et al. (2005) reported that an increase of vacuum time significantly affected the impregnation level of Mina cheese submitted to intermittent vacuum impregnation at 85.3 kPa. Chiralt et al. (2001), reviewing the scientific literature concerning the application of vacuum impregnation in salting process, reported that for meat fillets as longer the vacuum time as smaller the weight loss, in accordance with a greater pore volume impregnated from brine solution. Derossi et al. (2010) showed a significant positive effect of vacuum time on the pH reduction of pepper slices submitted to VI at a pressure of 200 mbar; instead a not clear behavior was observed for the experiments performed at 400 mbar. Moreover, in all cases the authors showed a positive effect of the increase of relaxation time on acidification level of the samples in a range of 10 and 30 minutes. Hironaka et al. (2011), studying the enrichment of whole potato with ascorbic acid through a vacuum impregnation treatment showed that an increase of vacuum time ranged between 0 and 60 minutes was effective in the improvement of ascorbic acid concentration of different varieties of whole potatoes. Apart the three above variables which are much strictly related to the vacuum process, the ensemble of the external solution characteristics with particular attention to its viscosity and temperature significantly affect the quality of vacuum impregnated foods. As reported from Barat et al. (2001), which studied the structural changes of apple tissues during vacuum osmotic dehydration, the impregnation level may be inhibited from a high viscosity of the hypertonic solution. The authors speculated that the relaxation of samples could occur when sample is taken out from the osmotic solution leading to an impregnation (regain) of gas if a liquid with very high viscosity is used. Also, the authors correlated the volume changes caused by capillary impregnation, ΔV^{LP-VI}, and the volume changes due to compression-relaxation phenomena, ΔV^{CR}, as a function of viscosity and temperature of the osmotic solution (figure 10).

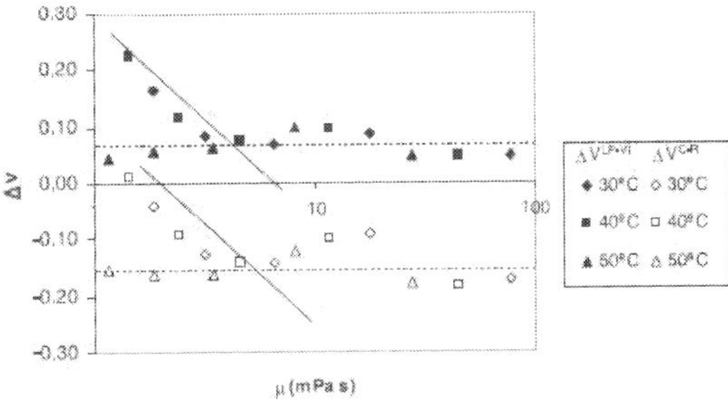

Fig. 10. Relationship between viscosity and temperature of osmotic solution and volume changes due to impregnation and compression-relaxation phenomena of apple samples submitted to vacuum osmotic dehydration (from Barat et al., 2001).

It is worth noting that for the experiments carried out at 40°C and 50°C the increase of viscosity did not show effects on impregnation and deformation. Instead, for samples treated at 30°C and for a viscosity of osmotic solution ranged between 0.5 and 3.5 mPa it is possible to observe that as viscosity increased as ΔV^{LP-VI} and ΔV^{CR} respectively decreased and raised. This behavior may be explained with a slow flow (because hindered from the high viscosity) of the osmotic solution into the pores; in this way more time for the relaxation process of the solid matrix is available which becomes the major contribution to the total volume changes of food. In line with this consideration, Barat et al. (2001) stated that: *"when viscosity is very low, impregnation tends to the theoretical value of a non-deformable porous matrix, and the deformation tends to zero".* Mujica-Paz et a (2003), studying the effects of osmotic solution concentration on water loss of apple, melon and mango samples, reported similar results. In general they showed that WL values increased in line with an increase of hypertonic solution concentration due to the greater osmotic dehydration; nevertheless, apple samples showed negative value of WL (a gain of water) for solution concentration lower than 50° Brix. In this condition the penetration of osmotic solution probably was the most important phenomenon which produced an increase of water content; also, it is important to take into account that the experiments performed from the authors were constituted from a vacuum period and a relaxation time of 10 minutes that represent a very short osmotic treatments to generate a high dehydration. However, for the experiments performed with a solution concentration > 50°Brix, WL values were positive because the OS has difficulty to penetrate into the pores and the pressure gradient promoted the outflow of water from the apple microstructure. Similar effect of the osmotic solution concentration was observed from other authors (Martinez-Monzo et al., 1998; Silva Paes et al., 2007).

The temperature of solution directly affects the mass transfer of process like osmotic dehydration, acidification, brining, etc., but also affects the viscosity of external liquid and the viscoelastic properties of solid matrix of foods. The latter properties which are on of the most important internal variable will be reviewed in the following paragraph; in general, as foods are soft as higher is the contribution of relaxation phenomena on the total volume changes which, in turn reduces the impregnation level.

3.2 Internal variables

Vacuum impregnation is a technique strictly dependent from the characteristic of the food structure both at mesoscopic and microscopic scale (Fito et al., 1996; Salvatori et al., 1998; Barat et al., 2001; Fito et al., 2001; Chiralt & Fito, 2003; Giraldo et al., 2003; Gras et al., 2003; Mujica-Paz et al., 2003; Zhao & Xie, 2004; Derossi et al., 2010; Derossi et al., 2011) . At first, the porosity fraction of biological tissue is the most important parameter for the application of VI because it presents the void space potentially available for the influx of the external solution. In general, fruits and vegetables show a higher porosity fraction in comparison with meats, fishes and cheeses making them more suitable for the use of VI techniques. In table 1 the porosity fraction values of several foods are reported.

In general, as greater is the porosity fraction as greater is the impregnation level. Fito et al. (1996) showed that for apple, banana, apricot, strawberry and mushroom samples the impregnation level was in the same order of the effective porosity values of fresh vegetables. Gras et al. (2003) studied the calcium fortification of eggplants, carrots and oyster mushroom by VI treatments. The authors reported the maximum X values for eggplants and oyster mushrooms which showed, in comparison with carrots, the greater intercellular spaces respectively of 54%±1 and 41%±2. Fito et al. (1996), studying the HDM and DRP on apple, banana, apricot, strawberry and mushroom samples submitted to VI, showed that the impregnation levels were directly correlated with the porosity values of the vegetables.

However, the porosity value is a not sufficient index to completely characterize and to predict the behavior of food during vacuum impregnation. As reported from Zhao & Xie (2004), during VI three main phenomena are involved: gas outflows, deformation-relaxation of solid matrix and the liquid influx. Since these phenomena simultaneously occur, the result of VI is a consequence of the equilibrium among their kinetics which, in turn, are affected by (Fito et al. 1996):

- Tissue structure (pores and size distribution)
- Relaxation time of the solid matrix, a function of the viscoelastic properties of the material;
- The rate of HDM, a function of porosity, size and shape of capillaries, their connectivity, the viscosity of solution;
- Size and shape of samples.

Food (variety, type)	Porosity fraction (ε, %)	Reference
Apple (Granny Smith)	23.8±1.0	Salvatori et al. (1998)
Apple (Golden Delisious)	27.3±1.1	Mujica-Paz et al. (2003b)
Apple (Gala)	18.3	Silvia Paes et al. (2007)
Mango (Tommy Atkins)	9.9±1.3	Salvatori et al. (1998)
Strawberry (Chandler)	6.3±1.6	Salvatori et al. (1998)
Kiwi fruit (Hayward)	2.3±0.8	Salvatori et al. (1998)
Pear (Passa Crassana)	3.4±0.5	Salvatori et al. (1997)
Plum (President)	2±0.2	Salvatori et al. (1997)
Peach (Miraflores)	2.6±0.5	Salvatori et al. (1998)
Peach (Criollo)	4.6±0.2	Mujica-Paz et al. (2003a)
Apricot (Bulida)	2.2 ± 0.2	Salvatori et al. (1997)
Pienapple (Espanola Roja)	3.7±1.3	Salvatori et al. (1997)
Banana (Giant Cavendish)	~ 9	Fito et al. (1996)
Banana (Macho)	1.6±0.3	Mujica-Paz et al. (2003a)
Orange peel (Valencia Late)	21±0.04	Chafer et al. (2000)
Mandarina peel (Satsuma)	25±0.11	Chafer et al. (2001)
Eggplant (Saroya)	64.1±2	Gras et al. (2001)
Zucchini (Blanco Grise)	4.4±0.9	Gras et al. (2001)
Zucchini	8.39±2.22	Derossi et al. (2011)
Mushroom (Albidus)	35,9±1.9	Gras et al. (2001)
Oyster mushroom	16.1±4	Gras et al. (2001)
Mango (Manila)	15.2±0.1	Mujica-Paz et al. (2003b)
Melon (Reticulado)	13.3±0.6	Mujica-Paz et al. (2003b)
Carrot (Nantesa)	13.7±2	Gras et al. (2001)
Beetroot	4.3±1.3	Gras et al. (2001)
Cherry	~ 30	Vursavus et al (2006)
Persimionn (Rojo Brillante)	4.9±0.8	Igual et al. (2008)
Cheese (Manchegu type)	~ 3	Chiralt et al. (2001)

Table 1. Porosity fraction values of several foods

Moreover, further internal variables such as the tortuosity of the internal pathway and the effect of temperature on the mechanical properties of solid matrix greatly affect the phenomena above reported. Furthermore, a synergistic (positive or negative) efferts of the internal variables were reported from several authors. Among the above variables the mechanical properties of biological tissues are one of the most important, because HDM is based on the pressure gradient generated by both vacuum pressure externally imposed and the deformation-relaxation of solid matrix. Salvatori et al. (1998) studied the effects of VI on pore volume impregnated from external (osmotic) solution and on deformation phenomena of several fruits. Strawberries, even tough showed a greater porosity fraction (6.3%) in comparison with kiwi fruit and peach, reported a negligible impregnation level (X = 0.2). It was hypothesized that the kinetic of liquid penetration was longer than the relaxation time promoting the deformation (compression) of pores rather than impregnation phenomena. These differences in the rate of HDM and DRP phenomena could be attributed to the microscopic properties of the strawberry tissues such as high tortuosity of the internal pathways and/or size and shape of pores which hindered the influx of the external solution. On the other hand strawberries being a soft material could be characterized from a fast compression rate which reduced the liquid penetration. In agreement with these hypothesis the authors showed a high deformation (compression) index (γ = - 4.0). Also, in the same paper it was reported that mango and peach samples showed positive γ values at the end of relaxation time, stating that an enlargement of the capillaries was still observed when the equilibrium was reached. In this case a high rigidity of the vegetable tissues could have reduced the rate of compression phenomena and increased the liquid penetration. However, among the studied fruits, strawberries and kiwi fruits were those with a lower impregnated pore volume (respectively of 0.2% and 0.89%) and between them kiwi fruits showed the lowest porosity fraction of ~ 2.3%. Mujica-Paz et al. (2003a) studied the effect of vacuum pressure on the pore volume impregnated from an osmotic solution. Figures 11a and 11b respectively show the prediction of X values of apple and others fruit samples as a function of the factor (1 - 1/r). Since *r* is defined as the ratio between atmospheric pressure and work pressure the *x*-axis reports an index of pressure gradient intensity.

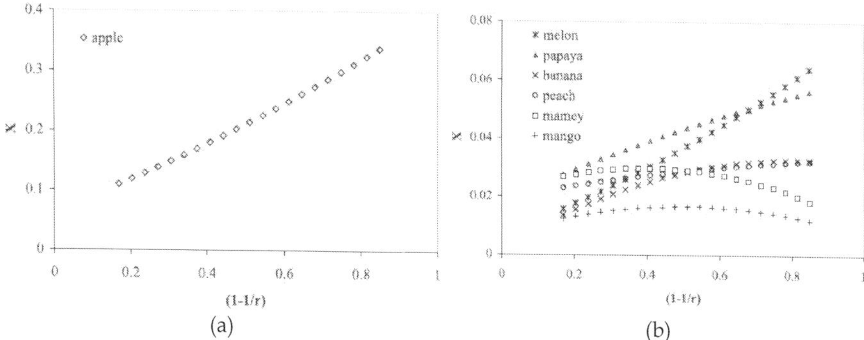

Fig. 11. Prediction of impregnation level value (X) as a function of (1-1/r) for some fruit samples submitted to VI for a vacuum time of 10 minutes and relaxation time of 25 minutes (From Mujica-Paz et al., 2002)

Mango, papaya, namey and peach samples showed a linear increase of X, stating that their vegetable tissues are subject to low deformation phenomena. Instead, banana, mango, namey, and peach showed a linear increase of X values until certain limits (about 400 mbar), after which impregnation level decreased. The authors supposed that these operative conditions could have produced a high vegetable tissue deformation (compression), which reduced the void phase available for liquid penetration. It was concluded that these fruits suffered of a great deformation when submitted to VI treatments in particular operative conditions. In their conclusion the authors highlighted the importance of the internal variables such as the number and the diameter of pores as well as the mechanical properties of solid matrix. Moreover, the spatial distribution of cells and their characteristics as well as the kind of fluid (liquid or gas) present in the intercellular space must be considered for the precise setting of vacuum impregnation treatments. Fito et al. (2001) deeply studied the vacuum impregnation and osmotic dehydration in matrix engineering. Among the obtained results, the authors reported that although orange peel showed a porosity fraction lower than eggplants, the fruit may be considered more suitable for VI treatments. This is because in orange peel the cells in flavedo zone are densely packed but the albedo zone shows large shape cells with a rupture of cell junctions that occurs during fruit ripening. This peculiar cell space distribution gives sponge-like properties with a high impregnation and swelling capacity. Taking into account the effect of the size of samples, it can be state that as thinner the food piece as greater the impregnation level because a greater void space is exposed to the surface. Gras et al. (2003) reported that carrot slices submitted to VI treatments showed an impregnation level of 5.8±0.3, 4.6±0.7 and 3.4±0.3 respectively for samples with a diameter of 10 mm, 15 mm and 25 mm. Derossi et al. (2011), studying the effect of a pulsed vacuum acidification treatment on zucchini slices in different operative conditions, showed that no significant differences were observed when the experiments were performed at 200 and 400 mbar in terms of pH reduction. It was hypothesized that the variability of vegetable microstructure coupled with the small diameter of the samples (~ 1.5 cm) favored the capillary impregnation reducing the effect of an increase of vacuum level. In accordance with this result the authors showed that no differences were observed in terms of porosity value reduction when fresh zucchini slices were submitted to VI at 200 mbar and 400 mbar. However the interaction of all these variables act simultaneously and could interact within them producing unexpected and/or complicated result. For instance, Chiralt et al. (1999) stated that samples with narrow pores treated with viscous liquid tend to show more deformation than impregnation effect.

4. Industrial applications

Back in the past, when HDM was proposed as an additional mass transfer mechanism occurring during osmotic dehydration, the first application of VI techniques had the aim to increase the rate of osmotic dehydration through the impregnation of capillaries with hypertonic solutions. Since then, vacuum osmotic dehydration (VOD) was subject to a great number of scientific papers; in addition to VOD, the application of an initial vacuum pulse followed by the restoration of atmospheric pressure and the maintaining of food immersed into the osmotic solution for a long time was proposed as pulsed vacuum osmotic dehydration (PVOD). In the last 10 years a great number of industrial applications have been studied (Javeri et al., 1991; Ponappa et al., 1993; Rastogi et al., 1996; Fito et al., 2001a; Fito et al., 2001b; Gonzalez-Martinez et al., 2002; Betoret et al., 2003; Giraldo et al., 2003; Gras et al., 2003; Mujica-Paz et al., 2003a ;Mujica-Paz et al., 2003b; Zhao & Xie, 2004; Hofmeister, et al., 2005; Corzo et al., 2007; Silva Paes et al , 2007; Igual et al., 2008; Cruz et al., 2009). In general it is possible to consider two principal aims: 1. The increase of mass transfer; 2. The introduction of chemical and biological compounds in food microstructure with the aim to improve their quality. On these basis, VI may be used as pre-treatment before drying or freezing, as innovative treatment inserted into a more complex processing with several industrial applications. In figure 12 is schematically reported some potential application of VI in food industry (Zhao & Xie, 2004).

4.1 VI as method to increase the rate of mass transfer of food during food processing
As known, several industrial processes, carried out with the aim to extend the shelf life or to improve/change their nutritional, functional or sensorial properties, are based on mass transfer toward foods or *vice versa*. If the purpose is to increase the rate of mass transfer

between food and an external solution as in the cases of osmotic dehydration acidification treatments, vacuum impregnation significantly increases the rate of the processes due to HDM and the increase of liquid-product contact area. A large literature concerning the effect of VI on the rate of osmotic dehydration recognized the effectiveness of VI (Shi & Fito, 1994; Fito et al., 1996; Rastogi et al., 1996; Tapia et al., 1999; Cunha et al., 2001;Moreno et al., 2004). Shi & Fito (1994) reported that water loss of apricot samples occurred much faster when vacuum was applied during the experiments. Shi et al. (1995) showed that osmotic dehydration under vacuum promotes a faster water diffusion when apricots, pineapples and strawberries were treated. Rastogi et al. (1996), studying the kinetics of osmotic dehydration of apple and coconut samples showed that the estimated rates were significantly higher in vacuum conditions (235 mbar). Fito et al. (2001b) stated that VI promotes the effective diffusion in the fruit liquid phase when impregnated with low viscosity solution.

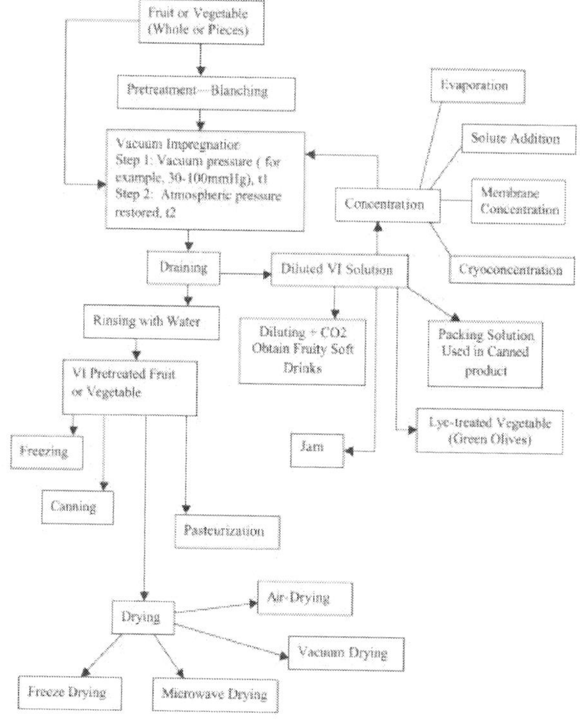

Fig. 12. Potential application of vacuum impregnation techniques in food industry (from Zhao & Xie, 2004)

Figure 13 shows the estimated diffusion coefficient values (De) of different type of fruits submitted to traditional osmotic dehydration and pulsed osmotic dehydration. It is possible to observe higher *De* values in the case of VI for vegetables which have high porosity values stating the effectiveness of the treatment as a consequence of the increase of mass transfer. Fito et al. (1994) found that water loss and solid gain obtained with a PVOD treatment performed at 70 mbar with a t_1 of 5 minutes were greater than those obtained through a traditional OD. Mujica-Paz et al. (2003a) performed osmotic dehydration treatments under vacuum (135-674 mbar, $t_1=t_2=10$ min) and at osmotic pressure measuring water (t = 20 min) activity depression (Da_w, %) for apple, melon and mango samples. The authors showed that in all cases the experiment carried out at sub-atmospheric conditions allowed to obtain the greater Da_w values stating that water loss and solid gain (the two main phenomena involved in the water activity depression) were faster than at atmospheric pressure. Giraldo et al. (2003) studied the kinetic of OD and PVOD treatments of mango samples performed with different osmotic solutions concentrations (35 – 65 °Bx). The authors, using a simplified Fickian equation, estimated De values of $5.9*10^{-10}$ m²/s, $6.4*10^{-10}$ m²/s and $9.7*10^{-10}$ m²/s for PVOD performed with osmotic solution concentration respectively of 65°Bx, 55°Bx and 35°Bx; in the same conditions were estimated De values of $1.8*10^{-10}$ m²/s, $5.9*10^{-10}$ m²/s and $5.9*10^{-10}$ m²/s for OD experiments. These results stated that in PVOD diffusion in the tissues is promoted, due to the removal of gas in the intercellular space and the introduction of hypertonic solution. Moreover, the application of VI as substituted of traditional brining of fishes, meats and cheeses or acidifying-dipping showed to be effective for the increase of mass transfer. Gonzales-Martinez et al. (2002) studied three type of brining methods applied to Machengo type cheese: at atmospheric pressure (BI), at vacuum pressure (VI) and with an initial vacuum pulse of 30 minutes (PVI). The authors estimated diffusion coefficients by using diffusional mass transfer mechanism. They obtained in the upper part of the samples D_e values of $4.4*10^{-10}$ m²/s, $6.1*10^{-10}$ m²/s and $9.5*10^{-10}$ m²/s respectively for BI (traditional), PVI and VI treatments. Also, similar results were obtained for the lower part of the samples. In their conclusion the authors stated that BI in vacuum condition greatly improved the kinetics of salting process as a consequence of hydrodynamic mechanism. Derossi et al. (2010), studying the application of an innovative vacuum acidification treatment (VA) on pepper slices, reported RpH values of 0.968, 0.929 and 0.894 for a traditional acidifying-dipping, a VA performed at 400 mbar (t_1 = 5 min, t_2 = 30 min) and a VA performed at 200 mbar (t_1=5 min, t_2=30 min) respectively. In agreement with the theoretical principles of VI the authors concluded that VA is able to increase the rate of pH reduction as a consequence of the capillary impregnation which improve the acid solution-product contact area.

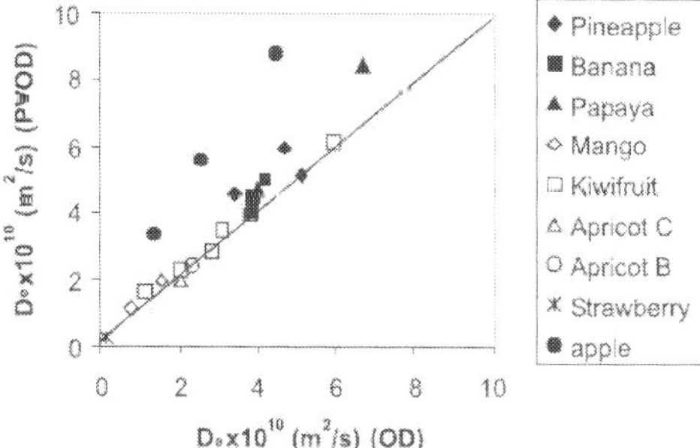

Fig. 13. Comparison of De values obtained from traditional osmotic dehydration (OD) and from pulsed vacuum osmotic dehydration (PVOD). Results obtained from several fruits (From Fito et al., 2001b)

Also Derossi et al. (2001) studied the kinetics of pulsed vacuum acidification treatments (PVA) of zucchini slices. The authors reported that an increase of vacuum did not affect the rate of pH reduction but in all cases the treatments were faster than acidification treatments performed at atmospheric pressure.

4.2 VI as method to improve food quality

As previous reported vacuum impregnation may be considered a technique by which is possible to introduce, in a controlled way, an external liquid in the void phase of foods. The term "external liquid" refers to a liquid obtained by dissolving in water any added chemical components. So, it is clear that VI may be used to introduce several compounds such as anti-browning agents, firming agents, nutritional compounds, functional ingredients, anti-microbial agents, anti-freezing, enzymes, etc. with several aims such as to prolong shelf life, to enrich fresh food with nutritional and/or functional substances, to obtain innovative food formulations as pre-treatments before drying or freezing, etc.

As well known, drying process performed with hot air, microwave, etc., is characterized by high energy consumption hence, high expenses; in this way, the possibility to reduce the initial amount of water in fresh foods (free water) by different techniques would allow to significantly reduce the cost of the process. Several papers showed the effectiveness of a traditional osmotic dehydration before air drying in the reduction of energy consumption. Instead, through a vacuum impregnation treatment applied before air drying two goals may be simultaneously obtained: 1. The reduction of energy costs; 2. The introduction of solutes such as antimicrobial, antibrowning and antioxidant to improve the final quality of dried foods (Sapers et al., 1990; Torreggiani, 1995; Barat et al., 2001b). Fito et al. (2001b) reported an improvement of the drying behavior of several fruits and vegetables submitted to VI pretreatments. Several authors reported that the stability of pigments was enhanced without the use of common compounds for color preservation when VI is applied before drying (Maltini et al., 1991; Torreggiani, 1995). Prothon et al. (2001) reported that water diffusivity of samples was increased when a pre-VI treatment was carried out. On the contrary, the same authors, studying the rehydration capacity of the samples, showed that this property was greater for non-treated samples, probably because both impregnation and DRP reduced the pores in which water flows during rehydration.

Among the stabilization processes, freezing is one of the most important because the use of low temperatures allows to better retain nutritional compounds in comparison with others traditional techniques such as dehydration, pasteurization/sterilization treatments, etc. However the formation of ice crystals leads to several physical damage and drip loss during thawing. Moreover, fluctuation of temperature along the cold chain may promote recrystallization phenomena leading to changes in size and shape of ice as well as their orientation (Zhao & Xie, 2004; Cruz et al., 2009). Vacuum impregnation was shown to be a useful method to incorporate in void phase of foods cryoprotectants and cryostabilizers, such as hypertonic sugar solution, antifreeze protein (AFP), high methoxyl pectin, etc. Martinez-Monzo et al. (1998) showed that as higher the concentration of sugar solution used during VI treatments as lower the freezable water content in food which reduces the drip loss during thawing. The same authors studied the potential application of VI treatment to introduce concentrate grape musts and pectin solution with the aim to decrease the damages of freezing on apple samples. It was observed that the both aqueous solutions were effective in the reduction of drip loss during thawing. In particular, grape must was able to reduce the freezable water content; instead pectin solution increased glassy transition temperature of liquid phase. Cruz et al. (2009) carried out several experiments to evaluate the potential application of vacuum impregnation with antifreeze protein (AFP) with the aim to preserve the quality of watercress leaves. Results stated that samples treated with AFP showed small ice crystals in comparison with water impregnated samples, due to the ability of the proteins to bound water molecules modifying the natural ice crystal formation. Figures 14a,14b and 14c show the results of wilting test on watercress leaves. It can be observed that samples impregnated with AFP showed a similar turgidity to the raw samples and higher than the control (water vacuum impregnated samples). This is in agreement with the formation of small ice crystals that during the thawing at room temperature led a reduction of cellular damage. Moreover, it is worth noting that this behavior could produce a better retention of nutrients traditionally lost during thawing. Ralfs et al. (2003) showed that carrot tissues vacuum impregnated with AFP had a higher mechanical strength in comparison with samples submitted to VI with ultrapure water.

(a)   (b)   (c)

Fig. 14. Wilting test of watercress leaves. (a) raw samples; (b) vacuum impregnated with water; (c) vacuum impregnated with AFP.

Xie and Zhao (2003) showed that vacuum impregnation of strawberries and marionberries with cryoprotectan solutions (HCFS and high methyl pectin) enriched with 7.5% of calcium gluconal highly enhanced the texture and the reduction of drip loss on frozen-thawed samples. In particular the authors reported an increase of compression force in a range of 50-100% and a reduction of drip loss of 20-50% in comparison with untreated samples. According to literature, the authors attributed these results to the reduction of freezable water content. Furthermore, it is worth noting that the reduction of water content as a consequence of VI treatment reduces energy consumption during freezing.

In the last years, on the basis of the consumer interest on the relation between the assumption of correct diet and its health benefits, the possibility to obtain foods whit high nutritional and functional properties exponentially increased the efforts of food scientists and industries in this research field. In this way, vacuum impregnation is a useful techniques to fill the void phase of foods with nutritional and functional ingredients. However, the scientific results concerning this application of VI are still few. Fito et al. (2001a) were the first scientists that evaluated the feasibility of VI to obtain innovative fresh functional foods (FFF). During their experiments the authors studied the impregnation of several vegetables with calcium and iron salt solution taking into account the solubility of these salts in water. In particular, the experiments had the aim to estimate the possibility to obtain fortified samples containing the 25% of the recommended daily intake (RDI) for each specific salts. As example, Figure 15 reports the weight of fruits that contain the 25% of RDI as a function of calcium lactate concentration.

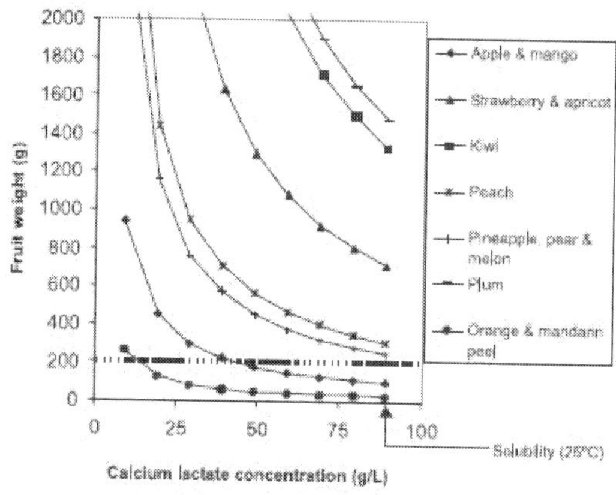

Fig. 15. Weight of some fruits containing the 25% of RDI of calcium salts.

The trends shown in the figure were estimated taking into account the porosity fraction of each type of fruit and by assuming the void phase completely impregnated with the external solution. For instance, by using a calcium lactate solution with a concentration of 50 g/L, less than 100g of orange peel should be consumed to assume the 25% of RDI. In the same way, about 600g and 1200g of peach and strawberry or apricot should be, respectively, assumed. Gras et al. (2003) studied the application of VI with the purpose to obtain eggplant, carrot and oyster mushroom fortified with calcium salts. The authors reported that eggplants and oyster mushroom may be considered as more appropriate to obtain FFF due to their high porosity in comparison with carrots. Hirinoka et al. (2011) studied the application of VI for the enrichment of whole potato with ascorbic acid. Also the authors compared the AA content of samples submitted to VI with the untreated one after a steam cooking treatment over boiling water for 25 minutes and during storage at 5°C for two weeks. Figures 16a, 16b, 16c and 16c show the visual aspect of potato samples submitted to VI and immersed at atmospheric pressure in red ink solution. The figures clearly show that the immersion of potato in solution at atmospheric pressure did not allow to impregnate capillaries of potatoes and the same results were obtained with a VI without apply any restoration time ($t_2 = 0$). Instead after a restoration time of 3 h a significant impregnation was observed.

As expected ascorbic acid content significantly decrease (about 42%) after steam cooking. Nevertheless, VI-cooked samples had an acid ascorbic concentration 22 time higher than samples only submitted to steam cooking (raw-cooking). Also, the VI-cooking samples showed an AA content of ~ 100 mg/100g which is twice of the FAO value (45mg/100g) and close to the RDA value. Furthermore, in another series of experiments Xie & Zhao (2003) used VI to enrich apple, strawberry and marionberry with calcium and zinc. The experiments performed with high corn syrup solution enriched with calcium and zinc showed that a 15-20% of RDI of calcium more than 40% RDI of zinc could be obtained in 200g of impregnated apple fresh-cut samples.

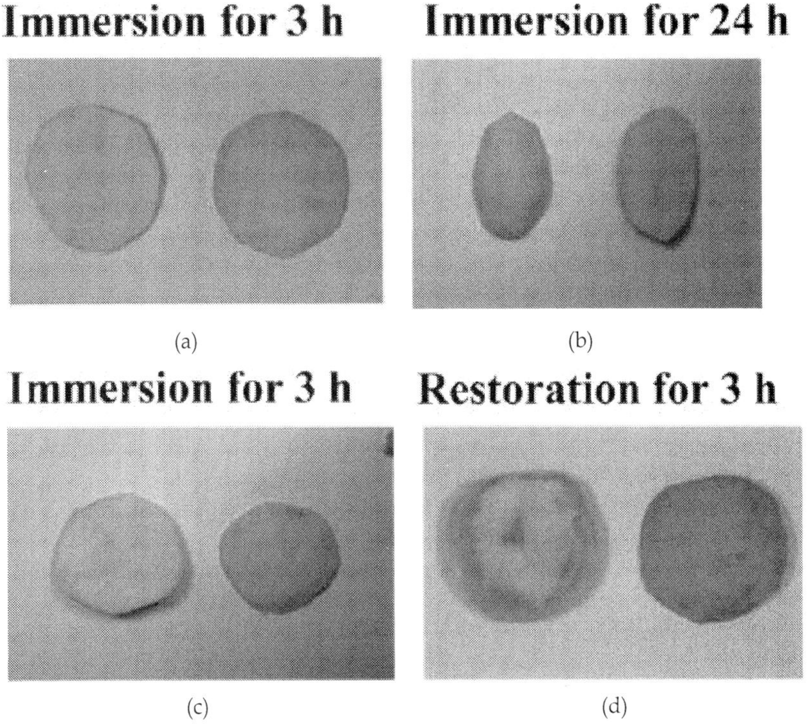

Immersion for 3 h   **Immersion for 24 h**

(a)   (b)

Immersion for 3 h   **Restoration for 3 h**

(c)   (d)

Fig. 16. Potato samples immersed in red ink solution without vacuum (a,b), after a vacuum time of 3 h (without restoration time) and after a restoration time of 3 h (From Hironaka et al., 2011).

Figure 17 reports the ascorbic acid content of whole potato submitted to VI and cooked over boiling water for 25 minutes and the controls (un-VI samples cooked).

Vacuum impregnation could be a method to produce a numerous series of innovative probiotic foods. For instance, Betoret et al. (2003) studied the use of VI to obtain probiotic enriched dried fruits. The authors performed VI treatments on apple samples by using apple juice and whole milk containing respectively *Saccharomyces cerevisiae* and *Lactobacillus casei* (*spp. Rhamnosus*) with a concentration of 10^7–10^8 cfu/ml. Results allowed to state that, combining VI and low temperature air dehydration, it was possible to obtain dried apples with a microbial content of 10^6–10^7 cfu/g. However, despite the wide number of the potential industrial application, shelf life extension is one of the most important. So, due to its unique advantage vacuum impregnation may be considered a useful methods to introduce inhibitors for microbial growth and/or chemical degradation reactions; nevertheless, the scientific literature concerning the application of VI in this field of research is still poor. Tapia et al. (1999) used a complex solution containing sucrose (40°Bx), phosphoric acid (0.6% w/w), potassium sorbate (100 ppm) and calcium lactate (0.2%) to increase the shelf life of melon samples. Results showed that foods packed in glass jars and covered with syrup maintained a good acceptance for 15 days at 25°C. Welty-Chanes et al. (1998), studying the feasibility of VI for the production of minimally processed oranges reported that the samples were microbiologically stable and showed good sensorial properties for 50 days when stored at temperature lower than 25°C. Derossi et al. (2010) and Derossi et al. (2011) proposed an innovative vacuum acidification (VA) and pulsed vacuum acidification (PVA) to improve the pH reduction of vegetable, with the aim to assure the inhibition of the out-grow of *Clostridium botulinum* spores in the production of canned food. The results stated the possibility to obtain a fast reduction of pH without the use of high temperature of acid solution as in the case of acidifying-blanching. However, the authors reported the effect of VI on visual aspect of vegetable that need to be considered for the industrial application, because the compression-deformation phenomena could reduce the consumer acceptability. Guillemin et al. (2008) showed the effectiveness of VI for the introduction of pectinmethylesterase which enhances fruit firmness.

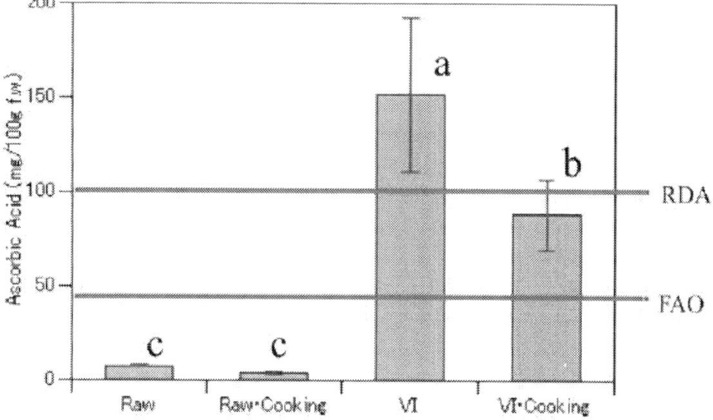

Fig. 17. Effect of steam cooking on ascorbic acid content of whole potato submitted to vacuum impregnation. VI solution: 10% AA, p = 70 cm Hg, t1=1h, t2= 3 h)

5. Conclusion

Although vacuum impregnation was for the first time proposed at least 20 years ago, it may be still considered an emerging technology with high potential applications. Due to its unique characteristics, VI is the first food processing based on the exploitation of three dimensional food microstructure. It is performed by immerging food in an external solution and applying a vacuum pressure (p) for a time (t_1). Then, the restoration of atmospheric pressure maintaining the foods into the solution for a relaxation time (t_2) allows to complete the process. During these steps three main phenomena occurs: the out-flow of native liquid and gases from the pores; the influx of external solution inside capillaries; deformation-relaxation of solid matrix. The influx of external liquid occurs under the action of a pressure gradient between the pores and the pressure externally imposed; this is known as hydrodynamic mechanisms (HDM). However, on the basis of its nature, VI is a very complex treatment and its results are affected from several external and internal variables. The former are the operative conditions above reported coupled with the temperature and viscosity of external solution. The latter are characterized from the microscopic and mesoscopic properties of food architecture such as length and diameter of pores, their shapes, the tortuosity of internal pathways, the mechanical (viscoelastic) properties of biological tissues, the high or low presence of gas and/or liquid inside capillaries, etc. VI has shown to be very effective in a wide number of industrial applications. The impregnation, causing a significant increase of the external solution/product contact area, is an important method to increase the mass transfer of several solid-liquid operation such as osmotic dehydration, acidification, brining of fish and meat products, etc. VI may be used as pretreatment before drying or freezing, improving the quality of final product and reducing cost operations due to the removal of native liquid (water) from the pores. Furthermore, the possibility to introduce, in a controlled way, an external solution enriched with any type of components catch light on a high number of pubblications. Indeed, VI has been used to extend shelf life, to produce fresh fortified food (FFF), to enrich food with nutritional/functional ingredients, to reduce the freezing damage, to obtain foods with innovative sensorial properties, to reduce oxidative reaction, to reduce browning, etc. Furthermore, from an engineering point of view some advantages may be considered: 1. it is a fast process (usually it is completed in few minutes); it needs low energy costs; it is performed at room temperature; the external solution may be reused many times. Nevertheless, the applications of VI at industrial scale are still poor. This problem may be attributed to the lack of industrial plants in which it is possible to precise control the operative conditions during the two steps of the process. Also, some technical problems need to be solved. For instance, as reported from Zhao & Xie (2004), the complete immersion of foods into the external solution is a challenge for the correct application of VI. Often, fruits and vegetables tend to float due to their low density in comparison with external solution as in the case of osmotic solution. The current VI is applied by stirring solution with the aim to keep food pieces inside solution with the drawback of an increase of energy costs and possible damages of foods. Furthermore, the lack of information for industries on the advantage of these techniques reduces its application at industrial scale.

6. References

Aguilera, J.M. (2005). Why food microstructure?. *Journal of Food Engineering*, Vol. 67, pp. 3-11.

Andres, I., Salvatori, D., Chiralt, A., & Fito, P. (2001). Vacuum impregnation viability of some fruits and vegetables. In: Osmotic dehydration and vacuum impregnation,

P. Fito, A. Chiralt, Barat, J.M., Spiess, W.E.L., Behsnilian, D., Barat, J.M. (eds.). Behsnilian.

Atares, L., Chiralt, A., & Gonzales-Martinez C. (2008). Effect of solute on osmotic dehydration and rehydration of vacuum impregnated apple cylinders (cv. Granny Smith). *Journal of Food Engineering*, Vol. 89, pp. 49-56.

Barat, J.M., Chiralt, A., & Fito, P. (1998). Equilibrium in cellular food osmotic solution systems as related to structure. *Journal of Food Science*, Vol. 63, pp. 836-840.

Barat, J.M., Chiralt, A., & Fito, P. (2000). Structural changes kinetics in osmotic dehydration of apple tissue. In *Proceeding of the 21th International Drying Simposium* IDS2000, Paper No. 416, Amsterdam, Elsevier.

Barat, J.M., Chiralt, A., Fito, P. (2001b). Effect of osmotic solution concentration, temperature and vacuum impregnation pretreatment on omostic dehydration kinetics of apple slices. Food Science and Technology International, Vol. 7, pp. 451-456.

Barat, J.M., Fito, P., & Chiralt, A. (2001a). Modeling of simultaneous mass transfer and structural changes in fruit tissues. *Journal of Food Engineering*, Vol. 49, pp. 77-85.

Betoret, N., Puente, I., Diaz, M.J., Pagan, M.J., Garcia, M.J., Gras, M.L., Marto, J. & Fito, P. (2003). Development of probiotic-enriched dried fruits by vacuum impregnation. *Journal of Food Engineering*, Vol. , No. (2-3), pp. 273-277.

Chafer, M., Gonzales-Martinez, C., Ortola, M.D., Chiralt, A., & Fito, P. (2000). Osmotic dehydration of mandarin and oragn peel by using rectified grape must. Proceedings of the 12th international drying symposium. IDS 2000, Elsevier Science, Amsterdam, Paper n. 103.

Chafer, M., Ortola, M.D., Martinez-Monzo, J., Navarro, E., Chiralt, A., & Fito, P. (2001). Vacuum impregnation and osmotic dehydration on mandarin peel. Proceedings of the ICEF8. Lancaster: Technomic Publishing Company.

Chiralt, A., & Fito, P. (2003). Transport mechanisms in osmotic dehydration: the role of the structure. *Food Science Technology International*, 9 (3), 179-186.

Chiralt, A., Fito, P., Andres, A., Barat, J.M., Martinez-Monzo, J, & Martinez-Navarrete, N. (1999). Vacuum impregnation: a tool in minimally processing of foods. In: *Processing of Foods: Quality Optimization and Process Assessment*. Oliveira F.A.R., & Oliveira, J.C. (eds), Boca Raton: CRC press, pp. 341-356.

Chiralt, A., Fito, P., Barat, J.M., Andres, A., Gonzales-Martinez, C., Escrichr, I., & Camacho, M.M. (2001). Use of vacuum impregnation in food salting process. *Journal of Food Engineering*, Vol. 49, pp. 141-151.

Corzo, O., Brancho, N., Rodriguez, J., & Gonzales, M. (2007). Predicting the moisture and salt contents of sardine sheets during vacuum pulse osmotic dehydration. *Journal of Food Engineering*, Vol. 80, pp. 781-780.

Cruz, R.M.S., Vieira, M.C., Silva, C.L.M. (2009). The response of watercress (*Nasturtium officinale*) to vacuum impregnation: effect of and antifreeze protein type I. *Journal of Food Engineering*, Vol. 95, pp. 339-345.

Cunha, L.M., Oliveira, F.A.R., Aboim, A.P., frias, J.M., & Pinheiro-Torres, A. (2001). Stochastic approach to the modelling of water losses during osmotic dehydration and improved parameter estimation. *International Journal of Food Science and Technology*, Vol. 36, pp. 253-262.

Datta, A.K. (2007a). Porous media approaches to studying simultaneous heat and mass transfer in food processes. I: Problem formulations. *Journal of Food Engineering,* pp. 80-95.

Datta, A.K. (2007b). Porous media approaches to studying simultaneous heat and mass transfer in food processes. II: Property data and representative results. *Journal of Food Engineering,* Vol. 80, pp. 96-110.

Derossi, A., De Pilli, T., La Penna, M.P., & Severini, C. (2011). pH reduction and vegetable tissue structure changes of zucchini slices during pulsed vacuum acidification. *LWT- Food Science and Technology,* Vol. 44, pp. 1901-1907.

Derossi, A., De Pilli, T., Severini, C. (2010). Reduction in the pH of vegetables by vacuum impregnation: A study on pepper. *Journal of Food Engineering,* Vol. 99, pp. 9-15.

Fito, P. (1994). Modelling of vacuum osmotic dehydration of foods. Journal of Food Engineering, 22, 313-318.

Fito, P., & Chiralt, A. (1994). An Update on vacuum osmotic dehydration. In *Food Preservation by Moisture Control: Fundamentals and Application,* G.V. Barbosa-Canovas and J. Welti-Chanes, eds., pp. 351-372, Technomic Pub. Co., Lancaster, PA.

Fito, P., & Chiralt, A. (2003). Food Matrix Engineering: The Use of the water-structure-functionality ensemble in dried food Product development. *Food Science Technology International,* Vol. 9(3), pp. 151-156.

Fito, P., & Pastor, R. (1994). *Non-diffusional mechanism occurring during vacuum osmotic dehydration (VOD).* Journal of Food Engineering, 21, 513-519.

Fito, P., Andres, A., Chiralt, A., & Pardo, P. (1996). *Coupling of Hydrodynamic mechanisms and Deformation-Relaxation Phenomena During Vacuum treatments in Solid Porous Food-Liquid Systems.* Journal of Food Engineering, 27, 229-241.

Fito, P., Andres, A., Pastor, R. & Chiralt, A. (1994). Modelling of vacuum osmotic dehydration of foods. In: *Process optimization and minimal processing of foods,* Singh, P., & Oliveira, F. (eds.), pp. 107-121, Boca Raton: CRC Press.

Fito, P., Chiralt, A., Barat, J.M., Andres, A., Martinez-Monzo, J., & Martinez-Navarrete, N. (2001b). Vacuum impregnation for development of new dehydrated products. *Journal of Food Engineering,* Vol. 49, pp. 297-302.

Fito, P., Chiralt, A., Betoret, M., Gras, M.C., Martinez-Monzo, J., Andres, A., & Vidal, D. (2001a). Vacuum impregnation and osmotic dehydration in matrix engineering. Application in functional fresh food development. *Journal of Food Engineering,* Vol. 49, pp. 175-183.

Giraldo, G., Talens, P., Fito, P., & Chiralt, A. (2003). Influence of sucrose solution concentration on kinetics and yield during osmotic dehydration of mango. *Journal of Food Engineering,* Vol. 58, pp. 33-43.

Gonzalez-Martinez, C., Chafer, M., Fito, P., Chiralt, A. (2002). Development of salt profiles on Machengo type cheese during brining. Influence of vacuum pressure. *Journal of Food Engineering,* Vol. 53. Pp. 67-73.

Gras, M., Vidal-Brotons, D., Betoret, N., Chiralt, A., & Fito, P. (2002). The response of some vegetables to vacuum impregnation. *Innovative Food Science and Emerging Technologies,* Vol. 3, pp. 263-269.

Gras, M.L., Fito, P., Vidal, D., Albors, A., Chiralt, A., & Andres, A. (2001). The effect of vacuum impregnation upon some properties of vegetables. *Proceedings of the ICEF8.* Technomic Publishing Company. Lancanster

Gras, M.L:, Vidal, D., Betoret, N., Chiralt, A., & Fito, P. (2003). Calcium fortification of vegetables by vacuum impregnation. Intercations with cellular matrix. *Journal of Food Engineering,* Vol. 56, pp. 279-284.

Guillemin, A., Degraeve, P, Noel, C., & Saurel, R. (2008). Influence on impregnation solution viscosity and osmolarity on solute uptake during vacuum impregnation of apple cubes (var. Granny Smith). *Journal of Food Engineering,* Vol. 86, pp. 475-483.

Halder, A., Dhall, A., & Datta, A.K. (2007). An improved, easily implementable porous media based model for deep-fat frying Part I: Model development and input parameters. *Food and Bioproducts Processing.*

Hironaka, K., Kikuchi, M., Koaze, H., Sato, T., Kojima, M., Yaamamoto, K., Yasuda, K., Mori, M., & Tsuda, M. (2011). Ascorbic acid enrichment of whole potato tuber by vacuum-impregnation. *Food Chemistry,* Vol 127, pp. 1114-1118.

Hofmeister, L.C., Souza, J.A.R., & Laurindo, J.B. (2005). Use of dye solutions to visualize different aspect of vacuum impregnation of Minas Cheese. *LWT – Food Science and Technology,* Vol. 38, pp. 379-386.

Igual, M., Castello, M.L., Ortola, M.D., & Andres, A. (2008). Influence of vacuum impregnation on respiration rate, mechanical and optical properties of cut persimmon. *Journal of Food Engineering,* Vol. 86, pp. 315-323.

Javeri, R.H., Toledo, R., & Wicker, L. (1991). Vacuum infusion of citrus pectinmethylesterase and calcium effects on firmness of peaches. *Journal of Food Science,* Vol. 56, pp. 739-742.

Lewis, M.J. (1993). *Propiedades fisicas de los alimentos y de los sistemas de procesado.* Ed. Acribia, Zaragoza, Espana.

Maltini, E., Pizzocaro, F., Torreggiani, D., & Bertolo, G. (1991). Effectiveness of antioxidant treatment in the preparation of sulfur free dehydrated apple cubes. In 8th World Congress: Food Science and Technology, Toronto, Canada, pp. 87-91.

Martinez-Monzo, J., Martinez Navarrete, N., Chiralt, A., & Fito, P. (1998). Mechanical and structural change in apple (var. Granny Smith) due to vacuum impregnation with cryoprotectans. *Journal of Food Science,* Vol. 63 (3), pp. 499-503.

Mebatsion, H.K., Verboven, P., Ho, Q.T., Verlinden, B.E., & Nicolai, B.M. (2008). Modelling fruit (micro)structures, why and how?. Trends in Food Science & Technolgy, 19, 59-66.

Moreno, J., Bugueno, G., Velasco, V., Petzold, G., & Tabilo-Munizaga, G. (2004). Osmotic dehydration and vacuum imprengation on physicochemical properties of Chilean Papaya (Carica candamarcensis). *Journal of Food Science,* Vol. 69, pp. 102-106.

Mujica-Paz, H., Valdez-Fragoso, A., Lopez- Malom A., Palou, E., & Welti-Chanes, J. (2002). Impregnation properties of some frutis at vacuum pressure. Journal of Food Engineering, Vol.56, pp. 307-314.

Mujica-Paz, H., Valdez-Fragoso, A., Lopez-Malo, A., palou, E., & Welti-Chanes (2003). Impregnation and osmotic dehydration of some fruits: effect of the vacuum pressure and syrup concentration. *Journal of Food Engineering,* Vol. 57, pp. 305-314.

Ponappa, T., Scheerens, J.C., & Miller, A.R. (1993). Vacuum infiltration of polyamines increases firmness of strawberry slices under various storage conditions. *Journal of Food Science*, Vol. 58, pp. 361-364.

Prothon, F., Ahrme, L.M., Funebo, T., Kidman, S., Langton, M., & Sjoholm, I. (2001). Effects of combined osmotic and microwave dehydration of apple on texture, microstructure and rehydration characteristics. *Lebensmittel-Wissenschaft und technologie*, Vol. 34, pp. 95-101.

Ralfs, J.D., Sidebottom, C.M., Ormerod, A.P. (2003). Antifreeze proteins in 444 vegetables. World intellectual property organization, patent WO 03/055320 A1, pp. 1-8.

Rastogi, N.K., & Raghavarao, K.S.M.S. (1996). Kinetics of Osmotic dehydration under vacuum. *Lebensm.-Wiss. U.-Technol.*, Vol. 29, pp. 669-672.

Salvatori, D. (1997). Osmotic dehydration of fruits: Compositional and structural changes at moderate temperatures. Ph.D. Thesis.

Salvatori, D., Andres, A., Chiralt, A., & Fito, P. (1998). The response of some properties of fruits to vacuum impregnation. *Journal of Food Process Engineering*, Vol. 21, pp. 59-73.

Sapers, G.M., Garzarella, L., & Pilizota, V. (1990). Application of browning inhibitors to cut apple and potato by vacuum and pressure infiltration. Journal of Food Science, Vol. 55, pp. 1049-1053.

Shi, X.Q., & Fito, M.P. (1994). Mass transfer in vacuum osmotic dehydration of fruits: A mathematical model approach. *Lebensm.-Wiss.-u.-Technol*, Vol. 27, pp. 67-72.

Shi, Z.Q., Fito, P., Chiralt, A. (1995). Influence of vacuum treatments on mass transfer during osmotic dehydration of fruits. *Food Research International*, Vol. 21, pp. 59-73.

Tapia, M.S., Ranirez, M.R., Castanon, X., & Lopez-Malo, A. (1999) Stability of minimally treated melon (*Cucumis melon, L.*) during storage and effect of the water activity depression treatment. No. 22D-13. *Presented at 1999 IFT annual meeting*, Chicago, IL.

Torquato, S. (2000). Modeling of physical properties of composite materials. *International Journal of Solids and Structures*, Vol. 37, pp. 411-422.

Torregiani, D. (1995). Technological aspect of osmotic dehydration in foods. In: Food Preservation by moisture control. Fundamentals and Applications, Barbosa-Canovas, G.V., & Welti-Chanes, J. (eds.), Lancaster: Technomic Publisher Co. Inc., pp 281-304.

Vursavus, K., Kelebek, H., & Selli, S. (2006). A study on some chemical and physico-mechanic properties of three sweet cherry varieties (Prunus avium L.). Journal of Food Engineering, Vol. 74, pp. 568-575.

Welti-Chanes, J., santacruz, C., Lopez-Malo, A., & Wesche-Ebeling, P. (1998). Stability of minimally processed orange segments obtained by vacuum dehydration techniques. *No. 34B-8. Presented at 1998 IFT annual meeting*, Atlanta, GA.

Xie, J., & Zhao, Y. (2003). Improvement of physicochemical and nutritional qualities of frozen Marionberry and by vacuum impregnation pretreatment with cryoprotectants and minerals. *Journal of Horticultural Science and Biotechnology*, Vol. 78, pp. 248-253.

Zao, Y., & Xie, J. (2004). Practical applications of vacuum impregnation in fruit and vegetable processing. *Trend in Food Science & Technology*, Vol. 15, pp. 434-451.

3

Freezing / Thawing and Cooking of Fish

Ebrahim Alizadeh Doughikollaee
University Of Zabol
Iran

1. Introduction

One of the greatest challenges for food technologists is to maintain the quality of food products for an extended period. Fish and shellfish are perishable and, as a result of a complex series of chemical, physical, bacteriological, and histological changes occurring in muscle, easily spoiled after harvesting. These interrelated processes are usually accompanied by the gradual loss or development of different compounds that affect fish quality. Fresh seafood has a high commercial value for preservation, and the sensory and nutritional loss in conventionally frozen/thawed fish is a big concern for producers and consumers. This chapter present the effect of Freezing/Thawing and Cooking on the quality of fish.

2. Freezing

Freezing is a much preferred technique to preserve food for long period of time. It permits to preserve the flavour and the nutritional properties of foods better than storage above the initial freezing temperature. It also has the advantage of minimizing microbial or enzymatic activity. The freezing process is governed by heat and mass transfers. The concentration of the aqueous phase present in the cell will increase when extra ice crystal will appear. This phenomenon induces water diffusion from surrounding locations. Of course, intra cellular ice induces also an increase of the concentration of the intra cellular aqueous phase. The size and location of ice crystals are considered most important factors affecting the textural quality of frozen food (Martino et al., 1998). It has been recognized that the freezing rate is critical to the nucleation and growth of ice crystals. Nucleation is an activated process driven by the degree of supercooling (the difference between the ambient temperature and that of the solid-liquid equilibrium). In traditional freezing methods, ice crystals are formed by a stress-inducing ice front moving from surface to centre of food samples. Due to the limited conductive heat transfer in foods, the driving force of supercooling for nucleation is small and hence the associated low freezing rates. Thus, the traditional freezing process is generally slow, resulting in large extracellular ice crystal formations (Fennema et al., 1973; Bello et al., 1982; Alizadeh et al., 2007a), which cause texture damage, accelerate enzyme activity and increase oxidation rates during storage and after thawing.

Pressure shift freezing (PSF) has been investigated as an alternative method to the existing freezing processes. The PSF process is based on the principle of water-ice phase transition under pressure: Elevated pressure depresses the freezing point of water from 0°C to -21°C at about 210 MPa (Bridgman, 1912). The sample is cooled under pressure to a temperature just above the melting temperature of ice at this pressure. Pressure is then fast released resulting in supercooling, which enhanced instantaneous and homogeneous nucleation throughout the cooled sample (Kalichevsky et al., 1995). Ice crystal growth is then achieved at atmospheric pressure in a conventional freezer. Pressure shift freezing (PSF), as a new technique, is increasingly receiving attention in recent years because of its potential benefits for improving the quality of frozen food (Cheftel et al., 2002; Le Bail et al., 2002). PSF process has been demonstrated to produce fine and uniform ice crystals thus reducing ice-crystal related textural damage to frozen products (Chevalier et al., 2001; Zhu et al., 2003; Otero et al., 2000; Alizadeh et al., 2007a). From a point of view of the tissue damage, pressure shift freezing seemed to be beneficial, causing a very smaller cell deformation than the classic freezing process.

2.1 Freezing process

Freezing is the process of removing sensible and latent heat in order to lower product temperature generally to -18 °C or below (Delgado & Sun, 2001; Li & Sun, 2002). Figure 1 shows a typical freezing curve for the air blast freezing (ABF). The initial freezing point was about -1.5 °C and was observable at the beginning of the freezing plateau (Alizadeh et al., 2007a). The temperature dropped slowly at follow because of the water to ice transition. This freezing point depression has been classically observed in several freezing trials (not always) and has been recognized to be due to the presence of solutes and microscopic cavities in the food matrix (Pham, 1987). The nominal freezing time was used to evaluate the freezing time. The nominal freezing time is defined by the International Institute of Refrigeration as the time needed to decrease the temperature of the thermal centre to 10 °C below the initial freezing point (Institut International du Froid, 1986).

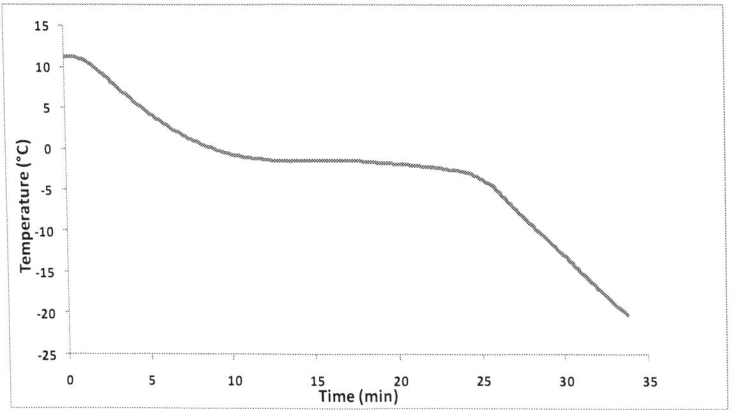

Fig. 1. A typical freezing curve of Atlantic salmon fillets obtained in air-blast freezing (Alizadeh, 2007).

Figure 2 shows a typical Pressure shift freezing curve. The process began when the unfrozen fish sample was placed in the high-pressure vessel. The temperature appeared to drop a little bit and a slight initiation of freezing can be detected at the surface of the sample after the sample was immersed into the ethanol/water medium (-18 °C) of the refrigerated bath (Alizadeh et al., 2007a). Pressurization (200 MPa) induced a temperature increase due to the adiabatic heat generated. It took about 57 min for the sample to be cooled to -18 °C without freezing which is close to the liquid-ice I equilibrium temperature (Bridgman, 1912).

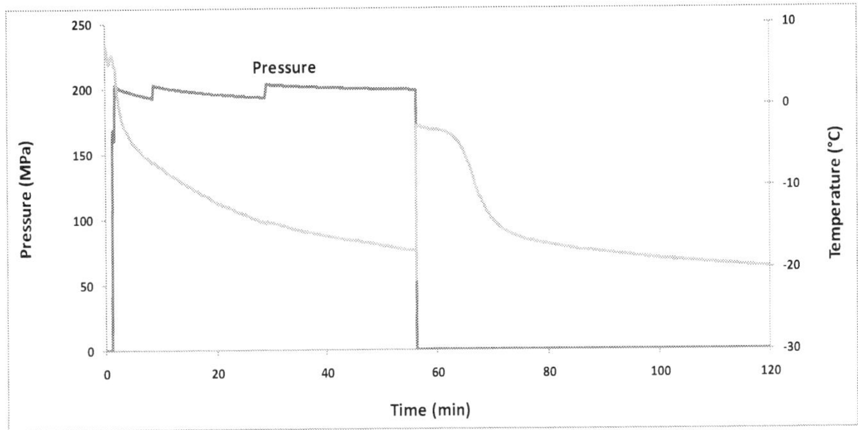

Fig. 2. A typical Pressure shift freezing curve of Atlantic salmon fillets (Alizadeh, 2007).

Then, the quick release of pressure created a large supercooling, causing a rapid and uniform nucleation, due to the shift in the freezing point back to the normal (-1.5°C) and the rapid conversion of the sensible heat (from -18 to -1.5 °C) to the latent heat. After depressurization, the temperature reached a stable temperature (-1.5 °C) for freezing at atmospheric pressure because of the latent heat release. The final step of the PSF process was similar to conventional freezing at atmospheric pressure.

2.2 Fish microstructure during freezing

Ice crystallization strongly affects the structure of tissue foods, which in turn damages the palatable attributes and consumer acceptance of the frozen products. The extent of these damages is a function of the size and location of the crystals formed and therefore depends on freezing rate. It is mentioned that slow freezing treatments usually cause texture damage to real foods due to the large and extracellular ice crystals formed (Fennema et al., 1973). Clearly, most area was occupied with the cross-section of the ice crystals larger than the muscle fibers. This means that the muscle tissue was seriously deformed after the air blast freezing at low freezing rate (1, 62 cm/h) which may cause an important shrinkage of the cells and formation of large extracellular ice crystals but it was very difficult to determine if these ice crystals were intra or extra-cellular (Figure 3). On the other hand, the intra and extracellular ice crystal have been seen during air blast freezing at high freezing rate (2, 51 cm/h). It is possible to observe the muscle fibers and analyse the size of intracellular ice crystal (Alizadeh, 2007).

The pressure shift freezing (PSF) process created smaller and more uniform ice crystals. A higher degree of supercooling should be expected during the pressure shift freezing experiments because of the rapid depressurization and the smaller ice crystals observed in the samples frozen by PSF at higher pressure. Burke et al. (1975) reported that there was a 10-fold increase in the rate of ice nucleation for each °C of supercooling. Thus, a higher pressure and lower temperature resulted in more intensive nucleation and formation of a larger number of small ice crystals. Moreover, PSF at a higher pressure is carried out at lower temperature, creating a larger temperature difference between the sample and the surrounding for final freezing completion after depressurization. This could also be a major factor affecting the final ice-crystal size in the PSF samples. Micrographs in Figure 3 also show well isotropic spread of ice crystals in the fish tissues, especially for the 200 MPa treatments. This is because the isostatic property of pressure allows isotropic supercooling and homogeneous ice nucleation. It is quite clear that the muscle fibers in the PSF treated samples (Figure 3) were well kept as compared with their original structures. Therefore, conventional freezing problems like tissue deformation and cell shrinkage could be much reduced or avoided using PSF process (Martino et al., 1998; Chevalier et al., 2000; Zhu et al., 2003; Sequeira Munoz et al., 2005; Alizadeh et al., 2007a).

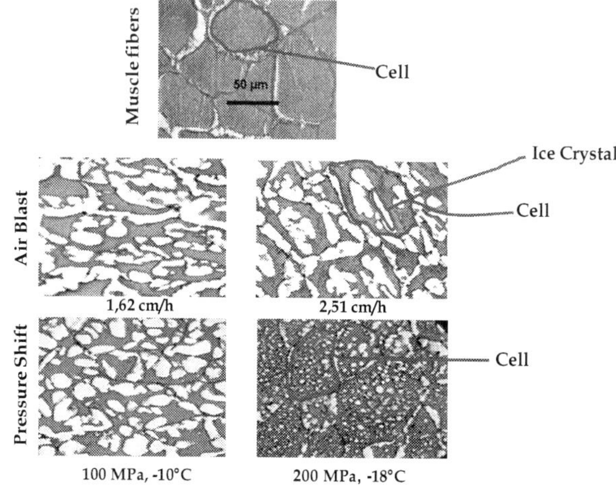

Fig. 3. Ice crystals formed in Atlantic salmon tissues during freezing (Alizadeh, 2007).

2.3 Ice crystal evolution during frozen storage

The evolution of the size of the ice crystal is important during frozen storage. It is difficult to evaluate the extracellular ice crystal for air blast freezing. But the size of high freezing rate extracellular ice crystals is smaller than low freezing rate ones. Alizadeh et al. (2007a) reported that the evolution of the intracellular ice crystal is not significant ($P<0.05$) during 6 months of storage for the air-blast (-30 °C, 4 m/s) and pressure (100 MPa) shift freezing. But for pressure shift freezing (200 MPa), the ice crystal size is changed after 6 months storage. Theoretically during frozen storage, small ice crystals have a tendency to melt and to aggregate to larger ones. It is known that the smallest ice crystals are the most unstable during storage. Indeed, the theory of ice nucleation permits to calculate the free energy of ice crystals as the sum of a surface free energy and of a volume free energy. The volume free energy increases faster than the surface free energy with increasing radius, explaining why the smaller ice crystals are more unstable. Thus the size of the ice crystals for pressure shift freezing (200 MPa) was stable for the first 3 months and then the size of the ice crystals tended to coarsen for longer storage (up to 6 months). In comparison, the size of the ice crystals obtained by pressure (100 MPa) shift freezing were much stable in size, demonstrating that a high pressure level is not necessarily required when prolonged frozen storage duration is envisaged (Alizadeh et al., 2007a).

3. Thawing process

The methodology and technique used for freezing and thawing processes play an important role in the preservation of the quality of frozen foods. Conventional thawing generally occurs more slowly than freezing, potentially causing further damages to frozen food texture. The thawing rate during conventional thawing processes is controlled by two main parameters outside the product: the surface heat transfer coefficient and the surrounding medium temperature. This medium temperature is supposed to remain below 15 °C during thawing, to prevent development of a microbial flora. The heat transfer coefficient then stays as the only parameter affecting the thawing rate at atmospheric pressure. Hence, the small temperature difference between the initial freezing point and room temperature does not allow high thawing rates (Chourot et al., 1996). Figure 4 shows a typical air blast thawing (ABT) curve. The temperature augmented to reach the melting point and temperature plateau appeared during this process.

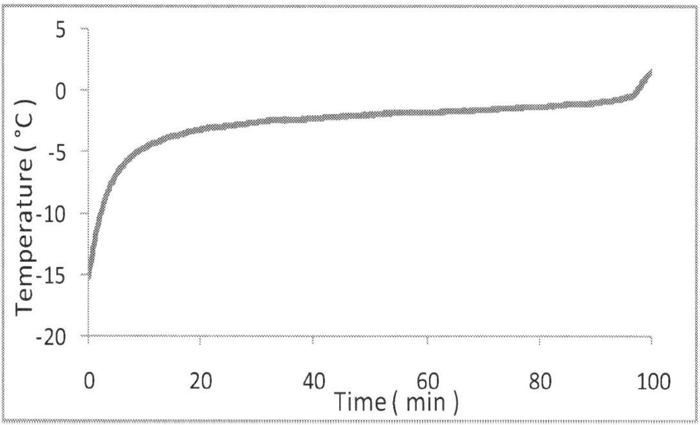

Fig. 4. A typical thawing curve of Atlantic salmon fillets obtained in air-blast thawing (Alizadeh, 2007).

Rapid thawing at low temperatures can help to prevent the loss of food quality during thawing process (Okamoto and Suzuki, 2002). This is obviously a challenge for traditional thawing processes, because the use of lower temperatures reduces the temperature difference between the frozen sample and the ambient, which is the principal driving force for the thawing process.

Pressure assisted thawing (PAT) may be attractive in comparison to conventional thawing when the quality and freshness are of primary importance. Figure 5 shows a typical pressure assisted thawing curve. Temperature increased slightly during the period of sample preparation (about 4 min) before pressurization due to the temperature difference between the sample and the medium in pressure chamber. During pressurisation the temperature decreases according to the depression of the ice-water transition under pressure (Bridgman, 1912). Then there was a temperature plateau due to the large amount latent heat needed for melting. The temperature rose quickly when thawing was completed. During the depressurization, the sample and the pressure medium were instantaneously cooled because of the positive coefficient of thermal expansion of water. To avoid ice crystal formation due to adiabatic cooling, sample temperature must be brought to a minimum level above 0 °C before releasing pressure (Cheftel et al., 2000).

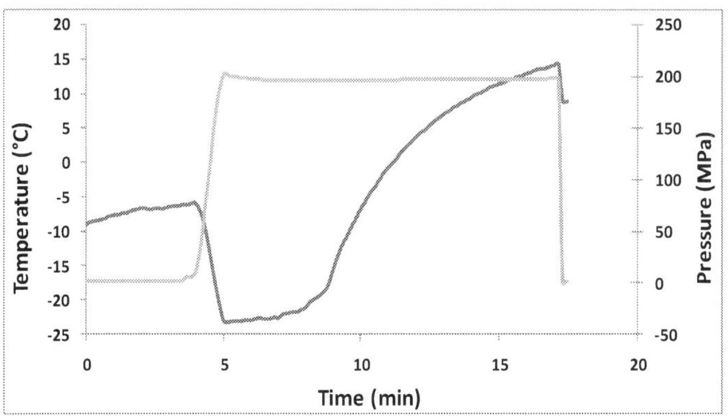

Fig. 5. A typical Pressure assisted thawing curve of Atlantic salmon fillets (Alizadeh, 2007).

3.1 Texture quality

Texture is an important quality parameter of the fish flesh. It is an important characteristic for consumer and also an important attribute for the mechanical processing of fillets by the fish food Industries. One critical quality factor influenced by freezing is food texture. Many foods are thawed from the frozen state and eaten directly, or cooked before consumption. In some cases, the texture of the thawed material is close to that of the fresh and unfrozen food. In other cases, the texture may be changed by the freezing process and yet result in a thawed product that is still acceptable to consumers. The texture of fish is modified after freezing and thawing (Figure 6). Pressure generally caused an increase in the toughness in comparison to conventional freezing and thawing (Chevalier et al., 2000; Zhu et al., 2004; Alizadeh et al., 2007b). This increase was attributed to the denaturation of proteins caused by high pressure processing. On the other hand, high pressure process was deleterious in some other aspects, mainly related to the effect of pressure on protein structures: high-pressure treatment (200 MPa) of Atlantic salmon muscle produced a partial denaturation with aggregation and insolubilization of the myosin (Alizadeh et al., 2007b). Freezing process is an important factor affecting textural quality of the fish. It is interesting to note that pressure shift freezing (200 MPa, -18 °C) induced formation of smaller and more regular ice crystals compared with air blast freezing (Chevalier et al., 2000; Alizadeh et al., 2007a; Martino et al., 1998). A tentative explanation could be that pressure shift freezing were less subjected to ice crystals injuries. Injuries involve a release of proteases (calpains and cathepsins) which are able to hydrolyse myofibrillar proteins and then to lead to quick textural changes (Jiang, 2000).

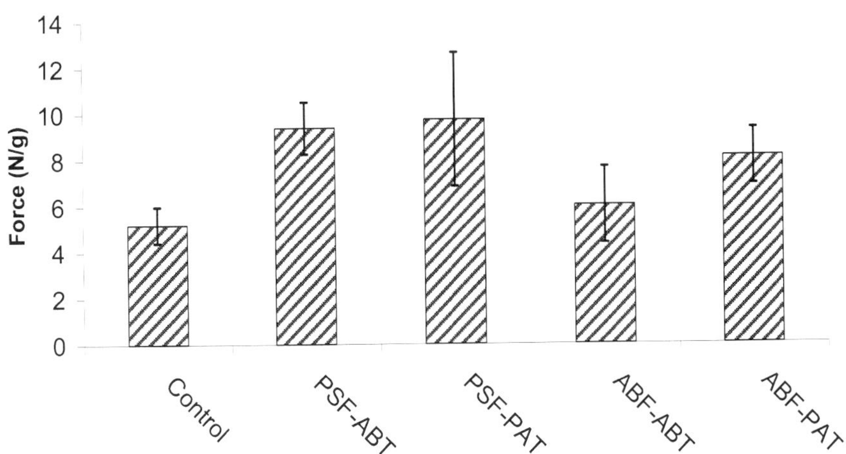

Fig. 6. Effect of Freezing (PSF, ABF) and thawing (PAT, ABT) on the texture of Atlantic salmon fillets (Alizadeh, 2007).

3.2 Colour changes

The first quality judgement made by a consumer on a food at the point of sale is its visual appearance. Appearance analyses of foods (colour and texture) are used in maintenance of food quality throughout and at the end of processing. Colour is one of the most important appearance attribute of food materials, since it influences consumer acceptability (Saenz et al., 1993).Various factors are responsible for the loss of colour during processing of food products. These include non-enzymatic and enzymatic browning and process conditions such as pH, acidity, packaging material and duration and temperature of storage.

The colour of fish is changed after freezing and thawing processes. This changes (assessed by very high colour differences ΔE) can be seen mainly caused by a strong increase in lightness (L*) and decrease for both redness (a*) and yellowness (b*) after pressure shift freezing. But this is opposite of those obtained for air blast freezing after thawing (Alizadeh et al., 2007b). Colour modifications and particularly modifications of lightness could be consequences of protein modifications. Changes in myofibrillar and sarcoplasmic proteins due to pressure could induce meat surface changes and consequently colour modifications (Ledward, 1998). The thawing process had little impact on overall colour change in fish after pressure shift freezing. But the discolouration of the flesh was visible with naked eyes after pressure assisted thawing (Alizadeh et al., 2007b). Murakami et al. (1992) also reported that an increase in all colour values (L*, a*, b*) of tuna when thawed by high pressure (50-150 MPa). This increase was stronger with increasing pressure. Furthermore, colour changes seem to be influenced by temperature, as lower temperatures caused stronger changes under the same pressure.

3.3 Drip loss

Drip loss is not only disadvantageous economically but can give rise to an unpleasant appearance and also involves loss of soluble nutrients. Drip loss during thawing is caused by irreversible damage during the freezing, storage (recrystallization), and thawing process (Pham & Mawson, 1997). Freezing can be considered as a dehydration process in which frozen water is removed from the original location in the product to form ice crystals. During thawing, the tissue may not reabsorb the melted ice crystals fully to the water content it had before freezing. This leads to undesirable release of exudate (drip loss) and toughness of texture in the thawed muscle (Mackie, 1993). Slow freezing produces larger extracellular ice crystals and resulting in more tissue damage and thawing loss. Thus, low freezing rate (air blast freezing) resulted in more drip than high freezing rate (pressure shift freezing) (Alizadeh et al., 2007b; Chevalier et al., 1999; Ngapo et al., 1999).

As shown in Figure 7, the freezing process was generally much more important than thawing for drip loss. Drip loss was reduced for all pressure shift freezing process irrespective to the thawing process but the air blast freezing resulted in a higher drip loss.

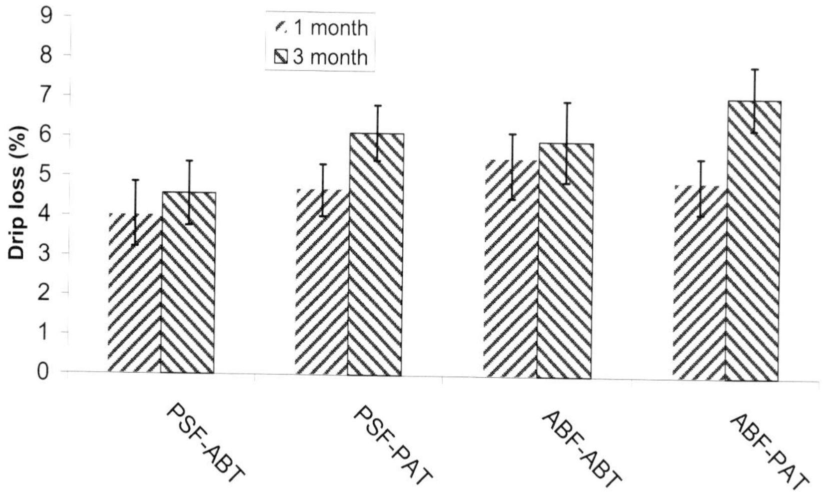

Fig. 7. Effect of Freezing (PSF, ABF) and thawing (PAT, ABT) on the drip loss of Atlantic salmon fillets (Alizadeh, 2007).

The pressure assisted thawing reduced the drip volume after conventional freezing. It can be assumed that during a slow thawing process, (corresponding to atmospheric pressure thawing), crystal accretion might occur leading to a mechanical damage of the cell membrane while thawing, and consequently in an increase of the drip volume. Pressure assisted thawing (PAT) reduced the thawing time and thus might have minimized the phenomenon of crystal accretion (Alizadeh et al., 2007b).

Few studies have reported the application of high pressure technology process for fish. Murakami et al. (1992) observed drip loss reduction in high pressure technology treated tuna meat. Chevalier et al. (1999) found that high freezing rate or high pressurization rate reduced thawing drip loss of whiting fillets (*Gadus merlangus*), but drip loss was reduced only by prolonging holding time of pressure as compared to atmospheric pressure thawing. Rouillé et al. (2002) found that high pressure technology processing (100, 150 and 200 MPa) of Spiny dogfish (*Squalus acanthias*) significantly reduced drip loss when compared with atmospheric thawing.

Crystal growth might enhance shrinkage of muscle fibers and even disrupt the cellular structure, resulting in a greater drip loss during frozen storage. Storage temperatures should be below -18 °C for optimum quality. Some studies suggest that drip loss may increase during extended frozen storage (Alizadeh et al., 2007b; Awonorin & Ayoade, 1992). Finally, drip loss seems to be a complicated process, and more studies are necessary for better understanding the phenomenon related to drip formation during pressure assisted thawing process.

4. Cooking process

Thermal processing techniques are widely used to improve eating quality and safety of food products, and to extend the shelf life of the products. Cooking is a heating process to alter the eating quality of foods and to destroy microorganisms and enzymes for food safety. Sous vide is a French term that means under vacuum. Sous vide involves the cooking of fish inside a hermetically sealed vacuum package. Cooking under vacuum (sous vide technology) defines those foods that are cooked in stable containers and stored in refrigeration. Because these products are processed at low temperatures (65 to 95 °C), the sensorial and nutritional characteristics are maximized in comparison with the sterilized products. The final product is not sterile and its shelf life depends on the applied thermal treatment and storage temperature. Figure 8 shows a typical cooking (Water bath) curve. The cooking was finished when the temperature reached at +80°C, then put in the ice water at 0°C for cooling.

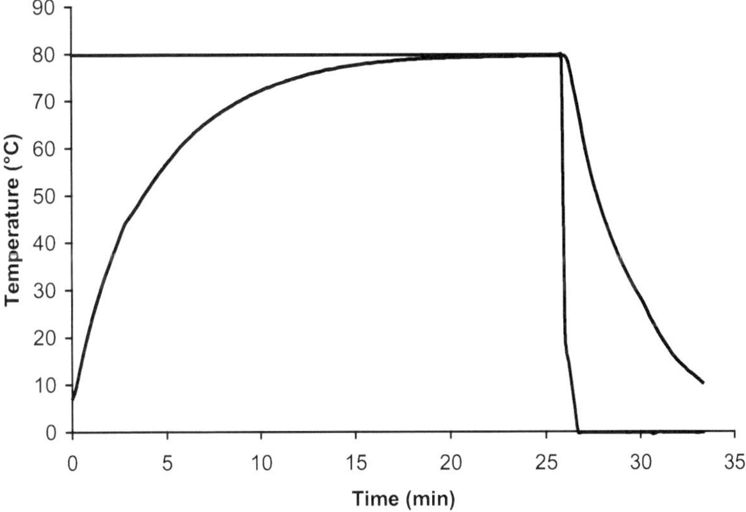

Fig. 8. Curve time/temperature during cooking under vacuum of Atlantic salmon fillets (Alizadeh, 2007).

Shaevel (1993) reported that the sous-vide process can also be used for cooking meat: this entails vacuum sealing the meat portions in plastic pouches, cooking in hot water vats for up to 4 h followed by rapid cooling at 1°C. Cooking time varied from food to food due to variation in heat transfer rates and the size of the food pieces.

4.1 Texture quality

Change in food texture was associated with heat treatment of the food such as cooking. It has been shown that thermal conditions (internal temperature) during meat cooking have a significant effect on all the meat texture profile parameters (cohesiveness, springiness, chewiness, hardness, elasticity). These reach their optimum level in the 70–80 °C range. As observed by Palka and Daun (1999), increasing the temperature to 100 °C causes the meat structure to become more compact due to a significant decrease in fiber diameter. During heating, at varying temperatures (37–75 °C), meat proteins denature and cause structural changes such as transversal and longitudinal shrinkage of muscle fibers and connective tissue shrinkage. Another effect is the destruction of cell membranes and the aggregation of sarcoplasmic proteins (Offer, 1984).

Pressure shift freezing (PSF) and cooking have an important effect on the quality of texture. Cooking process has more effect on texture than pressure shift freezing (Alizadeh et al., 2009). Meanwhile, the pressure shift freezing minimized the drip loss after cooking process. A partial cooking is a favorable fact for the pressure shift freezing, taking into account that a high proportion of fish is exposure to a cooking process before consuming. High cooking temperature can shorten cooking time and hence processing period, but it also causes high cooking loss and lower texture quality.

4.2 Protein denaturation

Denaturation can be defined as a loss of functionality caused by changes in the protein structure due to the disruption of chemical bonds and by secondary interactions with other constituents (Sikorski et al., 1976). Structural and spatial alterations can cause a range of textural and functional changes, such as the development of toughness, loss of protein solubility, loss of emulsifying capacity, and loss of water holding capacity (Miller et al., 1980; Awad et al., 1969; Dyer, 1951). In general during fish heating, sarcoplasmic and myofibrillar proteins are coagulated and denatured. The extent of these changes depends on the temperature and time and affects the yields and final quality of the fishery product.

Differential scanning calorimetry (DSC) can also be used to investigate the thermal stability of proteins and to estimate the cooking temperature of the seafood products. The proteins of salmon are denatured after freezing and cooking processes (Figure 9). Principal peaks are corresponding to myosin (42,5°C), sarcoplasmic proteins (55,5°C), collagen (64°C) and actin (73°C) (Schubring, 1999). Alizadeh et al. (2009) found that a partial denaturation of proteins, mainly to myofibrillar proteins denaturation, is induced by pressure shift freezing, similar to the effect of pressure on protein structures: high-pressure treatment (200 MPa) of sea bass muscle (Urrutia et al., 2007). As shown in Figure 9, cooking process was caused a total denaturation of proteins as comparison with pressure shift freezing. Bower (1987) showed the proteins were completely denatured under cooking process at 80°C.

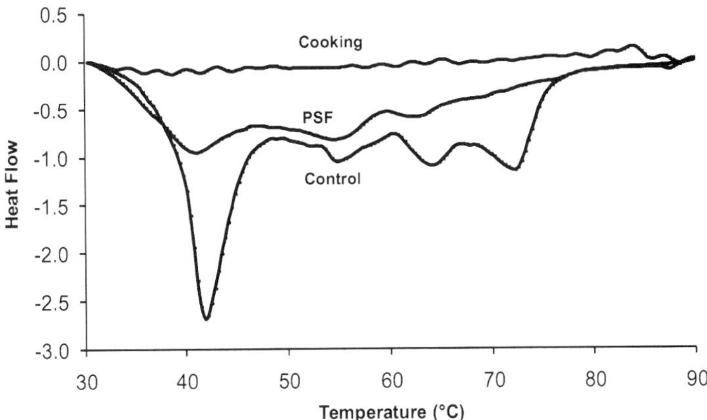

Fig. 9. Typical DSC thermograms of Atlantic salmon fillets (Alizadeh, 2007).

5. Conclusion

The quality of frozen foods is closely related to the size and distribution of ice crystals. Existence of large ice crystals within the frozen food tissue could result in mechanical damage, drip loss, and thus reduction in product quality. This chapter offers once again the advantage of pressure shift freezing process, which is widely used in the industry. Pressure shift freezing (200 MPa) process produced a large amount of small and regular intracellular ice crystals that can improved the microstructure of ice crystals (size, formation and location). The pressure shift freezing was responsible of a partial protein denaturation, which is reflected by an increase in texture. The total change of colour was observed after the freezing / thawing and cooking processes. The integration of results showed that the pressure shift freezing provides an interesting alternative compared to conventional freezing.

6. References

Alizadeh, E. (2007). Contribution à l'étude des procédés combinés de congélation – décongélation sur la qualité du saumon (*salmo salar*); impact des procédés haute pression et conventionnels. Thèse de Doctorat, Ecole Polytechnique de l'Université de Nantes, France, 77–150.

Alizadeh, E.; Chapleau, N.; De lamballerie, M. & Le bail, A. (2007a). Effects of different freezing processes on the microstructure of Atlantic salmon (*Salmo salar*) fillets. Innovative Food Science and Emerging Technologies, 8, 493–499.

Alizadeh, E.; Chapleau, N.; De lamballerie, M. & Le bail, A. (2007b). Effects of different freezing and thawing processes on the quality of Atlantic salmon (*Salmo salar*) fillets. Journal of Food Science: Food Engineering and Physical Properties, 72 (5), E279-E284.

Alizadeh, E.; Chapleau, N.; De lamballerie, M. & Le bail, A. (2009). Effect of Freezing and Cooking Processes on the Texture of Atlantic Salmon (*Salmo Salar*) Fillets.

Proceedings of the 5th CIGR Section VI International Symposium on Food Processing, Monitoring Technology in Bioprocesses and Food Quality Management (pp: 262-269), Potsdam, Germany, 31 August - 02 September 2009.

Awad, A.; Powrie, W. D. & Fennema, O. (1969). Deterioration of fresh-water whitefish muscle during frozen storage at -10°C. Journal of Food Science, 34, 1-9.

Awonorin, S. O. & Ayoade, J. A. (1992). Texture and eating quality of raw- and thawed-roasted turkey and chicken breasts as influenced by age of birds and period of frozen storage. Journal of Food Service System, 6, 241.

Bello, R. A.; Luft, J. H. & Pigott, G. M. (1982). Ultrastructural study of skeletal fish muscle after freezing at different rates. Journal of Food Science, 47, 1389-1394.

Bowers, J. A.; Craig, J. A.; Kropf, D. H. & Tucker, T. J. (1987). Flavor, color, and other characteristics of beef longissimus muscle heated to seven internal temperatures between 55 and 85°C. Journal Food Science, 52, 533-536.

Bridgman, P. W. (1912). Water in the liquid and five solid forms under pressure. Proceedings of the American Academy of Arts and Sciences, 47, 439-558.

Burke, M. J.; George, M. F. & Bryant R. G. (1975). Water in plant tissues and frost hardiness. In *Water relations of foods*, R. B. Duckworth, London, Academic Press, 111-135.

Cheftel, J. C.; Levy, J. & Dumay, E. (2000). Pressure-assisted freezing and thawing: principles and potential applications. Food Reviews International, 16, 453-483.

Cheftel, J. C.; Thiebaud, M. & Dumay, E. (2002). Pressure-assisted freezing and thawing: a review of recent studies. High Pressure Research, 22, 601–611.

Chevalier, D.; LeBail, A.; Chourot, J. M. & Chantreau, P. (1999). High pressure thawing of fish (Whiting): Influence of the process parameters on drip losses. Lebensmittel-Wissenschaft und-Technologie, 32, 25-31.

Chevalier, D.; Sequeira-Munoz, A. ; Le Bail, A. ; Simpson, B. K. & Ghoul, M. (2000). Effect of freezing conditions and storage on ice crystal and drip volume in turbot (*Scophthalmus maximus*): evaluation of pressure shift freezing vs. air-blast freezing. Innovative Food Science and Emerging Technologies, 1(3), 193-201.

Chevalier, D.; Sequeira-Munoz, A.; Le Bail, A.; Simpson, B. K. & Ghoul, M. (2001). Effect of freezing conditions and storage on ice crystal and drip volume in turbot (*Scophthalmus maximus*), evaluation of pressure shift freezing vs. air-blast freezing. Innovative Food Science and Emerging Technologies, 1, 193–201.

Chourot, J. M.; Lemaire, R.; Cornier, G. & LeBail, A. (1996). Modelling of high-pressure thawing. High Pressure Bioscience and Biotechnology, 439-444.

Delgado, A. E. & Sun, D. W. (2001). Heat and mass transfer models for predicting freezing processes – a review. Journal of Food Engineering, 47 (3),157–174.

Dyer, W. J. (1951). Protein denaturation in frozen and stored fish. Food Research, 16, 522–527.

Fennema, O. R.; Powrie, W. D. & Marth, E. H. (1973). Nature of the Freezing Process. In *Low Temperature Preservation of Foods and Living Matter*, M. Dekker, New York, 151-222.

Institut International du Froid (1986). Recommendations for the processing and handling of frozen foods. IIF, Paris, France, 32-39.

Jiang, S. T. (2000). Effect of proteinases on the meat texture and seafood quality. Food Science and Agricultural Chemistry, 2, 55-74.

Kalichevsky, M. T.; Knorr, D. & Lillford P. J. (1995). Potential food applications of high-pressure effects on ice-water transitions. Trends in Food Science & Technology, 6(8), 253-259.

Le Bail, A.; Chevalier, D. ; Mussa, D. M. & Ghoul, M. (2002). High pressure freezing and thawing of foods: a review. International Journal of Refrigeration, 25(5), 504–513.

Ledward, D. A. (1998). High pressure processing of meat and fish. Fresh novel foods by high pressure. Biotechnology and Food Research, 165-175.

Li, B. & Sun, D. W. (2002). Novel methods for rapid freezing and thawing of foods— a review. Journal of Food Engineering, 54 (3),175–182.

Mackie, I. M. (1993). The effect of freezing on fish proteins. Food Review International, 9, 575–610.

Martino, M. N.; Otero, L.; Sanz, P.D. & Zaritzky, N.E. (1998). Size and location of ice crystals in pork frozen by high-pressure-assisted freezing as compared to classical methods. Meat Science, 50(3), 303-313.

Miller, A. J.; Ackerman, S. A. & Palumbo, S. A. (1980). Effects of frozen storage on functionality of meat for processing. Journal of Food Science, 45, 1466–1471.

Murakami, T.; Kimura, I.; Yamagishi, T. & Fujimoto, M. (1992). Thawing of frozen fish by hydrostatic pressure. High Pressure Biotechnology, 224, 329-331.

Ngapo, T. M.; Babare, I. H.; Reynolds, J. & Mawson, R. F. (1999). Freezing and thawing rate effects on drip loss from samples of pork. Meat Science, 53, 149-158.

Offer, G. (1984). Progress in the biochemistry physiology and structure of meat. In R. Lawrie (Ed.), Proceedings of the 30th European meeting of meat science (4th ed.) (pp. 173–241). Oxford: Elsevier.

Okamoto, A. & Suzuki, A. (2002). Effects of high hydrostatic pressure-thawing on pork meat. Trends in High Pressure Bioscience and Biotechnology, 19, 571-576.

Otero, L. ; Martino, M. ; Zaritzky, N. ; Solas, M. & Sanz, P. D. (2000). Preservation of microstructure in peach and mango during highpressure- shift freezing. Journal of Food Science, 65(3), 466–470.

Palka, K. & Daun, H. (1999). Changes in texture, cooking losses, and miofibrillar structure of bovine M. semitendinosus during heating. Meat Science, 51, 237–243.

Pham, Q. T. (1987). Calculation of bound water in frozen food. Journal of Food Science, 52(1), 210-212.

Phan, Q. T. & Mawson, R. F. (1997). Moisture migration and ice recrystalization in frozen foods. In *Quality in Frozen Food,* M. C. Erickson, Y. C. Hung, New York, Chapman and Hall, 67–91.

Rouillé, J.; LeBail, A.; Ramaswamy, H. S. & Leclerc, L. (2002). High pressure thawing of fish and shellfish. Journal of Food Engineering, 53, 83-88.

Saenz, C.; Sepulveda, A. E. & Calvo, C. (1993). Color changes in concentrated juices of prickly pear (*Opuntia ficus indica*) during storage at different temperatures. Lebensmittel-Wissenschaft und-Technologie, 26, 417-421.

Schubring, R. (1999). DSC studies on deep frozen fishery products. Thermochimica Acta, 337, 89-95.

Sequeira-Munoz, A.; Chevalier, D.; Simpson, B. K.; Le Bail, A. & Ramaswamy, H. S. (2005). Effect of pressure-shift freezing versus air-blast freezing of carp (*Cyprinus carpio*) fillets: a storage study. Journal of Food Biochemistry, 29, 504-516.

Shaevel, M. L. (1993). Manufacturing of frozen prepared meals. In *Frozen Food Technology*, C. P. Mallett, London, Blackie Academic & Professional, 270–302.

Sikorski, Z.; Olley, J. & Kostuch, S. (1976). Protein changes in frozen fish. Critical Reviews in Food Science and Nutrition, 8, 97–129.

Urrutia, G.; Arabas, J.; Autio, K.; Brul, S.; Hendrickx, M.; Ka̱kolewski, A.; Knorr, D.; Le Bail, A.; Lille, M.; Molina-García, A. D.; Ousegui, A.; Sanz, P. D.; Shen, T. & Van Buggenhout, S. (2007). SAFE ICE: Low-temperature pressure processing of foods: Safety and quality aspects, process parameters and consumer acceptance. Journal of Food Engineering, 83, 293–315.

Zhu, S.; Le Bail, A. & Ramaswamy, H. S. (2003). Ice crystal formation in pressure shift freezing of Atlantic salmon (*Salmo salar*) as compared to classical freezing methods. Journal of Food Processing and Preservation, 27(6), 427–444.

Zhu, S.; Ramaswamy, H. S. & Simpson, B. K. (2004). Effect of high-pressure versus conventional thawing on colour, drip loss and texture of Atlantic salmon frozen by different methods. Lebensmittel-Wissenschaft und-Technologie, 37, 291-299.

Novel Fractionation Method for Squalene and Phytosterols Contained in the Deodorization Distillate of Rice Bran Oil

Yukihiro Yamamoto and Setsuko Hara
Seikei University
Japan

1. Introduction

To obtain the valuable constituents from natural products, "fractionation" is very important step in industrial processing. In addition, how it is efficient, how it is green for the environment, animals, nature and human, has been required for its standard. Especially in food industry, there is limitation of usage of organic solvent, which is well adopted for extraction of hydrophobic constituents from natural products. Although, hexane or ethanol has used for extraction in food industry at least in Japan, it is often less efficient than chloroform or methanol for extraction of low polar constituents such as lipids. In addition, it is difficult to remove residual solvent. Safe and simple extraction or fractionation methods have been required. In this point, supercritical fluid extraction method may be useful. In this chapter, after introducing about the key words of our study, we will present about - novel fractionation method for squalene and phytosterols contained in the deodorization distillate of rice bran oil.

2. Rice bran and rice bran oil

Rice bran, ingredient of rice bran oil (RBO), is co-product of milled rice containing pericarp, seed coat, perisperm and germ. In Japan, 1,000,000 t/year of rice bran have produced. One third is used for production of RBO and the residue is for feeding stuff, fertilizer, Japanese pickle and etc. Rice bran contains ~20% of RBO, and in RBO, unsaponifiables such as squalene and phytosterol are specifically highly contained compare to other seed oils (Table 1). Fatty acid in RBO is mainly composed by oleic acid (C18:1) and linoleic acid (C18:0). Palmitic acid (C16:0) is contained higher compare to other edible oils. In addition lower content of linolenic acid (C18:3) makes RBO stable for oxidation.

It is known that RBO exerts cholesterol lowering effect, for example, Chou et al. reported that RBO diet improves lipid abnormalities and suppress hyperinsulinemic responses in rats with streptozotocin/nicotinamide-induced type 2 diabetes (Chou et al., 2009). Physiological function of RBO is well documented in several reviews (Cicero and Gaddi, 2001; Jariwalla, 2001; Sugano et al., 1999).

In the process of produce RBO, crude RBO is steam-distillated for its flavor. The volatile component is called "deodorization distillate of RBO". The content of deodorization distillate of RBO is 0.15-0.45% varied with its condition of distillation. The deodorization distillate of RBO is viscous and typical smelled liquid. Besides diacylglyceride and free fatty acid, deodorization distillate of RBO has nearly 40% of unsaponifiable substances such as squalene, phytosterols and tocopherol. Particularly it contains ca 10% of squalene in deodorization distillate of RBO.

	RBO	Soy been	Canola	Corn
Unsaponifiables/Oil (%)	2.31	0.46	0.87	0.96
Fatty acid composition (%)				
C16:0	16.4	10.5	4.2	10.4
C18:0	1.6	3.9	2.0	1.9
C18:1	42.0	23.3	60.8	27.5
C18:2	35.8	53.0	20.6	57.2
C18:3	1.3	7.6	9.2	1.2
Others	2.9	1.7	3.2	1.8

Table 1. Chemical properties and fatty acid compositions of major edible oils.

3. Squalene

Squalene is widely found in marine animal oils as a trace component, and has been extensively studied their preventive effect in many diseases such as cardiovascular diseases and cancer (Esrich et al., 2011; Smith, 2000). Recently it is also attracted attention as food factor (Bhilwade et al., 2010). Since it has been well known that the liver oil of some varieties of shark (*Squalidae* family), especially those inhabiting the deep sea, is rich in squalene, this substance has been fractionated from shark liver oil. On the other hand, squalene obtained from shark liver oil has not been fully utilized recently on humane grounds, due to the unstable supply of shark liver oil as an industrial material, the characteristic smell of fish oil, and the large variation of the constituents. Then, attention has shifted to squalene of plant origin, and the application of this type of squalene to cosmetics, medicine and functional foods has been attempted. For example, squalene originating from olive oil is being produced in Europe. However, the production is poor to make up for the sagging production of squalene from shark liver oil. Therefore, the possibility of extracting squalene from RBO is being investigated. Furthermore, because squalene is easily oxidized owing to its structural character having a lot of carbon double bond (Fig. 1.), novel fractionation method conducting in mild condition is required.

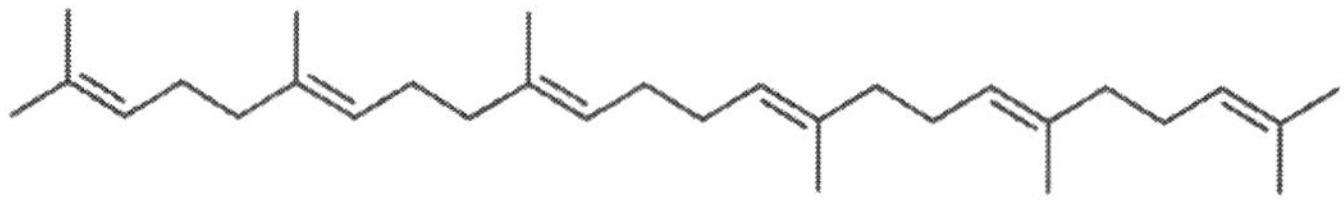

Fig. 1. Chemical structure of squalene.

4. Phytosterol

The main phytosterols in RBO are β-sitosterol, campesterol, stigmasterol and isofucosterol (Fig. 2.), and the content in phytosterols were ca 50%, 20%, 15%, 5%, respectively. It is well known that phytosterols inhibit cholesterol absorption on small intestine which results cholesterol-reducing activity of phytosterols (Gupta et al., 2011; Niijar et al., 2010; Malinowski & Gehret, 2010). In addition, the sterol content of RBO is specifically high compare to other major oils from plant- ca 1% in RBO, and 0.2-0.5% in soy bean, canola, and corn oil. These phytosterols are effective substances for utilization for functional foods called foods for specified health use.

Fig. 2. Chemical structures of main phytosterols in RBO; (A) β-sitosterol, (B) campesterol, (C) stigmasterol, (D) isofucosterol.

5. Supercritical fluid extraction and supercritical fluid chromatography

Supercritical fluid is any substance at a temperature and pressure above its critical point, where distinct liquid and gas phases do not exist. It can effuse through solids like a gas, and dissolve materials like a liquid. There are mainly two techniques, supercritical fluid extraction (SFE) and supercritical fluid chromatography (SFC). For high fractionation selectivity, combination of SFE and SFC is recently adopted. Following advantages of SFE or SFC are known.

- Carried at low temperature-thermal denaturation of substances is poorly occurred.
- Inactive gas (CO_2)-denaturation and oxidation are almost ignored.
- Supercritical fluid is whole diminished after extraction.
- Bland, innocuous and harmless- applicable for food industry.

Carbon dioxide and water are the most commonly used supercritical fluids, being used for decaffeination of coffee been (Khosravi-Darani, 2010), extraction of eicosapentaenoic acid and docosapentaenoic acid from fish oil (Higashidate et al., 1990) and fractionation of terpenes from lemon peel oil (Yamauchi & Saito, 1990), respectively. For the detail of the application of supercritical fluid for variety of field including food industry, several reviews are available (Herrero et al., 2010; Khosravi-Darani, 2010; Zhao & Jiang, 2010).

6. Novel fractionation method for squalene and phytosterols contained in the deodorization distillate of rice bran oil

Although there are already some reports on methods of concentrating squalene originating from plants, the thermal and oxidative deterioration of squalene poses significant problems for its use in foods and medicines. In addition, solvent fractionation bears some difficulties related to proper solvent selectivity during processing, the elimination of the residual solvent from the squalene fraction, and the disposal of the waste fluid. In this chapter, we presented a novel method of concentrating the squalene and phytosterols contained in the deodorization distillate of rice bran oil by combining SFE, SFC and solvent fractionation. Under the supercritical condition, gas acquires unique characteristics: upon a slight pressure change, it becomes more dense, less viscous and more soluble (Xiao-Wen, 2005). Therefore, supercritical fluid has the excellent ability to separate and extract specific components from a mixture. Since the critical temperature and pressure of carbon dioxide are 31.1$\circ$C and 72.8 kg/cm^2, respectively, there are many advantages to using supercritical carbon dioxide for extraction at room temperature: the extract undergoes no thermal and oxidative deterioration, no remnant gas is generated in the extract at atmospheric pressure and etc as described above. SFC is determined here that the method extracting from the solution mixed with silica gel as a stationary phase, not capillary column SFC or packed column SFC.

6.1 Experimental
6.1.1 Samples and reagents

The deodorization distillate of RBO used to concentrate squalene and phytosterols was supplied from Boso Oil & Fat Co., Ltd., (Funabashi, Japan). The composition of the deodorization distillate was as follows: squalene, 8.5%; phytosterols, 3.5% (sitosterol: 2.05%, stigmasterol: 0.69%, campesterol: 0.76%); tocopherol (Toc), 1.6%; triacylglycerol (TG), 58.0%; diacylglycerol (DG) + fatty acids (FA), 26.5%. Squalene standard of special grade and 2-propanol of HPLC grade were purchased from Wako Pure Chemical Industries, Ltd. (Osaka, Japan), and all other reagents used were of special grade or first grade, as commercially available, and used without further purification, unless otherwise specified.

6.1.2 Condensation of squalene from the deodorization distillate

6.1.2.1 Condensation of squalene by SFE

To fractionate squalene from the deodorization distillate, SFE with supercritical carbon dioxide was employed. SFE makes it possible to extract the target compound by changing the density of supercritical carbon dioxide under certain conditions of temperature and/or pressure. In the present experiment, SFE was carried out with an SFC apparatus, model 880-81 (JASCO, Tokyo, Japan), and a 10- or 50-mL stainless vessel. The extractor was kept in a thermostatic oven, model CH-201 (Scinics Corporation, Tokyo, Japan) and temperatures ranging from 30 to 80$\circ$C and pressures ranging from 80 to 220 kg/cm^2 were examined at an extraction time of 2 h and a supercritical carbon dioxide flow rate of 7 mL/min. The squalene concentration in the extract was determined by TLC-FID analysis with hexane as the primary developing solvent and benzene/ethyl acetate (9:1, v/v) as the secondary developing solvent.

6.1.2.2 Condensation of squalene by SFC

The optimal conditions for squalene concentration by SFE were determined to be as follows: extraction temperature, 30°C; supercritical carbon dioxide pressure, 100 kg/cm²; flow rate, 7 mL/min. However, squalene was only concentrated to 25%, and the other components in the extract had higher polarity, such as TG, DG and FA. Therefore, 1 to 3-fold higher quantities of activated silica gel as compared with the deodorization distillate were placed in the extraction vessel to adsorb the components with higher polarity, and then squalene was extracted and/or concentrated under the optimal conditions listed above by means of SFC as shown in Fig. 3.

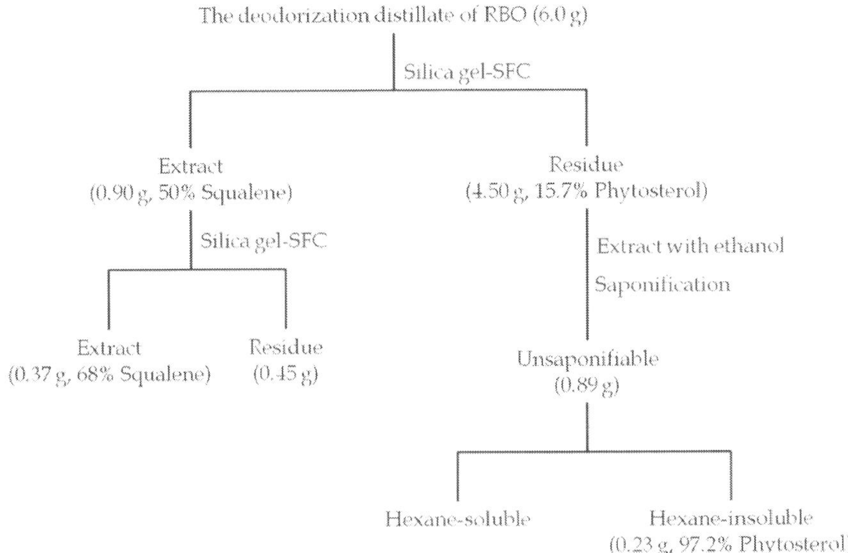

Fig. 3. Fractionation of squalene and phytosterols from deodorization distillate of RBO.

6.1.3 Condensation of phytosterols from the deodorization distillate

6.1.3.1 Condensation of phytosterols by SFC

Since phytosterols were concentrated in the residue after the extraction of squalene by SFC, further concentration of the phytosterols contained in 6 g of residue was attempted. In addition, the residue recovered as a result of SFE was also subjected to SFC to concentrate the phytosterols adsorbed onto the silica gel. In this case, the concentration ratio of phytosterols was compared with the amounts of residual phytosterols before and after the further extraction of squalene for 2 h, because phytosterols remained in the residue.

6.1.3.2 Condensation of phytosterols by solvent fractionation

The residue remaining in the vessel packed with silica gel were extracted with ethanol and then saponified by refluxing for 4 h with 3 mL of 25% potassium hydroxide aqueous solution and 40 mL each of ethanol and hexane. After the saponification, the reactant was separated into a hexane layer and a hydrated ethanol layer.

6.1.3.3 Condensation of phytosterols from the unsaponifiable components of the deodorization distillate

Another means of concentrating phytosterols was also examined, as shown in Figs. 4: the phytosterols were concentrated after the saponification of the deodorization distillate.

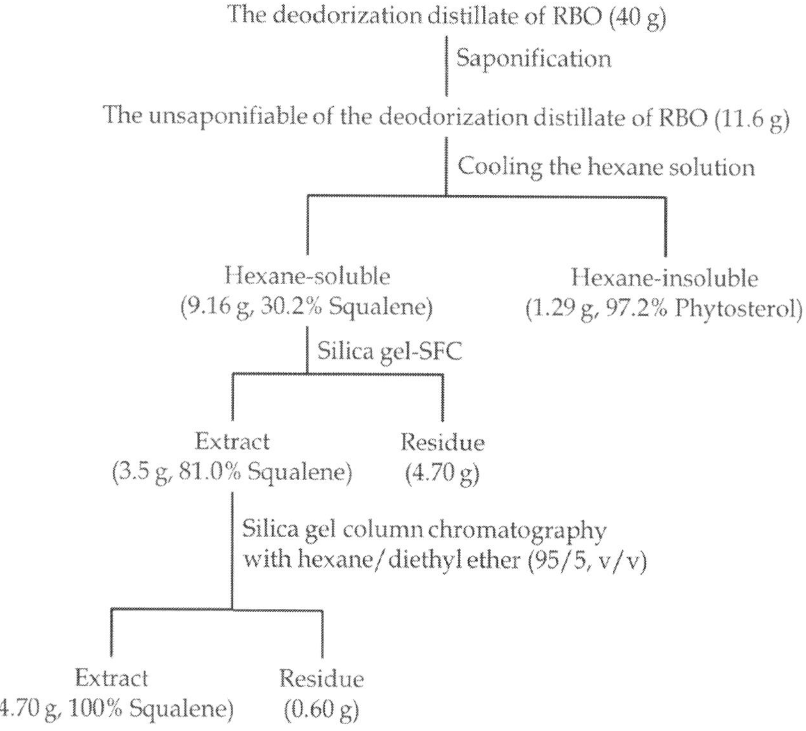

Fig. 4. Fractionation of squalene and phytosterols from unsaponifiable components of the deodorization distillate of RBO.

6.1.3.4 Preparation of highly purified squalene

To obtain more highly purified squalene, 6 g of the extract, which contained squalene having 81% purity and recovered by SFC of the unsaponifiable components, were column-chromatographed with hexane/diethyl ether (95:5, v/v) as an eluent by the procedure shown in Fig. 4.

6.1.4 Analysis

To confirm the composition of the deodorization distillate of rice bran oil, TLC-FID analysis was carried out on a Iatroscan, model MK-5 (Mitsubishi Kagaku Iatron Inc., Tokyo, Japan) equipped with a Chromatopac C-R6A (Shimadzu Corp., Kyoto, Japan). The deodorization distillate, squalene, canola oil, oleic acid, phytosterols and Toc as standard samples were loaded on silica gel sintering quartz rods (Chromarod S-III, Mitsubishi Kagaku Iatron Inc., Tokyo, Japan). Developing solvents consisting of hexane as a primary solvent to detect squalene and benzene/ethyl acetate at a ratio of 9:1 (v/v) as a secondary solvent to detect the other components were used. The fatty acid composition of the lipids was analyzed by GLC. The constituent fatty acids were methyl-esterified (Jham et al., 1982), and a fused silica capillary column, CBP1-S25-050 (0.32 mm x 25 m; Sinwa-kagaku Co., Ltd.), was connected to a GLC (model GC-18A; Shimadzu Corp., Kyoto, Japan) for constant-temperature GLC analysis. The column temperature was set at 200°C, helium was used as the carrier gas, and FID was used as the detector. A normal-phase column, Finepak SIL-5 (4.6 × 250 mm; JASCO Corp.), was connected to an HPLC (model BIP-1; JASCO), and a mobile-phase solvent that consisted of hexane and 2-propanol at a ratio of 124:1 was allowed to flow through the column at a rate of 0.7 mL/min. A fluorescence detector (model FP-2020 Plus, JASCO) with excitation at 295 nm and emission at 325 nm was used. For the quantitative analysis of Toc, 2,2,5,7,8-pentamethyl-6-chromanol (PMC) was used as the internal standard, and the calibration curves prepared beforehand for the Toc isomers [for a-Toc, y=0.4508x-0.0927 (R2=0.9682); for b-Toc, y=0.7795x-0.3017 (R2=0.9488); for g-Toc, y=1.2532x-0.7692 (R2=0.9887); and for δ-Toc, y=1.2454x-0.9934 (R2=0.9752)] were used.

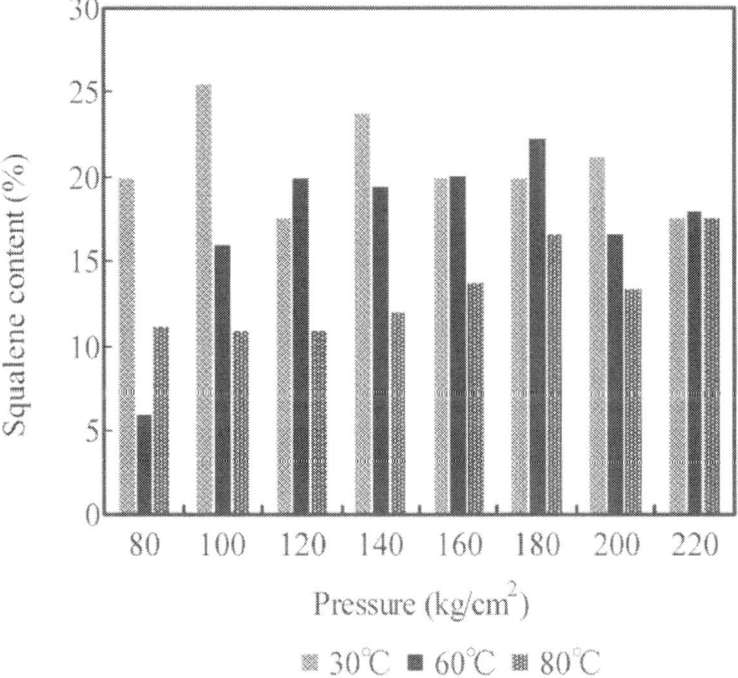

Fig. 5. Changes in squalene content under various conditions of SFE.

6.2 Results and discussion
6.2.1 Condensation of squalene from the deodorization distillate

6.2.1.1 Condensation of squalene by SFE

As shown in Fig. 5, squalene was concentrated to an average concentration of 25% with quantitative recovery under the following conditions: extraction temperature, 30°C; pressure, 100 kg/cm²; extraction time, 2 h. It was confirmed that squalene could be concentrated to 3 times the original content (8.5%) in the deodorization distillate with peroxide value (PV) of 3.0 meq/kg.

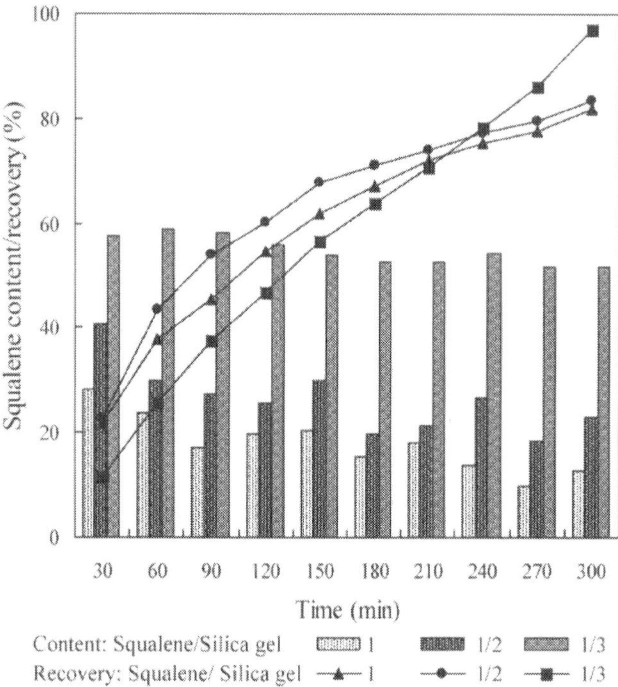

Fig. 6. Changes in the squalene content and recovery at various amounts of activated silica gel in SFC.

6.2.1.2 Condensation of squalene by SFC

As shown in Fig. 6, the higher squalene content was obtained with more than 95% recovery when 3 times more silica gel was added to the deodorization distillate and the squalene was extracted with supercritical carbon dioxide for 5 h. The extract obtained by silica-gel SFC contained, on average, 50% squalene and 38% components with higher polarity. Then, SFC of the extract obtained by the first SFC was repeated after mixing fresh silica gel. As a result, it was proved that squalene could be concentrated to 68% in the extract with PV of 3.1 meq/kg by repeating SFC with 3-fold more silica gel under the conditions stated earlier. The PV of intact squalene was 3.0 meq/kg as mentioned above.

From these results, it is suggested that the squalene was not oxidized under these conditions. Therefore, it was found that the present silica gel-SFC, with the addition of silica gel as a stationary phase into the supercritical vessel to create a chromatographic system, had a higher selectivity than mere SFE. The present silica gel-SFC is expected to become a very useful technique for concentrating squalene from the deodorization distillate of RBO as shown in Fig. 3.

6.2.2 Condensation of phytosterols from the deodorization distillate

6.2.2.1 Condensation of phytosterols by SFC

The composition of the residue with PV of 3.0 meq/kg recovered by SFC with silica gel packed as a stationary phase under the conditions of 30°C, 100 kg/cm^2, 7 mL/min and 5 h was 10.4% phytosterols, 3.9% Toc, 48.6% TG, and 37.1% DG + FA. In addition, the residue recovered from the procedure described in 6.1.2.1 contained 17.3% phytosterols under the following conditions: 30°C, 220 kg/cm^2, 7 mL/min and 7 h. From these results, it was considered that SFC did not suit the separation of phytosterols from a mixture of phytosterols, TG, DG and FA, which have nearly the same polarities, although SFC was suitable for the extraction of compounds with lower polarity, such as a squalene. Then, we examined solvent fractionation to concentrate the phytosterols from the deodorization distillate.

6.2.2.2 Condensation of phytosterols by solvent fractionation

The residual fraction shown in Fig. 3 contained 15.7% phytosterols in 4.5 g of recovered residue. The 4.5 g of residue remaining in the vessel packed with silica gel were extracted with ethanol and then saponified by refluxing for 4 h with 3 mL of 25% potassium hydroxide aqueous solution and 40 mL each of ethanol and hexane. After the saponification, the reactant was separated into a hexane layer and a hydrated ethanol layer. The unsaponifiable components thus obtained in the hexane layer were then cooled to obtain 0.23 g of crystalline phytosterols with 97.3% purity. As described above, squalene was fractionated by SFC with silica gel packed into the vessel, and phytosterols were highly concentrated from the residue by solvent fractionation. Therefore, it is considered that the combination of silica gel-SFC and solvent fractionation was a very effective means of obtaining both components with higher purity. This method, however, is rather time-consuming and costly, because SFC has to be repeated in order to concentrate the squalene, and the residue has to be extracted from the silica gel in the SFC column to concentrate the phytosterols.

6.2.2.3 Condensation of phytosterols from the unsaponifiable components of the deodorization distillate

After the saponification of the deodorization distillate (40 g) by refluxing for 4 h with 3 mL of 25% potassium hydroxide aqueous solution, 11.6 g of unsaponifiable components were recovered. Then, hexane was added to the components, and the crystalline phytosterols were recovered from the hexane-insoluble fraction under cooling. By a series of processes, 9.16 g of hexane-soluble fraction and 1.29 g of hexane-insoluble fraction were obtained and analyzed by TLC-FID. As a result, it was found that the phytosterols were concentrated to 97.2% in the hexane-insoluble fraction as shown in Table 2.

Fraction	Hexane soluble (9.16 g)	Hexane insoluble (1.29 g)
Less polar components	11.9	0
Squalene	30.2	0
Phytosterol	14.1	97.2
TG	17.5	2.8
DG	26.3	0
FA		
PV (meq/kg)	3.5	3.8

Table 2. Composition of the hexane-soluble and hexane- insoluble fractions by solvent fractionation (%).

In hexane soluble faction, saponifiables such as TG, DG and FA were contained. In this study, the condition of saponification was not finely examined. By controlling the reflux time and temperature or the concentration of potassium hydrate, TG, DG and FA could be well saponified.

It was confirmed that a combination of saponification and solvent fractionation of the deodorization distillate is an effective means of concentrating phytosterols. Since squalene was concentrated to 30.2% in the hexane soluble fraction, this fraction were subjected to SFC with silica gel under the following conditions to obtain higher purity squalene: flow rate of supercritical carbon dioxide, 3 or 7 mL/min; extraction pressure, 80-140 kg/cm^2. As results, it was found that higher squalene recovery tended to be obtained at faster flow rates and higher pressures. Furthermore, the squalene content in the extract reached 81.0%. From these results, it is considered that the deodorization distillate which is usually discarded as waste can be utilized for sources of functional components. In addition, the comparison of Fig. 3 with Fig. 4 indicates that the solvent fractionation of unsaponifiable components of the deodorization distillate is a practicable and convenient method of concentrating phytosterols and squalene. The combination of solvent fractionation and SFC developed in the present work is deemed to be an effective and safe means of fractionating squalene and phytosterols, which can then be used as additives in cosmetics and functional foods.

6.2.2.4 Preparation of highly purified squalene

The 3.50 g of extract containing 81.0% squalene obtained from the SFC were further purified by column chromatography with hexane/diethyl ether (95:5, v/v). As a result, 2.55 g of squalene with 100% purity and PV of 4.0 meq/kg could be obtained with 500 mL of eluate.

7. Conclusion

In this chapter, a novel method of fractionating squalene and phytosterols contained in the deodorization distillate of RBO without any oxidative rancidity was established by the combination of solvent fractionation and SFC after saponification of the deodorization distillate. Although there are some industrial production methods which are patented (Hirota & Ohta, 1997; Tsujiwaki et al., 1995; Ando et al., 1994) of squalene or squalane from the deodorization distillate of RBO, those methods have to perform many processes such as saponification, solvent fractionation, distillation, hydrogenation, and final molecular distillation to avoid the oxidative rancidity of squalene, or another is a cultivation method with yeast extracts for 6 days at 30°C. A Japan patent (Kohno, 2002) for the production for phytosterols are released from Kao Corporation, in which phytosterols are concentrated to

90-94% purity from crude phytosterols (purity: ca. 80%) with hydrocarbon solvents. Commercial squalenes obtained from shark liver oil, olive oil, and rice bran oil are now on sale as 1,000-1,500 yen/kg, 2,500 yen/kg, and 15,000 yen/kg, respectively. The market prices of phytosterols are 3,500-15,000 yen/kg based on their purities. Therefore, the present method has some merits such as a fewer operation process, time-saving, no oxidative rancidity and continuous production of the two functional components. In addition, there is a strong possibility of lower prices production than existent methods, since carbon dioxide used as a supercritical gas is costly but recyclable. It was found that the present method very safely and effectively fractionates the functional components contained in deodorization distillate, which is usually regarded as waste.

8. References

Ando, Y., Watanabe, Y. & Nakazato, M. (1994). Japan patent. 306387.

Bhilwade, HN., Tatewaki, N., Nishida, H. & Konishi, T. (2010). Squalene as Novel Food Factor. *Current Pharmaceutical Biotechnology*, Vol. 11 (No. 8): 29-36.

Chou, TW., Ma, CY., Cheng, HH. & Gaddi, A. (2009). A Rice Bran Oil Diet Improves Lipid Abnormalities and Suppress Hyperinsulinemic Responses in Rats with Streptozotocin/Nicotinamide-Induced Type 2 Diabetes. *Journal of Clinical Biochemistry and Nutrition*, Vol. 45 (No. 1): 29-36.

Cicero, AF. & Gaddi, A. (2001). Rice Bran Oil and Gamma-Oryzanol in the Treatment of Hyperlipoproteinaemias and Other Conditions. *Phytotheraphy research : (PTR)*, Vol. 14 (No. 4): 277-289.

Escrich, E., Solanas, M., Moral, R. & Escrich, R. (2011). Modulatory Effects and Molecular Mechanisms of Olive Oil and Other Dietary Lipids in Breast Cancer. *Current Pharmaceutical Design*, Vol. 17 (No. 8): 813-830.

Gupta, AK., Savopoulos, CG., Ahuja, J. & Hatzitolios, AI. (2011). Role of phytosterols in lipid-lowering: current perspectives. *QJM : Monthly Journal of the Association of Physicians*, Vol. 104 (No. 4): 301-308.

Herrero, M., Mendiola, JA., Cifuentes, A. & Ibáñez, E. (2010). Supercritical Fluid Extraction: Recent Advances and Applications. *Journal of Chromatography A*, Vol. 1217 (No. 16): 2495-2511.

Higashidate, S., Yamauchi, Y. & Saito, M. (1990). Enrichment of Eicosapentaenoic Acid and Docosahexaenoic Acid Esters from Esterified Fish Oil by Programmed Extraction-Elution with Supercritical Carbon Dioxide. *Journal of Chromatography A*, Vol. 515 (No. 31): 295-303.

Hirota, Y. & Ohta, Y. (1997). Japan patent. 176057.

Jarowalla, RJ. (2001). Rice Bran Products: Phytonutrients with Potential Applications in Preventive and Clinical Medicine. *Drugs Under Experimental and Clinical Research*, Vol. 27 (No. 1): 17-26.

Jham, GN., Teles, FFF. & Campos, LG (1982). Use of Aqueous HCl/MeOH as Esterification Reagent for Analysis of Fatty Acid Derived from Soybean Lipids. *Journal of the American Oil Chemists Society*, Vol. 59 (No. 3): 132-133.

Khosravi-Darani, K. (2010). Research Activities on Supercritical Fluid Science in Food Biotechnology. *Critical Reviews in Food Science and Nutrition*, Vol. 50 (No. 6): 479-488.

Khono, J. (2002). Japan patent. 316996.

Malinowski, JM. & Gehret, MM. (2010). Phytosterols for Dyslipidemia. *American Journal of Health-System Pharmacy : AJHP : Official Journal of the American Society of Health-System Pharmacists*, Vol. 67 (No. 14): 1165-1173.

Niijar, PS., Burke, FM., Bloesch, A. & Rader, DJ. (2010). Role of Dietary Supplements in Lowering Low-Density Lipoprotein Cholesterol: a review. *Journal of Clinical Lipidology*, Vol. 4 (No. 4): 248-258.

Smith. (2000). Squalene: Potential Chemopreventive Agent. *Expert Opinion on Investigational Drugs*, Vol. 9 (No. 8): 1841-1848.

Sugano, M., Koba, K. & Tsuji, E. (1999). Health Benefits of Rice Bran Oil. *Anticancer Research*, Vol. 10 (No. 5A): 3651-3657.

Tsujiwaki, Y., Yamamoto, H. & Minami, K. (1995). Japan patent. 327687.

Xiao-Wen, W. (2005). Leading Technology in the 21st Century "Supercritical Fluid Extraction".*Shokuhin to Kaihatsu*, Vol. 40: 68-69.

Yamauchi, Y. & Saito, M. (1990). Fractionation of Lemon-Peel Oil by Semi-Preparative Supercritical Fluid Chromatography. *Journal of Chromatography*, Vol. 505 (No. 1): 237-246.

Zhao, HY. & Jiang, JG. (2010). Application of Chromatography Technology in the Separation of Active Components from Nature Derived Drugs. *Mini Reviews in Medicinal Chemistry*, Vol. 10 (No. 13): 1223-1234.

5

Microorganism-Produced Enzymes in the Food Industry

Izabel Soares, Zacarias Távora,
Rodrigo Patera Barcelos and Suzymeire Baroni
Federal University of the Bahia Reconcavo / Center for Health Sciences
Brazil

1. Introduction

The application of microorganisms, such as bacteria, yeasts and principally fungi, by the food industry has led to a highly diversified food industry with relevant economical assets. Fermentation, with special reference to the production of alcoholic beverages, ethyl alcohol, dairy products, organic acids and drugs which also comprise antibiotics are the most important examples of microbiological processes.

The enzyme industry, as it is currently known, is the result of a rapid development of biotechnology, especially during the past four decades. Since ancient times, enzymes found in nature have been used in the production of food products such as cheese, beer, wine and vinegar (Kirk et al., 2002).

Enzymes which decompose complex molecules into smaller units, such as carbohydrates into sugars, are natural substances involved in all biochemical processes. Due to the enzymes' specificities, each substratum has a corresponding enzyme.

Although plants, fungi, bacteria and yeasts produce most enzymes, microbial sources-produced enzymes are more advantageous than their equivalents from animal or vegetable sources. The advantages assets comprise lower production costs, possibility of large-scale production in industrial fermentors, wide range of physical and chemical characteristics, possibility of genetic manipulation, absence of effects brought about by seasonality, rapid culture development and the use of non-burdensome methods. The above characteristics make microbial enzymes suitable biocatalysts for various industrial applications (Hasan et al., 2006).

Therefore, the identification and the dissemination of new microbial sources, mainly those which are non-toxic to humans, are of high strategic interest. Besides guaranteeing enzyme supply to different industrial processes, the development of new enzymatic systems which cannot be obtained from plants or animals is made possible and important progress in the food industry may be achieved.

2. Fungus of industrial interest

Owing to progress in the knowledge of enzymes, fungi acquired great importance in several industries since they may improve various aspects of the final product.

In fact, the fungi kingdom has approximately 200 species of *Aspergillus* which produce enzymes. They are isolated from soil, decomposing plants and air. *Aspergillus* actually produces a great number of extracellular enzymes, many of which are applied in biotechnology. *Aspergillus flavus, A. niger, A. oryzae, A. nidulans, A. fumigatus, A. clavatus, A. glaucus, A. ustus* and *A. versicolor* are the best known.

The remarkable interest in *Aspergillus niger*, a species of great commercial interest with a highly promising future and already widely applied in modern biotechnology, is due to its several and diverse reactions (Andersen et al, 2008).

Moreover, *A. niger* not only produces various enzymes but it is one of the few species of the fungus kingdom classified as GRAS (Generally Recognized as Safe) by the Food and Drug Administration (FDA). The species is used in the production of enzymes, its cell mass is used as a component in animal feed and its fermentation produces organic acids and other compounds of high economic value (Couto and Sanroman, 2006; Mulimania and Shankar, 2007).

2.1 Microbial enzymes for industries
2.1.1 Pectinase enzyme

Plants, filamentous fungi, bacteria and yeasts produce the pectinase enzymes group with wide use in the food and beverages industries. The enzyme is employed in the food industries for fruit ripening, viscosity clarification and reduction of fruit juices, preliminary treatment of grape juice for wine industries, extraction of tomato pulp (Adams et al., 2005), tea and chocolate fermentation (Almeida et al. 2005; da Silva et al., 2005), vegetal wastes treatment, fiber degumming in the textile and paper industries (Sorensen, et al. 2004; Kaur, et al. 2004, Taragano, et al., 1999, Lima, et al., 2000), animal nutrition, protein enrichment of baby food and oil extraction (Da Silva et al., 2005, Lima, et al., 2000).

The main application of the above mentioned enzyme group lies within the juice processing industry during the extraction, clarification and concentration stages (Martin, 2007). The enzymes are also used to reduce excessive bitterness in citrus peel, restore flavor lost during drying and improve the stableness of processed peaches and pickles. Pectinase and β-glucosidase infusion enhances the scent and volatile substances of fruits and vegetables, increases the amount of antioxidants in extra virgin olive oil and reduces rancidity.

The advantages of pectinase in juices include, for example, the clarification of juices, concentrated products, pulps and purees; a decrease in total time in their extraction; improvement in the production of juices and stable concentrated products and reduction in waste pulp; decrease of production costs; and the possibility of processing different types of fruit (Uenojo and Pastore 2007). For instance, in the production of passion fruit juice, the enzymes are added prior to filtration when the plant structure's enzymatic hydrolysis occurs. This results in the degradation of suspended solids and in viscosity decrease, speeding up the entire process (Paula, et al., 2004).

Several species of microorganisms such as *Bacillus, Erwinia, Kluyveromyces, Aspergillus, Rhizopus, Trichoderma, Pseudomonas, Penicillium* and *Fusarium* are good producers of pectinases (De Gregorio, et al., 2002). Among the microorganisms which synthesize pectinolytic enzymes, fungi, especially filamentous fungi, such as *Aspergillus niger* and *Aspergillus carbonarius* and *Lentinus edodes*, are preferred in industries since approximately 90% of produced enzymes may be secreted into the culture medium (Blandino et al., 2001).

In fact, several studies have been undertaken to isolate, select, produce and characterize these specific enzymes so that pectinolytic enzymes could be employed not only in food processing but also in industrial ones. High resolution techniques such as crystallography and nuclear resonance have been used for a better understanding of regulatory secretion mechanisms of these enzymes and their catalytic activity. The biotechnological importance of microorganisms and their enzymes triggers a great interest toward the understanding of gene regulation and expression of extracellular enzymes.

2.1.2 Lipases

Lipolytic enzymes such as lipases and esterases are an important group of enzymes associated with the metabolism of lipid degradation. Lipase-producing microorganisms such as *Penicillium restrictum* may be found in soil and various oil residues. The industries Novozymes, Amano and Gist Brocades already employ microbial lipases.

Several microorganisms, such as *Candida rugosa, Candida antarctica, Pseudomonas alcaligenes, Pseudomonas mendocina* and *Burkholderia cepacia,* are lipase producers (Jaeger and Reetz, 1998). Other research works have also included *Geotrichum* sp. (Burkert et al., 2004), *Geotrichum candidum* DBM 4013 (Zarevúcka et al., 2005), *Pseudomonas cepacia, Bacillus stearothermophilus, Burkholderia cepacia* (Bradoo et al., 2002), *Candida lipolytica* (Tan et al., 2003) *Bacillus coagulans* (Alkan et al., 2007), *Bacillus coagulans* BTS-3 (Kumar et al., 2005), *Pseudomonas aeruginosa* PseA (Mahanta et al., 2008), *Clostridium thermocellum* 27405 (Chinn et al., 2008), *Yarrowia lipolytica* (Dominguez et al., 2003) and *Yarrowia lipolytica* CL180 (Kim et al., 2007).

The fungi of the genera *Rhizopus, Geotrichum, Rhizomucor, Aspergillus, Candida* and *Penicillium* have been reported to be producers of several commercially used lipases.

The industrial demand for new lipase sources with different enzymatic characteristics and produced at low costs has motivated the isolation and selection of new lipolytic microorganisms. However, the production process may modify their gene expression and change their phenotypes, including growth, production of secondary metabolites and enzymes. Posterior to primary selection, the production of the enzyme should be evaluated during the growth of the promising strain in fermentation, in liquid medium and / or in the solid state (Colen et al., 2006). However, it is evident that each system will result in different proteins featuring specific characteristics with regard to reactions' catalysis and, consequently, to the products produced (Asther et al., 2002).

2.1.3 Lactase

Popularly known as lactase, beta-galactosidases are enzymes classified as hydrolases. They catalyze the terminal residue of b-lactose galactopiranosil (Galb1 - 4Glc) and produce glucose and galactose (Carminatti, 2001). Lactase's production sources are peaches, almonds and certain species of wild roses; animal organisms, such as the intestine, the brain and skin tissues; yeasts, such as *Kluyveromyces lactis, K. fragilis* and *Candida pseudotropicalis*; bacteria, such as *Escherichia coli, Lactobacillus bulgaricus, Streptococcus lactis* and *Bacillus* sp; and fungi, such as *Aspergillus foetidus, A. niger, A. oryzae* and *A. Phoenecia.*

The b-galactosidase may be found in nature, or rather, in plants, particularly almonds, peaches, apricots, apples, animal organs such as the intestine, the brain, placenta and the testis.

Lactase is produced by a widely diverse fungus population and by a large amount of microorganisms such as filamentous fungus, bacteria and yeast (Holsinger, 1997; Almeida and Pastore, 2001).

Beta-galactosidase is highly important in the dairy industry, in the hydrolysis of lactose into glucose and galactose with an improvement in the solubility and digestibility of milk and dairy products. Food with low lactose contents, ideal for lactose-intolerant consumers, is thus obtained (Mahoney, 1997; Kardel et al. 1995; Pivarnik et al., 1995). It also favors consumers who are less tolerant to dairy products' crystallization, such as milk candy, condensed milk, frozen concentrated milk, yoghurt and ice cream mixtures, (Mahoney, 1998; Kardel et al., 1995). It also produces oligosaccharides (Almeida and Pastore, 2001), the best biodegradability of whey second to lactose hydrolysis (Mlichová; Rosenberg, 2006).

2.1.4 Cellulases

Cellulases are enzymes that break the glucosidic bonds of cellulose microfibrils, releasing oligosaccharides, cellobiose and glucose (Dillon, 2004). These hydrolytic enzymes are not only used in food, drug, cosmetics, detergent and textile industries, but also in wood pulp and paper industry, in waste management and in the medical-pharmaceutical industry (Bhat and Bhat, 1997).

In the food industry, cellulases are employed in the extraction of components from green tea, soy protein, essential oils, aromatic products and sweet potato starch. Coupled to hemicellulases and pectinases they are used in the extraction and clarification of fruit juices. After fruit crushing, the enzymes are used to increase liquefaction through the degradation of the solid phase.

The above enzymes are also employed in the production process of orange vinegar and agar and in the extraction and clarification of citrus fruit juices (Orberg 1981). Cellulases supplement pectinases in juice and wine industries as extraction, clarification and filtration aids, with an increase in yield, flavor and the durability of filters and finishers (Pretel, 1997). Cellulase is produced by a vast and diverse fungus population, such as the genera *Trichoderma, Chaetomium, Penicillium, Aspergillus, Fusarium* and *Phoma*; aerobic bacteria, such as *Acidothermus, Bacillus, Celvibrio, pseudonoma, Staphylococcus, Streptomyces* and *Xanthomonas*; and anaerobic bacteria, such as *Acetovibrio, Bacteroides, Butyrivibrio, Caldocellum, Clostridium, Erwinia, Eubacterium, Pseudonocardia, Ruminococcus* and *Thermoanaerobacter* (Moreira & Siqueira, 2006; Zhang et al., 2006). *Aspergillus* filamentous fungi stand out as major producers of cellulolytic enzymes. It is worth underscoring the filamentous fungus *Aspergillus niger*, a fermenting microorganism, which has been to produce of cellulolytic enzymes, organic acids and other products with high added value by solid-state fermentation processes. (Castro, 2006, Chandra et. al., 2007, Castro & Pereira Jr. 2010)

2.1.5 Amylases

Amylases started to be produced during the last century due to their great industrial importance. In fact, they are the most important industrial enzymes with high biotechnological relevance. Their use ranges from textiles, beer, liquor, bakery, infant feeding cereals, starch liquefaction-saccharification and animal feed industries to the chemical and pharmaceutical ones.

Currently, large quantities of microbial amylases are commercially available and are almost entirely applied in starch hydrolysis in the starch-processing industries.

The species *Aspergillus* and *Rhizopus* are highly important among the filamentous fungus for the production of amylases (Pandey et al., 1999, 2005). In the production of amyloglucosidase, the species *Aspergillus niger, A. oryzae, A. awamori, Fusarum oxysporum, Humicola insolens, Mucor pusillus, Trichoderma viride* . Species Are producing α-amylase. *Aspergillus niger, A. fumigatus, A. saitri, A. terreus, A. foetidus foetidus, Rhizopus, R. delemar* (Pandey et al. 2005), with special emphasis on the species of the genera *Aspergillus* sp., *Rhizopus* sp. and *Endomyces* sp (Soccol et al. 2003).

In fact, filamentous fungi and the enzymes produced thereby have been used in food and in the food-processing industries for decades. In fact, their GRAS (Generally Recognized as Safe) status is acknowledged by the U.S. Food and Drug Administration in the case of some species such as *Aspergillus niger* and *Aspergillus oryzae*.

The food industry use amylases for the conversion of starch into dextrins. The latter are employed in clinical formulas as stabilizers and thickeners; in the conversion of starch into maltose, in confectioneries and in the manufacture of soft drinks, beer, jellies and ice cream; in the conversion of starch into glucose with applications in the soft-drinks industry, bakery, brewery and as a subsidy for ethanol production and other bioproducts; in the conversion of glucose into fructose, used in soft drinks, jams and yoghurts (Aquino et al., 2003, Nguyen *et al.*, 2002).

Amylases provide better bread color, volume and texture in the baking industry. The use of these enzymes in bread production retards its aging process and maintains fresh bread for a longer period. Whereas fungal α-amylase provides greater fermentation potential, amyloglucosidase improves flavor and taste and a better bread crust color (Novozymes, 2005). Amylases are the most used enzymes in bread baking (Giménez et al. 2007; Haros; Rosell, Leon; Durán).

Amylases have an important role in carbon cycling contained in starch by hydrolyzing the starch molecule in several products such as dextrins and glucose. Dextrins are mainly applied in clinical formulas and in material for enzymatic saccharification. Whereas maltose is used in confectioneries and in soft drinks, beer, jam and ice cream industries, glucose is employed as a sweetener in fermentations for the production of ethanol and other bioproducts.

The above amylases break the glycosidic bonds in the amylose and amylopectin chains. Thus, amylases have an important role in commercial enzymes. They are mainly applied in food, drugs, textiles and paper industries and in detergent formulas (Peixoto et al. 2003; Najafpour, Gupta et al., 2002; Asghar et al. 2006; Mitidieri et al., 2006).

Results from strains tested for the potential production of amylases, kept at 4°C during 10 days, indicated that the wild and mutant strains still removed the nutrients required from the medium by using the available substrate. This fact showed that cooling maintained intact the amylase's activities or that a stressful condition for the fungus caused its degradation and thus consumed more compounds than normal (Smith, et al., 2010).

The best enzyme activity of microbial enzymes occurs in the same conditions that produce the microorganisms' maximum growth. Most studies on the production of amylases were undertaken from mesophilic fungi between 25 and 37°C. Best yields for α-amylase were achieved between 30 and 37°C for *Aspergillus* sp.; 30°C for *A. niger* in the production of amyloglucosidase 30°C in the production of α-amylase by *A. oryzae* (Tunga, R.; Tunga B.S, 2003), 55°C by thermophile fungus *Thermomonospora*, and 50°C by *T. lanuginosus* in the production of α-amylase (Gupta et al., 2003). However, no reports exist whether increase in enzyme activity after growth of fungus in ideal conditions and kept refrigerated at 4°C for 10 days has ever been tested.

2.1.6 Proteases

Proteases are enzymes produced by several microorganisms, namely, *Aspergillus niger, A. oryzae, Bacillus amyloliquefaciens, B. stearotermophilus, Mucor miehei, M. pusillus.* Proteases have important roles in baking, brewing and in the production of various oriental foods such as soy sauce, miso, meat tenderization and cheese manufacture.

Man's first contact with proteases activities occurred when he started producing milk curd. Desert nomads from the East used to carry milk in bags made of the goat's stomach. After long journeys, they realized that the milk became denser and sour, without understanding the process's cause. Curds became thus a food source and a delicacy. Renin, an animal-produced enzyme, is the protease which caused the hydrolysis of milk protein.

Proteases, enzymes that catalyze the cleavage of peptide bonds in proteins, are Class 3 enzymes, hydrolases, and sub-class 3.4, peptide-hydrolases. Proteases may be classified as exopeptidases and endopeptidases, according to the peptide bond to be chain-cleaved.

Recently proteases represent 60% of industrial enzymes on the market, whereas microbial proteases, particularly fungal infections, are advantageous because they are easy to obtain and to recover (Smith et al, 2009).

An enzyme extract (Neves-Souza, 2005), which coagulates milk and which is derived from the fungus *Aspergillus niger* var. *awamori*, is already produced industrially.

Although the bovine-derived protease called renin has been widely used in the manufacture of different types of cheese, the microbial-originated proteases are better for coagulant (CA) and proteolytic (PA) activities (PA). The relationship AC / AP has been a parameter to select potentially renin-producing microbial samples. The higher the ratio AC / AP, the most promising is the strain. It features high coagulation activity, with fewer risks in providing undesirable characteristics from enhanced proteolysis (Melo et al, 2002).

The microbial proteases have also been important in brewery. Beer contains poorly soluble protein complexes at lower temperature, causing turbidity when cold. The use of proteolytic enzymes to hydrolyze proteins involved in turbidity is an alternative for solving this problem.

Most commercial serine proteinases (Rawling et al, 1994), mainly neutral and alkaline, are produced by organisms belonging to the genus *Bacillus*. Whereas subtilisin enzymes are representatives of this group, similar enzymes are also produced by other bacteria such as *Thermus caldophilus, Desulfurococcus mucosus* and *Streptomyces* and by the genera *Aeromonas* and *Escherichia coli*.

In their studies and observations on the activities of proteases from *Bacillus*, Singh and Patel (2005), Silva, and Martin Delaney (2007); Sheri and Al-Mostafa (2004) and others evaluated their properties for a better performance in pH and temperature ranges.

2.1.7 Glucose oxidase

Glucose oxidase [E.C. 1.1.3.4] (GOx) is an enzyme that catalyzes the oxidation of beta-D-glucose with the formation of D-gluconolactone. The enzyme contains the prosthetic group flavin adenine dinucleotide (FAD) which enables the protein to catalyze oxidation-reduction reactions.

Guimarães et al. (2006) performed a screening of filamentous fungi which could potentially produce glucose-oxidase. Their results showed high levels of GOx in *Aspergillus versicolor* and *Rhizopus stolonifer*. The literature already suggests that the genus *Aspergillus* is a major GOx producer.

The enzyme is used in the food industry for the removing of harmful oxygen. Packaging materials and storage conditions are vital for the quality of products containing probiotic microorganisms since the microbial group's metabolism is essentially anaerobic or microaerophilic (MattilaSandholm et al., 2002). Oxygen level during storage should be consequently minimal to avoid toxicity, the organism's death and the consequent loss of the product's functionality.

Glucose oxidase may be a biotechnological asset to increase stability of probiotic bacteria in yoghurt without chemical additives. It may thus be a biotechnology alternative.

2.1.8 Glucose isomerase

Glucose isomerase (GI) (D-xylose ketol isomerase; EC 5.3.1.5) catalyzes the reversible isomerase from D-glucose and D-xylose into D-fructose and D-xylulose, respectively. The enzyme is highly important in the food industry due to its application in the production of fructose-rich corn syrup.

Interconversion of xylose into xylulose by GI is a nutritional requirement of saprophytic bacteria and has a potential application in the bioconversion of hemicellulose into ethanol. The enzyme is widely distributed among prokaryotes and several studies have been undertaken to enhance its industrial application (Bhosale et al, 1996).

The isolation of GI in *Arthrobacter* strains was performed by Smith et al. (1991), whereas Walfridsson et al. (1996) cloned gene xylA of *Thermus themophilus* and introduced it into *Saccharomyces cerevisiae* to be expressed under the control of the yeast PGK1 promoter. The search for GI thermostable enzymes has been the target of protein engineering (Hartley et al., 2000).

In fact, biotechnology has an important role in obtaining mutants with promising prospects for the commercialization of glucose isomerase enzyme.

The development of microbial strains which use xylan with prime matters for the growth or selection of GI-constituted mutants should lead towards the discontinuation of the use of xylose as an enzyme production inducer.

2.1.9 Invertase

Invertase is an S-bD-fructofuranosidase obtained from *Saccharomyces cerevisiae* and other microorganisms. The enzyme catalyzes the hydrolysis from sucrose to fructose and glucose. The manufacture of inverted sugar is one of invertase's several applications. Owing to its sweetening effects which are higher than sucrose's, it has high industrial importance and there are good prospects for its use in biotechnology.

Invertase is more active at temperatures and pH ranging respectively between 40° and 60° C and between 3.0 and 5.0. When invertase-S is applied at 0.6% rate in a solution of sucrose 40% w / w at 40°C, it inverts 80% of sucrose after 4h. 20min.

When Cardoso et al. (1998) added invertases to banana juice to assess its sweetness potential, they reported an increase in juice viscosity besides an increase in sweetness.

Alternaria sp isolates from soybean seed were inoculated in a semi-solid culture and the microorganism accumulated large amounts of extracellular invertase, which was produced constitutively without the need for an inductor.

Microorganisms, such as filamentous fungi, are good producers of invertase with potential application in various industrial sectors.

Gould et al. (2003) cultivated the filamentous fungus *Rhizopus* sp in wheat bran medium, and obtained invertase identified as polyacrylamide gel. Another potentially producing fungus invertase is *Aspergillus casiellus*. It was inoculated in soybean meal medium and after 72 hours its crude extract was isolated (Novak et al., 2010). Since most invertases used in industry are produced by yeasts, underscoring the search for fungi that produce it in great amounts is a must.

3. Final considerations

Perspectives for biotechnological production of enzymes by microorganisms.

Biotechnology is an important tool for a more refined search for microorganisms with commercial assets. Microorganisms have existed on the planet Earth during millions of years and are a source of biotechnological possibilities due to their genetic plasticity and adaptation.

The isolation of new species from several and different habitats, such as saltwater and freshwater, soils, hot springs, contaminated soils, caves and hostile environments is required. Microorganisms adapted to these conditions may have great biotechnological potential.

Methods such as the selection of mutants are simple ways to obtain strains or strains with enzymatic possibilities and these methods are widely used by researchers in academic pure science laboratories.

Geneticists also employ the recombination and selection of mutants, which feature promising characteristics, in new strains. This method consists of transferring genetic material among contrasting genotypes, obtain recombinants and use the selection for the desired need.

The recombinant DNA technology (TDR) is a very useful method under three aspects: it increases the production of a microbial enzyme during the fermentation process; it provides enzymes with new properties suitable for industrial processes, such as thermostability and ability to function outside the normal pH range; it produces enzymes from animal- and vegetable-derived microorganisms.

Extremophile microorganisms are potentially producing enzymes with useful characteristics for high temperature industrial processes.

Microorganisms that grow at low temperatures have important biotechnological assets since their enzymes are more effective at low temperatures and enables contamination risks in continuous fermentation processes. This will shorten fermentation time and enhance energy saving.

The DNA sequencing technologies have advanced greatly in recent years and important progress on genes that synthesize proteins and thus determine their function in organisms has been achieved.

Genomes of several microorganisms have been sequenced, including those which are important for the food industry, such as *Saccharomyces cerevisiae*, *Bacillus subtilis*, *Lactococcus Latis*, *Lactobacillus acidophilus*, *Lactobacillus* sp, and *Streptococcus thermophilus*. These genomes have revealed several new genes, most of which codify enzymes.

Microorganisms are potential producers of enzymes useful for the food industry. Biotechnological tools are available for the selection and obtaining of strains and for strains which increase enzymes' production on a large scale.

Progress and achievements in this area will bring improvements in the food industry and, consequently, a better health quality for mankind.

4. References

Alkan, H., Baisal, Z., Uyar, F., Dogru, M. (2007). Production of Lipase by a Newly Isolated *Bacillus coagulans* Under Solid-State Fermentation Using Melon Wastes. *Applied Biochemistry and Biotechnology,* 136, 183-192.

Almeida, C.; Brányik, T.; Moradas-Ferreira, P.; Teixeira, J. (2005). *Process Biochem.,* 40, 1937.

Almeida, M.M. de; PASTORE G.M.(2001). Galactooligossacarídeos-Produção e efeitos benéficos, Ciência e tecnologia de Alimentos, Campinas, SBCTA, Vol. 35, No. 1/2, p.12-19.

Al-Sheri, M. A.; Mostafa, S.Y. (2004). Production and some properties of protease produced by *Bacillus licheniformis* isolated from Tihamet Aseer, Saudi Arabia. *Paquistan Journal of Biological Sciences,* Vol.7, p.1631-1635.

Andersen,M R; Nielsen, M L; Nielsen, J. (2008). Metabolic model integration of the bibliome, genome, metabolome and reactome of *Aspergillus niger, Molecular, Systems Biology,* Vol. 4, No.178, 1-13.

Aquino, A.C.M.M.; jorge, J.A.; terenzi, H.F.; polizeli, M.L.T.M. (2003). Studies on athermostable a-amylase from thermophilic fungus Scytalidium *thermophilum. Appl.Microbiol. Biotechnol.,* 61: 323-328, 2003.

Asghar, M.; Asad M. J.; Rehman, S.; Legge, R. L. A. (2006). Thermostable α-amylase from a Moderately Thermophilic *Bacillus subtilis* Strain for StarchProcessing. *Journal of Food Engineering,* Vol.38, p. 1599-1616.

Asther, M., Haon, M., Roussos, S., Record, E., Delattre, M., Meessen-Lesage, L., Labat, M., Asther, M. (2002). Feruloyl esterase from *Aspergillus niger* a comparison of the production in solid state and submerged fermentation. *Process Biochemistry,* Vol. 38, 685-691.

Bhat, M. K. (2000). *Biotechnol. Adv., 18,* 355.*Biochem. 37,* 497.

Bhosale, S.H.; Rao, M.B.; Deshpande,V.V.(1996). Molecular and industrial aspects of glucose isomerase. *Microbiol. Rev,* Vol.60, No.2, p.280-300.

Blandino, A.; Dravillas, K.; Cantero, D.; Pandiella, S. S.; Webb, C.(2001). *Process*

Bradoo, S., Rathi, P., Saxena, R.K., Gupta, R. (2002). Microwave-assisted rapid characterization of lipase selectivities. *Journal of Biochemistry,* 51, 115-120.

Cardoso, M.H.; Jackix, M.N.H.; Menezes, H.C.; Gonçalves, E. B.; Marques, S.V.B.(1998). Efeito da associação de pectinase, invertase e glicose isomerase na qualidade do suco de banana. Ciênc. Tecnol. Aliment. Vol. 18,No.3, ISSN 0101-2061.

Castro. A. M.; Pereira Jr, N.(2010). Produção, propriedades e aplicação de celulases na hidrólise de resíduos agroindustriais. *Quimica. Nova,* Vol 33, No.1, PP.181-188.

Chandra, M. S.; Viswanath, B.; Rajaseklar Reddy, B. (2007). Cellulolytic enzymes on lignocellulosic substrates in solid state fermentation by *Aspergillus niger. Indian Journal of Microbiology,* Vol.47, pp.323-328.

Chinn, M.s., Nokes, S.E., Strobel, H.J. (2008). Influence of moisture content and cultivation duration on *Clostridium thermocellum* 27405 endproduct formation in solid substrate cultivation on Avicel. *Bioresource Technology,* Vol. 99, 2664-2671.

Colen, G., Junqueira, R.G., Moraes-Santos, T. (2006). Isolation and screening of alkaline lipase-producing fungi from Brazilian savanna soil.*World Journal of Microbiology & Biotechnology,* Vol. 22, 881-885.

Couto, S.R.; Sanroman M A. (2006). Application of solid state fermentation to food industry – A review, *Journal of Food Engineering,* Vol.76, No.3, pp.291-302.

Da Silva, E. G.; Borges, M. F.; Medina, C.; Piccoli, R. H.; Schwan, R. F.; (2005) *FEMS Yeast Res.* 5, 859.

De Gregorio, A.; Mandalani, G.; Arena, N.; Nucita, F.; Tripodo, M. M.; Lo Curto, R. B. (2002). SCP and crude pectinase production by slurry-state fermentation of lemon pulps. *Bioresour. Technol.*, Vol.83, No.2, p. 8994.

Dillon, Aldo. Celulases. *In:* SAID, S.; PIETRO, R. C. L. (2004). *Enzimas como agentes biotecnológicos*. Ribeirão Preto: *Legis Summa*, p. 243-270.

Dominguez, A.; Costas, M.; Longo, M.A.; Sanromán, A. (2003). "A novel application of solid culture: production of lipases by *Yarrowia lipolytica*", *Biotechnology Letters* , Vol.25, p. 1225-1229.

Giménez, A.; Varela, P. Salvador, A.; Ares, G; Fiszman, S.; and Garitta, L. (2007). Shelf life estimation of brown pan bread: A consumer approach. *Food Quality and Preference*, Barking, Vol.18, No.2, pp. 196-204.

Glazer, A. N.; Nikaido, H. *Microbiol biotechnology*. (1995). New York: W.H.Freeman, 662 p.

Goesaert, H; Brijs, K.; Veraverbeke, W.S. Courtin, C.M. Gebruers, K. and Delcour, J.A. (2005). Wheat flour constituents: how they impact bread quality, and how to impact their functionality. *Trends in Food Science & Technology*, Cambridge, Vol.16, No.1-3, pp. 12-30.

Gomes, E. Guez, M. A. U, Martin, N, Silva, R. (2007). Enzimas termoestáveis: fontes, produção e aplicação industrial, *Química Nova*, Vol.30, No.1,pp 136-145.

Goto, C. E; Barbosa, E. P.; Kistner, L. C. L.; GANDRA, R. F.; ARRIAS,V. L.; , R. M. (1998). Production of amylases by *Aspergillus fumigatus*. *Revista de Microbiologia*, v.29, p.99-103.

Goulart, A. J.; Adalberto, P.R.; Monti, R. (2003). Purificação parcial de invertase a partir de *Rhizopus sp* em fermentação semi-sólida. *Alim. Nutr.*, Vol.14, No.2, pp. 199-203.

Gupta, R. et al. Microbial α-Amylases. (2003). Biotechnological Perspective. *Process Biochemistry*, Vol.38, No.11, p. 1-18.

Haros, M.; ROSELL, C. M.; BENEDITO, C. (2002). Effect of different carbohydrases on fresh bread texture and bread staling. *Eur. Food Res. Technol.*, Berlin,Vol.215, No. 5, pp. 425-430.

Hartley, B.S.; Hanlon, N. Robin, J.; Rangarajan, J.; Ragaranjan M. (2000). Glucose isomerase: insights into protein engineering for increased thermostability. *Biochimica et Biophysica Acta (BBA) - Protein Structure and Molecular Enzymology*, Vol. 1543, Issue 2, p.294-335.

Hasan, F.; Shah, A. A.; Hameed, A.; (2006), "Industrial application of microbial lipase." *Enzyme and Microbial technology*, Vol.39, No.2, pp. 235-251.

Holsinger, V. H.; kilgerman, K. H. (1991). Application of lactose in dairy foods and other foods containing lactose. *Food Technology*, Vol.45, No.1, pp. 94-95.

Houde, A.; Kademi, A.; Leblanc, D. (2004).Lipases and their industrial applications: an overview. *Appl. Biochem. Biotechnol.*, Clifton, Vol.118, No.1-3, pp. 155–170.

Jaeger, K.E & Reetz, M.T., (1998). Microbial lipases form versatile tools for biotechnology. *Tibtech*, Vol.16, pp 396-403.

Kashyap, D. R.; Chandra, S.; Kaul, A.; Tewari, R. (2000). *World J. Microbiol. Biotechnol.* 16, 277.

Kardel, G.; Furtado, M.M.; Neto, J.P.M.L. (1995). Lactase na Indústria de Laticínios (Parte 1). *Revista do Instituto de Laticínios "Cândido Tostes"*. Juiz de Fora, Vol.50, No.294, pp.15-17.

Kaur, G.; Kumar, S.; Satyanarayama, T. (2004). *Bioresour. Technol. 94*, 239.

Kim, J.-T., Kang, S. G., Woo, J.-H., Lee, J.-H., Jeong, B.C., Kim, S.-J., (2007). Screening and its potential application of lipolytic activity from a marine environment:characterization of a novel esterase from *Yarrowia lipolytica* CL180. *Applied Microbiology and Biotechnology*, Vol.74, pp 820-828.

Kirk, O.; Borchert, T. V.; Fuglsang, C. C. (2002). Industrial enzyme applications. *Current Opinion Biotechnology*, Vol. 13, pp. 345 – 351.

Kumar, S., Kikon, K., Upadhyay, A., Kanwar, S.S, Gupta, R., (2005). Production, purification, and characterization of lipase from thermophilic and alkaliphilic *Bacillus coagulans* BTS-3. *Protein Expression Purificati*, Vol. 41, 38-44.

León, A. E.; Durán, E.; Barber, C. B. (2002).Utilization of enzyme mixtures to retard bread crumb firming. *Journal of Agricultural and Food Chemistry*, Easton, Vol..50, No.6, p. 1416-1419.

Lima, A. S.; Alegre, R. M.; Meirelles, A. J. A. (2000). *Carbohydr. Polym. 50*, 63.

Mattila-Sandholm, T.; Crittenden, R.; Mogensen,G.; Fondén, R.; Saarela, M. Technological challenges for future probiotic foods. *Int. Dairy J.* Vol:12, pp. 173-182.

Mahoney, R.R. (1997), Lactose: Enzymatic Modification. In: *Lactose, water, salts and vitamins*, London, *Advanced Dairy Chemistry*, Vol.3, p.77-125.

Melo, I.S.; Valadares-Inglis, M.C.; Nass, L.L.; Valois, A.C.C. (2002). *Recursos Genéticos & Melhoramento- Microrganismos*. (1 ed), Embrapa, ISBN 85-85771-21-6- Jaguariauna-São Paulo- SP.

Milichová, Z.; Rosenberg, M. (2006). Current trends of β-galactosidase application in food techonology. *Journal of Food an Nutrition Research*, Vol.45, No.2, p. 47-54.

Mitidieri, S.; Martinelli, A. H. S.; SCHRANK, A.; VAINSTEIN, M. H. (2006). Enzymatic detergent formulation containing amylase from *Aspergillus niger*: A comparative study with commercial detergent formulations. *Bioresource Technology*, Vol.97, p. 1217–1224.

Najafpour, G. D.; Shan, C. P. (2003). Enzymatic hydrolysis of molasses. *Bioresource Technology*, v. Vol.86, p. 91-94.

Nguyen, Q.D.; Rezessy-SZABO, J.M.; Claeyssens, M.; STALS, I.; Hoschke, A. (2002). Purification and characterisation of amylolytic enzymes from thermophilic fungus*Thermomyces lanuginosus* strain ATCC 34626. *Enzimes. Microbial Technol.*, Vol.31, pp.345-352.

Novaki, L.; Hasan, S.D.M.; Kadowaku, M.K. Andrade, D. (2010). Produção de invertase por fermentação em estado sólido a partir de farelo de soja Engevista, Vol.12, No.2, pp. 131-140.

Pandey, A., Benjamin, S., Soccol, C.R., Nigam, P., Kriger, N., Soccol, V.T. (1999). The realm of microbial lipases in biotechnology.*Biotechnology Applied Biochemist*, Vol. 29, 119-131.

Patel, R. M.; Singh, S. P. (2005). Extracellular akaline protease from a newly isolated haloalkaliphilic *Bacillus* sp.: Production and Optimization. *Process Biochemistry*, Vol.40, pp.3569-3575.

Paula, B.; Moraes, I. V. M.; Castilho, C.C.; Gomes, F. S.; Matta, V. M.; Cabral, L. M. C. (2004). Melhoria na eficiência da clarificação de suco de maracujá pela combinação dos processos de microfiltração e enzimático. Boletim CEPPA, Vol.22, No.2, pp. 311-324.

Peixoto, S.C.; jorge, J.A.; Terenzi, H.F.; Polizeli, M.L.T.M. (2003). Rhizopus microsporus var. rhizopodiformis: a thermotolerant fungus with potential for production of thermostable amylases. *Int. Microbiol.*, Vol.6, pp.269-273.

Pretel, M.T.(1997). Pectic enzymes in fresh fruit processing: optimization of enzymic peeling of oranges. *Process Biochemistry.* Vol.32, No.1, pp. 43-49.

Pivarnik, L. F.; Senegal, A.G.; Rand, A.G. (1995). Hydrolytic and transgalactosil activities of commercial β-galactosidase (lactase) in food processing. *Advances in Food and Nutrition Research*, New York, Vol.38, p. 33.

Rawling, N.D., Barret, A. (1994). Families of serine peptidases. *Meth. Enzymol.*, Vol.244, pp.18-61.

Shankar, S.K.; Mulimania, V.H. (2007). β-Galactosidase production by *Aspergillus* oryzae *Bioresource Technology*, Vol.98, No.4, pp. 958-961.

Silva, C.R.; Delatorre, A. B.; Martins, M. L. L. (2007). Effect of the culture conditions on the production of an extracellular protease by thermophilic *Bacillus* sp. and some properties of the enzymatic activity. *Brazilian Journal of Microbiology*, Vol.38, pp.253-258.

Silva, G.A.B; Almeida, W.E.S; Cortes, M.S; Martins, E.S. (2009). Produção e caracterização de protease obtida por *gliocladium verticilloides* através da fermentação em estado sólido de subprodutos agroindustriais. *Revista Brasileira de Tecnologia Agroindustrial*, Vol.03, No.01, pp. 28-41. ISSN: 1981-3686.

Smith, C. A.; Rangarajan, M.; Hartley, B. S. (1991). D-Xylose (D-glucose) isomerase from Arthrobacter strain N.R.R.L. B3728. Purification and properties. *Biochemestry Journal*, Vol.1; No.277, pp. 255–261.

Soccol, C.R. & Vandenberghe, L.P.S., (2003). Overview of applied solidstate fermentation in Brazil. *Biochemical Engineering Journal*, Vol. 13, pp. 205-218.

Sorensen, J. F.; Krag, K. M.; Sibbesen, O.; Delcur, J.; Goesaert, H.; Svensson, B.; Tahir, T. A.; Brufau, J.; Perez-Vendrell, A. M.; Bellincamp, D.; D'Ovidio, R.; Camardella, L.; Giovane, A.; Bonnin, E.; Juge, N. (2004). *Biochim. Biophys. Acta, 1696*, 275.

Systems Biology, Vol.4, No.178, 1-13.

Tan, T., Zhang, M., Wang, B., Ying, C., Deng, Li. (2003). Screening of high lipase producing *Candida* sp. And production of lipase by fermentation. *Process Biochemistry*, Vol.39, pp.459-465.

Taragano, V. M.; Pilosof, A. M. R. (1999). *Enzyme Microb. Technol. 25*, 411.

Tunga, R.; Tunga, B. S. (2003).Extra-cellular Amylase Production by *Aspergillus oryzae* Under Solid State Fermentation. Japan: *International Center for Biotechnology*, Osaka University, 12 p.

Uenojo, M., Pastore, G. M. (2007). Pectinases: Aplicações Industriais e Perspectivas. *Química Nova*, Vol.30, No. 2, pp. 388-394.

Walfridsson, M.; Bao, X.; Anderlund, M.; Lilius, G. Bulow, L.; Hahn-Hagerdal, B. (1996). Ethanolic fermentation of xylose with Saccharomyces cerevisiae harboring the *Thermus thermophilus* xylA gene, which expresses an active xylose (glucose) isomerase. *Appl. Environ. Microbiol.*, Vol.62, No.12, pp. 4648-4651.

Zarevúcka, M., Kejík, Z., Saman, D., Wimmer, Z., Demnerová, K. (2005). Enantioselective properties of induced lipases from *Geotrichum*. *Enzyme and Microbial Technology*, Vol.37, pp.481-486.

6

Nanotechnology and Food Industry

Francisco Javier Gutiérrez, Mª Luisa Mussons,
Paloma Gatón and Ruth Rojo
*Centro Tecnológico CARTIF. Parque Tecnológico de Boecillo,
Valladolid
España*

1. Introduction

Human population will reach 9,100 million by 2,050, which supposes an increase of 34% respect present situation. This growth will occur in emerging countries mainly. As a consequence of that, there will be an increase in global demand for foods, feed and energy. Initial estimation on the increment of this world demands are in the order of 70%. Accordingly, the pressure on resources (in special water and crops) will be higher. World surface devoted to crop production will be increased, in order to meet demands on food energy and other industrial uses and so, the environmental impact is some areas could be high.

In order to obtain commodities and other feedstock in a sustainable way it is necessary to improve the current working methods and control the environmental impact, acting on:

- Water management and use,
- Agriculture,
- Animal exploitation and, in general,
- Food processing.

In a broad context, some factors affecting living standards are the water availability and food. In fact, life expectancy and health are determined by both. Considering the expected demand on food (and water) as well as the global situation, several technological challenges should be overcome to make a rational use of resources possible. In this sense, nanotechnology could suppose a great tool in solving that situation.

Where is the interest of nanotechnology? Which are the possibilities of nanotechnology in relation to foods and food production? Is it possible to improve crops using nanotechnology? Which are the advantages concerning water and use management? How is it envisaged to achieve those potential benefits at a global scale?

A general view on the nanotechnology concerning foods and water is presented in this chapter and some answers to the former questions are proposed.

According to Chaudhry et al., 2011, in food nanotechnology different categories can be specified:

- When food ingredients are processed to form nanostructures.
- When additives are used in nanocapsules (or reduced somehow to nanometric size).
- When nanomaterials are used in surfaces, contact surface development, packaging materials intelligent packaging and nanosensors.

- When nanomaterials are used in the development of new pesticides, veterinarian drugs or other agrochemical aimed at production improvement.
- When nanomaterials are used in the removal of unwanted substances from foods and water.

The state of the nanotechnology in relation with food production and water use is presented in this chapter through some examples. Why the nanotechnology shows a great application potential will be explained presenting some of the most recent findings. Therefore, a general view on how nanotechnology could be a solution for the improvement of methods used in the food production and water uses is provided.

1.1 Basic notions on nanotechnology

Nanotechnology can be briefly defined as 'the engineering of very small systems and structures'. Actually, nanotechnology consists in a set of technologies than can be developed and used in several activities and agro-food sector it is not an exception.

The main feature of nanotechnology is defined by the size of the systems of work: the 'nanoscale'. At this scale the matter presents properties which are different (and new) than the observed at macroscopic level. These properties emerge as a consequence of the size of the structures that produces the material and their interactions. Somehow, in nanotechnology, atomic and molecular forces turn on determinant factors over other effects of relevance at higher scales. Other properties and possibilities of nanotechnology, which have great interest in food technology, are high reactivity, enhanced bioavailability and bioactivity, adherence effects and surface effects of nanoparticles.

According to the former, which is the magnitude order defining nanoscale?. Although there is a consensus about considering this as a range between 10 and 100 nm, there is growing evidence on particles and systems of several hundred of nanometres with activity that can be related with a typical behaviour of nanomaterial (Ashwood et al., 2007).

The definition 'working under 100 nm' could be considered a common approach when defining nanotechnology, but this approach could be too restrictive in the agro-food sector, as well as when the effects on health and environment are considered.

At present several definitions about the term 'nano' are presented by organisations at international level. For instance, Commonwealth Scientific and Industrial Research Organisation (CSIRO) or the Food and Drug Administration (FDA) define nanomaterial as that material under 1,000 nm.

According to the ISO/TS 8004-1:2010 norm, nanotechnology is: 'The application of scientific knowledge to control and utilize matter in the nanoscale, where properties and phenomena related to size or structure can emerge'. And 'nanoscale' is defined as: 'Size range from approximately 10 nm to 100 nm'.

Regardless of the former, in this chapter nanoscale will be considered when the working range is less than 1,000 nm. This decision comes from the fact that a clear distinction can be made when considering the production of foods and water use:

- Application could be devoted to be a part of the food that will be not absorbed. (i.e. packaging)
- Application is designed to be an ingredient or additive that will be consumed.

So, since the final aim is the search for a kind of functionality (mainly physiological type) through the control of the size of the matter and, considering the fact that particles of several hundred of nanometres showing activity can be found, it seem reasonable to extent the definition up to 1,000 nm.

Despite that the potential of nanotechnology has been already recognised and applied in other industrial sectors (electronic, medical, pharmaceutical energetic sector and material sciences), the application on the agro-food sector has been limited up to now. The most promising applications of nanotechnology in foods includes: Enhancement of activity and bioavailability of nutrients and activity principles of foods, improvement of organoleptic features (colour, flavour), better consistence of food matrix, new packaging development, food traceability, safety and food monitoring during transport and storage.

In the case of the water use, the high nanoparticle activity allows the use of new purification techniques and removal of unwanted substances. In agriculture, the increase of bioavailability and the particles behaviour could boost the reduction of the side-effects in environment.

2. Potential applications of nanotechnology in the agro-food industries

2.1 Can nanotechnology enhance the access to the crops?

The pressure on crops for agricultural production will increase in the future (both the area available for cultivation and the consumption of water required for this purpose). This is related to the fact that soil is used not only for food production but also for other products of industrial interest such as biofuels. To give some figures, the annual cereal production will have to increase from 2.1 billion to 3 billion while the annual meat production will have to double.

The scientific community trust in nanotechnology as a tool which could help to solve the challenge faced by the farmer: get highly productive crops while minimizing the use of synthetic chemicals. Despite the promising use of this technology in agriculture, most applications are still under research: nanotechnology and the genetic improvement of crops and production of more selective, effective and easier to dose plant protection products (FAO / WHO, 2009; Nair et al., 2010). Nanotechnology will likely have something to say about the search of alternative energy sources and cleaning and decontamination of water or soil resources. Researches also work in the application of nanotechnology in surveillance and control systems to determine when is the best moment to harvest or to monitor crop safety (Joseph & Morrison, 2006).

But the crops will not only be benefited from the potential advantages of this technology, but also could be used as organic producers of nanoparticles.

2.1.1 Nanotechnology and crops genetic improvement

This technology, combined with others such as biotechnology, can make genetic manipulation of plants easier. It allows that nanoparticles, nanofibers or nanocapsules are used as vectors of new genetic material instead of conventional viral vectors. These new vehicles could carry a larger number of genes as well as substances able to trigger gene expression (Miller & Senjen, 2008; Nair et al., 2010) or to control the release of genetic material throughout time.

Chitosan is one of the most studied non-viral gene vectors since its positive charge density allows the condensation of DNA by electrostatic interaction, protecting the entrapped genes from nuclease action. However, chitosan nanocarriers have still not replaced conventional vectors. These vehicles are often associated with low transfection efficiencies. Authors like Zhao et al., 2011 are recently working on the enhancement of this property through modification of chitosan nanocomplexes with an octapeptide. They have condensed DNA into spherical nanoparticles of around 100-200 nm size with higher transfection efficiencies and lower cytotoxicity than those of DNA complexed with unmodified chitosan. Their transfection capacity varies depending on cell type. Authors like Mao et al., 2001 when they used as a genetic vehicle chitosan nanoparticles, found higher levels of gene expression in human kidney cells and bronchial cells than in cancerous cells. Wang et al., 2011 worked with *Jatropha curcas* callus cells and demonstrated the integration of DNA carried by this type of nanoparticles in their genome.

Chitosan nanoparticles are quite versatile, as well as their transfection efficiency can be modified, they can be PEGylated in order to control the release of genetic material as time goes by (Mao et al., 2001). This effect of time controlled genetic material release can be achieved by encapsulating pDNA into poly (DL-lactide-co-glycolide) particles. Specifically speaking, Cohen et al., 2000 achieved sustained release of pDNA over a month. Nonetheless the major advantage of nanobiotechnology is the simultaneous delivery of both DNA and effector molecules to specific sites. This effect has been achieved in intact tobacco and maize tissues when using gold-capped mesoporous silica nanoparticles (MSNs) as plasmid DNA transfers (Nair et al., 2010).

2.1.2 Production of nano-agrochemicals

Nanotechnology can help significantly to improve crop management techniques and it seems to be particularly useful in the use and handling of agrochemicals. Usually, only a very small amount of a given active compound is needed for treatment. Nowadays larger amount of chemicals, which are applied several times, are used in the field and nanotechnology could help to increase efficacy and to prevent losses. These losses can be explained for several reasons: UV degradation, hydrolysis, microbiota interaction or leaching. This way of handling chemicals brings negative effects to the environment (soil degradation, water pollution and side effects in other species). It is postulated that active substances in their nano form will allow to formulate more effective products, with time controlled action, active only under certain environmental conditions and against specific organisms, or able to reach and act on specific sites inducing changes in plant metabolism.

Research in this field is carried out by groups such as Wang et al., 2007 who formulated beta-cypermethrin nanoemulsions. They were more effective than commercial micro-emulsions of the same compound. The surface-functionalized silica nanoparticle (SNP) developed by Debnath et al., 2011 (about 15-30 nm) caused 90% mortality in *Sitophilus oryzae* when comparing its effectiveness with conventional silica (> 1 μm). Qian et al., 2011 achieved 2 weeks of validamycin sustained release when nano-sized calcium carbonate (nano-CC) was used. Guan et al., 2010 encapsulated imidacloprid with a coating of chitosan and sodium alginate through layer-by-layer self-assembly, increasing its speed rate in soil applications.

In the synthesis of products active only under specific environmental conditions work among others, Song et al., 2009 group, who showed that triazophos can be effectively protected from hydrolysis in acidic and neutral media by including it in a nano-emulsion, while its release was very easily achieved in alkaline media. Other examples of selective chemicals are the nanoparticles functionalized with the aim of being absorbed through insects' cuticle. The protective wax layer of insects is damaged and leads them to death by desiccation. This approach is safe for plants and entails less environmental damage. Nair et al., 2010 reported that certain nanoparticles are able to reach sap, and Corredor et al., 2009 found that they can move towards several sites in pumpkins. In other words, nanoparticles have the ability to act in sites different from their application point.

Substances in their nano form can affect cellular metabolism in a specific manner; for instance Ursache-Oprisan et al., 2011 reported an inhibition in chlorophyll biosynthesis caused by magnetic nanoparticles. But nanomaterials not only influence plants, but also animals like *Eisenia fetida* earthworms which avoid silver nanoparticles enriched soils as was evidenced by Shoults-Wilson et al., 2011.

2.1.3 Alternative energy sources and nanotechnology

One of the biggest challenges in this century is the search for new and feasible energy sources different from fuel, nuclear or hydroelectric power. Alternative energy should maintain socio-economic development without jeopardizing the environment. According to U.S. Energy Information Administration, more than 90% of the total energy produced during the first eight months of 2010 in the USA was obtained from coal, gas, nuclear, oil and wood. This energetic mix finds explanation in the fact that renewable energies (solar, wind, geothermal and tidal) are still non cost-effective.

Although, it is not the aim of this chapter to talk about the connection between energy and agriculture, it is worth to mention that agricultural sector requires important energy inputs (direct but also indirect). We think any step forward in the production; storage or use of new energy sources could be easier and faster with the help of nanotechnology. Then society will be moving towards a more sustainable, affordable and less dependent on fuel agriculture.

2.1.4 Water and soil resources remediation

Furthermore being vehicle for active substances (pesticides, plant growth regulators or fertilizers), nanoparticles can also be synthesized with a catalytic oxidation-reduction objective. The latter application would reduce the amount of these active substances present in the environment and also the time during which it is exposed to their action (Knauer & Bucheli, 2009). This section focuses on the application of reduction-oxidation catalytic nanoparticles as soil decontaminants, while the role of nanotechnology in water remediation is developed in section 2.3. Nanotechnology and water supplies.

The research focus is twofold in this issue: first, try to accelerate the degradation of residual pesticides in the soil, and secondly, improving these pollutants' detection and quantification methods.

Between those who try to accelerate the decomposition of these contaminants in soil, are Shen et al., 2007 who synthesized magnetic Fe3O4-C18 composite nanoparticles (5-10 nm) more effective than conventional C18 materials as cleaner substances of organophosphorous pesticides. Zeng et al., 2010 proved that TiO2 nanoparticles can enhance organophosphorous and carbamates' degradation rate (30% faster) in crop fields.

Currently, several authors are working in pesticide analysis methods' optimization. Magnetic composite nanoparticle-modified screen printed carbon electrodes (Gan et al.,

2010), electrodes coated with multiwalled carbon nanotubes (Sundari & Manisankar, 2011) or nano TiO2 (Kumaravel & Chandrasekaran, 2011) are some examples of nanomaterials which have being successfully used. They all act as sensors of pesticides. As opposed to more conventional ones, these methods are more sensitive and selective.

2.1.5 Crop monitoring systems

Nanotechnology is contributing very much in the development of sensors with applications in fields such as: agriculture, farming or food packaging. In agriculture, nano-sensors could make real-time detection of: humidity, nutrient status, temperature, pH or pesticides, pollutants and pathogens presence in air, water, soil or plants. All these collected data could help to save agro-chemicals and to reduce waste production (Baruah & Dutta, 2009; Joseph & Morrison, 2006).

2.1.6 Particle farming

One of the cornerstones of nanotechnology is the synthesis of nanoparticles and their self-assembly (Gardea et al., 2002) since the methods used until now are very expensive and some of them involve the use of hazardous chemical reagents.

Alternative nanoparticles production processes are continually sought in order to make them more easily scalable and affordable. One of these routes of synthesis under study is known as "particle farming" and involves the usage of living plants or their extracts as factories of nanoparticles. This process opens up new opportunities in the recycling of wastes and could be useful in areas such as cosmetics, food or medicine.

The latest research in this field focus on the synthesis of gold and silver nanoparticles with various plants: *Medicago sativa* (Bali & Harris, 2010; Gardea et al., 2002), *Vigna radiata, Arachis hypogaea, Cyamopsis tetragonolobus, Zea mays , Pennisetum glaucum, Sorghum vulgare* (Rajani et al., 2010), *Brassica juncea* (Bali & Harris, 2010; Beattie & Haverkamp, 2011) or extracts from *B. juncea* and *M.sativa* (Bali & Harris, 2010), *Memecylon edule* (Elavazhagan & Arunachalam, 2011) or *Allium sativum* L. (Ahamed et al., 2011).

Depending on the nanoparticle's nature, specie of plant or tissue in which they are stored, metal nanoparticles of different shapes and sizes can be obtained. However, all these processes share the advantages of being simple, cost-effective and environmentally friendly.

Apart from the potential benefits of nanotechnology in agricultural sector (described throughout this section), it also involves some risks. Farmers' chronic exposure to nanomaterials, unknown life cycles, interactions with the biotic or abiotic environment and their possible amplified bioaccumulation effects, should be seriously considered before these applications move from laboratories to the field.

2.2 Nanotechnology and animal production

Livestock contribute 40 percent of the global value of agricultural output and support the livelihoods and food security of almost a billion people (FAO, 2009). Rapidly rising incomes and urbanization, combined with underlying population growth, are driving demand for meat and other animal products in many developing countries, being the annual growth rate 0,9% in developed countries and 2,7% annual worldwide rate. In the last decade, per capita consumption in developing countries is nearly twice and this tendency keeps on growing surpassing 100 million tonnes of produced meat in Asia in 2007 for instance (FAO, 2009).

Therefore, some of the challenges the animal production sector will have to deal with are:

- Look for an environmentally friendly sustainable production system that contributes to maintain environment preventing further degradation (livestock is responsible for 18 percent of global greenhouse gas emissions, (FAO, 2009) and livestock grazing occupies 26 percent of the earth's ice-free land surface), this joint to the increasing supply of meat products to a growing population (food demand is foreseen to be twice by 2050 due to the socio-economic and population growths, FAO, 2009). However, this growing population has not always higher economic resources (FAO estimates that between 2003–05, 75 million more people were added to the total number of undernourished and high food prices share part of the blame and this number increased in 2007 up to 923 million (FAO,2008).

- Diseases control in animal-food production sector, since these diseases are extended quickly now than in the past due to the market globalisation, so this point is a key to preserve human health and food safety (according to a study carried out in USA, UK and Ireland during the last decade approximately 20% of the retired food for food safety came from the meat sector, 12% from food processed products and 11% from vegetable and fruits sectors (Agromeat, 2011). Animal diseases reduce production and productivity, disrupt local and national economies, threaten human health and exacerbate poverty being essential its minimisation.

- An accurate traceability of products derived from the meat sector in an international growing market, which would warrant the product identity and avoid possible food fraud in this sector.

In this way, nanotechnology will be able to solve these problems in the animal production sector. In order to explain this issue, we will speak of the different areas within the animal production where nanotechnology can give support and provide some important solutions. These fields can be classified within 5 categories (Kuzma, 2010) being currently all of them under research and development.

- Pathogen detection and removal
- Veterinary medicine
- Feed improvement and waste remediation
- Animal breeding and genetics
- Identity preservation and supply-chain tracking

2.2.1 Pathogen detection and removal

This first category includes the use of nanodetectors not only to detect pathogens but to bind and remove them. An example will be the case of S. *typhi* detection in chickens skins, which uses magnetic particles functionalised with antibodies that after binding the pathogen, removes them by means of the introduction of magnetic forces. Other case is the study against foot-and mouth disease (FMD) virus in chickens, with the development of surfaces based on nanostructured gold films with topography matched to that of the size of FMDV. These surfaces are functionalized with a single chain antibody specific for FMDV in such a way that liquid crystals will uniformly anchor on these surfaces. FMDV will bind to these surfaces in such a way that it will give rise to easily visualized changes in the appearance of liquid crystals anchored on these surfaces and so virus presence is detected (Platypus Technologies LLC, 2002).

The benefit of this application would allow improving human health reducing the risk of diseases derived from food consumption and allow great benefits for animal health, likewise an economic advantage when producing. However, this technology is not still very developed and it is necessary studies to guarantee the lack of toxicity of these nanoparticles. Also, nanotechnology is used in the diagnosis of animal diseases control based on microfluidic, microarray, electronic and photo-electronic, integrated on-chip and nanotechnology together with analytical systems, which enable the development of point-of-care analysers (Bollo, 2007).

2.2.2 Veterinary medicine

In the veterinary medicine field, nanoparticles to deliver animal drugs like growth hormones (somatotropine) in pigs through PGLA nanocapsules and the improvement in animal vaccination are two of the application examples. Among vaccination products, some polystyrene nanoparticles bound to antigens are experimentally being proved in sheep (Scheerlinck et al., 2005) and vaccines against *Salmonella enteriditis* based on an immunogenic subcellular extract obtained from whole bacteria encapsulated in nanoparticles made with the polymer Gantrez (Ochoa et al., 2007).

Animal production and food safety will be increased and this will lead to cheaper and greater availability of meat products. However, some of these measures could not be successfully applied in some countries, as it would be the case of hormones supply to animals, which is not accepted as good practice by consumers or even banned in some countries.

2.2.3 Feed Improvement and waste remediation

In the category of feed improvement and waste remediation are being studied the use of nanoparticles in feed for pathogen detection. These particles would bind pathogens in the gut of poultry preventing its colonization and growth and avoiding its presence in waste.

Other example would be the use of nanoparticles to detect chemical and microbiological contamination in feed production. Advantages of both applications are clear: Farms would be more environmentally friendly and alternatives for using of antibiotics to face pathogens will be proposed. The former option could be very interesting due to the growth of antibiotic resistance found in animals and humans. However, life cycle of these particles must be verified in order to warrant that these particles are not harmful for the animal and that they don't end up on animal-derived food products or food chain.

Toxicology must be studied and environmental exposure, likewise the effect that animal wastes containing these particles would have. In feed production case, workers safety must be studied and the effect of leftovers remaining in farms and fields.

Within this part, we can also mention the use of polymeric nanocapsules to carry bioactive compounds (such as proteins, peptides, vitamins, etc.) in such a way that they will be able to pass through the gastrointestinal mucus layer and/or epithelial tissue providing or increasing the absorption of these bioactive compounds (Luppi et al., 2008; Plapied et al., 2011), improving the availability of these nutrients and therefore, the quality of animal production.

2.2.4 Animal breeding and genetics

Genetic animal therapy is one of the possible applications in a near future. Currently, studies of genes delivery in animal cells for selection and temporal expression are being carried out in order to determine specific features of livestock (Mc Knight et al., 2003).

Animal health and breeding would be improved as well as animal production. For instance, some kind of genetic diseases could be avoided and genetic configuration could be determined in a way that animals will be more fertile. However, before more studies about toxicity, safety and legislation must be made. Moreover, public perception and current legal framework related to genetic modification would have to change a lot, since society opposition is still rather difficult to save.

2.2.5 Identity preservation and supply-chain tracking

Concerning traceability and identity preservation chips of DNA are being studied in order to detect genes in feed and food, likewise nanobarcodes to get feed (Jayarao et al., 2011) and animal traceability from farm to fork. This will aid to maintain animals healthy but also to improve public health, since feed traceability would help to prevent and avoid possible food related illnesses, being able to find quickly the source of such diseases and even to prevent them.

In conclusion, the potential benefit of these technologies would allow a diminution cost and an improvement in the animal production sector as much in quality as in quantity, the waste reduction, increasing efficiency in the use of resources and rising safety in animal derived food products.

Possible risks related to human and animal health and also the environmental impact have to be overcome before applying of these technologies. Other matters such as exploitation rights, intellectual property and legal issues derived from the use of these inventions and patents must be considered too.

2.3 Nanotechnology and water supplies

In this decade, the demand of water supplies is rapidly growing and the competition for water resources is being every day a bigger concern. Currently, 70% of the obtained fresh water in the entire world is used in agriculture, while 20% is used in the industry and 10% are devoted to municipal uses (FAO, Agotamiento de los Recursos de Agua dulce).

The access to inexpensive and clean water sources is an overriding global challenge mainly due to:

- The global climate change that will contribute to fresh water scarcity.
- The constant growth in world population. Water sources demand has exceeded by a factor of two or more the world population growth, (FAO, Agotamiento de los Recursos de Agua dulce valoración ambiental: Indicadores de Presión Estado Respuesta), this means that the expected population growth will be pressing water resources. In order to have a rough idea, the daily drinking-water requirements per person are 2-4 litres. However, it takes 2000 - 5000 litres of water to produce a person's daily food, if this is added to the increasing population growth... (FAO. FAOWATER. Water at a Glance: The relationship between water, agriculture, food security and poverty.)
- The raising industrial and municipal water waste. Every day, 2 million tons of human waste is disposed of in water courses. Worldwide total water waste volume was estimated in more than 1.500 km3 in 1995, knowing that every litre of waste water contaminates 8 litres of freshwater, 12.000 km3 of water resources from the planet are estimated not being available for its use. If this figure maintains its growth at the same

rate than the population is increasing (by 2050, population will reach 9.000 million people), the planet will lose every year around 18.000000 km3 of water resources (UNESCO, 2007).

- The large quantities of water used to produce increasing amounts of energy from traditional sources.
- The intensive water use in agricultural (i.e., it takes 1-3 tonnes of water to grown 1kg of cereal; FAO. FAOWATER. Water at a Glance: The relationship between water, agriculture, food security and poverty) and livestock activities. Added to this, it is the fact that the foreseen urban population growth will increase pressure on local water resources, while food production demand is rising in remote areas to support the industrial animal production systems connected to cities.

These are some of the reasons why alternatives should be thought for obtaining and recycling water and in this point is where nanotechnology could provide innovative solutions.

Currently, there are several nanotechnology research lines concerning water treatment. Some of these examples are mentioned next:

- Magnetic nanoadsorbents, which can transform high polluted water in water for human consumption, sanitary uses and irrigation. Magnetic nanoadsorbents can be used to bind pollutants such as arsenic or petroleum and then could be easily eliminated by a magnet (Mayo et al., 2007; Wolfe, 2006).
- Water desalination by means of carbon nanoelectrodes. Water for human consumption could be obtained in areas, which are rich in high salinity water, using less economic and energy resources. An example would be the use of capacitive deionization (CDI), which is based on high-surface-area electrodes and that when are electrically charged can adsorb ionic components from water (Oren, 2008).
- Biofilms, which create surfaces that avoid bacterial adhesion and which would be self-cleaning, preventing fouling and hindering biocorrosion and pathogens in water distribution, connexion and storage systems. (Brame et al., 2011).
- Nanomaterials with catalytic and phototacalytic activities able to eliminate bacteria, pesticides, antibiotics and hormonal disruptors, likewise chlorine of organic solvents like trichloroethylene (hypercatalysis) from the environment. Remediation of groundwater contaminated with organochlorine compounds and other compounds can be done, having impact not only on water purification systems for drinkable water but on human health. Among some of the applications would be nanocomposites of dioxide of titanium, which in presence of UV-visible and solar light have bactericide and anti-viral properties. Other application would be the hypercatalysis with palladium covered by gold nanoparticles that would eliminate organic solvents like trichloroethylene (Clean Water Project: WATER DETOXIFICATION USING INNOVATIVE vi-NANOCATALYSTS, FP7 Collaborative Project, ENV NMP227017, 2009-2012; Nutt et al., 2005; Senthilnathan & Ligy, 2010).
- Membranes systems based on metal nanoparticles, which prevent fouling to purify water in places where water quality is not good for human consumption and in water waste, reducing the number of treatment steps; nevertheless several drawbacks must be solved before applying it. Among these problems to be solved would be the use of metal-based nanoparticles, which implies a potential selection for antibiotic resistance. These multi-functional membranes would be self-cleaning, antifouling

performance and could inactivate virus and eliminate organic matter. An example would be reverse osmosis membranes and its spacers with nanosilver coating, which would improve bacteria removal and antifouling performance (Yang at al., 2009; Zodrow et al., 2009).

- Nanosponge for rainwater harvesting, a combination of polymers and glass nanoparticles that can be printed onto surfaces like fabrics to soak up water. This application can be very beneficial for tropical climate countries where humidity is high and there could be a more efficient water mist catching (Grimshaw, 2009).

- Nanomesh waterstick, a straw-like filtration device made of carbon nanotubes placed on a flexible and porous material. This stick cleans water as someone drinks and eliminates virus and bacteria. It has been already used in Africa (Seldom laboratories) and is currently available in the market. Speaking about active carbon, this one has been used together with nanofibres, in point-of-use water purifying systems. They are some kind of biodegradable tea-bags-like to filtrate water. Once, they have been used the nanofibres will disintegrate in liquids after a few days (developed by E. Cloete; Universidad de Pretoria).

- Nanoparticles with DNA able to track hydrological flows, which will allow to know water sources and if the water is polluted or not (Walter et al., 2005).

The key point is that these technologies would allow to treat processing and residual water from food industry and recycle and reuse them, contributing to the economy, environmental improvement and water security of local communities and industries.

This is essential in areas where water scarcity is a problem, drinkable water is difficult to obtain and water is considered as a precious resource.

Although the operation cost of some of these installations and facilities would be higher than the operation costs of the traditional facilities, the initial investment would be cheaper (construction costs), moreover life cycle of some of these materials is unknown and these are only some of issues that must be studied before trying to implement anything, so there is still a long way to cover before some of these applications can be real facts.

2.4 Nanotechnology in food processing

Up to now in this chapter, the nanotechnology applications have been referred to water use and commodity production but, nanotechnology can be a way to get better manufacturing food process (or even a completely new and different one). Actually, food sector could be facing a paradigm shift. As commented by Pray & Yaktine, 2009, nowadays foods are structured using a recipe (o formulation) whereas two simultaneous processes occur:

- Formation the structure (i.e: by means of phase creation, reactions, biopolymer transformation ...)
- Stabilization of the system (i.e. vitrification, crystallization, network formation...)

On the other side, nanotechnology affords the possibility of structure foods from the base elements (in an approach derived from its constituents self-assembly). Rather than the use of a recipe, nanotechnology raises the opportunity of using molecules as starting material, so that new interaction are achieved which in turn generates the required properties. The paradigm shift resides in the fact the food formulation will based on the use of "food matrix precursors or structural elements". To achieve this possibility it will be necessary the development of new knowledge and techniques.

Some examples on the potentiality of the former approach is the development of new textures and flavours, the possibility of the design of low dense, low calorie foods but nutritionally enhanced. Those foods can be aimed at a nutrition adapted to different life-style and consumer conditions (i.e. obesity cases).

Two points are commented here to give a general view of the multiple possibilities of nanotechnology use in food processing:

- Food processing issues: mixing, component stability, safety, etc.
- Intrinsic food features: texture, flavour, taste masking, availability and delivery, etc.

Among the processing issues the most of the applications are related with the use of nanoparticles and nanocapsules. These particles enhance the products functionality, and they are responsible of that enhancement because they perform a protection function on:

- The contained active principle. Nanoparticles avoid the degradation produced by the surrounding environment of the food or by the manufacturing process. Gökmen et al., 2011, have reported that omega 3 was successfully nanoencapsulated and used in bread making. In addition of the positive effect in taste masking, thermooxidation during baking was reduced. As a consequence of that, the production of other further degradation by-products (i.e. acrylamide) was strongly reduced.
- The food itself. This happens when the active principle is used with a technological purpose. For instance nanoparticles containing essential oil have been used to improve antimicrobial activity in juices (Donsì et al., 2011). In this case, the addition of small amounts of nanoparticles containing terpenes to an orange and pear juice delayed or avoided the microbial growth (depending on the concentration). Organoleptic properties were maintained.

One advanced aspect of the use of nanotechnology is the possibility of acting straight onto the food structure. Nanostructured foods aimed at producing better texturized, flavoured and other properties can be obtained. In particular it can be highlighted the possibility of producing cream-like foods such as: ice-creams, spreads and low fat dressings. Leser et al., 2006, comment how polar lipids (monoglycerides and phospholipids) can be used with this purpose. Some of those polar lipids are capable of create nanostructures that finally result in crystalline phases. These structures can be used as a base for the development of spreads taking advantage on the texture properties emerging from interaction between nanostructures clusters. The manufacturing process will depend on water relative quantity, stirring, and the system temperature evolution.

The contribution of particles to the stability of foams and emulsions is another aspect of interest in the use of nanotechnology in foods. Most of the foods are (or have been during processing) dispersions like emulsions or foams. Several examples could be derived from bakery, confectionery, meat products, dressings and spreads. Another example is the prepared foods. In general, foam and emulsion stability can be improved in the presence of nanoparticles and nanostructures.

The macroscopic properties of dispersion can be improved by controlling the formation of nanostructures in the interphase (Rodríguez et al., 2007). The application of new techniques to food development such as Brewster angle microscopy (BAM), atomic force microscopy (AFM) y imaging ellipsometry (IE), will allow the study and application of those possibilities to processing.

Dickinson, 2010 presents a review about the use of inorganic nanoparticles (silica), fat crystals and protein nanoparticles. Main conclusions to highlight are:

- "Natural biopolymer structural assemblies are obviously attractive as (nano)particle building blocks"
- Polysaccharides (i.e. starch or cellulose) represent can be considered as a cheap and a quickly source of nanomaterial for food uses. Anyhow, it is required some type of modification to increase hydrophobicity.
- Proteins represent the best potential derived from the amphiphilic behaviour as well as due to their carrying capacity (i.e. casein nanotubes)
- The interaction between protein and polysaccharides is of great interest also. For instance, the interaction obtained from the interaction of caseinate and Arabic gum or, the heat degraded beta-lactoglobulin aggregates and stabilised by pectin. One important implication of this type of interaction could be the complexing between casein and a charged polysaccharide which can be used in the maintenance of solubility and functionality under acidic pH conditions

Another relevant aspect in food processing is the time. In this case, when considering the interaction between food components in a structure, the time required to reach a position should be considered. "In order to interact, different components of a food structure must come into position at the right time." (Pray & Yaktine, 2009) For example, the foam formation is a kinetic process that is initiated at the nanoscale in miliseconds, after that foam is stabilised at macroscopic level in the order of minutes. The capability of nanoparticles and nanosystems to arrange in time is then another issue to consider. In this sense, Sánchez & Patino, 2005 have reported how the diffusion speed of caseinate to the liquid-air interphase affects to the foam formation and, how depending on the concentration (above 2%), interphase saturation occurs and appear a micelle formation for sort times.

Besides processing features, a possible breakthrough in food manufacturing is the enhancement of the bioavailability. Nanoemulsions and nanoparticles suppose a delivery vehicle capable of override two great problems:

- Conservation in the food previous to the consumption and the stomach degradation
- The absorption of the compounds by the organism

In the last case, it should be considered that the main absorption route of foods is the intestine and nanotechnology could improve that somehow. Direct absorption of nanoparticles is controlled by size and surface chemistry of the particle. Generally nanoparticles can be done by an active transport mechanism or by a passive transport (Acosta, 2009).

In the first case (active transport) absorption happens by means of specific channels that present the epithelial intestinal cells. In this case absorption is controlled by the hormonal system and by homeostatic effect derived from blood levels of compounds. According to the last control mechanism, once a level is surpassed for an ingredient, it will be not absorbed and the excess is accumulated or excreted. Selective absorption can be conditioned by specific receptors in the cell surface (enterocytes and M cells). Those nanoparticles with a specific surface composition can be absorbed by this way. Generally, the M cells present a better permeability. For example, there is a great interest in the development of lecithin nanoparticles since it could be an improvement in the absorption of isoflavones by this route (Acosta, 2009).

Passive transport happens by means of diffusion though epithelial tissue. The particle has to be fixed to intestinal mucosa, and from there, to contact the cell. The absorption speed is determined by a concentration gradient. In an interesting work, Cattania et al., 2010, studied the use *in vivo* of lipidic core nanocapsules. Those nanocapsules are fixed to gastrointestinal lumen and act as active principle reservoirs. Nanocapsules delivered the active principle once they were fixed by diffusion. As relevant conclusion of this work it is pointed that there is important differences between evaluation models used (*in vivo* and *ex vivo*), and the live model system complexity cannot be predicted by a simpler model, at least at this time.

In fact, nanomaterial dosimetry it is not clear at all. Despite this, there is a general consensus about the relation of particle size and bioavailability in the sense that, as far the size decreases, the bioavailability is improved. A size reduction under 500 nm produces a higher absorption of the active principle and more particle uptake regardless of the system composition. However, above 500 nm, the bioavailability depends on the system (Acosta, 2009). So, it is possible to obtain more effect from a food additive with the use of nanotechnology (with lower content in an active principle) than with the use of the traditional approach (i.e. microencapsulation).

Food properties can be enhanced this way but there is a necessity for new test and assay methods to evaluate the state of the particle in the food (or structure) and the effect on the organism and health.

Ultimately, nanotechnology application on foods supposes a great challenge but requires solving several aspects in the near future. Once those barrier have been overcome, manufacturing of foods adapted to especial necessities will be obtained, as well as the improvement in efficiency, action in the organism and even getting a better experience in the consume.

2.5 Food packaging

Most of the nanotechnology applications in food industry are in the packaging sector. It is expected that by 2015 this sector will be 19% of nanotechnology food applications.

This is mainly due to the big development of nanotechnology in this field, to a higher acceptance by consumers of the use of this technology in packaging than in food as ingredients and to the normative requirements, which are less restrictive than for the enforced current food legislation. All this contributes to the fact that there are many more applications in this area than in others of food sector, together with the necessity of a more sustainable, lighter and stronger at the same time, efficient and intelligent packages. These packages would be able to provide safer products with more quality and at the same time maintaining the products in the best possible conditions and with a longer shelf-life.

During the last decade, the use of polymers as material for food packaging has incredibly increased due to its advantages over the use of traditional materials.

In polymers world market, which has been increased of about 5 million tons in 1950 to almost 100 million tons at the present time, 42% corresponds to packaging and containers (Silvestre et al., 2011).

In the following section, some nanotechnology derived applications of materials in contact with food are mentioned. These applications are currently available or at research level.

2.5.1 Nanomaterials to improve packaging properties

As it was previously mentioned, polymers provide packaging with strength, stiffness, oxygen barrier, humidity, protection against certain food compounds attacks and flexibility (Sánchez-García et al., 2010b).

The possibility of polymers improving these features in food packaging by means of nanoparticles addition has allowed the development of a huge variety of polymers with nanomaterials in its composition (Azeredo, 2009; Bradley et al., 2011; Lagaron & López-Rubio, 2011).

The use of nanocomposites for food packaging not only protects food but also increases shelf-life of food products and solves environmental problems reducing the necessity of using plastics. Most of packaging materials are not degradable and current biodegradable films have poor barrier and mechanical properties, so these properties need to be considerably improved before these films can replaced traditional plastics and aid to manage worldwide waste problem (Sorrentino et al., 2007).

With the introduction of inorganic particles as clay in the biopolymer matrix (Bordes et al., 2009; 48, Utracki et al., 2011), numerous advantages are reached.

Natural structure of clay in layers at nanoscale level makes that when clay is incorporated to polymers, gas permeation will be restricted and product will be anti UV radiation proof. In addition, mechanical properties and thermal stability of package are improved.

Polymers made up of nanoclay are being made from thermoplastics reinforced with clay nanoparticles. At present, there are a huge number of nanoclay polymers available on the market. Some well-known commercial applications are beer and soft drinks bottles and thermoformed containers.

Other examples that can be mentioned are nitrure of naotitanium, which is use to increase mechanical strength and as aid in processing and dioxide of titanium to protect against UV radiation, in transparent plastics. Among these applications the oxide of nanozinc and nanomagnesium are expected to be affordable and safer solutions for food packaging in a near future (Lepot et al., 2011; Li et al., 2010).

Also carbon nanotubes (Sánchez-García et al., 2010a) or nanoparticles of SiO_2 have been used for improving mechanical and barrier properties of several polymeric matrices (Vladimiriov et al., 2006).

The use as reinforcement elements of biodegradable cellulosic nanowhiskers and nanostructures obtained by electrospinning (Goffin et al., 2011; López-Rubio et al., 2007; Siqueira et al., 2009; Torres-Giner et al., 2008) must be highlighted too. Finally, to mention the use of biological nanofillers to strength bioplastics has the added value of generating formulations of complete biological base. These nanofillers have a high surface-mass ratio, an excellent mechanical strength, flexibility, lightness and in some cases, even they are edible, since they can be made from food hydrocolloids.

2.5.2 Active packaging

Active packaging is thought to incorporate components that liberate or absorb substances in the package or in the air in contact to food. Up to now, active packaging has being mainly developed for antimicrobiological applications, nevertheless other promising applications include oxygen captation, ethylene elimination, CO_2 absorption / emission, steam resistances and bad odours protection, liberation of antioxidants, preservatives addition, additives or flavours.

Nanoparticles more used in active packaging development are nanomaterials of metals and oxide of metals in antimicrobial packaging. Nanosilver use in packaging helps to maintain healthy conditions in the surface of food avoiding or reducing microbial growth. However, its action is not as a preservative even though, it is a biocide (Morones et al, 2005; Travan et al., 2009). Based on these properties, a big number of food contact materials, which inhibit microorganisms' growth have been created (i.e. plastic containers and bags to store food).

2.5.3 Intelligent packaging

Nanotechnology can be also applied in coatings or labels of packaging providing information about the traceability and tracking of outside as well as inside product conditions through the whole food chain. Some examples of these applications are: leak detections for foodstuffs packed under vacuum or inert atmosphere (when inert atmosphere has been ruptured some compounds change of colour warming consumers that air has come inside in where should be an inert atmosphere) (Mills & Hazafy, 2009); temperature changes (freeze–thaw–refreezing, monitoring of cold chain by means of silicon with nanopores structure), humidity variations through the product shelf-life or foodstuffs being gone off (unusual microbial presence).

Currently, sensors based on nanoparticles incrusted in a polymeric matrix isolated to detect and identify pathogens transmitted by food are being studied. These sensors work producing a specific pattern of answer against each microorganism (Yang et al., 2007). Technology called "Electronic tongue" must be underlined, too. It is made up of sensor arrays to signal condition of the foodstuffs. The device consists of an array of nanosensors extremely sensitive to gases released by spoiling microorganisms, producing a colour change which indicates whether the food is deteriorated.

DNA-based biochips are also under development, which will be able to detect the presence of harmful bacteria in meat, fish, or fungi affecting fruit (Heidenreich et al., 2010).

3. Nanotechnology challenges

As described throughout part 2, the implementation of nanotechnology in the food industry offers a wide range of opportunities to improve farm management, livestock waste, processing and food packaging. According to Helmut Kaiser Consultancy, this market was valued at USD 2.6 bn in 2003, doubled in 2005 and is expected to soar to USD 20.4 bn in 2015 (Groves & Titoria, 2009).

Despite these figures, nanotechnology has a lot of work to do in the food industry compared to its implementation in other fields such as health and fitness, home and garden or automotive. These were the three categories with the largest number of nanoproducts in March 2011 (Project on Emerging Nanotechnologies, 2011).

According to the same source, in 2010 there were sold a total of 1317 different products based on nanotechnology. This figure is small compared with the R & D investment and shows that nanotechnology commercialization is still in its infancy not only in the food sector.

The main common themes addressed by all company surveys related to commercialising nanotechnology, are: high processing costs, problems in the scalability of R & D for prototype and industrial production and concerns about public perception of environment, health and safety issues (Palmberg et al., 2009).

At the same time, as research on new and different applications of nanotechnology is carried out, others should be done with the aim of developing reliable and reproducible instrumental techniques for detecting, quantification and characterization of new materials in environmental, food and human samples. It will be necessary to study: different absorption pathways, exposure levels, metabolism, acute and chronic toxicity and its short or long term bioaccumulation. The knowledge gained in all these areas is essential to sketch a realistic and effective nanotechnology regulatory framework.

3.1 Scientific and technological challenges

There is currently a research boom in nanotechnology; both companies and universities are increasing their efforts to study the human health and environmental effects of exposure to nanomaterials. During last years, it has been shown that these materials can affect biological behaviours at the cellular, sub-cellular and protein levels due to its high potential to cross cell membranes. Some of these effects are not at all desirable, turning to be even toxic. Despite the efforts, conventional toxicity studies need to be updated to nanoscale. These new methods must define scenarios and routes of human exposure (so far there are only few studies involving oral routes), consider the behaviour of nanomaterials in watery environment and conditions that may influence its aggregation state and association with toxicants. Also, they must select a model organism to test toxicity.

In order to carry out such toxicity studies, it is necessary to implement new analytical methods able to detect the presence of very small quantities of nanomaterials in both environmental and food samples. This issue arouses so much interest that scientific journals such as Trends in Analytical Chemistry have published two special issues about characterization, analysis and risks of nanomaterials in environmental and food samples. Its papers emphasize that these are complex samples and therefore their analysis often involves four stages: (1) Sample preparation, (2) Imaging by means of different microscope techniques, (3) Separation and (4) Characterization by measuring size, size distribution, type, composition or charge density by, between others, light scattering techniques. Anyway it is very important to take into account that nanoparticles can change its structure and composition as a function of the medium and treatment. That is why resulting sample after its preparation may differ from the original one determining the reliability and conclusions of the whole analysis (Peters et al., 2011).

3.2 Socio-economic challenges

After years of public and private economic investments in R & D, nanotechnology in return is thought to develop new and more environmental friendly and efficient production methods in order to supply a growing population with commodities, and new and safer products with enhanced properties, and to generate qualified jobs as well as scientific advances.

In other words, one of the biggest challenges which nanotechnology faces up to is the ability to create an industrial and business scope. It is thought that nanotechnology will have an important impact on employment sooner than later, despite the fact that not all consultancies agree in their expectations. It depends, for instance, on nanomaterials' definition. The American NSF (National Science Foundation) estimated in 2001 that 2 million workers will be needed in nanotechnology based companies by 2015. According to LuxResearch in its 2004 report, 10 million jobs related to nanotechnology will have been created worldwide by 2014 (Palmberg et al., 2009). Although all long-term forecasts share that they were made in a buoyant and optimistic economic scenario (before 2008 crisis), and they should be considered with caution.

Related to its geographical distribution and according to NSF, by 2015 45% of new jobs will be generated in USA and 30% in Japan. Other agencies think that job layout will change and countries like China, India or Russia will become more important (Seear et al., 2009).

Another aspect to consider when studying nanotechnology influence on employment is the workers´ health. Professionals are not immune to the new materials' effects on health and could show symptoms related to chronic expositions (Seear et al., 2009). This fact would have repercussion on economic status of health systems.

In order to achieve private initiative investment in this sector, it is strictly necessary that a stable and effective regulatory framework exists, but also channels to inform and educate the public about what this technology is, its advantages, disadvantages but also the risks it may involve.

3.2.1 Development of an effective and specific legislation

The current implementation of nanotechnology in food industry does not count with a specific legislation. Although it does not mean that new food products, ingredients, surfaces or materials intended to come into contact with food are not obliged to pass safety controls before entering the market.

According to European Commission, the scope of the current legislation is wide enough to deal with new technologies (Commission of the European Communities, 2008). It should be applied what is established in one or other normative depending on the nanotechnology implementation and on the resultant product (ingredients, additives, packages...).

Against public agencies' opinion, other society sectors like Friends of the Earth defend that it is necessary to develop new nano-specific legislation which consider engineered nanomaterials as new substances with characteristic risks and properties and different from those associated with the same substance in its bulk form (Miller & Senjen, 2008).

These rules should be observed by any food producer, whether using nanotechnology or not. Although it is true that this technology implementation in the agro-food sector goes further than the observance of the regulations, codes or acts which appears in the table. It also concerns such different topics as: workplace health and security, water quality, wastes management, pesticides or animal health.

Implementation of this current legal framework is quite complex. It is necessary for a nanotechnology regulation to work properly in food industry that public agencies define precisely: (1) Nanomaterials, (2) An international regulatory body, (3) Detection, characterization and quantification methods and (4) Exposure and risk assessment of the new products.

The main problem is the lack of agreement between the most important agencies and international bodies in the legal definition of engineered nanomaterials. Council of the European Union defines it as follows: "any intentionally produced material that has one or more dimensions of the order of 100 nm or less or is composed of discrete functional parts, either internally or at the surface, structures, agglomerates or aggregates, which may have a size above the order of 100 nm to the nanoscale include: (i) those related to the large specific surface area of the materials considered; and or (ii) specific physico-chemical properties that are different from those of the nanoform of the same material" in the proposal for the novel foods amending Regulation (EC) No. 258/97 (Council of the European Union, 2009).

Use	European Union	United States of America	Australia & New Zealand.
General Food Safety	Regulation (EC) No.178/2002.	Federal Food, Drug and Cosmetic Act (the FDC Act).	Australian and New Zealand Food Standards Code (the Food Standards Code)
Novel Foods and Novel Food Ingredients	Regulation (EC) No.258/97.		Part 1.3. of the Food Standards Code
Food additives	Regulation (EC) No.1331/2008 Regulation (EC) No.1332/2008. Regulation (EC) No.1333/2008. Regulation (EC) No.1334/2008.		
Packaging and Food Contact Materials (FCMs)	Regulation (EC) No.1935/2004. Regulation (EC) No.450/2009.	Federal Food, Drug and Cosmetic Act (the FDC Act) in its Title 21, Chapter 9.	Standard 1.4.3 of the Food Standards Code

Table 1. Summary of the food related legislation in the UE, USA, Australia and New Zealand.

Institutions like International Union of Food Science & Technology (IUFoST) or House of Lords advice that engineered nanomaterials' legal definition shouldn't be based on size alone and recommend that it should refer in an explicit manner to its functionality. Other way, if the size threshold is fixed in 100 nm, producers could declare that their goods only contain particles with dimensions of 101 nm, avoiding the established safety controls (House of Lords, Science and Technology Committee, 2010a; Morris, 2010).

Overlapping between the different international regulatory entities or agencies is another difficulty related to nanotechnology implementation control. It is because this technology can be used in many and different fields and also because its resulting products should compete in a global market, this is why it is so important to define what body is going to organize the trade. Once this challenge has been overcome, trade barriers will be reduced and there will be a free movement of goods. Codex Alimentarius organized an expert meeting in June 2009 where they thought about the use of nanotechnology in the food and agriculture sectors and its potential food safety implications (FAO/WHO, 2009). On the other hand, entities such as Organisation for Economic Co-operation and Development (OECD) can also act as arbitrator in this issue on an international scale. Regarding this body,

Friends of the Earth Australia pointed out that many countries are not represented at the OECD, in particular developing nations (House of Lords, Science and Technology Committee, 2010a). There are also people who think that an international regulatory agency is unnecessary and agreements between countries are enough.

Apart from establishing nanomaterials' definition and deciding which organization is going to coordinate the international trade of nanotechnology food products, it is also necessary to standardize protocols and reliable detection, characterization and quantification methods of nanomaterials in food samples. Otherwise, any written regulation would be limited *per se* (Institute of Food Science & Technology, 2009).

Currently, food safety legislation in western countries is designed with the aim of offering the highest health guaranties. Each agency has established its own pre-market approval assessments for new products, additives, flavourings, enzymes and materials intended to come into contact with food. Any approved application will be included in a positive list which authorizes its use under certain concentrations and foods. If the application fails the assessment, (neither the company nor the authorities in charge are able to prove the substance's safety), its marketing will be denied.

Normally, every substance approved for human consumption is associated with a tolerable intake (expressed in concentration units). The nanomaterials inclusion in those lists will make necessary to change this units, because their effects are quite different from those in the macro-scale (Gergely et al., 2010). This is exactly the reason why it is strictly necessary the exposure and risk assessment of every new nanotechnology implementation in food industry.

3.2.2 Public perception generation from a critical approach

Public perception is a key point in the development of any new technology, since without public acceptation any opportunity for development, even if scientific-technological perspectives are gorgeous, would be vanished. The good news is that public perception can be created, changes and evolutions (an example of this would be genetically modified food products).

Nevertheless, for a true and lasting public perception, this one has to be created by and inside consumers of the mentioned technology, has not to be something only imposed from outside and for it some requirements must be accomplished (Magnuson, 2010; Yada, 2011).

Among the requirements would be the simple and transparent access to information, education related to nanotechnology to acquire the necessary knowledge for allowing society benefits and risks identification, as well as management of risk control by independent and reliable organisations, knowing cost impact and who will pay for its implementation, assessment of environmental impact and finally and more important, freedom of choice. This means allowing users of this technology to choose and decide consciously if they want to consume products in which nanotechnology has been used or not and for it again, here we are, the right to be informed.

Citizens' participation in committees and forums, where people give their opinion and can be informed in these matters and "nano" mandatory labelling would be some of the pending issues (Miller & Senjen, 2008), happily the path starts to be tracked in this direction.

Currently public perception of nanotechnology faced two problems: on one hand, the technical unknowledge of the subject and on the other hand the exaggerated expectatives arisen, which show it as the solution for all the problems together with the rejection to this excessive idealized vision.

The repulse to these exaggerated views, fear of nanotechnology being uncontrolled and becoming a threat, the fact that nanotechnology is a difficult concept to understand for consumers due to its complex nature and the small size of what is being treated (it's not something that can be seen just looking at it) contribute to the fact that there is still too many things to be done in this field.

According to a public perception document of nanotechnology published by the Food Standard Agency (Food Standards Agency, 2011), its success is conditioned by several factors. Next, some of the factors mentioned in this and other reports are shown:

- Use: On one side, consumers in general are more favourable to use nanotechnology in other sectors than Food sector (House of Lords, Science and Technology Committee, 2010b).This is due to food is not only perceived as from the functional point of view but related to health, environment, science, etc. not to mentioning personal and familiar habits in each home. On the other side, within Food sector it seems that public is more likely favourable to accept nanotechnologies in fat and salt reduction meals (issues that directly affect to its health) without taste and texture damage, than to accept it for new tastes and textures development (Joseph & Morrison, 2006).

- Physical proximity with food: As some authors have said consumers are more favourable to accept the use of nanotechnology in packaging than the use of the same as an additional food ingredient. Moreover, they better understand the advantages related to packaging use of nanotechnology (extended shelf-life of the product, intelligent packaging that shows when the product has gone off, etc.) (Harrington & Dawson, 2011). This was shown in a study carried out by Siegrist et al., 2007, 2008 that evaluated public perception of different kinds of food products. Results in 153 people interviewed were that packaging derived from nanotechnology was perceived as more beneficial that food modified with nanotechnology. The use of this technology is perceived as more acceptable if it is outside food. Furthermore, even if age didn't seem to be a determining factor, it was observed that people older than 66 years old was more favourable to consider the use of the nanotechnology in packaging, not being significant differences with respect to the other age groups.

- Concerns about unknowns in some issues: its effectiveness, human health risks, and regulation (40% people interviewed in a study carried out by the Woodrow Wilson Centre for Scholars, WWCS show their concerns with these issues), testing and research for safety (12% show their worries), environmental impact (10%), short and long term side-effects in food and food chain (7%), control and regulatory concerns, etc. As reported by the WWCS (Macoubrie, 2005) 40% of the study participants relieves that regulatory agencies shouldn't be trust, from 177 participants 55% thought that voluntary standards applied by industry weren't enough to assess nanotechnology risks. However, after receiving a bit more of information, when they were asked to ban this technology until more studies of potential risks were carried out 76% of people considered that this measure would be exaggerated. When they were asked about how government and industry could increase their trust in nanotechnology, 34% answered that increasing safety tests before the product will be on the market and 25% said that providing information to consumers supporting them to make an informed choice. Other suggestion was tracking risks of products on the market. These proposals of improvement of public information and consumers' education would allow them to make better choices and gain trust on industry and government, since lack of information is one the main mechanisms that breed suspicions and lack of trust. This coincides with the opinions shown in other reports, for example the one of FSA "Nanotechnology and food. TNS-BMRB Report" in which the consumers' acceptance is conditioned by transparent information transmission and the reliability in the involved authorities (Food Standards Agency, 2011).

- Information sources (Dudo et al., 2011; House of Lords, Science and Technology Committee, 2010a): Sources and means from which public obtain information conditions in part social perception, being more likely favourable to accept or defeat a new technology. Means from which consumers have obtained more information are mainly television and radio (22%) and from other people (20%) (Macoubrie, 2005). This probably will mean that those that had acquired their knowledge through television and radio have a general knowledge about nanotechnology and not a view as much scientific-technological as if they had acquired this knowledge through journals or specialised papers.

- Socio-demographic and cultural factors (Rollin et al., 2011): According to these authors women are less optimistic than men (33% vs. 49%), and slightly less supportive (53% vs. 59%); religious people were less likely to a favourable perception; the age is not a significant influencing factor in perception, although older people (more than 66 years old) are less likely favourable to the use of nanotechnology.

- Finally, different results were found when nanotechnology perception was studied in different countries. For instance, in 2005 in Europe, 44% of Europeans had heard about nanotechnology. In Europe, acceptance seems to be increasing. In 2002, only 29% agreed on the future positive impact of nanotechnology, and 53% answered "don't know", while in 2005, almost half (48%) considered that nanotechnology will have positive effects on their way of life in the next 20 years. In 2006, over half of Europeans interviewed (55%) support the development of nanotechnology as they perceived this technology as useful to society and morally acceptable. However, in USA in 2002 consumers were more optimistic about nanotechnology (50% optimistic) than Europeans (29% optimistic). Nevertheless, by 2005, European, US and Canadian citizens were equally optimistic about nanotechnology. Europeans were more concerned about the impact of nanotechnology on the environment and were less confident in regulation than North Americans.

- Public general interests in nanotechnology applications. Among these interest must be cited the following ones, 31% in medical applications, 27% better consumers products (i.e. less toxic paint coatings, rubbish bags that biodegrade, etc..), 12% general progress (i.e. qualitative and quantitative advance in human knowledge, improvement in communications, etc.), 8% environmental protection, 6% in food safety and 4% in energy, economy, and electronics, finally 3% shows interest in army and militar security (Macoubrie, 2005).

- Attitude towards risks-benefits balance. This point had a big influence in the consumers' acceptance or not of this technology. When expected risks were rather lower than benefits public were more likely favourable to accept this technology, this could explain packaging issue too. Personal situation influences perception of the risks-benefits balance, for instance people with diseases like obesity, hypertension or diabetes usually prone to see higher benefits in the applications to mitigate these diseases than risks in their possible applications (Food Standards Agency, 2011).

None previous studies have been found about how nanotechnology´s perception conditions behaviours regarding eating and food buying. These are very important facts, if a more useful and commercial approach of the real acceptance of this technology wants to be known, but to get it, first the access to this information must be simple and transparent.

4. Conclusions

Throughout this chapter a review on how the world is facing a situation where in the near future access to foods and water will be one of the main problems for a great part of the world has been described. The pressure on environment, efficiency on production systems and population growth will require new and imaginative solutions to answer those problems. A great part of these solutions could require a technological leap or a breakthrough to achieve the final result. In this sense, nanotechnology could result a great opportunity for that.

As previously mentioned, nanotechnology shows solutions for foods manufacturing and production as well as for the water management. These approaches raise technical possibilities that could help to solve the situation in real context. Despite this, it will be needed to invest more time, financial resources and technological means to achieve widespread nanotechnology application in the sectors described here. For this reason some years will be still required to see a general market application.

Challenges of the nanotechnology are technological and social. Technological challenges will require new analytical techniques so that we can understand how the things are actually working at the nanoscale. On the other hand, and according to the evidence, new evaluation methods for the determination of potential environmental and health effects of using nanotechnology are required. This way the responsible and safe use of the nanotechnology will be possible.

Concerning the social aspects, the achievement of a global consensus about the use of nanotechnology will be necessary so that some limits were respected, (at least from a health and safety point of view).

So, technological and scientific considerations are not enough for the development of nanotechnology. A great part of the possibilities and potential applications of nanotechnology in the near future will depend upon public acceptance. Society (as a whole) must evaluate critically and objectively nanotechnology.

5. Acknowledgement

This work has been granted by the Science and Innovation Office of Spanish Government (2008-2011) within National Programme of Applied Research, Applied Research-Technology Centre Subprogramme (DINAMO Project: *Development of Nanoencapsulates for Nutritional Use* , AIP-600100-2008-23).

6. References

Acosta, E. (2009). Bioavailability of nanoparticles in nutrient and nutraceutical delivery. *Current Opinion in Colloid & Interface Science,* Vol. 14, pp. 3–15, ISSN 1359-0294

Agromeat (2011). Calidad Alimentaria: Cuáles son las causas principales para retirar alimentos del mercado., In: *Agromeat website,* 4 July 2011, Available from: <http://www.agromeat.com/index.php?idNews=115714>

Ahamed, M.; Khan, M. A. M.; Siddiqui, M. K. J.; AlSalhi, M. S. & Alkorayan, S. A. (2011). Green synthesis, characterization and evaluation of biocompatibility of silver nanoparticles. *Physica E-Low-Dimensional Systems & Nanostructures,* Vol. 43, No. 6, (April 2011), pp. 1266-1271, ISSN 1386-9477

Ashwood, P.; Thompson, R.; Powell, J. (2007). Fine Particles That Adsorb Lipopolysaccharide Via Bridging Calcium Cations May Mimic Bacterial Pathogenicity Towards Cells. *Experimental Biology and Medicin,* Vol. 232, pp. 107-117, ISSN 1535-3702

Australia New Zealand Food Standards Code. (2003). In: *Food Standards Australia New Zealand (FSANZ) website,* 8 June 2011, Available from: <http://www.foodstandards.gov.au/foodstandards/foodstandardscode.cfm>

Azeredo, H. M. C. (2009). Nanocomposites for food packaging applications. *Food Research International,* Vol. 42, pp. 1240-1253, ISSN 0963-9969

Bali, R. & Harris, A. T. (2010). Biogenic Synthesis of Au Nanoparticles Using Vascular Plants. *Industrial & Engineering Chemistry Research,* Vol. 49, No. 24, (December 2010), pp. 12762-12772, ISSN 0888-5885

Baruah, S. & Dutta, J. (2009). Nanotechnology applications in pollution sensing and degradation in agriculture: a review. *Environmental Chemistry Letters,* Vol. 7, No. 3, (September 2009), pp. 191-204, ISSN 1610-3653

Beattie, I. R. & Haverkamp, R. G. (2011). Silver and gold nanoparticles in plants: sites for the reduction to metal. *Metallomics,* Vol. 3, No. 6, (2011), pp. 628-632, ISSN 1756-5901

Bollo, E. (2007). Nanotechnologies applied to veterinary diagnostics. Conference 60th Annual Meeting of the Italian-Society-of-Veterinary-Sciences, Terrasini, Italy, Aug-2007 ,*Veterinary Research Communications,* Vol. 31, Supplement 1, pp. 145-147

Bordes, P.; Pollet, E. & Avérous, L. (2009). Nano-biocomposites: biodegradable polyester/nanoclay systems. *Progress in Polymer Science,* Vol. 34, pp. 125-155, ISSN 0079-06700

Bradley, E. L.; Castle, L. & Chaudhry, Q. (2011). Applications of nanomaterials in food packaging with a consideration of opportunities for developing countries. *Trends in Food Science & Technology,* doi:10.1016/j.tifs.2011.01.002

Brame, J.; Li, Q. & Álvarez, P. J. J. (2011). Nanotechnology-enabled water treatment and reuse: emerging opportunities and challenges for developing countries. *Trends in Food Science & Technology,* doi:10.1016/j.tifs.2011.01.004

Cattania, V. B.; Almeida, Fiel L. A.; Jager, A.; Jager, E.; Colomé, L. M.; Uchoa, F.; Stefani, V.; Dalla, Costa T.; Guterres, S. S. & Pohlmann A. R. (2010). Lipid-core nanocapsules restrained the indomethacin ethyl ester hydrolysis in the gastrointestinal lumen and wall acting as mucoadhesive reservoirs. *European Journal of Pharmaceutical Sciences,* Vol. 39, No. 1-3, pp. 116–124, ISSN 0928-0987

Chaudhry, Q. & Castle, L. (2011). Food applications of nanotechnologies: an overview of opportunities and challenges for developing countries. *Trends in Food Science & Technology*, doi:10.1016/j.tifs.2011.01.001

Clean Water Project: Water Detoxification Using Innovative vi-Nanocatalysts, FP7 Collaborative Project, ENV NMP227017. (2009-2012), In: *Project website*, 04 August 2011, Available from: <http://www.photocleanwater.eu/>

Cohen, H.; Levy, R. J.; Gao, J.; Fishbein, I.; Kousaev, V.; Sosnowski, S.; Slomkowski, S. & Golomb, G. (2000). Sustained delivery and expression of DNA encapsulated in polymeric nanoparticles. *Gene Therapy*, Vol. 7, No. 22, (November 2000), pp. 1896-1905, ISSN 0969-7128

Commission of the European Communities. (2008). Communication from the Commission to the European Parliament, the Council and the European Economic and Social Commmittee. Regulatory Aspects of Nanomaterials. [SEC (2008) 2036], In: *European Commission website* , 20 June 2011, Available from: <http://ec.europa.eu/nanotechnology/pdf/comm_2008_0366_en.pdf>

Corredor, E.; Testillano, P. S.; Coronado, M. J.; González-Melendi, P.; Fernández-Pacheco, R.; Marquina, C.; Ibarra, M .R.; de la Fuente, J. M.; Rubiales, D.; Pérez-de-Luque, A. & Risueño, M.C. (2009). Nanoparticle penetration and transport in living pumpkin plants: in situ subcellular identification. *BMC Plant Biology*, Vol. 9, (April 2009), ISSN 1471-2229

Council of the European Union (June 2009). Proposal for a Regulation of the European Parliament and of the Council on novel foods an amending Regulation (EC) No XXX/XXXX [common procedure] (LA) (First reading), In: *Council of the European Union website*, 8 June 2011, Available from: <http://register.consilium.europa.eu/pdf/en/09/st10/st10754.en09.pdf>

Debnath, N.; Das, S.; Seth, D.; Chandra, R.; Bhattacharya, S. C. & Goswami, A. (2011). Entomotoxic effect of silica nanoparticles against Sitophilus oryzae (L.). *Journal of Pest Science*, Vol. 81, No. 1, (March 2011), pp. 99-105, ISSN 1612-4758

Dickinson, E. (2010). Food emulsions and foams: stabilization by particles. *Current Opinion in Colloid & Interface Science*, Vol. 15, pp. 40–49, ISSN 1359-0294

Donsì, F.; Annuziata, M.; Sessa, M. & Ferrari, G. (2011). Nanoencapsulation of essential oils to enhance their antimicrobial activity in foods / LWT. *Food Science and Technology*, Vol. 44, pp. 1908-1914

Dudo, A.; Choi, D-H. & Scheufele, D. A. (2011). Food nanotechnology in the news: coverage patterns and thematic emphases during the last decade. *Appetite*, Vol.56, No.1, pp. 78 89, ISSN 0195-6663

Elavazhagan, T. & Arunachalam, K.D., (2011). Memecylon edule leaf extract mediated green synthesis of silver and gold nanoparticles. *International Journal of Nanomedicine*, Vol. 6, (2011), pp. 1265-1278, ISSN 1178-2013

FAO (2008). El estado de la inseguridad alimentaria en el mundo 2008: Los precios elevados de los alimentos y la seguridad alimentaria: amenazas y oportunidades, In: *Depósito de documentos de la FAO website*, 01 August 2011, Available from: <ftp://ftp.fao.org/docrep/fao/011/i0291s/i0291s00.pdf>

FAO (2009). El Estado mundial de la agricultura y la alimentación 2009: La ganadería, a examen, In: *Depósito de documentos de la FAO website*, 01 August 2011, Available from: <http://www.fao.org/docrep/012/i0680s/i0680s.pdf>

FAO. Agotamiento de los Recursos de Agua Dulce Valoración Ambiental: Indicadores de Presión Estado Respuesta, In: *FAO website*, 04 August 2011, Available from: <http://www.fao.org/ag/againfo/programmes/es/lead/toolbox/Indust/DepWatEA.htm>

FAO. Agotamiento de los Recursos de Agua Dulce, In: *FAO website*, 04 August 2011, Availablefrom:<http://www.fao.org/ag/againfo/programmes/es/lead/toolbox/Indust/DFreWat.htm>

FAO. FAOWATER. Water at a Glance: The relationship between water, agriculture, food security and poverty. In: *FAO website*, 04 August 2011, Available from: <http://www.fao.org/nr/water/docs/waterataglance.pdf>

FAO/WHO (2009). Expert Meeting on the Application of Nanotechnologies in the Food and Agriculture Sectors: Potential Food Safety Implications, In: *FAO website,* April 2011, Available from: <http://www.fao.org/ag/agn/agns/files/FAO_WHO_Nano_Expert_Meeting_Report_Final.pdf>

Federal Food, Drug, and Cosmetic Act (FD&C Act) Chapter IV: Food. (March 2005). In: *FDA U.S. Food and Drug Administration website*, 8 June 2011, Available from: <http://www.fda.gov/RegulatoryInformation/Legislation/FederalFoodDrugandCosmeticActFDCAct/default.htm>

Food Standards Agency (2011). FSA Citizens Forums: Nanotechnology and food. In: *TNS-BMRB Report. JN 219186*, April 2011, Available from: <http://www.food.gov.uk/multimedia/pdfs/publication/fsacfnanotechnologyfood.pdf>

Gan, N.; Yang, X.; Xie, D. H.; Wu, Y. Z. & Wen, W. G. (2010). A Disposable Organophosphorus Pesticides Enzyme Biosensor Based on Magnetic Composite Nano-Particles Modified Screen Printed Carbon. *Sensors*, Vol. 10, No. 1, (January 2010), pp. 625-638, ISSN 1424-8220

Gardea-Torresdey, J. L.; Parsons, J. G.; Gomez, E.; Peralta-Videa, J.; Troiani, H. E.; Santiago, P. & Yacaman, M. J. (2002). Formation and growth of Au nanoparticles inside live alfalfa plants. *Nano Letters*, Vol. 2, No. 4, (April 2002), pp. 397-401, ISSN 1530-6984

Gergely, A.; Chaudry, Q. & Bowman, D. M., (2010). Regulatory perspectives on nanotechnologies in foods and food contact materials, In: *International Handbook on Regulating Nanotechnologies*, Hodge, G.A., Bowman, D.M., & Maynard, A.D., pp. 321-341, Edward Elgar Publishing Limited, Retrieved from <http://www.steptoe.com/assets/attachments/Gergely%20et%20al%202010%20Regulatory%20perspectives%20on%20nanotech%20in%20foods.pdf>

Goffin, A. L.; Raquez, J. M.; Duquesne, E.; Siqueira, G.; Hbibi, Y.; Dufresne, A. & Dubois, P. (2011). From interfacial ring-opening polymerization to melt processing of cellulose nanowhisker-filled polylactide-based nanocomposites. *Biomacromolecules*, Vol. 12, No. 7, pp. 2456-2465, ISSN 1525-7797

Gökmen, V.; Mogol, B.; Lumaga, R.; Fogliano, V.; Kaplun, Z. & Shimoni, E. (2011). Development of functional bread containing nanoencapsulated omega-3 fatty acids. *Journal of Food Engineering*, Vol. 105, pp. 585–591, ISSN 0260-8774

Grimshaw, D. (2009). Nanotechnology for clean water: facts and figures, In: *SciDevNet. Science and Development Network website*, 02 August 2011, Available from: <http://www.scidev.net/en/new-technologies/nanotechnology/features/ nanotechnology-for-clean-water-facts-and-figures.html>

Groves, K. & Titoria, P. (2009). Nanotechnology for the food industry. *Nano the Magazine for Small Science*, Vol. 13, (August 2009), pp. 12-14, ISSN 1757-2517

Guan, H. A.; Chi, D.F.; Yu, J. & Li, H. (2010). Dynamics of residues from a novel nano-imidacloprid formulation in soyabean fields. *Crop protection*, Vol. 29, No. 9, (September 2010), pp. 942-946, ISSN 02161-2194

Harrington, R. & Dawson, F. (2011). Nano-based packaging more acceptable to consumers, In: *FoodProductionDaily.com website*, Available from: <http://www. foodproductiondaily.com/Quality-Safety/Nano-based-packaging-more-acceptable-to-consumers>

Heidenreich, B.; Pohlmann, C.; Sprinzi, M. & Gareis, M. (2010). Detection of Escherichia coli in meat with an electochemical biochip. *Journal of Food Protection*, Vol. 73, No. 11, pp. 2025-2033, ISSN 0362-028X

House of Lords, Science and Technology Committee. (2010a). Nanotechnologies and Food. 1st Report of Session 2009-10 Volume I: Report, In: *UK Parliament website*, 9 June 2011, Available from: <http://www.publications.parliament.uk/pa/ld200910 /ldselect/ldsctech/22/22i.pdf>

House of Lords, Science and Technology Committee. (2010b). Nanotechnologies and Food. 1st Report of Session 2009–10 Volume II: Minutes of Evidence, In: *UK Parliament website*, 9 June 2011, Available from: <http://www.publications.parliament. uk/pa/ld200910/ldselect/ldsctech/22/2202.htm>

Institute of Food Science & Technology (2009). The House of Lords Science and Technology Committee Inquiry into the Use of Nanotechnologies in the Food Sector 2009. Written evidence submitted by the Institute of Food Science & Technology (IFST), In: *Institute of Food Science & Technology website*, 10 June 2011, Available from: <http://www.ifst.org/documents/submissions/holselectcttenanotechnologyinqui r.pdf>

Jayarao, B.; Wolfgang, D. R.; Keating, C. D.; Van Saun, R. J. & Hovingh, E. (2011). Nanobarcodes-based bioassays for tracing consignments of meat and bone meal wiyhin the project use nanobar technologies to trace commodities for contamination and/or purity, In: *Veterinary and Biomedical Sciences*, 03 August 2011 Available from: <http://vbs.psu.edu/extension/research/projects/nanobarcodes-based-bioassays-for-tracing-consignments-of-meat-and-bone-meal, College of Agricultural Sciences web 2011>

Joseph, T. & Morrison, M. (2006). Nanotechnology in Agriculture and Food. Nanoforum Report, In: *Nanoforum.org European Nanotechnology Gateway website*, April 2011, Available from: <http://www.nanoforum.org/dateien/temp/nanotechnology %20in%20agriculture%20and%20food.pdf>

Knauer, K. & Bucheli, T. (2009). Nano-materials-the need for research in agriculture. *Agrarforschung*, Vol. 16, No. 10, (October 2009), pp. 390-395, ISSN 1022-663X

Kumaravel, A. & Chandrasekaran, M. (2011). A biocompatible nano TiO(2)/nafion composite modified glassy carbon electrode for the detection of fenitrothion. *Journal of Electroanalytical Chemistry*, Vol. 650, No. 2, (January 2011), pp. 163-170, ISSN 1572-6657

Kuzma, J. (2010). Nanotechnology in animal production—Upstream assessment of applications. *Livestock Science*, Vol. 130, pp. 14–24, ISSN 1871-1413

Lagaron, J. M. & López-Rubio, A. (2011). Nanotechnology for bioplastics: opportunities, challenges and strategies. *Trends in Food Science & Technology*, doi:10.1016/j.tifs.2011.01.007

Lepot, N.; Van Bael, M. K.; Van den Rul, H.; D'Haen, J.; Peeters, R.; Franco, D. & Mullens, J. (2011). Influence of incorporation of ZnO nanoparticles and biaxial orientation on mechanical and oxygen barrier properties of polypropylene films for food packaging appliations. *Journal of Applied Polymer Science*, Vol. 120, No. 3; pp. 1616-1623, ISSN 0021-8995

Leser, M. E.; Sagalowicz, L.; Michel, M. & Watzke, H. J. (2006). Self-assembly of polar food lipids. *Advances in Colloid and Interface Science,* pp. 123– 126, 125– 136, ISSN 0001-8686

Li, X. H.; Xing, Y. G.; Li, W. L.; Jiang, J. H. & Ding, Y. L. (2010). Antibacterial and physical properties of poly(vinylchoride)-based film coated with ZnO nanoparticles. *Food Science and Technology International*, Vol. 16, No. 3, pp. 225-232, ISSN 1082-0132

Lopez-Rubio, A.; Lagaron, J. M.; Ankerfors, M.; Lindstrom, T.; Nordqvist, D. & Mattozzi, A. (2007). Enhanced film forming and film properties of amylopectin using micro-fibrillated cellulose. *Carbohydrate Polymers*, Vol. 68, No. 4, pp. 718-727

Luppi, B.; Bigucci, F.; Cerchiara, T.; Mandrioli, R.; Di Pietra, A. M. & Zecchi, V. (2008). New environmental sensitive system for colon-specific delivery of peptidic drugs. *International Journal of Pharmaceutics*, Vol. 358, pp. 44–49, ISSN 0378-5173

Macoubrie, J. (2005). Informed Public Perceptions of Nanotechnology and Trust in Government. The Project on Emerging Nanotechnologies at the Woodrow Wilson International Center for Scholars, In: *Woodrow Wilson International Center for Scholars website*, 05 August, 2011, Available from: <http://www.pewtrusts.org/ uploadedFiles/wwwpewtrustsorg/Reports/Nanotechnologies/Nanotech_0905.pdf>

Magnuson, B. (2010). State of the science on safety/toxicological assessment of nanomaterials for food applications Presentation. Cantox (Health Science International). IFT International Food Nanoscience Conference, Chicago, Illinois (USA), Ontario, Canada. July 17, 2010

Mao, H-Q.; Roy, K.; Troung-Le, V.L.; Janes, K. A.; Lin, K. Y.; Wang, Y.; August, J. T. & Leong, K. W. (2001). Chitosan-DNA nanoparticles as gene carriers: synthesis, characterization and transfection efficiency. *Journal of Controlled Release*, Vol. 70, No. 3, (February 2001), pp.399-421, ISSN 0168-3659

Mayo, J. T.; Yavuz, C.; Yean, S.; Cong, L.; Shipley, H.; Yu, W.; Falkner, J.; Kan, A.; Tomson, M. & Colvin, V.L., (2007). The effect of nanocrystalline magnetite size on arsenic

Removal. *Science and Technology of Advanced Materials*, Vol. 8, pp. 71–75, ISSN 1468-6996

Mc Knight, T. E.; Melechko, A. V.; Griffin, G.D.; Guillorn, M.A.; Merkulov, V.I.; Serna, F.; Hensley, D.K.; Doktycz, M.J.; Lowndes, D.H. & Simpson, M.L. (2003). Intracellular integration of synthetic nanostructures with viable cells for controlled biochemical manipulation. *Nanotechnology*, Vol. 14, No. 5, pp. 551-556, ISSN 0957-4484

Miller, G. & Senjen, R. (2008). Out of the laboratory and on to our plates. Nanotechnology in food & agriculture, In: Friends of the Earth, Australia, Europe & U.S.A., In: *Friends of the Earth Europe website*, 31 May 2011, Available from: <http://www.foeeurope.org/activities/nanotechnology/Documents/Nano_food_report.pdf>

Mills, A. & Hazafy, D. (2009) Nanocrystalline SnO_2-based, UVB-activated, colourimetric oxygen indicator. *Sensors and Actuators B-Chemical*, Vol. 136, No. 2, pp. 344–349, ISSN 0925-4005

Morones, J. R., Elechiguerra, J. L., Camacho, A., Holt, K., Kouri, J. B., Ramirez, J. T. & Yacaman, M.J. (2005). The bactericidal effect of silver nanoparticles. *Nanotechnology*, Vol. 16, No. 10, pp. 2346–2353, ISSN 0957-4484

Morris, V. (2010). Nanotechnology and Food, In: *International Union of Food Science & Technology IUFoST Scientific Information Bulletin August 2010 Update*, 9 June 2011, Available from: http://www.iufost.org/sites/default/files/docs/IUF.SIB.NanotechnologyandFoodupdate.pdf>

Nair, R.; Varguese, S. H.; Nair, B. G.; Maekawa, T.; Yoshida, Y. & Kumar, D. S. (2010). Nanoparticulate material delivery to plants. *Plant Science*, Vol. 179, pp. 154-163

Nutt, M. O.; Hughes, J. H. & Wong, M. S. (2005). Designing Pd-on-Au bimetallic nanoparticle catalysts for trichloroethene hydrodechlorination. *Environmental Science & Technology*, Vol. 39, pp. 1346-1353, ISSN 0013-936X

Ochoa, J.; Irache, J. M.; Tamayo, I.; Walz, A.; DelVecchio, V.G. & Gamazo, C. (2007). Protective immunity of biodegradable nanoparticle-based vaccine against an experimental challenge with Salmonella Enteritidis in mice , *Vaccine*, Vol. 25, No. 22, pp. 4410-4419, ISSN 0264-410X

Oren, Y. (2008). Capacitive deionization (CDI) for desalination and water treatment — past, present and future (a review). *Desalination*, Vol. 228, pp. 10–29, ISSN 0011-9164

Palmberg, C.; Dernis, H. & Miguel, C. (2009). Nanotechnology: an overview based on indicators and statistics. statistical analysis of science, technology and industry, In: *Organisation for Economic Co-operation and Development (OECD) website*, 18 July 2011, Available from: <http://www.oecd.org/dataoecd/59/9/43179651.pdf>

Peters, R.; Dam, G.; Bouwmeester, H.; Helsper, H.; Allmaier, G.; Kammer, F.; Ramsch, R.; Solans, C.; Tomaniová, M.; Hajslova, J. & Weigel, S. (2011). Identification and characterization of organic nanoparticles in food. *Trends in Analytical Chemistry*, Vol. 30, No. 1, (January 2011), pp. 100-112, ISSN 0167-2940

Plapied, L.; Duhem, N.; des Rieux, A. & Préat, V. (2011). Fate of polymeric nanocarriers for oral drug delivery. *Current Opinion in Colloid & Interface Science*, Vol. 16, pp. 228–237, ISSN 1359-0294

Platypus Technologies LLC (2002). Detection of FMDV by Anchoring Transitions of Liquid Crystals. Program: 2002/SBIR, Award ID:57257, In: *SBIR/STTR Small Business Innovation Research Small Business Technology Transfer*, 05 August 2011, Available from: <http://sba-sbir-qa.reisys.com/sbirsearch/detail/277343>

Pray, L. & Yaktine, A. (2009). Nanotechnology in Food Products: Workshop Summary. *Food Forum; Institute of Medicine*, ISBN 978-0-309-13772-0

Project on Emerging Nanotechnologies (2011). Inventory of nanotechnology-based consumer products, In: *Project on Emerging Nanotechnologies website*, July 2011, Available from

http://www.nanotechproject.org/inventories/consumer/analysis_draft/>

Qian, K.; Shi, T. Y.; Tang, T.; Zhang, S. L.; Liu, X. L. & Cao, Y. S. (2011). Preparation and characterization of nano-sized calcium carbonate as controlled release pesticide carrier for validamycin against Rhizoctonia solani. *Microchimica Acta*, Vol. 173, No. 1-2, (April 2011), pp. 51-57, ISSN 0026-3672

Rajani, P.; SriSindhura, K.; Prasad, T. N. V. K. V.; Hussain, O. M.; Sudhakar, P.; Latha, P.; Balakrishna, M.; Kambala, V.; Reddy, K. R.; Giri, P. K.; Goswami, D. K.; Perumal, A. & Chattopadhyay, A. (2010). Fabrication of Biogenic Silver Nanoparticles Using Agricultural Crop Plant Leaf Extracts. *Proceedings of International Conference on Advanced Nanomaterials and Nanotechnology*, ISBN 978-0-7354-0825-8, Guwahati, December 2009

Rodríguez, J. M.; Lucero, A.; Rodríguez, M. R.; Mackie, A. R.; Gunning, A. P.; Morris, V. J. (2007). Some implications of nanoscience in food dispersion formulations containing phospholipids as emulsifiers. *Food Chemistry*, Vol. 102, pp. 532–541, ISSN 0308-8146

Rollin, F.; Kennedy, J. & Wills, J. (2011). Review: Consumers and new food technologies. *Trends in Food Science & Technology*, Vol. 22, Issue. 2-3, pp. 99-111, ISSN 0167-7799

Sánchez, C. C. & Patino, J. M. R. (2005). Interfacial, foaming and emulsifying characteristics of sodium caseinate as influenced by protein concentration in solution. *Food Hydrocolloids*, Vol.19 407–416, ISSN 0268-005X

Sánchez-García, M. D.; Lagaron, J. M. & Hoa, S. V. (2010a). Effect of addition of carbon nanofibers and carbon nanotubes on properties of thermoplastic biopolymers. *Composites Science and Technology*, Vol. 70, No. 7, pp. 1095-1105, ISSN 0266-3538

Sánchez-García, M. D.; López-Rubio, A. & Lagaron, J. M. (2010b). Natural micro and nanobiocomposites with enhanced barrier properties and novel functionalities for food biopackaging applications. *Trends in Food Science & Technology*, Vol. 21, pp. 528-536, ISSN 0167-7799

Scheerlinck, J. P. Y.; Gloster, S.; Gamvrellis, A.; Mottram, P.L. & Plebanski, M. (2005). Systemic immune responses in sheep, induced by a novel nano-bead adjuvant. *Vaccine*, Vol.24, No. 8, pp. 1124-1131, ISSN 0264-410X

Seear, K.; Petersen, A. & Bowman, D. (2009) The Social and Economic Impacts of Nanotechnologies: A Literature Review, In: *Department of Innovation, Industry, Science and Research of Australian Government website*, 25 July 2011, Available from:<http://www.innovation.gov.au/Industry/Nanotechnology/NationalEnabli

ngTechnologiesStrategy/Documents/SocialandEconomicImpacts_LiteratureRevie
w.pdf>

Seldon laboratoires. Personal Water treatment: Waterstick, In: *Source: Seldon website*, 04 August 2011, Available from: <http://www.seldontechnologies.com/products. htm>

Senthilnathan, J. & Ligy, P. (2010). Removal of mixed pesticides from drinking water system using surfactant-assisted nano-TiO_2. *Water Air and Soil Pollution*, Vol. 210, pp. 143-154, ISSN 0049-6979

Shen, H. Y.; Zhu, Y.; Wen, X. E. & Zhuang, Y.M. (2007). Preparation of Fe3O4-C18 nano-magnetic composite materials and their cleanup properties for organophosphorus pesticides. *Analytical and Bioanalytical Chemistry*, Vol. 387, No. 6, (March 2007), pp. 2227-2237, ISSN 1618-2642

Shoults-Wilson, W. A.; Zhurbich, O. I.; McNear, D. H.; Tsyusko, O. V.; Bertsch, P. M. & Unrine, J. M., (2011). Evidence for avoidance of Ag nanoparticles by earthworms (Eisenia fetida). *Ecotoxicology*, Vol. 20, No. 2, (March 2011), pp. 385-296, ISSN 0963-9292

Siegrist, M.; Cousin, M-E.; Kastenholz, H. & Wiek, A. (2007). Public acceptance of nanotechnology foods and food packaging: The influence of affect and trust. *Appetite*, Vol. 49, pp. 459–466, ISSN 0195-6663

Siegrist, M.; Stampfli, N.; Kastenholz, H. & Keller, C. (2008). Perceived risks and perceived benefits of different nanotechnology foods and nanotechnology food packaging. *Appetite*, Vol. 51, pp. 283–290, ISSN 0195-6663

Silvestre, C.; Duraccio, D. & Cimmino, S. (2011). Food packaging based on polymer nanomaterials. Progress in Polymer Science,
doi:10.1016/j.progpolymsci.2011.02.003

Siqueira, G.; Bras, J. & Dufresne, A. (2009). Cellulose whiskeys versus microfibrils: influence of the nature of the nanoparticle and its surface functionalization on the thermal and mechanical properties of nanocomposites. *Biomacromolecules*, Vol. 10, No. 2, pp. 425-432, ISSN 1525-7797

Song, S. L.; Liu, X. H.; Jiang, J. H.; Qian, Y. H.; Zhang, N. & Wu, Q. H. (2009). Stability of triazophos in self-nanoemulsifying pesticide delivery system. *Colloids and Surfaces A-Physicochemical and Engineering*, Vol. 350, No. 1-3, (October 2009), pp. 57-62, ISSN 0927 7757

Sorrentino, A.; Gorrasi, G. & Vittoria, V. (2007). Potential perspectives of bionanocomposites for food packaging applications. *Trends in Food Science & Technology*, Vol. 18, pp. 84–95, ISSN 0167-7799

Sundari, P. A. & Manisankar, P. (2011). Development of Nano Poly (3-methylthiophene) /Multiwalled Carbon Nanotubes Sensor for the Efficient Detection of Some Pesticides. *Journal of the Brazilian Chemical Society*, Vol. 22, No. 4, (2011), pp. 746-755, ISSN 0103-5053

The European Parliament and the Council. (December 2008). REGULATION (EC) No 1331/2008 OF THE EUROPEAN PARLIAMENT AND OF THE COUNCIL of 16 December 2008 establishing a common authorisation procedure for food additives, food enzymes and food flavorings, In: *Official Journal of the European Communities*, 8

June 2011, Available from: <http://eur-lex.europa.eu/LexUriServ/LexUriServ.do?uri=OJ:L:2008:354:0001:0006:EN:PDF>

The European Parliament and the Council. (December 2008). REGULATION (EC) No 1332/2008 OF THE EUROPEAN PARLIAMENT AND OF THE COUNCIL of 16 December 2008 on food enzymes and amending Council Directive 83/417/EEC, Council Regulation (EC) No 1493/1999, Directive 2000/13/EC, Council Directive 2001/112/EC and Regulation (EC) No 258/97, In: *Official Journal of the European Communities*, 8 June 2011, Available from: <http://eur-lex.europa.eu/LexUriServ/LexUriServ.do?uri=OJ:L:2008:354:0007:0015:EN:PDF>

The European Parliament and the Council. (December 2008). REGULATION (EC) No 1333/2008 OF THE EUROPEAN PARLIAMENT AND OF THE COUNCIL of 16 December 2008 on food additives, In: *Official Journal of the European Communities*, 8 June 2011, Available from: <http://eur-lex.europa.eu/LexUriServ/LexUriServ.do?uri=OJ:L:2008:354:0016:0033:en:PDF>

The European Parliament and the Council. (December 2008). REGULATION (EC) No 1334/2008 OF THE EUROPEAN PARLIAMENT AND OF THE COUNCIL of 16 December 2008 on flavorings and certain food ingredients with flavoring properties for use in and on foods and amending Council Regulation (ECC) No 1601/91, Regulations (EC) No 2232/96 and (EC) No 110/2008 and Directive 2000/13/EC, In: *Official Journal of the European Communities*, 8 June 2011, Available from:
<http://eur-lex.europa.eu /LexUriServ/ LexUriServ.do?uri=OJ:L:2008:354 :0034: 0050:en:PDF>

The European Parliament and the Council. (January 1997). REGULATION (EC) No 258/97 OF THE EUROPEAN PARLIAMENT AND OF THE COUNCIL of 27 January 1997 concerning novel foods and novel food ingredients, In: *Official Journal of the European Communities*, 8 June 2011, Available from:
<http://eur-lex.europa.eu/LexUriServ/LexUriServ.do?uri=OJ:L:1997:043:0001: 0006:EN:PDF>

The European Parliament and the Council. (January 2002). REGULATION (EC) No 178/2002 OF THE EUROPEAN PARLIAMENT AND OF THE COUNCIL of 28 January 2002 laying down the general principles and requirements of food law, establishing the European Food Safety Authority and laying down procedures in matters of food safety, In: *Official Journal of the European Communities*, 8 June 2011, Available from:
<http://eur-lex.europa.eu/LexUriServ/LexUriServ.do?uri=OJ:L:2002:031:0001 :0024:EN:PDF>

The European Parliament and the Council. (May 2009). COMMISSION REGULATION (EC) No 450/2009 of 29 May 2009 on active and intelligent materials and articles intended to come into contact with food, In: *Official Journal of the European Union*, 8 June 2011, Available from:
<http://eur-lex.europa.eu/LexUriServ/LexUriServ.do?uri=OJ:L:2009:135:0003 :0011:EN:PDF>

The European Parliament and the Council. *(October 2004). REGULATION (EC) No 1935/2004 OF THE EUROPEAN PARLIAMENT AND OF THE COUNCIL of 27 October 2004 on materials and articles intended to come into contact with food and repealing Directives 80/590/EEC and 89/109/EEC, In: Official Journal of the European Communities, 8 June 2011, Available from: <http://eur-lex.europa.eu/LexUriServ/LexUriServ.do?uri=OJ:L:2004:338:0004:0017:en :PDF>*

Torres-Giner, S., Gimenez, E., & Lagaron, J. M. (2008). Characterization of the morphology and thermal properties of zein prolamine nanostructures obtained by electrospinning. *Food Hydrocolloids*, Vol. 22, pp. 601-614

Travan, A.; Pelillo, C.; Donati, I.; Marsich, E.; Benincasa, M. & Scarpa, T. (2009). Non-cytotoxic silver nanoparticle-polysaccharide nanocomposites with antimicrobial activity. *Biomacromolecules*, Vol. 10, No. 6, pp. 1429-1435, ISSN 1525-7797

UNESCO. (2007). Boletín Semanal del Portal del Agua de la UNESCO nº 184: Las Aguas Residuales, Hechos y Cifras sobre Aguas Residuales, In: *UNESCO website*, 08 August 2011, Available from: <http://www.unesco.org/water/news/newsletter /184_es.shtml>

Ursache-Oprisan, M.; Focanici, E.; Creanga, D. & Caltun, O. (2011). Sunflower chlorophyll levels after magnetic nanoparticle supply. *African Journal of Biotechnology*, Vol. 10, No.36, (July 2011), pp. 7092-7098, ISSN 1684-5315

Utracki, L. A.; Broughton, B.; González-Rojano, N.; de Carvalho, L. H. & Achete, C. A. (2011). Clays for polymeric nanocomposites. *Polymer Engineering and Science*, Vol. 51, Nº 3, pp. 559-572, ISSN 0032-3888

Vladimiriov, V., Betchev, C., Vassiliou, A., Papageorgiou, G., & Bikiaris, D. (2006). Dynamic mechanical and morphological studies of isotactic polypropylene /fumed silica nanocomposites with enhanced gas barrier properties. *Composites Science and Technology*, Vol. 66, pp. 2935–2944, ISSN 0266-3538

Walter, M.T., Luo, D., Regan, J.M., 2005. Using nanotechnology to identify and characterize hydrological flowpaths in agricultural landscapes. USDA-CRIS grant proposal 2004-04459

Wang, L. J.; Li, X. F.; Zhang, G. Y.; Dong, J. F. & Eastoe, J. (2007). Oil-in-water nanoemulsions for pesticide formulations. *Journal of Colloid and Interfacial Science*, Vol. 314, No. 1, (October 2007), pp. 230-235, ISSN 0021-9797

Wang, Q.; Chen, J.; Zhang, H.; Lu, M.; Qiu, D.; Wen, Y. & Kong, Q. (2011). Synthesis of Water Soluble Quantum Dots for Monitoring Carrier-DNA Nanoparticles in Plant Cells. *Journal of Nanoscience and Nanotechnology*, Vol. 11, No. 3, (March 2011), pp. 2208-2214, ISSN 1533-4880

Wolfe, J. (2006). Top Five Nanotech Breakthroughs of 2006, In: *Forbes/Wolfe Nanotech Report 2006*, 04 August 2011, Available from: http://www.forbes.com/2006/12/26/nanotech-breakthroughs-ibm-pf-guru-in_jw_1227soapbox_inl.html>

Yada, R. Y. (2011). Nanotechnology. How Science of the Small Could Have Large Implications for Your Food. The 18th Annual FFIGS Educational Workshop: Food Safety Today –A Micro and Macro Perspective!. 4 May, 2011, Available from:

<http://ffigs.org/2011%20R.%20Yada%20Food%20Safety%20Forum%20Ingersoll
%20May%201%202011.pdf>

Yang, H. L.; Lin, J. C. & Huang, C. (2009). Application of nanosilver surface modification
    to RO membrane and spacer for mitigating biofouling in seawater desalination.
    Water Research, Vol. 43, pp. 3777-3786, ISSN 0043-1354

Yang, L.; Chakrabartty, S. & Alocilja, E. (2007). Fundamental building blocks for
    molecular biowire based forward error-correcting biosensors. *Nanotechnology*,
    Vol. 18, N°. 42, pp. 1-6, ISSN 0957-4484

Zeng, R.; Wang, J. G.; Cui, J. Y.; Hu, L. & Mu, K. G. (2010). Photocatalytic degradation of
    pesticide residues with RE(3+)-doped nano-TiO(2). *Journal of Rare Earths*, Vol. 28,
    No. 1, (December 2010), pp. 353-356, ISSN 1002-0721

Zhao, X.; Zhaoyang, L.; Wenguang, L.; Wingmoon, L.; Peng, S.; Richard, Y. T. K.; Keith, D.
    K. L. & William, W.L. (2011). Octaarginine-modified chitosan as a nonviral gene
    delivery vector: properties and in vitro transfection efficiency. *Journal of Naparticle
    Research*, Vol. 13, No. 2, (February 2011), pp. 693-702, ISSN 1572-896X

Zodrow, K.; Brunet, L.; Mahendra, S.; Li, D.; Zhang, A.; Li, Q. & Alvarez, P. J. J. (2009).
    Polysulfone ultrafiltration membranes impregnated with silver nanoparticles
    show improved biofouling resistance and virus removal. *Water Research*, Vol. 43,
    pp. 715-723, ISSN 0043-1354

7

Micro and Nano Corrosion in Steel Cans Used in the Seafood Industry

Gustavo Lopez Badilla[1],
Benjamin Valdez Salas[2] and Michael Schorr Wiener[2]
[1]UNIVER, Plantel Oriente, Mexicali, Baja California
[2]Instituto de Ingenieria, Departamento de Materiales, Minerales y Corrosion,
Universidad Autonoma de Baja California
Mexico

1. Introduction

The use of metal containers for food preservation comes from the early nineteenth century, has been important in the food industry. This type of packaging was developed to improve food preservation, which were stored in glass jars, manufactured for the French army at the time of Napoleon Bonaparte (XVIII century), but were very fragile and difficult to handle in battlefields, so it was decided the produce metal containers (Brody et al, 2008). Peter Durand invented the metallic cans in1810 to improve the packaging of food. In 1903 the English company Cob Preserving, made studies to develop coatings and prevent internal and external corrosion of the cans and maintain the nutritional properties of food (Brody, 2001). Currently, the cans are made from steel sheets treated with electrolytic processes for depositing tin. In addition, a variety of plastic coatings used to protect steel from corrosion and produce the adequate brightness for printing legends on the outside of the metallic cans (Doyle, 2006). This type of metal containers does not affect the taste and smell of the product; the insulator between the food and the steel, is non-toxic and avoid the deterioration of the food. The differences between metal and glass containers, as well as the negative effects that cause damage to the environment and human health are presented in Table 1.

The wide use of steel packaging in the food industry, from their initial experimental process, has been very supportive to keep food in good conditions, with advantages over other materials such as glass, ceramics, iron and tin. The mechanical and physicochemical properties of steel help in its use for quick and easy manufacturing process (Brown, 2003). At present, exist a wide variety of foods conserved in steel cans, but in harsh environments, they corrode. Aluminum is used due to its better resistance to corrosion, but is more expensive. With metal packaging, the food reaches to the most remote places of the planet, and its stays for longer times without losing its nutritional properties, established and regulated for health standards by the Mexican Official Standards (NOM). The difference between using metal cans to glass (Table 1) indicate greater advantages for steel cans (Finkenzeller, 2003). In coastal areas, where some food companies operate, using steel cans, three types of deterioration are detected: atmospheric corrosion, filiform corrosion and microbiological corrosion. Even with the implementation of techniques and methods of protection and use of metal and plastic coatings, corrosion is still generated, being lower with the use of plastics (Lange et al, 2003). Variaitons of humidity and temperature deteriorate steel cans (Table2).

1.1 Steel

Steel is the most used metal in industrial plants, for its mechanical and thermal properties, and manufacturing facility. It is an alloy of iron and carbon. Steel manufacturing is a key part of the Mexican economy. Altos Hornos is the largest company in Mexico, with a production of more than 3,000000 tons per year, located in Monclova, Coahuila, near the U.S. border (AHMSA, 2010). Steel is used in the food industry, especially in the packaging of sardines and tuna (Lord, 2008).

PROPERTIES AND UNPROPERTIES		NEGATIVE EFFCTS	
METAL	**GLASS**	**METÁL**	**GLASS**
Resist the irregular handling and transport	Fragile and easily broken	Generation of filiform and microbiological corrosion	Cause spots of black color
Hermetically sealed	Not sealed; air enters	Bad sealed, creates rancidity by microbiological corrosion	High percentage of microorganisms by poor seal
Good shelf life without refrigeration at room temperature	Necessity of refrigeration of marine food	At warm and cold temperatures, foods lose their nutritional properties	At warm and cold temperatures, foods lose their nutritional properties
Accessibility manufacture	Manufacturing process complex by its fragility	By bad handling and the internal deterioration of the coating, generates filiform corrosion	Cover deformation generates gas food deterioration
No frequent supervision	Frequent supervision	Susceptible to atmospheric corrosion in indoor and outdoor environments	Broken pieces of glass are mixed with food, generating health damage
Easy recycling	Difficult to recycle	Sterilization time is 20 minutes	In sterilizing process, glass cans remains in hot water for 10 to 15 minutes and can generate bacteria

Table 1. Differences between metallic cans and glass containers in the food industry and their effect on health and environment

1.2 Metallic cans

The steel cans consist of two parts: body and ring or three parts: body, joint and ring (Figures 1a and 1b).

When a steel can is not properly sealed, it is damaged by drastic variations of humidity and temperature creating microorganisms, which cause an injury on the health of consumers (Cooksey, 2005). Every day millions of cans are produced, the companies express their interest in research studies to improve their designs. There are two main types of steel cans: tin plated and plastic coated. Plastic coatings have good resistance to compression, and the resistance to corrosion is better than the tin plate. Since the oxide layer that forms on the container surface is not completely inert, the container should be covered internally with a health compatible coating (Nachay, 2007).

Corrosion	Climatic factors	Coatings	
		Metallic	Plastic
Atmospheric External	High levels of humidity and temperature	In aggressive environments, is generated external and internal damage of steel cans	Originates stains and bad appearance without damage
Filiform Internal	Low levels of humidity	In harsh environments, are generated cracks under the coatings and, is formed the filiform corrosion	No formation of cracks in coatings as in the steel cans
Microbiological Internal	High levels of humidity	Dense black spots are formed by OH- and rancidity	Isolated black spots

Table 2. Corrosion types in coated metallic cans used in the food industry

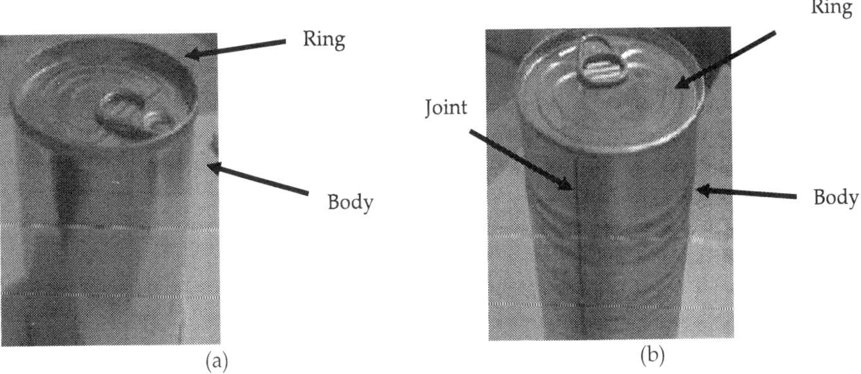

Fig. 1. (a) Aluminum cans without seams, in two parts: ring and body (b) Steel cans with seams: body, joint and rings

1.3 Production stages

The manufacture stages in a food industry are shown in Figure 2 (Avella, 2005):

Washing: Cans are cleaned thoroughly to remove the bacteria that could alter the food nutritional value.

Blanching: The product is subjected to hot water immersion to remove the enzymes that produce food darkening and the microorganisms that cause rancidity.

Preparation: Before placing food in the can the non-consumable parts of the sardine and tuna are removed, then the ingredients to prepare the food in accordance with the consumption requirements are added.

Packaging: The food is placed in the can, adding preservatives such as vinegar, syrup, salt and others to obtain the desired flavor.

Air removal: The can pass through a steam tunnel at 70 ° C, to avoid bad taste and odor.

Sealing: by soldering or with seams.

Sterilization: It is of great importance for the full elimination of microorganisms that might be left over from the previous stages, when the can is treated at temperatures of 120 ° C.

Cooling. Once sterilized the cans are cooled under running cold water or cold water immersion, from the outside without affecting the food quality.

Labeling. On the can label are placed legends with product ingredients, expiration dates and lot numbers of production.

Packaging, is made to organize the food steel cans in boxes.

Food technology specialists considers, that an adequate manufacturing process of canned foods, helps to keep certain products up to several months and years, as the case of milk powder to nine months, some vegetables and meat foods two and up to five years. A diagram summarizing all these stages is displayed in Figure 2.

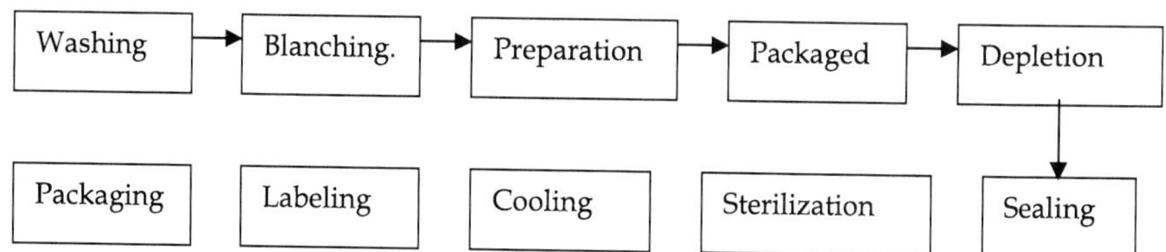

Fig. 2. Manufacturing steps in a food industry.

1.4 Sea food industry in Mexico

The main coastal cities in Mexico, with installed companies that fabricate metallic cans for sardines and tuna conservation are Acapulco, Guerrero, Ciudad del Cabo in the State of Baja California Sur; Ensenada, Baja California, Campeche, Campeche, Mazatlan, Sinaloa, Veracruz, Veracruz (Bancomext, 2010). The sardine is a blue fish with good source of omega-3, helping to lower cholesterol and triglycerides, and increase blood flow, decreasing the risk of atherosclerosis and thrombosis. Due to these nutrition properties, its widely consumed in Mexico; it contains vitamins B12, niacin and B1, using its energy nutrients (carbohydrates, fats and proteins) as a good diet. This food is important in the biological processes for formation of red blood cells, synthesis of genetic material and production of sex hormones. Tuna is an excellent food with high biological value protein, vitamins and minerals. It has minerals such as phosphorus, potassium, iron, magnesium and sodium and vitamins A, D, B, B3 and B12, which are beneficial for the care of the eyes and also provides folic acid to pregnant women. Fat rich in omega-3, is ideal for people who suffer from cardiovascular disease (FAO, 2010).

1.5 Atmospheric corrosion

Atmospheric corrosion is an electrochemical phenomenon that occurs in the wet film formed on metal surfaces by climatic factors (Lopez et al, 2011, AHRAE, 1999). One factor that determines the intensity of damage in metals exposed to atmosphere is the corrosive chemical composition in the environments. The sulphur oxides (SO_X), nitrogen oxides (NO_X), carbon oxide (CO) and sodium chloride (NaCl) that generates chloride ions (Cl^-), are the most common corrosive agents. The NaCl enters to the atmosphere from the sea; SO_X, NO_X and CO, is emitted by traffic vehicle. The joint action of the causes of pollution and weather determine the intensity and nature of corrosion processes, acting simultaneously, and increasing their effects. It is also important to mention other factors such as exposure conditions, the metal composition and properties of oxide formed, which combined, have an influence on the corrosion phenomena (Lopez, 2008). The most important atmospheric feature that is directly related to the corrosion process is moisture, which is the source of electrolyte required in the electrochemical process. In spite of existing corrosion prevention and protection systems as well as application of coatings in steel cans the corrosion control, is not easy in specific climatic regions, especially in marine regions. Ensenada which is a marine region of Mexico on the Pacific Ocean has a marine climate with cold winter mornings around 5 °C and in summer 35° C. Relative humidity (RH) is around 20% to 80%. The main climate factors analyzed were humidity, temperature and wind to determine the time of wetness (TOW) and the periods of formation of thin films of SO_X and Cl^- which were analyzed to determine the corrosivity levels (CL) in outdoors and indoors of seafood industry plants (Lopez et al, 2010).

1.6 Corrosion of steel cans

Corrosion of tinplate for food packaging is an electrochemical process that deteriorates the metallic surfaces (Ibars et al, 1992). The layer of tin provides a discontinuous structure, due to their porosity and mechanical damage or defects resulting from handling the can. The lack of continuity of the tin layer allows the food, product to be in contact with the various constituents of the steel, with the consequent formation of galvanic cells, inside of the cans. The presence of solder alloy used in the conventional container side seam is a further element in the formation of galvanic cells. Corrosion of tin plate for acidic food produces the dissolution of tin and hydrogen gas formation resulting from the cathodic reaction that accumulate in the cans. At present, the problems arising from the simultaneous presence of an aggressive environment, mechanical stress and localized corrosion (pitting) are too frequent (CGB, 2007).

1.7 Coatings

The food in steel can is protected by a metallic or plastic coating regulated by the FDA (Food Drugs Administration, USA) that does not generate any health problems in consumers (Weiss et al, 2006). The coating is adhered on the metal plate and its function is due to three main features:

- Thermal and chemical resistance assures the protection of the steel surface when a food produces a chemical attack by rancidity, changing the food taste.
- Adherence. The coating is easily attached to the inside can surface.
- Flexibility. Resistance to mechanical operations that modify the structure of the can, in the manufacturing process, such as molding shapes and bad handling.

Currently, new materials and coatings are analyzed to fit them to food variety, beverages and other canned products (Table 3). The coatings used in the food industry are organosol type, with high solids content, creating dry films with thickness 10 to 14 g/m^2, for manufacturing or recycling, allowing large deformation (Soroka, 2002). To improve the strength of steel, two layers of epoxy-phenolic are applied, in the organosol film. If the food suffers decomposition, it generates deformation in the can (Yam et al, 2005). Coatings are applied on the cans on the inside and outside. Since the early twentieth century, coatings manufacturers have supported the food and beverage industries, using oleoresins resins, phenolic and later in 1935, was applied vinyl coating in the beer cans. Later comes the epoxy-phenolic coatings, organosoles, acrylic and polyester (Ray, 2006).

CLASIFICACTION	
COATINGS	**DEFINITION**
Protection in indoors of cans	They are in contact with the packaged product and are called "health coatings"
Exterior coatings pigmented	Are used to underlie the decorative printing of packaging, called "white enamel" or "white lacquer. "
Exterior coatings transparent	Are used to underlie print, called "coatings of hitch. " Protects the printing inks of defectives manipulations, known as "clear coatings".
FUNCTIONS	
Protects the metal of the food	
Protects the product from contamination when steel parts from the can, are detached	
Facilitates the production	
Provides a basis for decoration	
Acts as a barrier against external corrosion and abrasion	
CHARACTERISTICS	
Must be compatible with the packaged product and to resist their aggressiveness	
Have a high adhesion to the tin or other metal	
Are free of toxic substances	
Not affect the organoleptic characteristics of the packaged product	
Not contain any items prohibited by the health legislation	
Is resistant of the sterilization and / or treatment, that is subject to packaging the product	
Adequately support to the welding of the body in two and three pieces of containers	
TYPES	
METALLIC	**PLASTICS**
Tin compound	Oleoresins, phenolic, epoxy, vinyl, acrylic, polyester
COMPOSITION	
COATING MATERIAL	**COMPOSITION**
Acrylic Polyester Copolymer	The polyester is inserted in the acrylic
Polyester resin	The resin is unsaturated
Crosslinker	The polyester contains acrylic acid, or maleic anhydride with styrene.

Table 3. Coatings used in the food industry

2. Materials and methods

2.1 Climate factors

The climate is composed of several parameters; RH and temperature are the most important factors in the damage of steel cans. Scientists that analyze the atmospheric corrosion, consider that the grade of deterioration of steel cans is due to the drastic changes in the humidity and temperature in certain times of the year, as expressed in ISO 9223 (ISO 9223, 1992). Managers and technicians of companies and members of health institutions in Mexico are concerned in some periods of the year, by the quality of seafood contained in steel cans (Moncmanova; 2007).

2.2 Corrosion testing

Pieces of steel rolls were prepared for corrosion testing simulating steel cans, which were exposed at indoor conditions of seafood plants for periods of one, three, six and twelve months in Ensenada, following the ASTM standards G 1, G 4, G 31 (ASTM, 2000). The results were correlated with RH, TOW and temperature parameters. The concentration levels of SO_X and Cl- were evaluated with the sulfatation technique plate (SPT) and wet candle method (WCM), (ASTM G 91-97, 2010; ASTM G140-02, 2008). The industrial plants of seafood in this city are located at distances at 1 km to 10 from the sea shore. Steel plates used to fabricate steel cans with dimensions of 3 cm. X 2 cm. and 0.5 cm of width, were cleaned by immersion in an isopropyl alcohol ultrasound bath for 15 minutes (ISO 11844-1, ISO 11844-2, Lopez et al, 2008). Immediately after cleaning the steel probes were placed in sealed plastic bags, ready to be installed in the test indoor and outdoor sites. After each exposure period the steel specimens were removed, cleaned and weighed to obtain the weight loss and to calculate the corrosion rate (CR).

2.3 Examination techniques

The corrosion products morphology was examined by the scanning electron microscope (SEM) and the Auger Electron Spectroscopy (AES) techniques.

- SEM. Used to determine the morphology of the corrosion products formed by chemical agents that react with the steel internal an external surface. The SEM technique produces very high-resolution images of a sample surface. A wide range of magnifications is possible, from about 10 times to more than 500,000 times. The SEM model SSX-550 was used; revealing details less than 3.5 nm, in size from 20 to 300,000 magnifications and 0.5 V to 30kV by step.

- AES. It determines the chemical composition of elements and compounds in the steel cans and rolls, and analyzes the air pollutants deposited on the steel. With this technique we knew in detail, quickly and with a good precision, the structural form and location of corrosion at surface level which determined the type of corrosion (Clark,et al, 2006). AES analysis was performed in Bruker Quantax and ESCA / SAM 560 models, and the bombarding were obtained when samples with a beam of electrons with energy of 5keV. We made a clean surface of steel specimens analyzed with an ion beam with energy Ar $^+$ 5keV and current density of 0.3 uA / cm^3 to remove CO_2 from the atmosphere (Asami et al, 1997). The sputtering process indicates the type of film formed on the metallic surface of steel and the corrosion on separated points such as pitting corrosion.

2.4 Numerical analysis

A mathematical correlation was made applying MatLab software to determine the CL in indoors of seafood industry in Ensenada in summer and winter (Duncan et al, 2005). With this simulation we find out the deterioration grade of steel probes, correlating the climate factors (humidity and temperature) and air pollutants (CO, NO_X and SO_X), with the corrosion rate (CR).

3. Results

The generation of corrosion in steel cans is promoted by the formation of the thin film of corrosion products in their surface and the exposition of chlorides and sulfides. The seafood industry is concerned with the economic losses caused by bad appearance of the containers and the loss of nutritional properties of sardine and tuna.

3.1 Deterioration of steel cans

Levels of humidity and temperature bigger than 75% and 35 °C accelerated the CR. In summer the CR was higher after one year. For temperatures in the range from 25 °C to 35 °C, and RH level of 35% to 75%, the CR was very high. Furthermore, in winter, at temperatures around 10 °C to 20 °C and RH levels from 25% to 85%, water condensates on the metal surface and the CR increases very fast. Variations of RH in the range from 25% to 75% and temperatures from 5 °C to 30 °C, and the concentration levels of air pollutants such as sulfides and chlorides, which exceeds the permitted levels of the air quality standard (AQS), increase the corrosion process. In the autumn and winter, corrosion is generated by a film formed uniformly on the steels (Lopez et al, 2010). Exposition to SO_2 indicates more damage, compared with the effect of the chlorides on the steel surface. The maximum CR representing the deterioration with steel exposed to SO_2 was in winter for the high concentration levels of RH and the minimum was in spring. The major effect of Cl- on the deterioration of metallic surface occurred in winter and the minimum was in spring, same with the exposition of SO_2 (Table 4).

Climate factors	Sulphur oxide (SO_2) RH[a] T[b] C[c] CR[d]				Chloride (Cl-) RH[a] T[b] C[c] CR[d]			
Spring								
Max	76.3	21.4	0.24	68.8	73.5	25.6	234	59.3
Min	24.3	13.5	0.20	35.7	24.3	13.3	122	24.8
Autumn								
Max	82.7	29.8	0.31	145.7	78.9	30.6	267	136.7
Min	28.4	20.7	0.18	109.8	16.7	15.8	187	99.8
Winter								
Max	88.9	23.4	0.51	205.6	84.3	31.2	299	178.9
Min	23.2	15.6	0.36	144.6	22.1	13.7	197	122.3

[a] RH. Relative Humidity (%), [b] T. Temperature (°C), [c] C. Concentration Level of Air Pollutant (ppm), [d] CR- Corrosion rate (mg/m2.year).; Source. TPS and WCM.

Table 4. Effect of RH, temperature and air pollutants on the CR of steel (2010)

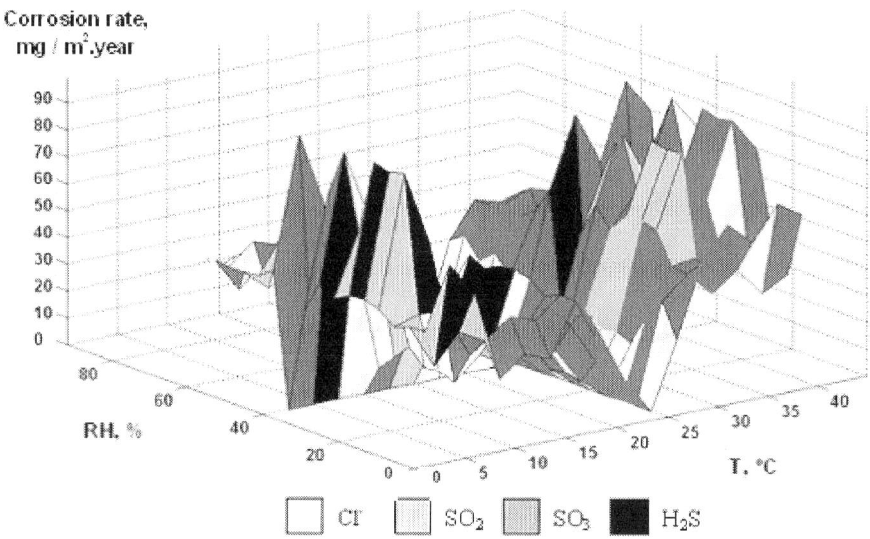

Fig. 3. CL of steel during summer exposition in Ensenada.

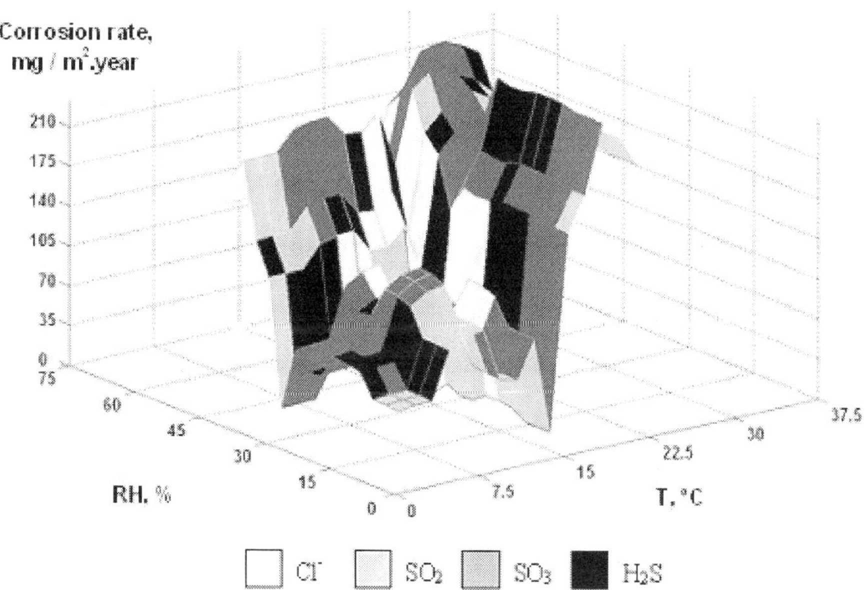

Fig. 4. CL of steel during winter exposition in Ensenada.

3.2 Corrosivity analysis

A computer model of atmospheric corrosion has been used to simulate the steel exposed to air pollutants: Cl⁻, SO_2, NO_2, O_3 and H_2S from a thermoelectric station located between Tijuana and Ensenada. RH was correlated with the major CR was 35% to 55% with temperatures of 20°C to 30°C. In summer CR was different than in winter, and in both environments (Figures 3 and 4). Air pollutants such as Cl⁻, NO_2 and sulfide penetrate through defects of the air conditioning systems. Figure 3 shows the CL analysis of indoors in summer, indicating the level 1, as the major aggressive environment and levels 4 the low aggressiveness grade which generate high deterioration grade of this type of materials. Some sections of the Figure 4, represents the different grades of aggressiveness, with high areas of level 1 and 2 but levels 3 and 4 exists in less percentage. RH and temperature ranges were from 25% to 80% and 20 ° C to 30° C with CR from 30 mg / m².year to 100 mg / m².year with RH and temperatures from 40% a 75% and 20 ° C to 35 ° C, with CR from 10 to 160 mg / m².year.

3.3 SEM analysis

The steel samples of 1, 3, 6 and 12 months show localized corrosion with small spots during the summer period and more corroded areas with uniform corrosion in the winter. Air pollutants that react with steel surface form corrosion products, in some zones of steel cans and rolls with chloride ions (light color) and other with sulfides (dark color), as shown in the AES analysis. Some corrosion products in the internal of steel cans appeared on the surface contaminating the sardine (Figures 5 and 6). Various microorganisms and microbial metabolites are human pathogens in sardine and tuna conserved in steel cans were detected (Figures 7 and 8). According to the most common source of these organisms, they can be grouped as follows:

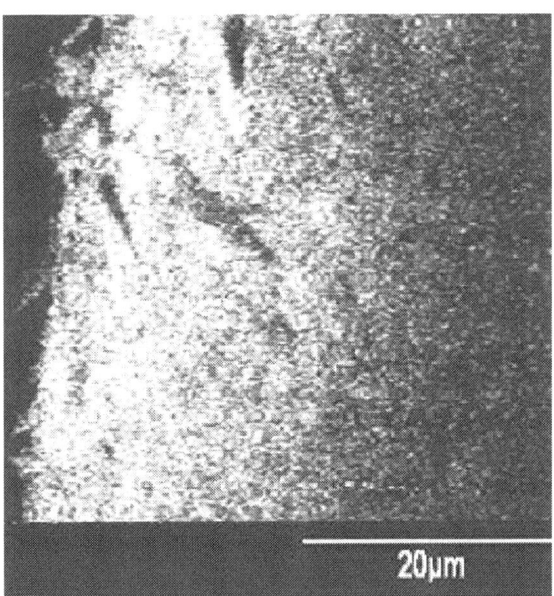

Fig. 5. Sardine contaminate with tin plate steel corrosion products.

Fig. 6. Filiform corrosion formed in internal of tin plate steel cans

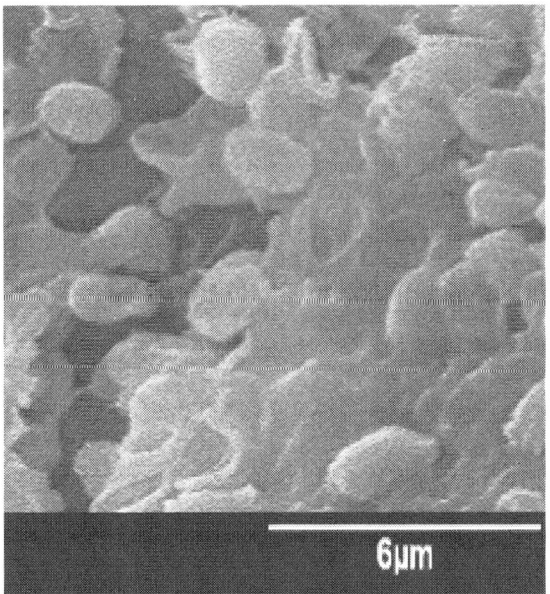

Fig. 7. Microbiological corrosion in internal of steel cans with plastic coatings.

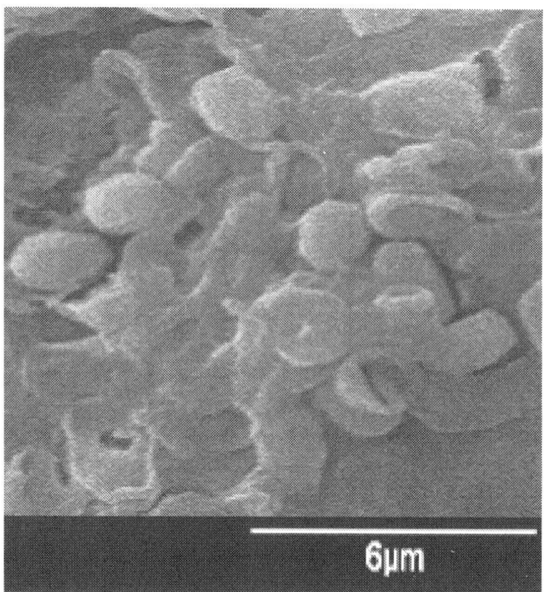

Fig. 8. Microbiological corrosion in internal of steel cans with tin coatings.

1. Endogenous. Originally present in the food before collection, including food animal, which produces the zoonoses diseases, transmitted from animals to humans in various ways, including through the digestive tract through food.
2. Exogenous. Do not exist in the food at the time of collection, at least in their internal structures, but came from the environment during production, transportation, storage, industrialization. Fungi are uni-or multicellular eukaryotic type, their most characteristic form is a mycelium or thallus and hyphae that are like branches.

3.4 AES examination

AES analyses were carried out to determine the corrosion products formed in indoor and outdoor of the steel cans. Figure 9a show scanning electron micrograph (SEM) images of areas selected for AES analysis covered by the principal corrosion products which are rich in chlorides and sulfides in tin plate steel cans evaluated. The Auger map process was performed to analyze punctual zones, indicating the presence of Cl- and S^{2-} as the main corrosive ions present in the steel corrosion products. The Auger spectra of steel cans was generated using a 5keV electron beam (Clark et al, 2006), which shows an analysis of the chemical composition of thin films formed in the steel surface (Figure 9b). The AES spectra of steel cans in the seafood plants show the surface analysis of two points evaluated in different zones of the steel probes. The peaks of steel appear between 700 and 705 eV, finding the chlorides and sulfides. In figure 10, the spectra reveals the same process as in figure 9 wit plastic coatings, with variable concentration in the chemical composition. In the two regions analyzed, where the principal pollutant was Cl- ion. In the region of steel surface were observed different concentrations of sulfide, carbon and oxygen, with low levels concentrations of H_2S, which damage the steel surface.

The standard thickness of 300 nm of tin plate and plastic coatings of internal and external of steel cans was determined by the AES technique with the sputtering process.

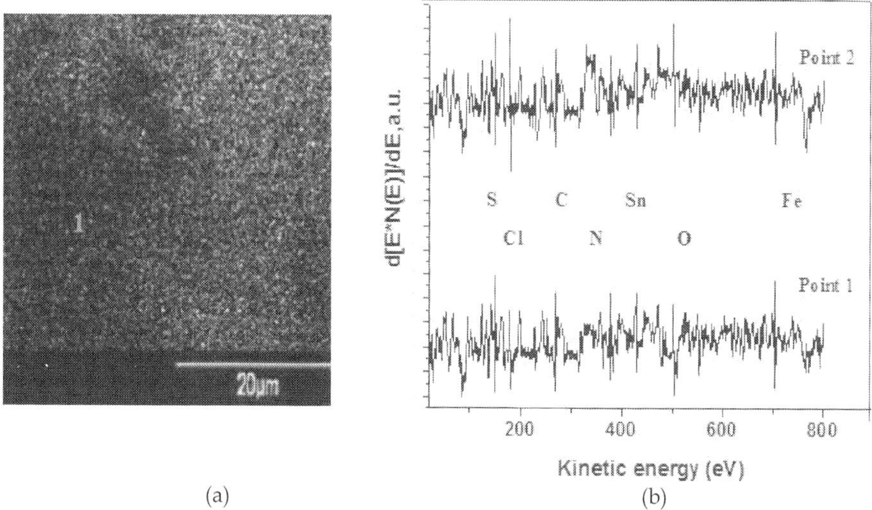

Fig. 9. Corrosion products of tin plated steel: (a) SEM microphotograph and (b) AES analysis, three months exposure.

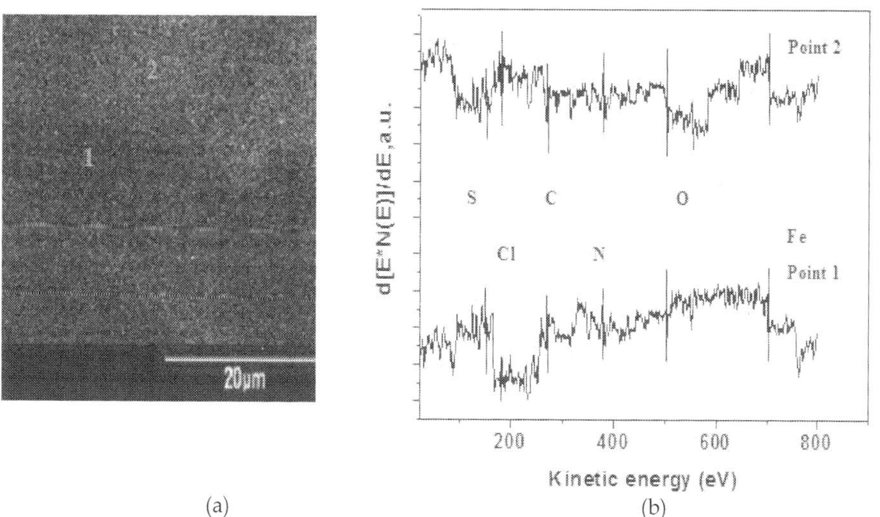

Fig. 10. Corrosion products of plastic coatings: (a) SEM microphotograph and (b) AES analysis, three months exposure.

4. Conclusions

Corrosion is the general cause of the destruction of most of engineering materials; this destructive force has always existed. The development of thermoelectric industries, which generates electricity and the increased vehicular traffic, has changed the composition of the atmosphere of industrial centers and large urban centers, making it more corrosive. Steel production and improved mechanical properties have made it a very useful material, along with these improvements, but still, it is with great economic losses, because 25% of annual world steel production is destroyed by corrosion. The corrosion of metals is one of the greatest economic losses of modern civilization. Steel used in the cannery industry for seafood suffer from corrosion. The majority of seafood industries in Mexico are on the coast, such as Ensenada, where chloride and sulfide ions are the most aggressive agents that promote the corrosion process in the steel cans The air pollutants mentioned come from traffic vehicles and from the thermoelectric industry, located around 50kms from Ensenada. Plastic coatings are better than tin coating because, on the plastic coatings do not develop microorganisms and do not damage on the internal surface.

5. References

AHRAE; Handbook; Heating, Ventilating and Ari-Conditioning; applications; *American Society of Heating*, Refrigerating and Air-Conditioning Engineers Inc.; 1999.

Altos Hornos de Mexico, Acero AHMSA para la industria petrolera y de construccion; www.ahmsa.com, consulted, may 2011.

Annual Book of ASTM Standards, 2000, Wear and Erosion: Metal Corrosion, Vol. 03.02.

Asami K., Kikuchi M. and Hashimoto K.; An auger electron spectroscopic study of the corrosion behavior of an amorphous $Zr_{40}Cu_{60}$ alloy; *Corrosion Science*; Volume 39, Issue 1, January 1997, Pages 95-106; 1997.

ASTM G140-02; Standard Test Method for Determining Atmospheric Chloride Deposition Rate by Wet Candle Method; 2008.

ASTM G91–97; Standard Practice for Monitoring Atmospheric SO2 Using the Sulfation Plate Technique (SPT); 2010.

Avella M, De Vlieger JJ, Errico ME, Fischer S, Vacca P, Volpe MG.; Biodegradable starch/clay nanocomposite films for food packaging applications. Food Chem; 93(3):467–74; 2005.

BANCOMEXT, Datos de producción pesquera en México;

Brody A, Strupinsky ER, Kline LR. Odor removers. In: Brody A, Strupinsky ER, Kline LR, editors. Active packaging for food applications. Lancaster, Pa.: Technomic Publishing Company, Inc. p 107–17; 2001.

Brody Aaron L., Bugusu Betty, Han Jung h., Sand Koelsh, Mchugh Tara H.; Innovative Food Packing Solutions; Journal of Food Science; 2008.

Brown H, Williams; Packaged product quality and shelf life. In: Coles R, McDowell D, Kirwan MJ, editors. Food packaging technology. Oxford, U.K.: Blackwell Publishing Ltd. p 65–94; 2003.

Canning Green Beans (CGB); Ecoprofile of Truitt Brothers Process; Institute for Environmental Research and Education; 2007.

Clark A. E., Pantan C. G, Hench L. L; Auger Spectroscopic Analysis of Bioglass Corrosion Films; *Journal of the American Ceramic Society*; Volume 59 Issue 1-2, Pages 37–39; 2006.

Cooksey K.; Effectiveness of antimicrobial food packaging materials. Food Addit Contam 22(10):980-7; 2005.

Doyle ME. ; Nanotechnology: a brief literature review. Food Research Institute Briefings [Internet]; http://www.wisc.edu/ fri/briefs/FRIBrief Nanotech Lit Rev.pdf; 2006.

FAO, Corporate Repository Report; consulted in

Finkenzeller K.; RFID handbook: fundamentals and applications. 2nd ed. West Sussex, U.K.: JohnWiley & Sons Ltd. 452 p.; 2003.
  http://www.fao.org/documents/en/Fisheries%20and%20aquaculture%20manage ment%20and%20conservation/topicsearch/3, 2011.
  http://www.financierarural.gob.mx/informacionsectorrural/Documents/Sector% 20pesquero/SectorPesqueroM%C3%A9xicoFR07.pdf, consulted, june 2011.

Ibars JR, Moreno DA, Ranninger C.; Microbial corrosion of stainless steel; Microbiologia. Nov;8(2):63-75; 1992.

ISO 11844-1:2006. Corrosion of metals and alloys - Classification of low corrosivity of indoor atmospheres- Determination and estimation of indoor corrosivity. ISO, Geneva, 2006.

ISO 11844-2:2005. Corrosion of metals and alloys - Classification of low corrosivity of indoor atmospheres - Determination and estimation attack in indoor atmospheres. ISO, Geneva, 2005.

ISO 9223:1992, Corrosion of metals and alloys, Corrosivity of Atmospheres, Classification.

Lange J, Wyser Y.; Recent innovations in barrier technologies for plastic packaging—a review. Packag Technol Sci 16:149–58.; 2003.

Lopez B. Gustavo, Valdez S. Benjamin, Schorr W. Miguel, Zlatev R., Tiznado V. Hugo, Soto H. Gerardo, De la Cruz W.; AES in corrosion of electronic devices in arid in marine environments; *AntiCorrosion Methods and Materials*; 2011.

Lopez B.G.; Ph.D. Thesis; Caracterización de la corrosión en materiales metálicos de la industria electrónica en Mexicali, B.C., 2008 (Spanish).

Lopez B.G.; Valdez S. B.; Schorr M. W.; "Spectroscopy analysis of corrosion in the electronic industry influeced by Santa Ana winds in marine environments of Mexico"; INTECH Ed. INDOOR AND OUTDOOR POLLUTON, 4; Edited by Jose A. Orosa, Book, 2011.

Lord JB.; The food industry in the United States In: Brody AL., Lord J, editors. Developing new food products for a changing market place. 2nd ed. Boca Raton, Fla.: CRS Press. p 1–23; 2008.

Moncmanova A. Ed. ; Environmental Deterioration of Materials, WITPress, pp 108-112; 2007.

Nachay K. ; Analyzing nanotechnology. Food Tech 61(1):34–6; 2007.

Ray S, Easteal A, Quek SY, Chen XD; The potential use of polymer-clay nanocomposites in food packaging. Int J Food Eng 2(4):1–11; 2006

Soroka, W, "Fundamentals of Packaging Technology", Institute of Packaging Professionals (IoPP), ISBN 1-930268-25-4; 2002.

Walsh, Azarm, Balachandran, Magrab, Herold & Duncan Engineers Guide to MATLAB, Prentice Hall, 2010, ISBN-10: 0131991108.

Weiss J, Takhistov P, McClements J.; Functional materials in food nanotechnology; J. Food Science; 71(9):R107–16; 2006.
Yam KL, Takhistov PT,Miltz J.; Intelligent packaging: concepts and applications; J Food Sci 70(1):R1–10; 2005.

8

Characteristics and Role of Feruloyl Esterase from *Aspergillus Awamori* in Japanese Spirits, '*Awamori*' Production

Makoto Kanauchi
Miyagi University
Japan

1. Introduction

Feruloyl esterases (EC 3.1.1.73), known as ferulic acid esterases, which are mainly from *Aspergillus* sp. (Faulds & Williamson, 1994), can specifically cleave the (1→5) ester bond between ferulic acid and arabinose. The esterases show high specificity of hydrolysis for synthetic methyl esters of phenyl alkanoic acids (Kroon and others, 1997). The reaction rate increases markedly when the substrates are small soluble feruloylated oligosaccharides derived from plant cell walls (Faulds and other, 1995; Ralet and others, 1994).

These enzymes have high potential for application in food production and other industries. Ferulic acid links hemicellulose and lignin. In addition, cross-linking of ferulic acids in cell wall components influences wall properties such as extensibility, plasticity, and digestibility, as well as limiting the access of polysaccharides to their substrates (Borneman et al., 1990).

Actually, feruloyl esterase is used for *Awamori* spirit production. *Awamori* spirits are Japanese spirits with a distinctive vanilla-like aroma. Feruloyl esterase is necessary to produce that vanilla aroma. Actually, lignocellulosic biomass is one means of resolving energy problems effectively. It is an important enzyme that produces bio-fuel from lignocellulosic biomass.

As explained in this paper, *Awamori* spirit production is described as an application of feruloyl esterase. The vanillin generating pathway extends from ferulic acid as precider, with isolation of *Aspergillus* producing feruloyl esterase, which is characteristic of the enzyme. Moreover, the application of feruloyl esterase for beer production and bio-fuel production is explained.

2. *Awamori* spirits

2.1 *Awamori* spirit characteristics

Awamori spirits have three important features. First, mash of *Awamori* spirit is fermented using *koji*, *Aspergillus* sp. are grown on steamed rice, which is the material and saccharifying agent used in *Awamori* spirit production. That fermentation is done in a pot still. Mash used in *Awamori* spirit processing is different from beer brewing, in which fermenting is done with saccharified mash by malt. Their fermentative form is call 'parallel fermentation' which progresses simultaneously with saccharification and fermentation. The resultant fermentative yeast can produce high concentrations of ethanol, approximately 16–18%, from mash of *Awamori* without osmotic injury.

Secondly, highly concentrated citric acid is produced in this process from *koji* made by black *Aspergillus* sp., classified as *Aspergillus awamori*. Because of this acid, the mash maintains low pH. It is usually made in the warm climate of Okinawa, with average temperatures of 25°C in all seasons. Koizumi (1996) describes that spoiling bacteria are able to grow in mash under pH 4.0 conditions. Moreover, although amylase from *Aspergillus oryzae* is inactivated at less than pH 3.5, that from *Aspergillus awamori* reacts stably at pH 3.0. Furthermore, the mash ferments soundly under those warm conditions.

Finally, aging is an important feature of *Awamori* spirits, which have a vanilla aroma that strengthens during aging. The *Awamori* spirit is aged in earthen pots for three years or more. Particularly, the spirit aged for more than three years, called '*Kusu*', is highly prized. The vanilla aroma in Scotch whisky, bourbon, or brandy is produced from lignin in barrel wood during aging. *Kusu* is not aged in barrels, but it does have a vanilla aroma resembling those of aged Scotch whisky, bourbon, and brandy.

Differences between *Awamori* spirit and other beverages are shown in the table. History and production methods of *Awamori* spirits are described below.

	Awamori	Sake	Whisky	Brandy
Type	Distilled beverage	Brewed beverage	Distilled beverage	Distilled beverage
Place	Okinawa	Mainly Japan	Worldwide	Worldwide
Production Temperature	All seasons average annual temperature (25°C)	Mainly winter 0–4°C	Room temperature (10–15°C)	Room temperature (15–20°C)
Material	*Indica* rice	*Japonica* rice	Barley, corn	Grapes
Mash	Parallel Fermentation Containing citric acid produced by	Parallel fermentation, Containing lactic acid produced by Lactic acid	Single Fermentation Not Containing Acid	Single Fermentation containing Malic acid from Material
Saccharifying agent	*Koji*	*Koji*	Malt	-
Microorganisms	*Aspergillus awamori* *Awamori* yeast	*Aspergillus oryzae* Sake yeast Lactic acid	Whisky yeast	Wine yeast

Fermentative Temperature	High temperature (27–30°C)	Low temperature (10–15°C)	Middle temperature (15–25°C)	Middle temperature (15–26°C)
Alcohol concentration	25–30%	15–16%	40–50%	40–50%
Aging period	Approximately 3 years or more	Very short term	More than 3 years	More than 3 years
Aging vessel	Mainly earthen pot	Mainly stainless tank	Barrel	Barrel
Taste and aroma	Vanilla like	Estery, fruit-like	Vanilla like	Vanilla like

Table 1. *Awamori* spirit and other alcoholic beverages

2.2 History of *Awamori*

Awamori spirits are traditionally produced in Okinawa, which has 47 production sites. *Awamori* spirits are produced from long-grain rice and rice imported from Thailand. Partly because it uses long grain rice imported from Thailand for production, it is believed that *Awamori* spirit production methods were brought from Thailand (Koizumi, 1996).

According to one account (Koizumi, 1996) of Okinawa's history, 'Ryukyu' was an independent country ruled by king Sho in 1420, which traded with the countries of Southeast Asia. At the time, the port of Naha bustled as a junction port between Japan and the South China Sea Islands, Indonesia, Cambodia, Vietnam, the Philippines, and Thailand. *Awamori* spirits were brought from there and also traded. In 1534, 'Chen Kan's Records', reported to his home country, China, noted that *Awamori* spirits have a clear aroma and were delicious; he noted also that *Awamori* spirits had been brought from Thailand.

Moreover, it was written that long-grain rice harvested in Thailand was used in *Awamori* spirit production, and the distilled spirits were aged in earthen pots. Their ancient technology of *Awamori* spirit production is followed by the present technology. Distillation technology was brought also via Thailand from China, as it was with *Awamori* spirits. Furthermore, they transported the technology eventually to the main islands of Japan.

The cradle of distillation technology is actually ancient Rome. In that era, distillation methods were used to produce essential oils in the following manner: plant resin was boiled in a pan on which a wool sheet had been placed. After boiling, the upper wool sheet was pressed to obtain the essential oil. That is a primitive distillation method.

The distillation method brought to Okinawa was superior to the Roman method, but the efficiency of distillation was low, according to Edo period accounts: 360 mL of distillate was obtained from 18 l of fermented alcohol beverages. Eventually, 72 ml of spirits were distilled from the first distillate (Koizumi, 1996). We can infer the alcohol concentration experimentally: fermentative alcoholic beverages (*Sake*) have approx. 10% alcohol concentration, the first distillate has approximately 20% alcohol concentration, and final spirits have approximately 30% alcohol concentration. The distillate yield by the original method was lower than that of the present method because the condenser was not a water-cooled system.

2.3 *Awamori* spirits production method

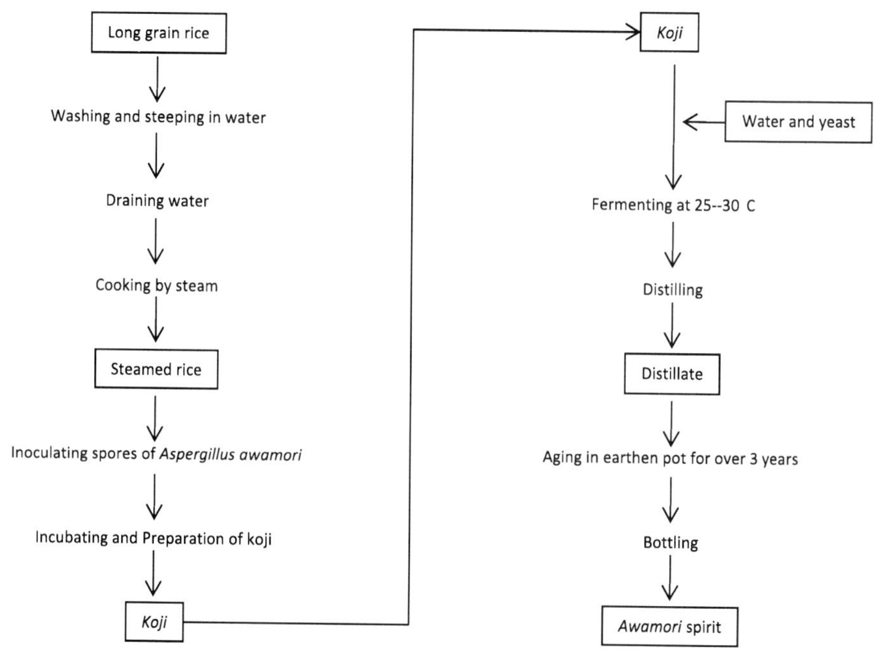

Fig. 1. Schemes of *Awamori* spirit production.

2.3.1 Rice

Awamori spirits are produced using long-grain rice imported from Thailand.

Oryza sativa is a perennial plant. Kato (1930) reported rice taxonomy. He reported some differences in *Indica* rice strains (imported from Thailand) and Japonica rice strains (grown in Japan): rice grains of *Indica* strain are longer than those of Japonica rice strain. Its leaves are light green. Moreover, *Indica* rice grains are longer than Japonica rice grains. The two strains sterilize in mating with each other. Furthermore, they relate to each other as subspecies, *Oryza sativa* subsp. *japonica* and *Oryza sativa* subsp. *indica*.

Some merits exist for the use of imported Thailand rice as the material for production.

1. The rice material is cheaper than Japanese rice. 2. Because the indica rice is not sticky, it is easy to work with during *koji* preparation. 3. Mash temperatures of mash using *Indica* rice are easy to control because this rice is hard and saccharifies slowly. 4. The alcohol yield from *Indica* rice is higher that from *Japonica* strains. 5. Indica rice has been used to produce *Awamori* spirit since it was brought from Chiame, Thailand (Nishiya, 1991).

The rice strains differ not only in grain size and shape but also in starch characteristics. Rice contains starches of two types: amylose and amylopectin. The structure is shown in the figure.

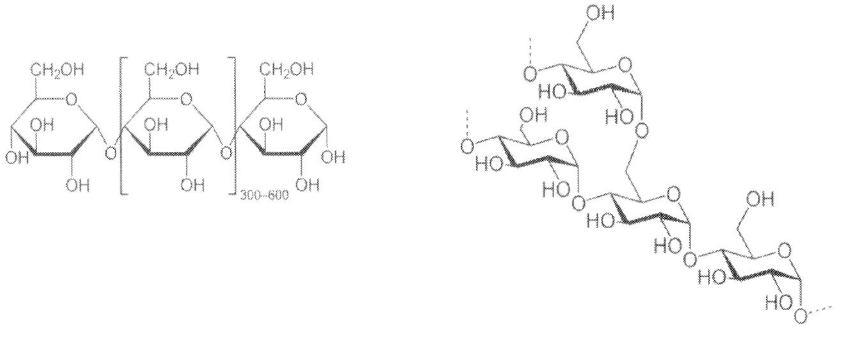

Fig. 2. Structures of amylose and amylopectin
Wikipedia (http://www.wikipedia.org)

Generally, *Japonica* rice strains contain 10–22% amylose, and amylase is not present in *Japonica* waxy strains. The starch is almost entirely composed of amylopectin (Juliano, 1985). In contrast, *Indica* rice contains about 18–32% amylase (Juliano, 1985). High concentrations of amylopectin make cooked rice sticky. Therefore, *Indica* rice is not sticky and is therefore suitable for preparation of *koji* for *Awamori* spirit production. Moreover, Horiuchi and Tani (1966) reported amylograms of nonwaxy starch prepared from some *Japonica* rice and *Indica* rice. The gelatinization temperature of *Indica* rice is 71.5°C (69.5–73.5°C), although the mean pasting temperature for *Japonica* varieties is 63.5°C (59–67%). No definite differences were found in the values of maximum viscosity and breakdown between *Japonica* and *Indica* varieties. Juliano et al. (1964a) obtained narrower pasting temperature ranges for *Japonica* rice flour (62–67°C) than for *Indica* (62–76.5°C). Gelatinization temperature of *Indica* rice was the highest among major cereals as waxy (*Japonica*) 62°C, 66°C for maize, and 62°C for wheat. (Dendy and Dobraszczyk, 2000).

2.3.2 Preparation of '*koji*' growing *Aspergillus awamori* on steamed rice

A. awamori have a role of saccharification during fermentation. Furthermore, the strain has a black area (conidia), and it differs completely from *A. niger*. *A. awamori* was isolated by Inui from *koji* for *Awamori* spirit production and named *A. luchuensis*, Inui (Inui, 1991). The strain was later renamed *A. awamori*. The strain is an important strain for alcohol production in Japan. The mold strain prevents contamination during fermentation processing by produced citric acid and acid tolerance α-amylase supplied by the strain saccharified starch as substrate under acid conditions during fermentation processing. Raper and Fennell (1965) investigated Black *Aspergillus*. Murakami (1982) compared the size of conidia, lengths of conidiophores, and characteristics of physiology, and classified two groups as *Awamori* Group and *Niger* Group in Black *Aspergillus*. *A. awamori* can grow at 35°C. It assimilates nitrite, and it has a short conidiophore: less than 0.9–1.1 mm.

2.3.3 Preparation of *koji*

Koji is used for alcohol beverage production or production of Asian seasonings such as miso paste and soy sauce. However, *koji* for *Awamori* spirits is prepared with a unique method (Nishiya, 1991). After steaming the rice, the seed mash is inoculated to the steamed rice.

Furthermore, the inoculated rice is incubated at 40°C and reduced step-by-step. Generally *koji* for seasoning or *sake* production is prepared at 30–35°C or with temperatures raised step-by-step. At high temperatures, amylase is produced and other enzymes such as feruloyl esterase break the cell walls. At the same time, fungi grow in rice grains. After 30 hr, the temperature is decreased gradually. At low temperatures of less than 35°C, *A. awamori* produces very much citric acid. The characteristic vanilla aroma of *Awamori* spirits is produced from ferulic acid as a precursor. Fukuchi (1999) reported that *A. awamori* grown at 37–40°C has high feruloyl esterase activity.

2.3.4 Fermentation of *Awamori* mash

Awamori spirit yeast, '*Awamori* yeast', that had been isolated from *Awamori* brewery was used during fermentation process and it might be peopling in brewer's house. The yeast is tolerant of low pH and high alcohol contents, and it can grow and ferment at low pH and high citric acid concentrations, producing very high alcohol concentrations (Nishiya, 1991). Nishiya (1972) and Suzuki (1972) reported that yeast growth was inhibited in conditions of greater than 11% alcohol and 35°C temperatures. However, the yeast ferments up to approximately 17% alcohol.

Awamori mash is prepared with *koji* and water only, and the water as 170% of koji weight is added. Fermented *Awamori* mash has 3.8–4.8% citric acid. The range of mash temperatures is 23–28°C. The yeast grows rapidly and fermentation finally ceases at temperatures higher than 30°C. At temperatures less than 20°C, the yeast grows slowly, and the mash is contaminated by bacteria. Generally, after 3 days, the mash has more than 10% of alcohol concentration. After 4 days, it has approx. 14%. After 7 days, it has more than 17%.

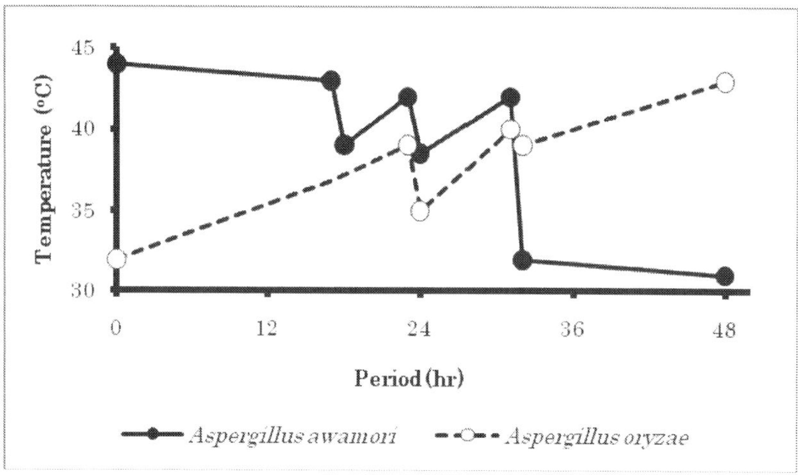

Fig. 3. Temperatures during two *Koji* preparations.

2.3.5 Mash distillation

After fermentation, the mash is distilled quickly because the mash aroma sours after yeast is digested by high alcohol and high temperatures. Two distillation systems exist as atmospheric distillation systems and reduced pressure distillation systems. *Awamori* spirits distilled with atmospheric distillation systems have three features (Nishiya, 1991).

1. Spirits have many components and a rich taste.
2. The spirit quality is good after aging.
3. Spirits have a distinctive aroma like scorching as furfural, which is produced by heat during distillation.

The distillation system with reduced pressure has no scorching aroma or higher alcohol, and it has a softer taste. It requires no long period of aging. During distillation, aldehydes and esters are distilled in the initial distillate. Then ethanol is distilled. The continuation of distillation decreases ethanol. The scorching aroma, that of furfural, is increased by heat during distillation. A direct heat distillation system is used traditionally: mash is heated in a kettle directly by fire. It has a scorched aroma. The compounds in distillate are shown during distillation of spirits.

2.3.6 Aging

Awamori spirits are aged in earthen pots using the *Shitsugi* method. The spirits show accelerated aging in earthen pots. Fatty acids or volatile acids as acetic acid in spirit are neutralized by calcium or magnesium released from earthen pots during aging. The acid or fatty smell in spirit is removed by aging. Recently, consumers favor a dry taste *Awamori* spirit aged in stainless steel tank after filtrate by ion exchange resin or activated charcoal.

The *Shitsugi* method is conducted by aging in earthen pots as follows. The distilled spirits are aged in earthen pots for 3–5 years. The spirits in the oldest earthen pots are consumed, and the consumed volume is supplied from the second oldest pot, and second oldest pot is supplied from third oldest pot is supplied from fourth oldest pot. Finally, the distilled spirit is poured in the newest pot. It resembles the solera system of sherry wine production. It is a specific aging method that has been used for *Kusu Awamori* spirits.

3. Mechanisms of making vanillin via ferulic acid in *'Awamori'* during aging within an earthen pot

3.1 Vanilla aroma of *'Awamori* spirit'

Awamori spirits have a vanilla aroma. Koseki (1998) reported mechanisms of vanilla aroma production in *Awamori* spirits. They found that not only vanillin but also phenol compounds as 4-vinyl guaiacol and ferulic acid from sufficient aged *Awamori* spirit. It does not age in barrels containing lignin. Therefore, they consider that the phenol compound is extracted from the material as rice.

3.2 Vanilla from ferulic acid in rice cell walls

Ferulic acid is contained in cell walls of rice. Then the ferulic acid is converted to vanilla during production. Koseki (1998) reported that ferulic acid dissolved in citric acid buffer was distilled, and that 4-vinyl guaiacol was detected in the distillate. Furthermore, the vanillin was detected from aged distillate containing 4-vinyl-guaiacol. Koseki mentions that the phenol compound was converted by heat. The ferulic acid is released from rice, and converts 4-vinyl guaiacol by heat when distilled. Furthermore, 4-vinyl guaiacol converts vanillin during aging. 4-Vinyl guaiacol is a well known compound as an off-flavor of beer and orange juice. Also, it is known as a characteristic flavor in weizen beer, or wheat beer. It is a distinct flavor.

Recently however, ferulic acid is converted to 4-vinyl guaiacol by microorganisms as *Saccharomyces cerevisiae* (Huang and others, 1993), *Pseudomonas fluorescens* (Huang and others, 1993), *Rhodotorula rubra* (Huang and others, 1994), *Candida famata* (Suezawa, 1995), *Bacillus pumilus* (Degrassi and others, 1995) and *Pseudomonas fluorescens* (Zhixian and others, 1994) via ferulic acid decarboxylase.

Furthermore, lactic acid bacteria converting ferulic acid to guaiacol were isolated from *Awamori* mash, and their ferulic acid decarboxylase was purified and compared (Watanabe, 2009). The lactic acid bacteria identified *Lactobacillus paracasei,* and their enzyme was the protein inferred as 47.9 kDa. The *p*-coumaric acid decarboxylase in *Lactobacillus plantarum* was 93 kDa; ferulic acid decarboxylase disagreed with those for *p*-coumaric acid decarboxylase (Cavin, 1997).

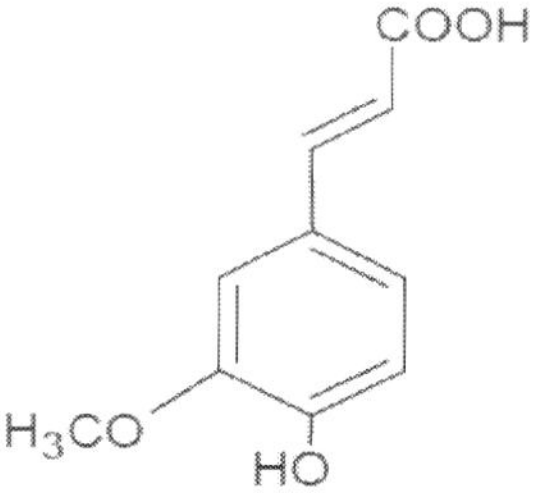

Ferulic acid

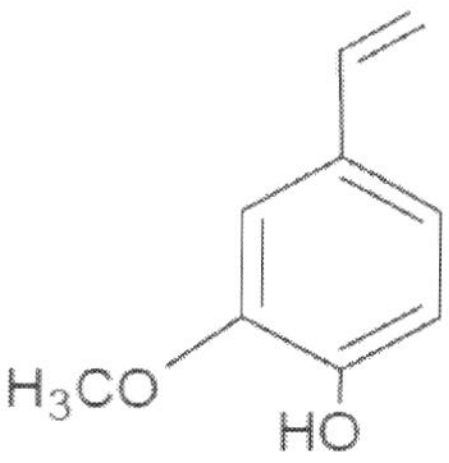

4-Vinyl guaiacol

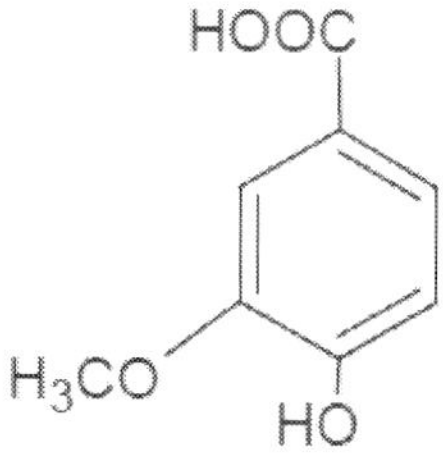

Vanillin

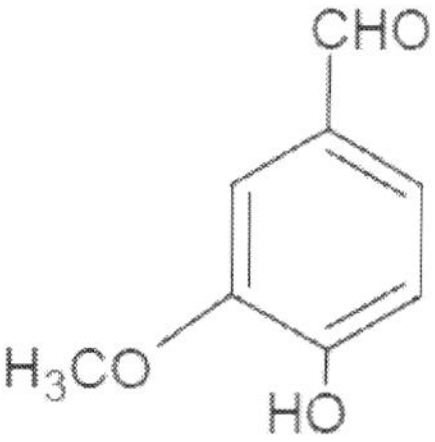

Vanillic acid

Fig. 4. Structure of phenol compound.

The optimum temperature was 60°C, and the optimum pH was approximately pH 3.0. The *p*-coumaric acid decarboxylase from *Lactobacillus plantarum* disagreed with that reported by Cavin (1997).

The optimum temperature was 30°C; the optimum pH was 5.5–6.0. The *Awamori* mash has an acid condition that is maintained by citric acid produced using *A. awamori.* Consequently, the mash is not contaminated by other microorganisms. The enzyme was regarded as having been grown optimally at pH 3.0: high acidity and low pH conditions. The vanillin is produced not only by chemical conversion but also through bio-conversion in *Awamori* spirit or the mash during *Awamori* spirit production.

3.3 Feruloyl esterase is an important enzyme for high-quality *A. awamori*

Ferulic acid is supplied from combining xylan on rice cell walls. The ferulic acid is released by feruloyl esterase. In this awamori production, *A. awamori* on *koji* provides the enzyme.

According to their reports, microorganisms are important to produce vanillin in *Awamori* spirits. The released ferulic acid is converted to 4-vinyl-guaiacol by lactic acid bacteria. Then vanillin is produced from guaiacol.

The amount of ferulic acid is obtained in relation to feruloyl esterase activity: high concentration ferulic acid gives *Awamori* spirits a high vanilla aroma. Namely, the amount of vanillin in *Awamori* spirits is determined by the feruloyl esterase activity.

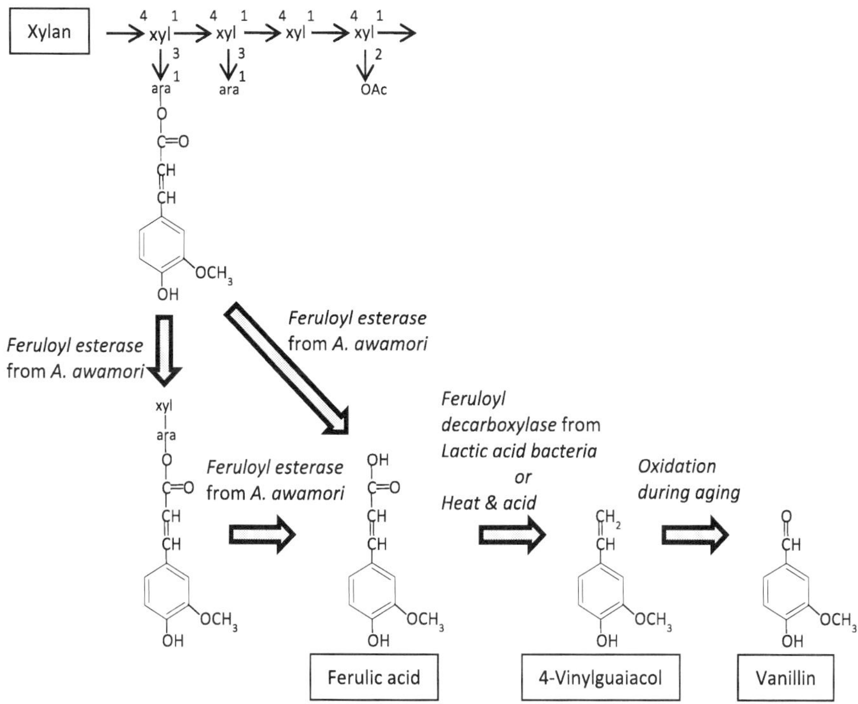

Fig. 5. Mechanisms of production of vanillin from ferulic acid during *Awamori* spirit

4. Selection of *Aspergillus awamori* strain producing the highest feruloyl esterase and characteristics of their enzymes

4.1 Screening of *Aspergillus awamori* producing the highest feruloyl esterase

We next describe the importance of feruloyl esterase during *Awamori* spirit production. We isolated and screened *A. awamori* to give high feruloyl esterase activity for *Awamori* spirits production.

Mold strains were isolated and selected from some locations near *Awamori* spirits breweries in Okinawa. Consequently, black *Aspergillus* of many kinds was isolated. Then the strains were cultivated on xylan plate medium (1.5% xylan, 0.5% yeast extract, 0.5% polypeptone and 1% agar) for the first screening. The strain with a clear zone on the plate medium was screened and the strain was assayed for feruloyl esterase activity.

Results show that the largest clear zone was made by G-2. Its feruloyl esterase was 394 U/ml, which was 3–4 times higher than the others. As presented in Table 2, feruloyl esterase contents of G-2 were higher than those of either *Aspergillus awamori* NRIC1200 (118 U/ml) or *Aspergillus usami* IAM2210 (50 U/ml) as a standard strain.

The mold strains that produced feruloyl esterase formed a black colony. Many reports have described feruloyl esterase from the black *Aspergillus* group. Results of this study concur with those previously reported results. Conidiophores were observed using a microscope.

The conidia were observed using scanning electric microscopy (SEM). Conidiophores of G-2 were similar to those of *Aspergillus* sp.: their conidia were black with a smooth surface.

production.In addition, nitrite was not assimilated by G-2. Generally, black conidia were *A. niger* or *A. awamori*; other *Aspergillus* were brown, yellow, or green (Pitt and Kich, 1988).

Strains	Feruloyl esterase (unit/ml)
A. awamori NRIC1200	118
A. saitoi IAM2210	50
G-2 strain	394
Other selected strains	130–293

Table 2. Activity of feruloyl esterase from the strains

Moreover, *A. awamori* were not assimilated and the surfaces of their conidia were smooth. In contrast, *A. niger* was assimilated; its conidia surfaces were rough (Murakami, 1979). For those reasons, G-2 was identified as *Aspergillus awamori*, used to produce Awamori spirits as Japanese traditional spirits produced in Okinawa. The G-2 strain does not produce ochratoxin or aflatoxin as a mycotoxin. This *A. awamori* is applicable of food production and Awamori production.

4.2 Feruloyl esterase characteristics

The enzyme was purified using ion-exchange, size-exclusion, and HPLC chromatography. After purification, the specific activity of the enzyme was 20-fold higher and the yield was 16% higher than that of the crude enzyme solution. The enzyme solution was then analyzed using sodium dodecyl sulfate poly acrylamide gel electrophoresis (SDS-PAGE).

The molecular weight of feruloyl esterase in *A. awamori* was 35 kDa (Koseki, 1998b). The enzyme was 78 kDa, which was higher than that of *A. awamori*. The molecular weight of feruloyl esterase was measured using size-exclusion chromatography. The molecular weight was estimated as 80 kDa. The feruloyl esterase was inferred to be a monomer protein.

The optimum pH of the feruloyl esterase was pH 5.0 and the optimum temperature was 40°C. The activity stabilized at pH 3.0–5.0. The feruloyl esterase of *A. awamori* is reportedly (Koseki and others, 1998b) unstable at pH 3.0, which was 30–50% of non-treated enzyme activity. However, the enzyme was stable in acid. The enzyme was stable at 50°C. Feruloyl esterase of *A. niger* showed tolerance at 80°C (Sundberg and others, 1990). The enzyme is therefore acid-tolerant, but not heat-tolerant. Use of the enzyme is expected to be advantageous for food production under acid conditions, in which *A. awamori* produces citric acid very well.

Mercury ion (Hg^{2+}) inhibited feruloyl esterase activity completely; Fe^{2+} inhibited 80% of the activity. Both PMSF and DFP completely inhibited feruloyl esterase activity. In general, feruloyl esterase, produced by black *Aspergillus* group, was classified into the serin-esterase group enzyme, which was inhibited by Hg^{2+} and Fe^{2+} (Koseki and others, 1998b). They reported that the active center of feruloyl esterase from *A. awamori* had Ser-Asp-His. It was therefore considered that the feruloyl esterase had similar catalysis.

The feruloyl esterase activity was the highest among four substrates with 1-naphthyl acetate. The activity of the feruloyl esterase had 40% of the activity related to 1-naphthyl acetate, against 1-naphthyl propionate (composed of three carbons). In addition, 1-naphthyl butyrate (comprising four carbons) showed 5% activity related to 1-naphthyl acetate. The activity of feruloyl esterase was decreased against substrates containing more than three carbons; 2-naphthyl acetate did not react completely. The reported feruloyl esterase showed activity against 1-naphthyl propionate, as did 1-naphthyl acetate (Ishihara and others, 1999). The feruloyl esterase was inferred to be substrate-specific.

Adsorption of the feruloyl esterase to cell walls was not observed. The enzyme was not absorbed by xylan or starch that were present in the supernatant. However, it was absorbed by cellulose in the supernatant. It was also found in the precipitant with cellulose.

Strain	Mycelium color	Conidium color	Conidiophore (mm)	Conidial head diameter (μm)	Conidium diameter (μm)	Conidium surface	Assimilation of $NaNO_2$
A. awamori NRIC1200	White	Black	0.61	220	3.8	Smooth	-
A. niger NRIC1221	White	Black	0.86	170	3.4	Spinky	+
A. saitoi IAM2210	White	Black	0.91	190	4.1	Rough	-
G-2	White	Black	0.68	200	3.4	Smooth	-

+, positive; -, negative

Table 3. Morphological and physiological characteristics of black *Aspergillus* spp.

Feruloyl esterase from *A. niger* and *A. awamori* reportedly adsorbed specifically to cellulose (Ferreira and others, 1993), which agrees with the results shown for this study. The intensity of absorbing the enzyme with cellulose was conducted to wash cellulose by boiling for 10 min in SDS solution (1–11%). The enzyme had a cellulose binding domain, as reported also by Koseki and others (2006). The protein was unbound by 10% and 11% SDS solution. It is considered that they were bound strongly.

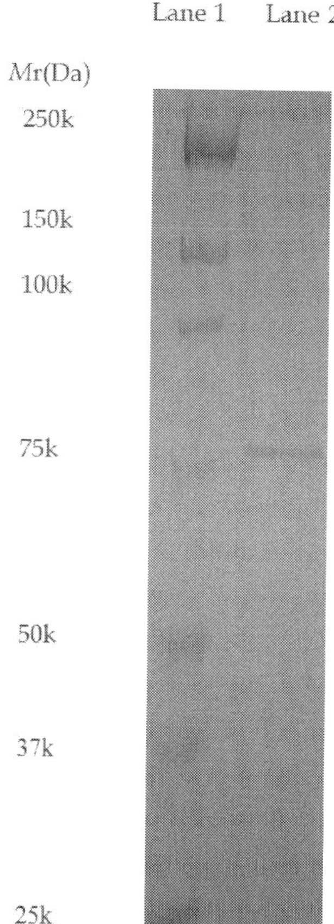

Fig. 6. SDS–PAGE of feruloyl esterase.

We investigated Michaelis constants of the feruloyl esterase. A Lineweaver–Burk plot is shown to calculate the Km of the feruloyl esterase: its Km was 0.0019% (0.01 mM). Koseki reported that Km was 0.26–0.66 mM (Koseki and others, 2006). The feruloyl esterase has higher affinity to 1-naphthyl acetate than that described by Koseki. The feruloyl esterase from *A. awamori* G-2 was more stable under the acid condition than the other black *Aspergillus* groups. In general, the esterase was as unstable under the acid condition as *Awamori* mash or shochu mash during fermentation. However, results suggest that the feruloyl esterase survived under the acidic conditions. The ferulic acid was related to the aroma of *Awamori* spirits or shochu spirits. The ferulic acid released from the cell wall was converted to vanilla via 4-vinyl guaiacol by yeast or lactic acid bacteria during aging (Gabe and Koseki, 2000; Watanabe and others, 2007). In addition, results showed that the enzyme-stabilized activity under acid and heat conditions of pH 3.0 and 50°C is applicable to food production.

Substrates	Relative activity (%)
1-naphthyl acetate	100
1-naphthyl propionate	40
1-naphthyl butyrate	5
2-naphthyl acetate	0

Table 4. Substrate specificity of feruloyl esterase

5. Application of feruloyl esterase to beer brewing

In beer brewing processes, the mash is used differently from the mash of *Awamori* spirits. Beer mash is saccharified and then fermented. In the mashing process, barley malt produces not only maltose as base material of ethanol for fermentation. The malt also supplies β-glucan in mash. β-glucan (called (1-3) (1-4)-β-D-glucans) are components of endosperm cell walls in barley, occupying 75% of the cell wall (MacGregor and Fincher, 1993). The amount of β-glucans is negatively correlated with yields of the amount of wort in the mashing process (Bourne and others,1982; Kato and others, 1995) . The amount of β-glucan in grain shows levels of decomposition of cell wall in barley endosperm. The malt contained a great amount of β-glucan, showing insufficiency in decomposing β-glucan in malt during germination. The malt insufficiently decomposes β-glucans and does not have sufficient starch or protein in endosperm. Therefore the yields fall. Furthermore, the β-glucans in malt used to make mash have high viscosity, which reduces the mash filtration speed (MacGregor and Fincher, 1993). Speed is a limiting factor of beer brewing. The slow filtration of beer presents problems for breweries. Moreover, glucan is a factor of invisible haze or pseudohaze (Jacson and Bamforth, 1983) (1-3), (1-4)−β-D-glucanase is usually added to elevate the activity (Bamforth, 1985).

Kanauchi and Bambort reported that β-glucan was decomposed not only by $(1-3)$, $(1-4)$-β-D-glucanase but also by xylanase, arabinofuranosidase, and feruloyl esterase. At least, feruloyl esterase decomposed xylan or feruloylxylan layer covering β-glucan. The enzyme helps to access β-glucanase to b-glucan.

6. Application of feruloyl esterase for biomass processing

Global crude oil production is 25 billion barrels, but human societies are expected to reduce the use of fossil fuels after bio-fuel development (Benoit et al., 2006). Particularly in the US, these gasoline fuels contain up to 10% ethanol by volume. In the future, automobiles can use a blend of 85% ethanol and 15% gasoline by volume (Fazary and Ju, 2008).

Under this situation, lignocellulosic and other plant biomass processing methods have been developed recently. The material is not only a renewable resource but it is also the most abundant source of organic components in high amounts on the earth: the materials are cheap, with a huge potential availability.

Plant lignocellulosic biomass comprises cellulose, hemicellulose, and lignin. Its major component is cellulose (35–50%) with hemicellulose (20–35%) and lignin (25%) following, as shown in Table 5 (Noor and others, 2008). Proteins, oils, and ash make up the remaining fraction of lignocellulosic biomass (Wyman, 1994b). Cellulose is a high-molecular-weight linear polymer of b-1,4-linked D-glucose units which can appear as a highly crystalline material (Fun and others, 1982). Hemicelluloses are branched polysaccharides consisting of the pentoses D-xylose and L-arabinose, and the hexoses D-mannose, D-glucose, D-galactose and uronic acids (Saka, 1991). Lignin is an aromatic polymer synthesized from phenylpropanoid precursors (Adler, 1977).

It is noted throughout the world that ethanol is producible from biomass material. For their bioconversion, pretreatment is an important procedure for practical cellulose conversion processes (Bungay, 1992; Dale and Moreira, 1982; Weil and others, 1994; Wyman, 1994a). Ferulic acid is found in the cell walls of woods, grasses, and corn hulls (Rosazza and others, 1995). It is ester-linked to polysaccharide compounds, where it plays important roles in plant cell walls including protein protection against pathogen invasion and control of extensibility of cell walls and growth (Fry, 1982).

	Composition (% dry weight basis)		
	Cellulose	Hemicellulose	Lignin
Corn fiber	15	35	8
Corn cob	45	35	15
Rice straw	35	25	12
Wheat straw	30	50	20
Sugarcane bagasse	40	24	25

Table 5. Composition of some agricultural lignocellulosic biomass (Noor and other, 2008)

In bio-ethanol production, complete enzymatic hydrolysis of hemicelluloses as arabinoxylan requires both depolymerizing and sidegroup cleaving enzyme activities such as FAEs.

Any hemicellulose-containing lignocellulose generates a mixture of sugars upon pretreatment alone or in combination with enzymatic hydrolysis. Fermentable sugars from cellulose and hemicellulose will fundamentally be glucose and xylose, which can be released from lignocellulosics through single-stage or two-stage hydrolysis. In Europe, portable alcohol manufacturing plants are based on wheat endosperm processing, with the hemicellulosic by-product remaining after fermentation consisting of approximately 66% (W/W) arabinoxylan (Benoit et al., 2006). A synergistic action between cellulases, FAEs, and xylanases might prove to be more effective when applied at a critical concentration in the saccharification of steam-exploded wheat straw (Borneman and others, 1993; Kennedy and others, 1999; Tabka and others, 2006).

7. References

Adler, E. (1977). Lignin chemistry: Past present and future. *Wood Sci. Technol.*, Vol. 11, pp. 169-218

Bamforth, C.W. (1985). Biochemical approaches to beer quality. *J. Instit. Brew.*, Vol. 91, pp. 154-160

Benoit, I., Navarro, D., Marnet, N., Rakotomanomana, N., Lesage-Meessen, L., Sigoillot, J-C. & Asther, M. (2006). Feruloyl esterases as a tool for the release of phenolic compounds from agro-industrial byproducts. *Carbohydr. Res.*, Vol. 341, pp. 1820-1827

Borneman, W.S., Hartley, R.D., Morrison, W.H., Akin, D.E. & Ljungdahl, L.G. (1990). Feruloyl and *p*-coumaroyl esterase from anaerobic fungi in relation to plant cell wall degradation. *Appl. Micro. Biotech.*, Vol. 33, pp. 345-351

Bourne, D.T., Powlesland, T. & Wheeler, R.E. (1982). The relationship between total b-glucan of malt and malt quality. *J. Instit. Brew.*, Vol. 88, pp. 371-375

Bungay, H. (1992). Product opportunities for biomass refining. *Enzyme Microb. Technol.*, Vol. 14, pp. 501-507

Cavin, J.F., Barthelmebs, L.G., Van Beeumen, J., Samyn, J., Travers, B. & Diviées, J.-F. (1997). Charles Purification and characterization of an inducible *p*-coumaric acid decarboxylase from *Lactobacillus plantarum*. *FEMS Microbiology Letters*, Vol. 147, pp. 291-295

Dale, B.E. & Moreira, M.J. (1982). *A freeze-explosion. Biotechnol Bioeng Symp*, Vol. 12, pp. 31-43

Degrassi, G., De Laureto, P.P. & Bruschi, C.V. (1995). Purification and characterization of ferulate and *p*-coumarate decarboxylase from *Bacillus pumilus*. *Appl. Environ. Microbiology*, Vol. 61, No. 1, pp. 326-332

Dendy, D. A. V., Dobraszczyk, B. J. (2000). Cereals and cereal products: chemistry and technology, Aspen publishers, Inc., Gaithersburg, Maryland, USA, pp. 305-306

Faulds, C.B. & Williamson, G. (1994). Purification and characterization of a ferulic acid esterase (FAE-III). from *Aspergillus niger*: specificity for the phenolic moiety and binding to micro-crystalline cellulose. *Microbiol.*, Vol. 144, pp. 779-787

Faulds, C.B., Kroon, P.A., Saulnier, L., Thibault, J.F. & Williamson, G. (1995). Release of ferulic acid from maize bran and derived oligosaccharides by *Aspergillus niger* esterases. *Carbohydr. Polym.*, Vol. 27, pp.187-190

Fazary, A.E. & Ju, Y.H. (2008). The large-scale use of feruloyl esterases in industry. *Biotech. Molecular Biology Reviews*, Vol. 3, No.5, pp. 95-110

Fry, S.C. (1982). Phenolic components of the primary cell wall. *Biochemistry Journal*, Vol. 203, pp. 493-504

Fukuchi, K., Tamura, H. & Higa, K. (1999). The studies on preparation of koji for *Kusu Awamori* spirits. *Report Indust. Tech. Center in Okinawa.*, Vol. 2, pp. 77-83 (in Japanese)

Fun, L.T., Lee, Y.H. & Gharpuray, M.M. (1982). The nature of lignocellulosics and their pretreatment for enzymatic hydrolysis. *Adv. Biochem. Eng.*, Vol. 23, pp. 158-187

Horiuchi, H. & Tani, T. (1966). Studies of cereal starches. Part V. Rheological properties of the starch of rices imported into Japan. *J. Agr. Biol. Chem.*, Vol. 30, pp. 457-465

Huang, Z., Dostal, L. & Rosazza, J.P.N. (1993). Microbial transformations of ferulic acid by *Saccharomyces cerevisiae* and *Pseudomonas fluorescens*. *Appl. Environ. Microbiol.*, Vol. 59, pp. 2244-2250

Huang, Z., Dostal, L. & Rosazza, J.P.N. (1994). Mechanism of ferulic acid *Rhodotorula rubra*. *J. Biol. Chem.*, Vol. 268, pp. 23594-23598

Inui, T. (1901). Untersuchungen uber die der Zubereitung des alkoholischen Getrankes "Awamori Betheiligen. *J. Coll. Sci. Imp. Univer. Tokyo*, Vol. 15, pp. 465-476

Ishihara, M., Nakazato, N., Chibana, K., Tawata, S. & Toyama, S. (1999). Purification and characterization of feruloyl esterase from *Aspergillus awamori*. *Food Sci. Tech. Res.*, Vol. 5, No.3, pp. 251-254

Jacson, G. & Bamforth, C.W. (1983). Anomalous haze readings due to β-glucan. *J. Instit. Brew.*, Vol. 89, pp. 155-156

Juliano, B.O. (1985). Criteria and test for rice grain quality. Rice chemistry and technology. American Association of Cereal Chemists, Inc., St. Paul, Minnesota, USA, pp. 443-513

Juliano, B.O., Bautista, G.M., Lugay, J.C. & Reyes, A.C. (1964). Studies of the physicochemical properties of rice. *J. Agric. Food Chem.*, Vol. 12, pp. 131-138

Kato, S., Kosaka, H., Hara, S., Maruyama, Y. & Takigushi, Y. (1930). On the affinity of the cultivated varieties of rice plants, *Oryza sativa* L. *J. Depart. Agri., Kyushu Imperial University*, Vol. 2, pp. 241-276

Kato, T., Sasaki, A. & Takeda, G. (1995). Genetic Variation of β-glucan Contents and β-glucanase Activities in Barley, and their Relationships to Malting Quality. *Breeding Science*, Vol. 45, No. 4, pp. 471-477 (in Japanese)

Kennedy, J.F., Methacanon, P. & Lloyd, L.L. (1999). The identification and quantification of the hydroxycinnamic acid substituents of a polysaccharide extracted from maize bran. *J. Sci. Food Agric.*, Vol. 79, pp. 464-470

Koizumi, T. (1996). Birth of Quality Liquor Tokyo: Kodansha

Koseki, T. & Iwano, K. (1998). A mechanism for formation and meaning of vanillin in Awamori. *Soc. Brew. Japan*, Vol. 93, No. 7, pp. 510-517 (in Japanese)

Koseki, T., Furuse, S., Iwano, K. & Matsuzawa, H. (1998). Purification and characterization of a feruloyl esterase from *Aspergillus awamori*. *Biosci. Biotechnol. Biochem.*, Vol. 62, No.10, pp. 2032-2034

Koseki, T., Takahashi, K., Hanada, T., Yamane, Y., Fushinobu, S. & Hashizume, K. (2006). N-Linked oligosaccharides of *Aspergillus awamori* feruloyl esterase are important for thermostability and catalysis. *Biosci. Biotechnol. Biochem.*, Vol. 70, No. 10, pp. 2476-2480

Kroon, P.A., Faulds, C.B., Brezillon, C. & Williamson, G. (1997). Methyl phenylalkanoates as substrates to probe the active sites of esterases. *Eur. J. Biochem.*, Vol. 248, pp. 245-251

MacGregor, A.M. & Fincher, G.B. (1993). Carbohydrates of the barley grain, pp. 73-130, In: A.W. MacGregor, & R.S. Bhatty, (eds.), Barley: Chemistry and technology. The American Association of Cereal Chemists, Inc., St. Paul, Minnesota, USA

Murakami, H. (1979). The taxonomic studies of koji mold (26th), The methods of physiological examination and characteristics of cultivation on black *Aspergillus* sp. *J. Brew Soc. Japan*, Vol. 74, No.5, pp. 323-327 (in Japanese)

Murakami, H. (1982). Classification of the black *Aspergillus* sp. *J. Brew. Soc. Japan*, Vol. 77, No. 2, pp. 84 (in Japanese)

Nishiya, S., Suzuki, A. & Shigaki, K. (1972). Kinetic studies of yeast growth (part 4), *J. Soc. Brew. Japan*, Vol. 67, pp. 445-448 (in Japanese)

Nishiya, T. (1991). *Honkaku Syochu-seizo-gijyutsu*, Brewing Society of Japan, Tokyo, pp. 1-279 (in Japanese)

Noor, M.S., Zulkali, M.M.D. & Ku Syhudah, K.I. (2008). Proceedings of Malaysian University Conferences on Engineering and Technology, Ferulic acid from lignocellulosic biomass: Review, Malaysian Universities Conferences on Engineering and Technology (MUCET2008)

Pitt, J.Y. & Kich, M.A. (1988). A laboratory guide to common *Aspergillus* species and their teleomorphs. CSIRO, Division of Food Research Sydney. Academic Press Inc., Australia

Ralet, M.-C., Faulds, C.B., Williamson, G. & Thibault, J.F. (1994). Degradation of feruloylated oligosaccharides from sugar-beet pulp and wheat bran by ferulic acid esterases from *Aspergillus niger*. *Carbohydr. Res.*, Vol. 263, pp. 257-269

Raper, K.B. & Fennell, D.I. (1965). The Genus *Aspergillus*, pp. 1-686, Baltimore: Williams & Wilkins

Rosazza, J.P.N., Huang, Z., Dostal, L., Vom, T. & Rousseau, B. (1995). Biocatalytic transformations of ferulic acid: an abundant aromatic natural product. *J. Ind. Microbiol.*, Vol. 15, pp. 457-471

Saka, S. (1991). Chemical composition and Distribution. Dekker, New York pp. 3-58

Suezawa, Y. (1995). Bioconversions of Ferulic Acid and *p*-Coumaric Acid to Volatile Phenols by Halotolerant Yeasts.: Studies of Halotolerant Yeasts in Soy Sauce Making (Part I). *J. Agri. Chem. Soci. Japan*, Vol. 69, pp. 1587-1596 (in Japanese)

Sundberg, M., Poutanen, K., Makkanen, P. & Linko, M. (1990). An extracellular esterase of *Aspergillus awamori*. *Biotechnol. Appl. Biochem.*, Vol. 12, pp. 670-680

Suzuki, A., Nishiya, T. & Shigaki, K. (1972). Kinetic studies of yeast growth (part 5). *J. Soc. Brew. Japan*, Vol. 67, pp. 449-452 (in Japanese)

Tabka, M.G., Herpoël-Gimbert, I., Monod, F., Asther, M. & Sigoillot, J.C. (2006). Enzymatic saccharification of wheat straw for bioethanol production by a combined cellulase xylanase and feruloyl esterase treatment, *Enzyme and Microbial. Tech.*, Vol. 39, No.4, pp. 897-902

Watanabe, S., Kanauchi, M., Kakuta, T. & Koizumi, T. (2007). Isolation and characteristics of lactic acid bacteria in Japanese spirit awamori mash. *J. Am. Soc. Brew Chem.*, Vol. 65, pp. 197-201

Watanabe, S., Kanauchi, M., Takahashi, K. & Koizumi, T. (2009). Purification and characteristics of feruloyl decarboxylase produced by lactic acid bacteria from *Awamori* mash, *J. Am. Soc. Brew. Chem.*, Vol. 65, pp. 229-234

Wei, J., Westgate, P., Koholmann, K. & Ladisch, M.R. (1994). Cellulose pretreatments of lignocellulosic substrate, *Enzyme Microb. Technol.*, Vol. 16, pp. 1002-1004

Wyman, C.E. (1994a). Ethanol from lignocellulosic biomass: technology, economics, and opportunities, *Bioresour. Technol.*, Vol. 50, pp. 3-16

Wyman, C.E. (1994b). Alternative fuels from biomass and their impact on carbon dioxide accumulation. *Appl. Biochem. Biotechnol.*, Vol. 45, No. 46, pp. 897-915

Zhixian, H., Dostal, L. & Rosazza, J.P.N. (1994). Purification and characterization of a ferulic acid decarboxylase from *Pseudomonas fluorescens*. *J. Bacteriol.*, Vol. 176, No. 19, pp. 5912-5918

Part 2

Social and Economic Issues

9

Functional Foods in Europe: A Focus on Health Claims

Igor Pravst
Nutrition Institute, Ljubljana
Slovenia

1. Introduction

The functional foods concept started in Japan in the early 1980s with the launch of three large-scale government-funded research programs on *systematic analyses and development of functional foods, analyses of physiological regulation of the functional food* and *analyses of functional foods and molecular design* (Ashwell 2002; Pravst et al. 2010). In 1991, in an effort to reduce the escalating cost of health care, a category of foods with potential benefits was established (Foods for Specified Health Use – FOSHU) (Ashwell 2002). In the USA, evidence-based health or disease prevention claims have been allowed since 1990, when the *Nutrition Labelling and Education Act* was adopted; claims have to be approved by the Food and Drug Administration (FDA) (Arvanitoyannis and Houwelingen-Koukaliaroglou 2005). *Codex Alimentarius Guidelines for the use of nutrition and health claims* were accepted in 2004, and amended in 2008 and 2009, followed by *Recommendations on the scientific basis of health claims* (Grossklaus 2009). In the European Union, harmonisation was achieved in 2006 with *Regulation (EC) No 1924/2006 on nutrition and health claims made on foods,* which requires authorization of all health claims before entering the market.

The definition of functional foods is an ongoing issue and many variations have been suggested (Arvanitoyannis and Houwelingen-Koukaliaroglou 2005). A consensus on the functional foods concept was reached in the European Union in 1999, when a working definition was established whereby a food can be regarded as functional if it is satisfactorily demonstrated to beneficially affect one or more target functions in the body beyond adequate nutritional effects in a way that is relevant to either an improved state of health and well-being or a reduction of disease risk. Functional foods must remain foods and demonstrate their effects when consumed in daily amounts that can be normally expected (Ashwell 2002). In practice, a functional food can be: an unmodified natural food; a food in which a component has been enhanced through special growing conditions, breeding or biotechnological means; a food to which a component has been added to provide benefits; a food from which a component has been removed by technological or biotechnological means so that the food provides benefits not otherwise available; a food in which a component has been replaced by an alternative component with favourable properties; a food in which a component has been modified by enzymatic, chemical or technological means to provide a benefit; a food in which the bioavailability of a component has been modified; or a combination of any of the above (Ashwell 2002). Regardless of the various definitions, the main purpose of functional food should be clear – to improve human health and well-being. However, health claims are a very convenient tool when it comes to marketing functional foods. The latest analyses show that in some European countries health claims are now used in over 15% of commonly eaten packaged foods (Lalor et al. 2010). Consumers are very sensitive to health-related communications and the use of health claims is unfortunately often connected with intentions to mislead them (Pravst 2011b). While this is clearly prohibited by law (Colombo 2010), companies are employing innovative strategies to avoid these rules so as to make a greater profit. Sophisticated regulation of this area is needed to provide consumers with high quality foods and non-misleading information. In this review the assessment of functional foods and their use in Europe will be critically discussed.

2. Consumer acceptance of functional foods and nutrition labelling

Health-related information on food labels may play a crucial role in informing consumers and influence their purchase decisions (Pothoulaki and Chryssochoidis 2009). However, consumer acceptance is far from unconditional and depends on beliefs about functional foods being viewed as a marketing stunt, as well as on familiarity and perceptions related to the perceived fit of ingredients and carrier or base products rather than on classical demographic characteristics (Verbeke 2010). In addition, it is unlikely one can count on consumer willingness to compromise on the taste of functional foods for health (Verbeke 2006). Recent studies show that the consumer might have a stronger preference for simple health statements compared to the implied benefits that result from consuming a functional food product (Bitzios et al. 2011). Indeed, consumers' attitudes to functional foods and the use of nutrition information or claims on food labels have been studied extensively in recent years (Bitzios et al. 2011; Hoefkens et al. 2011; Krutulyte et al. 2011; Pothoulaki and Chryssochoidis 2009; Urala and Lahteenmaki 2004; Verbeke 2005; Verbeke 2006; Verbeke et al. 2009) and some useful reviews of these topics are available (Bech-Larsen and Scholderer 2007; Pothoulaki and Chryssochoidis 2009; Verbeke 2010; Wills et al. 2009).

A recent multi-country study confirmed that consumers perceive some nutrients as qualifying (fibre, vitamins and minerals) or disqualifying (energy, fat, saturated fat, salt, sugars) (Hoefkens et al. 2011) and only small differences are observed between countries. Overall, consumers perceive the nutritional value of foods as important when selecting foods, particularly when it comes to qualifying nutrients (Hoefkens et al. 2011). Such a finding is quite interesting since public health nutrition awareness campaigns across Europe mainly target disqualifying nutrients, i.e. salt (Celemin et al. 2011). The marketing campaigns of the food industry can obviously have very strong effects, particularly if consumers perceive them as trustworthy. It must be noted that the labelling of nutrition information on foods in the EU is currently not mandatory, except for foods with nutrition and health claims. This is about to change under the accepted *Regulation on the provision of food information to consumers*. However, most food labels already contain at least back-of-pack nutrition labelling or related information, although there are significant differences between various food categories and countries (Bonsmann et al. 2010).

3. Assessment of functionality

The effects of particular functional foods on human health must be based on scientific evidence of the highest possible standard. However, because a direct measurement of the effect a food has on health is not always possible, the key issue is to identify critical biomarkers that can be used to monitor how biological processes are being influenced (Howlett 2008). If appropriate biomarkers are chosen it is possible to study the effect of the food on the final endpoints by measuring the biomarkers. The markers could be chosen to reflect either some key biological function or a key stage in development that is clearly linked to the study endpoint (markers of an intermediate endpoint) (Fig. 1) (Howlett 2008). In such a way we can study the long-term effects in the shorter term. Nonetheless, the biomarkers must be very carefully selected.

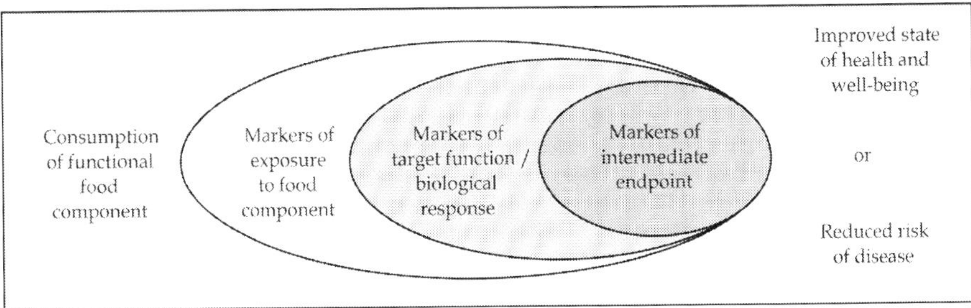

Fig. 1. The use of markers to link food consumption to health outcomes. Reproduced with permission of ILSI Europe (Howlett 2008).

In Europe the consensus on the criteria for the scientific support for claims on foods was reached in 2005 within PASSCLAIM – the Process for the Assessment of Scientific Support for Claims on Foods (Aggett et al. 2005). It delivered criteria to assess the scientific support for claims on foods. The set of criteria developed were derived through an iterative continuous improvement process that involved developing and evaluating criteria against key health claim areas of importance in Europe. These areas included diet-related cardiovascular disease, bone health, physical performance and fitness, body-weight regulation, insulin sensitivity and diabetes risk, diet-related cancer, mental state and performance and gut health and immunity. Expert groups reviewed the availability of indicators of health and disease states within their respective areas of expertise. They have demonstrated the limitations of existing biomarkers and have identified the need for better biomarkers (Aggett et al. 2005).

3.1 Food vs. medicine

In a time when the role of a healthy diet in preventing non-communicable diseases is well accepted, the borderline between foods and medicine is becoming very thin. The concept of foods should obviously go beyond providing basic nutritional needs. While products intended to cure diseases are classified as medicine, a healthy diet consisting of foods with functional properties can help promote well-being and even reduce the risk of developing certain diseases (Howlett 2008). However, a diet can only be healthy if the combination of individual foods is good. Limiting certain components (e.g. salt and saturated or trans fatty acids) or simply delivering an intake of nutrients cannot be regarded as a healthy diet. While it is easy to give advice on a *healthy diet*, it is much harder to define it for a particular need.

The borderline between food and medicine is shown in Table 1. As with traditional foods, functional foods should also not have serious side effects when consumed in reasonable amounts, although some functional foods can be intended for specific populations. This is contrary to medicine, which is intended for curing a disease or treating its symptoms, and used in exactly defined quantities to minimize side effects and assure therapeutic action. Food legislation also defines *food supplements* and *foods for special medical purposes*. *Food supplements* are concentrated sources of nutrients or other substances with a nutritional or physiological effect and are meant for supplementing a normal diet. Supplements are usually marketed as capsules, pastilles, tablets, pills and other similar forms of liquids and powders designed to be taken in measured small unit quantities. In contrast, *foods for special medical purposes* are specifically formulated, processed and intended for the dietary management of diseases, disorders or medical conditions of individuals who are being treated under medical supervision. These foods are intended for the exclusive or partial feeding of people whose nutritional requirements cannot be met by normal foods.

	Functional food	**Medicinal food**	**Medicine**
Usage	Providing nutritional needs, weight management, supporting health of digestive system, skeleton, heart, reducing risk of development of certain diseases.	Food management of some diseases or certain medical conditions, in which there are specific nutritional needs (i.e. problems with swallowing, lost appetite, postoperative nutrition)	Curing disease or treating its symptoms
To be consumed:	As wanted	As needed	As prescribed

Table 1. The borderline between food and medicine (Raspor 2011)

4. Functional foods in the context of regulation

To promote the use of any particular functional food its beneficial effects must be communicated to consumer. This is usually done through the use of nutrition and health claims in the labelling and advertising of foods. In this context, functional foods in Europe are probably most critically affected by the *Regulation (EC) No 1924/2006 on nutrition and health claims made on foods* (EC 2006). The regulation applies to all nutrition and health claims made in commercial communications, including trademarks and other brand names which could be construed as nutrition or health claims. The general principle is that claims should be substantiated by generally accepted scientific data, non-misleading and pre-approved on the EU level. Additionally, claims shall not give rise to doubt about the safety of other foods or encourage excessive consumption of a food.

4.1 Nutrition claims

The regulation defines a *nutrition claim* as any claim stating, suggesting or implying that a food has particular beneficial nutritional properties due to the calorific value or composition with respect to the presence (or absence) of specific nutrients or other substances. Only nutrition claims which are listed in the regulation are allowed, providing they are in accordance with the set conditions of use. Some examples of authorised nutrition claims and the conditions applying to them are listed in Table 2.

Nutrition claim	Conditions of use
low energy	less than 170 kJ/100 g for solids or 80 kJ/100 ml for liquids
low sodium	less than 0.12 g of sodium per 100 g or per 100 ml (different limits for waters)
source of fibre	at least 3 g of fibre per 100 g or at least 1.5 g of fibre per 100 kcal
high fibre	at least 6 g of fibre per 100 g or at least 3 g of fibre per 100 kcal
source of vitamins or minerals	a significant amount[1] of vitamins or minerals
high in vitamins or minerals	at least twice the amount of 'source of vitamins or minerals'
source of omega-3 fatty acids	at least 0.3g alpha-linolenic acid per 100g and per 100kcal, or at least 40mg of the sum of eicosapentaenoic acid and docosahexaenoic acid per 100g and per 100kcal.
high omega-3 fatty acids	at least 0.6g alpha-linolenic acid per 100g and per 100kcal, or at least 80mg of the sum of eicosapentaenoic acid and docosahexaenoic acid per 100g and per 100kcal.

Note: [1]As a rule, 15 % of the recommended allowance specified in the Annex of Directive 90/496/EEC on nutrition labelling for foodstuffs supplied by 100 g or 100 ml (or per package if the package contains only a single portion) should be taken into consideration in deciding what constitutes a significant amount.

Table 2. Examples of nutrition claims and the conditions applying to them (EC 2006)

Labelling of the content of vitamins and minerals must also comply with the *Council Directive 90/496/EEC on nutrition labelling for foodstuffs*, which was amended in 2008. In this directive vitamins and minerals which may be declared on labels and their recommended daily allowances (RDAs) are set (Table 3).

Vitamins	RDA	D-A-CH RI/AI[1]	Minerals and trace elements	RDA	D-A-CH RI/AI[1]
Vitamin A (µg)	800	1,000	Potassium (mg)	2,000	2,000
Vitamin D (µg)	5	5	Chloride (mg)	800	830
Vitamin E (mg)	12	14	Calcium (mg)	800	1,000
Vitamin K (µg)	75	70	Phosphorus (mg)	700	700
Vitamin C (mg)	80	100	Magnesium (mg)	375	350
Thiamin (mg)	1.1	1.2	Iron (mg)	14	10
Riboflavin (mg)	1.4	1.4	Zinc (mg)	10	10
Niacin (mg)	16	16	Copper (mg)	1	2.0-5.0
Vitamin B6 mg)	1.4	1.5	Manganese (mg)	2	1.0-1.5
Folic acid (µg)	200	400	Fluoride (mg)	3.5	3.8
Vitamin B12 µg)	2.5	3	Selenium (µg)	55	30-70
Biotin (µg)	50	30-60	Chromium (µg)	40	30-100
Pantothenic acid (mg)	6	6	Molybdenum (µg)	50	50-100
			Iodine (µg)	150	150-200

Notes: [1]RI- Recommended intake, AI- Estimated value for adequate intake (provided value for adult men, 25-51 years).

Table 3. Vitamins and minerals which may be declared and their recommended daily allowances (RDAs) as set in Council Directive 90/496/EEC on nutrition labelling for foodstuffs (amended in 2008) in comparison to D-A-CH reference values (D-A-CH 2002)

4.2 Health claims

Regulation (EC) 1924/2006 defines *health claim* as any claim which states, suggests or implies that a relationship exists between a food category, a food or one of its constituents and health. All health claims should be authorised and included in the list of authorised claims. The quantity of the food and pattern of consumption required to obtain the claimed beneficial effect should be included in the labelling and reasonably expected to be consumed in context of a varied and balanced diet. The claim must be specific, based on generally accepted scientific data and well understood by the average consumer. Reference to general, non-specific benefits of the nutrient or food for overall good health or health-related well-being may only be made if accompanied by a specific health claim (EC 2006). All authorised health claims are listed in a public *Community Register of nutrition and health claims made on food* which includes the wordings of claims and the conditions applying to them, together with any restrictions. Basically, the regulation distinguishes three categories of health claims (Table 4). General function claims describe the role of a food in body functions, including the sense of hunger or satiety and not referring to children, while disease risk reduction claims state that the consumption of a food or food constituent significantly reduces a risk factor in the development of a human disease. The regulation also mentions claims referring to children's development and health, for which no further definition is given.

General function health claims	Claims referring to children's development and health	Reduction of disease risk claims
Claims not referring to children and describing the role of a nutrient or other substance in growth, development and the functions of the body; or psychological and behavioural functions; or slimming or weight control or a reduction in the sense of hunger or an increase in the sense of satiety or to the reduction of the available energy from the diet.	No further definition provided.	Claims that state, suggest or imply that the consumption of a food category, a food or one of its constituents significantly reduces a risk factor in the development of a human disease.

Table 4. Types of health claims according to EC regulations (EC 2006)

5. Evaluation and authorisation of health claims

The regulation requires authorization of all health claims by the European Commission through the Comitology procedure, following scientific assessment and verification of a claim by the European Food Safety Authority (EFSA) (Pravst 2010). Claims referring to children's development and health, reduction of disease risk claims and general function claims which are based on newly developed scientific evidence are submitted directly by companies. In cases where health claim substantiation is based on (unpublished) proprietary data and if a claim cannot be substantiated without such data, the applicant can request 5 years of protection of such data.

In addition to the described procedure of health claims submission by applicants, all health claims which were on the market prior to 2006 were included in the so-called *Article 13 evaluation process*. Lists of general function claims on the market were provided by EU member states in collaboration with the industry and included in a consolidated list which formed the basis for the EFSA evaluation. In 2009 it became clear that the process of evaluating existing general function health claims would be much more demanding than expected. After examining over 44 000 claims supplied by the EU member states, the EFSA has received a consolidated list of over 4600 general function claims which entered the evaluation process. The evaluation of most claims was finished in 2011 with 341 published scientific opinions, and providing scientific advice for 2758 general function health claims, about 20% being favourable. While some claims were withdrawn, about 1500 claims on so-called *botanicals* have been placed on hold by the European Commission pending further consideration on how to proceed with these.

Scientific and technical guidance for the preparation and presentation of an application for authorisation of a health claim (EFSA 2011c) and *General guidance for stakeholders on the evaluation of health claims* (EFSA 2011a) were prepared by the EFSA. Scientific substantiations of claims are performed by taking into account the totality of the available pertinent scientific data and by weighing up the evidence, in particular whether:

- the effect is relevant for human health;
- there is an established cause-and-effect relationship between the consumption of the food and the claimed effect in humans;
- the effect has been shown on a study group which is representative of the target population; and
- the quantity of the food and pattern of consumption required to obtain the claimed effect could reasonably be achieved as part of a balanced diet.

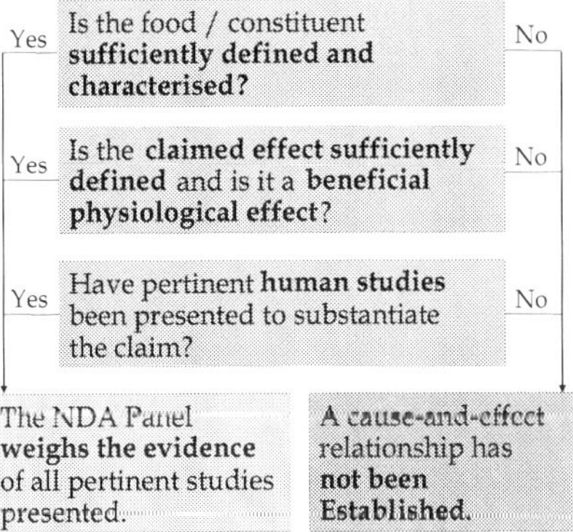

Fig. 2. Key questions addressed by the EFSA in the scientific evaluation of health claims (EFSA 2011a)

The process of the implementation of health claims legislation has involved a steep learning curve which is far from complete. While the EFSA has had to cope with an unprecedented and unforeseen workload, coupled with very short deadlines, the industry is financing expensive trials which are often still not being performed using standards that would enable successful substantiation. More specific guidance is therefore being released to help industry to substantiate health claims, starting with *Guidance for health claims related to gut and immune function* (EFSA 2011b). Guidance for health claims related to (1) antioxidants, oxidative damage and cardiovascular health, (2) bone, joints and oral health, (3) appetite ratings, weight management and blood glucose concentrations, (4) physical performance, and (5) cognitive function is also being developed. The key questions addressed by the EFSA in the scientific evaluation of health claims are presented in Fig. 2.

5.1 The wording of a health claim

The wording of a health claim must be specific, it must reflect the scientific evidence and it must be well understood by the average consumer. In the process of the scientific evaluation the EFSA can propose changing the wording to reflect the scientific evidence, but such wording is sometimes hard to understand by the general population. An example of such a procedure can be shown with a health claim for specific water-soluble tomato concentrate and maintenance of a healthy blood flow (Table 5). The EFSA considered that the wording which was proposed by the applicant did not reflect the scientific evidence because only measures of platelet aggregation have been used, whereas blood flow and circulation also depend on many other factors. It changed the wording to reflect the scientific evidence, although the Commission later reworded it to be understood by consumers (Pravst 2010).

	Claim wording
Applicant proposal:	Helps to maintain a healthy blood flow and benefits circulation
EFSA's proposal:	Helps maintain normal platelet aggregation
Authorised by the EC:	Helps maintain normal platelet aggregation, which contributes to healthy blood flow

Table 5. The wording of a health claim for specific water-soluble tomato concentrate and maintenance of a healthy blood flow (EFSA 2009; Pravst 2010).

When discussing the wording of health claims it should also be noted that, for reduction of disease risk claims, the wording should refer to the specific risk factor for disease and not to disease alone.

5.2 Food characterisation

The sufficient characterisation of foods or food constituents is important for the proper scientific evaluation of a health claim application and must enable authorities to control the market. In the scientific evaluation process the characterisation is needed to identify the food or food constituent, to define the appropriate conditions of use, and to connect it with the provided scientific studies. These studies should be performed using the same food or food constituent (which should also be sufficiently characterised). The lack of characterisation was one of the most common reasons for the EFSA's non-favourable opinions regarding general function claims (Pravst 2010).

In relation to characterisation it is necessary to distinguish between a specific formulation, specific constituent and a combination of constituents. All combinations must be characterised in detail, particularly in relation to the active constituents. Beside the physical and chemical properties and the composition, the analytical methods must also be provided (EFSA 2011a). In cases where variations in composition could occur, variability from batch to batch should be addressed together with stability with respect to storage conditions during shelf life (EFSA 2011a). Where applicable it is useful to show that a constituent is bioavailable and provide a rationale of how the constituent reaches the target site.

For microorganisms genetic typing should be performed at the strain level with internationally accepted molecular methods and the naming of strains according to the International Code of Nomenclature. Depositing samples in an internationally recognised culture collection for control purposes has been suggested (EFSA 2011a). The stability of the microorganisms and the influence of the food matrix on their activity should be studied.

For plant products the scientific name of the plant should be specified, together with the part of the plant used and details of the preparation used, including details of the extraction, drying etc. It is beneficial when the applicant can show that the composition of the plant-derived product can be controlled by analyses of specific chemical ingredients. For example, the above mentioned specific tomato concentrate was characterised on the basis of a clearly described production manufacturing process of tomatoes (*Lycopersicum esculentum*) together with a detailed chemical specification of the most important components and demonstrated batch-to-batch reproducibility (EFSA 2009). Chemical compounds which have been shown to have a beneficial effect *in vitro* were identified and quantified using the HPLC-MS technique, and the presence of unspecified constituents was limited. Several chemical and physical characteristics were assessed during stability testing, including breakdown products and the microbiological status. The bioactive components were shown to survive and to retain their activity *in vitro* over typical product shelf lives and when the product was included in specified matrixes.

Study population	Function
Subjects with untreated hypertension	maintaining normal blood pressure (for general population)
Overweight and obese subjects	weight reduction (for general population)
Patients with functional constipation	bowel function (for general population)
Irritable Bowel Syndrome (IBS) patients	bowel function (for general population)
Irritable Bowel Syndrome (IBS) patients	gastro-intestinal discomfort (for general population)
Subjects with immunosuppression	immune function (for population groups considered to be at risk of immunosuppression)

Table 6. Some specific study populations which were found as appropriate for scientific substantiation of health claims (EFSA 2011b; Pravst 2010)

5.3 Specific conditions of use

The quantity of the food or food constituent and pattern of consumption must be specified together with possible warnings, restrictions on use and directions for use. It is important that the consumer can consume enough food as part of a balanced diet to obtain the claimed effect (EFSA 2011a). The target population must be specified and in relation to this, it is critical that the specific study group in which the evidence was obtained is also representative of the target population for which the claim is intended (Pravst 2010). A key question is whether the results of studies on patients can be extrapolated to the target population. A judgement on this is made on a case-by-case basis, but in most cases patients were not found to be an appropriate study group. Usually, such studies are not considered as pertinent. In cases where studies are not performed on a representative of the target population, evidence must be provided that the extrapolation can be performed. No scientific conclusions can be drawn from studies on patients with genetically and functionally different cells and tissues. Some examples of specific study populations which were found as appropriate for scientific substantiation of health claims are listed in Table 6.

5.4 Relevance of the claimed effect

The claimed effect should be clearly defined and relevant to human health (Pravst 2010). This can be demonstrated with the example of the *Caralluma fimbriata* extract, and its effect on one's waist circumference (EFSA 2010c). Studies showed a statistically significant reduction in waist circumferences, but a reduction in one's waist circumference is not a beneficial physiological effect if it is not accompanied by an improvement in the adverse health effects of excess abdominal fat. On the contrary, only a slight improvement in parameters which are widely accepted as important to human health can be recognised as key evidence, such as in the case of a general functional claim for the role of omega-3 in maintaining normal blood pressure (EFSA 2009i). The EFSA concluded that high doses (3 g per day) of docosahexaenoic (DHA) and eicosapentaenoic acid (EPA) may have smaller, but statistically significant, effects in normotensives of about 1 mmHg; better results were observed in subjects with untreated hypertension.

5.5 Scientific substantiation of the claimed effect

Human data are critical for substantiating a claim and particular attention is paid to whether such studies are pertinent to the claim. Studies need to be carried out with the subject product with similar conditions of use in a study group representative of the population group and using an appropriate outcome measure of the claimed effect (EFSA 2011a). Using appropriate outcome measures can be a challenge because a limited number of validated biomarkers is available (Cazaubiel and Bard 2008). Biomarkers are characteristics that are objectively measured and evaluated as indicators of normal biological processes, pathogenic processes or pharmacologic responses to therapeutic intervention. Well-performed human intervention trials are particularly important for successful substantiation. Double-blind, randomised, placebo-controlled trials are considered the *gold standard* not only for the substantiation of disease risk reduction claims but also for general function claims (Pravst 2010). During the scientific evaluation such trials are assessed critically to assure there are no weaknesses. A good study design, proper performance, well-defined statistics and appropriate statistical power (enough subjects) are key issues in this context. In some cases, non-blind studies are also acceptable, particularly in the case of non-processed foods where blinding is not possible. This was confirmed recently in the case of a general function claim application for dried plums and their laxative effect (EFSA 2010h) – a study in which subjects free of gastrointestinal and eating disorders were randomised to consume either dried plums or grape juice was found to be pertinent. Nevertheless, taking the results of other studies into account there was insufficient evidence to establish a cause-and-effect relationship.

Human observational studies and data from studies in animals or model systems are considered only as supporting evidence.

6. Components of functional foods

As already mentioned, limiting certain food components or simply delivering nutrient intake cannot be regarded as a healthy diet. However, the functional effects of foods are usually studied in relation to their composition and bioactive components (in some cases also with isolated components). In this chapter over 300 scientific opinions on general function health claims (for the ones which are currently present on the European market) were reviewed. On the basis of these opinions and discussion within member states the European Commission will prepare a list of substantiated health claims and their conditions of use.

6.1 Vitamins, minerals and trace elements

When talking about essential nutrients we need to consider that there is a well-established consensus among scientists on many functions of such nutrients. In the evaluation of health claims we may rely on such a consensus and in such cases it may not be necessary to review the primary scientific studies on the claimed effect of the food (EFSA 2011a). On such bases many general function health claims for vitamins, minerals and trace elements received favourable opinions in the assessment process (Table 7). In most cases the proposed condition of use of such a claim is that the food is at least a source of the nutrient (15% of the RDA specified in Table 3 per 100 g or 100 ml, or per package if the package contains only a single portion).

Several essential nutrients were recognised to possess antioxidant activity, which is commonly communicated on functional foods. While the current evidence does not support the use of antioxidant supplements in the general population or in patients with certain diseases (Bjelakovic et al. 2008), some food components are indeed included in the antioxidant defence system of the human body, which is a complex network including endogenous antioxidants and dietary antioxidants, antioxidant enzymes, and repair mechanisms, with mutual interactions and synergetic effects among the various components (EFSA 2009f). For example, vitamin C functions physiologically as a water-soluble antioxidant and plays a major role as a free radical scavenger. On such a basis a cause-and-effect relationship has been established between the dietary intake of vitamin C and the protection of DNA, proteins and lipids from oxidative damage. Other antioxidants include vitamin E, riboflavin, copper, manganese and selenium (Table 7).

A series of vitamins and trace elements is involved in the functioning of the human immune system, but their ability to promote immunity function is questionable, especially in populations with adequate intake. For example, zinc deficiency is associated with a decline in most aspects of immune function; lymphopaenia and thymic atrophy are observed, cell mediated and antibody mediated responses are reduced (EFSA 2009~). Additionally, zinc deficiency appears to induce apoptosis, resulting in a loss of B-cell and T-cell precursors within the bone marrow. Adequate zinc status is necessary for natural killer cell function and zinc deficiency renders people more susceptible to infections. A cause-and-effect relationship has been established between the dietary intake of zinc and the normal function of the immune system but it was noted that there is no evidence for inadequate intake of zinc in the general EU population (EFSA 2009~). A function in the immune system was also recognised for copper, selenium and various vitamins (vitamins A, D, B6, B12, C and folate) (Table 7). In the case of vitamin D it was also concluded that it contributes to healthy inflammatory response (EFSA 2010*f*).

Health realtionship[1]	Vitamins	Minerals and trace elements
Function as antioxidant	Vitamin E (EFSA 2010„) Riboflavin (EFSA 2010}) Vitamin C (EFSA 2009{)	Cu (EFSA 2009h) Mn (EFSA 2009r) Se (EFSA 2009v)
Function in immune system	Vitamin A (EFSA 2009x) Vitamin D (EFSA 2010*f*) Vitamin B6 (EFSA 2009z) Folate (EFSA 2009k) Vitamin B12 (EFSA 2009y) Vitamin C (EFSA 2009{)	Cu (EFSA 2009h) Fe (EFSA 2009n) Se (EFSA 2009v) Zn (EFSA 2009~)
Function in brain or nervous system	Thiamine (EFSA 2009c; EFSA 2010) Riboflavin (EFSA 2010}) Niacin (EFSA 2009s; EFSA 2010w) Pantothenate (EFSA 2009t; EFSA 2010x) Vitamin B6 (EFSA 2009z; EFSA 2010) Folate (EFSA 2010j) Vitamin B12 (EFSA 2010€) Biotin (EFSA 2010d) Vitamin C (EFSA 2009{; EFSA 2010,)	Cu (EFSA 2009h) Fe (EFSA 2009n; EFSA 2010n) I (EFSA 2010m) K (EFSA 2010{) Mg (EFSA 2009q; EFSA 2010r) Zn (EFSA 2009~)
Function in bone or teeth	Vitamin D (EFSA 2009g; EFSA 2009 |) Vitamin K (EFSA 2009})	Ca (EFSA 2009f; EFSA 2009g) Mg (EFSA 2009q) P (EFSA 2009u) Cu (EFSA 2009h) K (EFSA 2010{) Mn (EFSA 2009r) Zn (EFSA 2009~) F- (EFSA 2009j)
Function in skin, hair or connective tissues	Vitamin A (EFSA 2009x) Riboflavin (EFSA 2010}) Niacin (EFSA 2009s) Biotin (EFSA 2009e)	Cu (EFSA 2009h) I (EFSA 2009m) Se (EFSA 2009v; EFSA 2010~) Zn (EFSA 2010†)
Function in vision	Vitamin A (EFSA 2009x) Riboflavin (EFSA 2010})	Zn (EFSA 2009~)
Function in muscle	Vitamin D (EFSA 2010*f*)	Ca (EFSA 2009f) Cu (EFSA 2009h) K (EFSA 2010{) Mg (EFSA 2009q)

Health realtionship[1]	Vitamins	Minerals and trace elements
Function in blood, haemoglobin and oxygen transport	Vitamin A (EFSA 2009x) Vitamin K (EFSA 2009}) Riboflavin (EFSA 2010}) Vitamin B6 (EFSA 2009z) Folate (EFSA 2009k) Vitamin B12 (EFSA 2009y) Vitamin C (EFSA 2009{)	Ca (EFSA 2009f) Cu (EFSA 2009h) Fe (EFSA 2009n)
Function in cell division & differentiation	Vitamin A (EFSA 2009x) Vitamin D (EFSA 2009 |) Folate (EFSA 2009k) Vitamin B12 (EFSA 2009y)	Ca (EFSA 2010e) Fe (EFSA 2009n) Mg (EFSA 2009q) Zn (EFSA 2009~)
Function in regulation of hormones	Pantothenic acid (EFSA 2009t) Vitamin B6 (EFSA 2010)	I (EFSA 2009m) Se (EFSA 2009v) Zn (EFSA 2010†)
Function in metabolism of nutrients	Thiamine (EFSA 2009c) Riboflavin (EFSA 2010}) Niacin (EFSA 2009s) Pantothenate (EFSA 2009t) Vitamin B6 (EFSA 2009z; EFSA 2010) Vitamin B12 (EFSA 2009y) Biotin (EFSA 2009e) Vitamin C (EFSA 2009{)	Ca (EFSA 2009f) Cr (EFSA 2010f) Cu (EFSA 2009h) Fe (EFSA 2009n) I (EFSA 2009m) Mg (EFSA 2009q) Mn (EFSA 2009r) Mo (EFSA 2010v) P (EFSA 2009u) Zn (EFSA 2009~; EFSA 2010†)

Notes: [1] See references for details on specific health claims. A reference to general, non-specific benefits of the nutrient for overall good health or health-related well-being may only be made if accompanied by a specific health claim.

Table 7. Selection of general function health claims for vitamins, minerals and trace elements as assessed by the EFSA

According to a recent Irish study, brain function is most commonly communicated on soft drinks. In fact, all known water-soluble vitamins, and several minerals and essential elements (copper, iron, iodine, potassium, magnesium and zinc) are recognised as important in this context (Table 7). Functions which were favourably assessed by the EFSA include maintenance of normal function of the nervous system, cognitive function, psychological functions, neurological functions, and reduction of tiredness and fatigue (depending on the particular nutrient). The role of vitamin B6 in the maintenance of mental performance was also evaluated. Human studies have shown the effect of vitamin B6 on symptoms of depression, cognition, ageing, premenstrual syndrome and memory performance, however the daily doses for supplementation ranged from 40 - 600 mg (EFSA 2009z), well above the Tolerable Upper Intake Level (UL) (25 mg). Such a claim is not appropriate because it would encourage excess consumption of vitamin B6.

The effect of nutrition on bone health is well established. The maximum attainment of peak bone mass achieved during growth and the rate of bone loss with advancing age are the two principal factors affecting adult bone health. With increasing life expectancy the epidemiology of bone-health related conditions is changing drastically and represents a major public health threat in the Western world. The careful formulation of functional foods represents an important step in the promotion of bone health and consequently on the quality of one's life, but to minimise health risks for consumers both the positive and the negative effects of active ingredients should be considered when developing such products. Calcium and vitamin D are considered the most important constituents of functional foods for the support of bone health (Earl et al. 2010; Palacios 2006) and received favourable opinions for maintaining normal bone and teeth. Additionally, the role of vitamin D in maintaining normal blood calcium concentrations and absorption and utilisation of Ca and P was confirmed (EFSA 2009|). In the last decade, the role of vitamin K in γ-carboxylation of osteocalcin has also been recognised (Ikeda et al. 2006; Katsuyama et al. 2004) as about 10-30% of osteocalcin in the healthy adult population is in an under-carboxylated (and therefore inactive) state (Vermeer et al. 2004). Other nutrients recognised in maintaining normal bone health are magnesium, zinc, manganese, potassium, copper and phosphorus (Table 7). However, the enrichment of (functional) foods or drinks with phosphorus is controversial as its intake can easily exceed the recommendations and a bigger intake might have adverse effects on bone health (Pravst 2011b). Therefore, both health and ethical concerns arise as to whether such claims should be allowed, even though science is not yet clear on this issue. A useful solution in such cases would be to authorise the claim with more specific conditions of use.

Most of the nutrients with a function in bone health were also recognised as important in the maintenance of normal teeth (vitamin D, calcium, magnesium and phosphorus). Additionally, the beneficial role of fluoride for tooth health (by counteracting hydroxyapatite demineralisation and supporting remineralisation) is widely accepted as fluoride can replace hydroxyl ions in the hydroxyapatite crystal lattice of a tooth's tissues and make it more resistant to acid exposure (EFSA 2009j).

Biotin is the only vitamin that has received a favourable opinion for its role in the maintenance of normal hair (EFSA 2009e). While there were no studies available in which improvement in hair loss and hair quality were studied using objective methods, it is known that the symptoms of biotin deficiency include thinning hair and progression to a loss of all hair, including eyebrows and lashes (EFSA 2009e). Copper, selenium and zinc were also recognised as important in the maintenance of normal hair (Table 7). Other functions of these nutrients include the maintenance of normal nails, skin or mucous membranes, while a role in the normal formation of connective tissue was determined only for manganese (EFSA 2010s) and copper (EFSA 2009h).

Vitamin A and compounds with pro-vitamin A activity (i.e. beta-carotene) are recognised to have a function in maintaining normal vision. Retinal is required by the eye for the transduction of light into neural signals which are necessary for vision and without an adequate level of vitamin A in the retina night blindness occurs (EFSA 2009x). In a different way vitamin A deficiency leads to a reduction in mucus production by the goblet cells of the conjunctival membranes and the cornea becomes dry. Riboflavin (EFSA 2010}) and zinc (EFSA 2009~) also received favourable opinions.

With muscle weakness being a major clinical syndrome of vitamin D deficiency, vitamin D is the only vitamin with a favourable opinion about its role in the maintenance of normal muscle function (EFSA 2010f). Clinical symptoms of the deficiency include proximal muscle weakness, diffuse muscle pain and gait impairments such as a waddling way of walking.

Dietary intake of calcium, magnesium, potassium and copper was also connected with muscle function (Table 7). In a similar manner the importance of sodium was confirmed; however, current sodium intake in the general EU population is more than adequate and directly associated with a greater likelihood of increased blood pressure, which in turn has been directly related to the development of cardiovascular and renal diseases (EFSA 2011y). Promoting sodium intake with the use of health claims is in obvious conflict with current targets for a reduction in dietary sodium intakes (Cappuccio and Pravst 2011).

Several nutrients have received favourable opinions for their function in blood, formation of haemoglobin or oxygen transport (Table 7). In haemoglobin, iron allows reversible binding of oxygen and its transport in the erythrocytes to the tissues. The most common consequence of iron deficiency is microcytic anaemia. The role of iron in oxygen transport, normal formation of red blood cells and haemoglobin is obviously well established (EFSA 2009n). Interestingly, there is still a significant prevalence of iron deficiency in Europe, especially among children and women of reproductive age and during pregnancy. Because dietary iron is absorbed as Fe^{2+} (and not as Fe^{3+}), reducing agents, including vitamin C, promote non-haem iron absorption by keeping it reduced. The function of vitamin C and non-haem iron absorption therefore received a favourable opinion (EFSA 2009{). Other vitamins and minerals with a confirmed cause-and-effect relationship were vitamin A in the metabolism of iron, vitamin K and calcium in normal blood coagulation, folate in normal blood formation, copper in normal iron transport, and riboflavin, vitamin B6 and B12 in the normal formation of red blood cells.

Other specific functions of vitamins, minerals and trace elements which are rarely communicated on food labelling were also included in the evaluation process. The ones with favourable outcomes include a function in cell division & differentiation, regulation of hormones and metabolism of nutrients (Table 7). In addition, some of the following very specific health claims were assessed: selenium and the maintenance of normal spermatogenesis (EFSA 2009v); zinc and normal fertility and reproduction, normal DNA synthesis; normal vitamin A metabolism and normal acid-base metabolism (EFSA 2009~); vitamin C and normal collagen formation (EFSA 2009{) and regeneration of the reduced form of vitamin E (EFSA 2010,); vitamin B6 and normal glycogen metabolism (EFSA 2009z) and normal cysteine synthesis (EFSA 2010); folate and normal maternal tissue growth during pregnancy (EFSA 2009k); phosphorus and normal functioning of cell membranes (EFSA 2009u); magnesium and electrolyte balance (EFSA 2009q); and, calcium and normal functioning of digestive enzymes (EFSA 2009f).

Vitamin B6, B12 and folic acid were recognised as important for their contribution to normal homocysteine metabolism (EFSA 2009k; EFSA 2010€; EFSA 2010). Deficiencies in these vitamins lead to impaired homocysteine metabolism causing mild, moderate, or severe elevations in plasma homocysteine (depending on the severity of the deficiency) as well as the coexistence of genetic or other factors that interfere with homocysteine metabolism (EFSA 2010). In addition to these vitamins, a contribution to normal homocysteine metabolism has also been established for betaine (EFSA 2011g) and choline (EFSA 2011l). Choline can function as a precursor for the formation of betaine, which acts as a methyl donor in the remethylation of homocysteine in the liver by the enzyme betaine-homocysteine methyltransferase (EFSA 2011g). Choline can be biosynthesised in our liver and is therefore not a vitamin; however, most men and postmenopausal women need to consume it in their diets (Zeisel and Caudill 2010). The nutritional need for choline greatly depends on gender, age and genetic polymorphisms. For this reason the EFSA was unable to propose conditions of use; however, it was noted that a nutrient content claim has been authorised in the US, based on the adequate intake for adult males (550 mg of choline per day). Furthermore, choline has also received a favourable opinion for the maintenance of normal liver function and its contribution to normal lipid metabolism (EFSA 2011l). In relation to betaine it must be mentioned that the application for the use of betaine as a novel food ingredient in the EU was rejected, mainly due to concerns over the safety of betaine with long term use (EFSA 2005). Despite this, betaine as a constituent of traditional foods is not considered a novel food.

In practice, a functional food can also be a food from which a component has been removed to provide benefits not otherwise available (Ashwell 2002). Good examples of such foods are foods low or very low in sodium, whose consumption helps to maintain normal blood pressure (EFSA 2011p). In fact, sodium intake and blood pressure demonstrate a close and consistent direct relationship (Cappuccio and Pravst 2011) and there is extensive evidence to support this (He and MacGregor 2010). A 4.6g reduction in daily dietary salt intake decreases BP by about 5.0/2.7 mmHg in hypertensive individuals and by 2.0/1.0 mmHg in normotensive people. Dose-response effects have been consistently demonstrated in adults and children (He and MacGregor 2003). A 5g higher salt intake is associated with a 17% greater risk of total cardiovascular disease and a 23% greater risk of stroke (Strazzullo et al. 2009).

6.2 Fats, fatty acids and fatty acid composition

Fats are a major contributor to total energy intakes in most Western diets, supplying 35-40% of food energy through the consumption of 80-100g of fat per day (Geissler and Powers 2005). All fat sources contain mixtures of saturated, monounsaturated and polyunsaturated fatty acids, but the proportions vary significantly according to the source. Besides being a source of energy, fats also play other diverse roles in the human body. Two fatty acids, linoleic acid and α-linolenic acid are essential in the human diet as they cannot be synthesised in mammalian tissues. The human body can synthesise eicosapentaenoic acid (EPA) and docosahexaenoic acid (DHA) from α-linolenic acid, but the synthesis is not very efficient.

The balance between saturated and unsaturated fats in the diet is known to influence circulating cholesterol concentrations (WHO 2003). Current dietary recommendations for adults in the EU are to limit intake of fats to 20-35 energy % (E%) and to keep the intake of saturated fatty acid (SFA) and trans fatty acid as low as possible (EFSA 2010b). The proposed adequate intakes are: 4 E% for linolenic acid; 0.5 E% for α-linolenic acid (ALA); and 250 mg for EPA+DHA.

In Table 8 the proposed conditions of use of health claims related to fats, fatty acids and fatty acid composition are listed. Cholesterol related claims are one of the most common categories of health claims on the market (Lalor et al. 2010). The role of fatty acid composition in cholesterol management has been confirmed and the proposed conditions of use should stimulate a reduced intake of saturated fatty acids also through product reformulation (EFSA 2011o; EFSA 2011{). Such a product should contain at least 30% less saturated fatty acids compared to a similar product. The role of linolenic and α-linolenic acid in cholesterol management also received favourable opinions, where a food contains at least 15% of the proposed labelling reference intake.

Food component[1]	Function	Conditions of use	Reference
Fat	Normal absorption of fat-soluble vitamins	no conditions of use can be defined	(EFSA 2011n)
Saturated fatty acids	Management of cholesterol	reduced amounts of saturated fatty acids by at least 30% compared to a similar product	(EFSA 2011{) (EFSA 2011o)
Linoleic acid	Management of cholesterol	15% of the proposed labelling reference intake values of 10g linoleic acid per day	(EFSA 2009p)
ALA	Management of cholesterol	15% of the proposed labelling reference intake value of 2g ALA per day	(EFSA 2009b)
Plant sterols	Management of cholesterol	0.8g per day	(EFSA 2010z)
DHA/EPA	Management of blood triglycerides	2-4g per day	(EFSA 2009i)
	Management of blood pressure	3g per day	(EFSA 2009i)
	Heart health	250mg per day	(EFSA 2010i)
DHA	Vision	250mg per day	(EFSA 2010g)
	Brain function	250mg per day	(EFSA 2010g)

Notes: [1]ALA - α-linolenic acid; DHA - docosahexaenoic acid; EPA - eicosapentaenoic acid.

Table 8. Conditions of use health claims related to fats, fatty acids and fatty acid composition as assessed by the EFSA

Plant sterols have been known for several decades to cause reductions in plasma cholesterol concentrations (St Onge and Jones 2003). These are also well known by consumers and are one of the most well-known functional ingredients. Plant sterols are structurally very similar to cholesterol, except that they always contain a substituent at the C24 position on the sterol side chain (Ling and Jones 1995). It is generally assumed that cholesterol reduction results directly from the inhibition of cholesterol absorption through the displacement of cholesterol from micelles. Their concentrations in mammalian tissue are normally very low - primarily due to poor absorption from the intestine and faster excretion from the liver compared to cholesterol (Ling and Jones 1995). The LDL-cholesterol lowering effects of plant sterols and stanols have been reviewed several times, lately by Demonty and co-workers (Demonty et al. 2008). In the general health claim evaluation process plant sterols and stanols were confirmed to be helpful in maintaining normal blood cholesterol levels when the food provides at least 0.8g of plant sterols or stanols daily in one or more servings (EFSA 2010z). Such products may not be nutritionally appropriate for pregnant and breastfeeding women, or for children under the age of five (EFSA 2010z). It must be mentioned that the related reduction of disease risk claims were already authorised for both plant sterols and stanol esters. Lowering blood cholesterol and reference to cholesterol as a risk factor in the development of coronary heart disease can be communicated on foods which provide a daily intake of 1.5-2.4g plant sterols/stanols. Most studies of cholesterol lowering effects were conducted with plant sterols or stanols added to foods such as margarine-type spreads, mayonnaise, and dairy products. Studies are showing an increase in the consumption of plant sterols in Europe. The consumption of foods enriched with plant sterols was recently studied in Belgium; the results indicate that plant sterol-enriched food products are also consumed by the non-target group and that efficient communication tools are needed to better inform consumers about the target group of enriched products, the advised dose per day and alternative dietary strategies to lower blood cholesterol level (Sioen et al. 2011). Although intakes of plant sterols have not produced significant adverse effects, it is not known whether constant consumption at high levels would have any toxic effects (St Onge and Jones 2003). It is therefore logical to advise the food industry to also formulate cholesterol-management foods with other effective functional ingredients, particularly with specific dietary fibres, being a more common constituent of the human diet (see chapter *6.3 Carbohydrates and dietary fibre*).

Intervention studies have demonstrated the beneficial effects of preformed n-3 long-chain polyunsaturated fatty acids (DHA, EPA) on recognised cardiovascular risk factors, such as a reduction in plasma triacylglycerol concentrations and blood pressure, albeit in quite high daily dosages (EFSA 2010b). The proposed daily dose for the management of blood triglycerides and the management of blood pressure is 2-4g and 3g of DHA and EPA respectively (Table 8). The question remains, however, as to whether such claims will be authorised. Lower daily dosages of DHA and EPA are required to maintain normal cardiac function (250mg DHA/EPA), normal vision and brain function (250 mg DHA).

6.3 Carbohydrates and dietary fibre

According to the degree of polymerisation we categorise carbohydrates into sugars (monosaccharides and disaccharides), oligosaccharides (3-9 residues), and polysaccharides with over 9 monomeric residues.

6.3.1 Glycaemic carbohydrates

Glycaemic carbohydrates are digested and absorbed in the human small intestine and provide glucose to body cells as a source of energy. According to their degree of polymerisation, these are sugars and some oligosaccharides and polysaccharides. The main glycaemic carbohydrates in the diet are glucose and fructose (monosaccharides), sucrose and lactose (disaccharides), malto-oligosaccharides and starch (polysaccharides) (EFSA 2010a). Glycaemic carbohydrates provide about 40% of energy intake in average Western diets, with a desirable level at around 55% (Geissler and Powers 2005). Far more glycaemic carbohydrates are consumed in developing countries.

Glucose is the preferred energy source for most body cells and the brain requires glucose for its energy needs. An intake of 130g of dietary glycaemic carbohydrates per day for adults is estimated to cover the glucose requirement of the brain (EFSA 2010a). In the health claim evaluation process a cause-and-effect relationship has been established between the consumption of glycaemic carbohydrates and the maintenance of normal brain function (EFSA 2011r). However, when talking about carbohydrates as constituents of functional foods we are usually discussing either lowering their amount or the functional properties of indigestible polysaccharides (dietary fibre).

Sugar replacement in foods or drinks with xylitol, sorbitol, mannitol, maltitol, lactitol, isomalt, erythritol, D-tagatose, isomaltulose, sucralose or polydextrose was found to be efficient in the reduction of post-prandial blood glucose responses as compared to sugar-containing products (EFSA 2011 |). Lowering sugar levels is also considered beneficial in maintaining the mineralisation of teeth (EFSA 2011s; EFSA 2011|). However, excessive amounts of polyols may result in a laxative effect and this should be communicated to the consumer with an appropriate (and mandatory) advisory statement.

6.3.2 Dietary fibre

Dietary fibre is mostly derived from plants and is composed of complex, non-starch carbohydrates and lignin that are not digestible within the small intestine – since mammals do not produce enzymes capable of their hydrolyses into constituent monomers (Turner and Lupton 2011). Dietary fibre is considered a non-nutrient and contributes no calories to our diet as it reaches the colon intact. However, in the colon dietary fibres are available for fermentation by the resident bacteria, and the metabolites released can be used to meet some of the energy requirements.

Regulatory fibre is defined in *Council Directive 90/496/EEC on nutrition labelling for foodstuffs* as carbohydrate polymers with three or more monomeric units, which are neither digested nor absorbed in the human small intestine and belong to the following categories: (1) edible carbohydrate polymers naturally occurring in the food as consumed; (2) edible carbohydrate polymers which have been obtained from raw food material by physical, enzymatic or chemical means and which have a beneficial physiological effect demonstrated by generally accepted scientific evidence; (3) edible synthetic carbohydrate polymers which have a beneficial physiological effect demonstrated by generally accepted scientific evidence. The EFSA has evaluated several specific dietary fibres for their role in the management of cholesterol, glycaemic response and gut health (Table 9).

The beneficial health effects of water-soluble dietary fibre mainly relate to their ability to improve viscosity of the meal bolus in the small intestine and thus to delay the absorption of nutrients. A lowering of blood cholesterol was established for beta-glucans, chitosan, glucomannan, hydroxypropyl methylcellulose (HPMC), pectins and guar gum when at least 3-10g of fibre was consumed daily (Table 9). Among these, the lowest daily dosage (3g in one or more servings) is required for non-processed or minimally processed beta-glucans from specific sources (EFSA 2009d). The structural features of beta-glucans greatly influence their molecular shape (conformation) and the behaviour of the polysaccharide in a solution, including viscosity. The primary source of beta-glucans and the production processes therefore have a great impact on their functionality. Beta-glucans are widely distributed as non-cellulosic matrix phase polysaccharides in cell walls of the *Poaceae*, which consist of the grasses and commercially important cereal species (Burton and Fincher 2009). Chemically, these are (1,3;1,4)-β-D-Glucans - as they consist of unbranched and unsubstituted chains of (1,3)- and (1,4)-β-glucosyl residues. Their physicochemical and functional properties in cell walls are influenced by the ratio of (1,4)-β-D-glucosyl residues to (1,3)-β-D-glucosyl residues. An example of beta-glucans with a recognised cholesterol effect includes beta-glucans from barley (AbuMweis et al. 2010; Talati et al. 2009), a cereal grain derived from the *Hordeum vulgare*. The proportion between β-(1,3) and β(1,4) linkages is 30 and 70%, respectively (Jadhav et al. 1998). In the polymer chain the blocks of 2 or 3 contiguous (1,4) linkages are separated by single (1,3) linkages; however, blocks of 2 or more adjacent (1,3) linkages are absent (Jadhav et al. 1998). Because the (1,3) linkages occur at irregular intervals the overall shape of the polysaccharide is irregular, which reduces its tendency to pack into stable, regular molecular aggregates and enable the formation of stable viscous solutions. It must be noted that beta-glucans are useful particularly in the production of hard functional foods such as bread, toasts, pasta, extruded flakes, crisps etc. On the market there are also some beta-glucan-enriched liquid functional foods (i.e. yoghurts) and drinks with a labelled cholesterol management effect; however, the length of the polymeric chain of beta-glucans contained in such products is shortened with chemical or enzymatic procedures. Such an ingredient has a lower capacity to form viscous solutions, and consequently a lower efficiency. Thus, the use of a health claim is not allowed in these cases (EFSA 2009d).

Dietary fibre	Management of cholesterol	Management of glycaemic response	Gut health	Reference
Beta glucans	3g per day (non-processed beta-glucans)	4g for each 30g of available carbohydrate (oats and barley beta-glucans)	high in fibre[2]	(EFSA 2009d) (EFSA 2011f) (EFSA 2011u)
Chitosan	3g per day			(EFSA 2011k)
Glucomannan	4g per day			(EFSA 2009l)
HPMC[1]	5g per day	4g per meal		(EFSA 2010l)
Pectins	6g per day	10g per meal		(EFSA 2010y)
Guar gum	10g per day			(EFSA 2010k)
Arabinoxylan		8g of arabinoxylan-rich fibre per 100g of available carbohydrates		(EFSA 2011e)
Resistant starch		14% of total starch as resistant starch (high carbohydrate baked foods)		(EFSA 2011w)
Rye fibre			high in fibre[4]	(EFSA 2011x)
Wheat bran fibre			high in fibre[2] 10g per day[3]	(EFSA 2010…)

Notes: [1]HPMC - hydroxypropyl methylcellulose; [2]Increase in faecal bulk; [3]Reduction in intestinal transit time; [4]Changes in bowel function; Reference to general, non-specific benefits of the nutrient for overall good health or health-related well-being may only be made if accompanied by a specific health claim.

Table 9. Conditions of use for general function health claims of various dietary fibres as proposed by the EFSA

Decreasing the magnitude of elevated blood glucose concentrations after consuming carbohydrate-rich food is a critical target in the production of low glycaemic index (GI) foods. It is well established that the management of glycaemic responses is beneficial to human health, particularly for people with impaired glucose tolerance (which is common among the general population). Various specific dietary fibres were recognised as beneficial in the reduction of post-prandial glycaemic responses. In most cases, the rationale for such a function is, similar to cholesterol lowering, related to their ability to achieve improved viscosity of the meal bolus in the small intestine. This enables a delay in the absorption of sugars - which is considered beneficial as long as post-prandial insulinaemic responses are not disproportionally increased. In the evaluation process the EFSA found a cause-and-effect relationship between the mentioned effect and the consumption of beta-glucans from oats and barley, HPMC, pectins, arabinoxylan produced from wheat endosperm and resistant starch. Various conditions of use were proposed (Table 9).

Both water-soluble and water-insoluble dietary fibres are also known to support gut health through changes in bowel function. Reduced transit time, more frequent bowel movements, increased faecal bulk or softer stools may be a beneficial physiological effect, provided these changes do not result in diarrhoea (EFSA 2011x). On the basis of such changes the functional role of rye fibre was confirmed to contribute to normal bowel function in foods providing at least 6g of such fibre per 100g (or at least 3g per 100kcal), being *high in fibre* (EFSA 2011x). The ability to increase faecal bulk has also been confirmed for wheat, oat and barley grain fibre (Table 9). Similar to the previous case, the proposed conditions of use are that the food is *high in fibre*. It is well established that the bulking effect of dietary fibre is closely related to the physico-chemical properties of the fibre, and in that way to the degree of fermentation by the gut microbiota in the large intestine (EFSA 2011u). The insoluble components of fibre are minimally degraded by colonic bacteria and thus remain to trap water, thereby increasing faecal bulk. In contrast, the bulking effects of soluble dietary fibre are determined by the higher extent of fermentation, and thus an increase in the bacterial mass in faeces (EFSA 2011u). A somewhat different support of gut health can be based on the ability of dietary fibre to reduce intestinal transit time. Such a function was found for wheat bran fibre, which increases the water holding capacity of the contents of the intestine, increases intestinal and pancreatic fluid secretion and thus increases the velocity of chyme displacement through the intestine if at least 10g per day is consumed (EFSA 2010...). A similar effect, but with a different mechanism of action, was also confirmed for lactulose – a synthetic sugar used in the treatment of constipation (EFSA 2010p). In the colon, lactulose is broken down to lactic acid and to small amounts of acetic and formic acids by the action of beta-galactosidases from colonic bacteria. This process leads to an increase in osmotic pressure and a slight acidification of the colonic content, causing an increase in stool water content and a softening of stools (EFSA 2010p). However, due to the medicinal use of lactulose in some EU countries the authorisation of such a health claim on foods is questionable.

6.3.3 Prebiotics

Prebiotics were defined as non-digestible food ingredients that beneficially affect the host by selectively stimulating the growth or activity of one or a limited number of bacterial species already resident in the colon, and thus attempt to improve host health (Gibson and Roberfroid 1995). An intake of prebiotics can modulate the colonic microbiota by increasing the number of specific bacteria and thus changing the composition of the microbiota.

Consumers perceive prebiotics as having health benefits. The *Guidance on the implementation of regulation No 1924/2006 on nutrition and health claims made on foods* specifies that a claim is a health claim if, in the naming of the substance or category of substances, there is a description or indication of functionality or an implied effect on health. The examples provided include the claim *contains prebiotic fibres*. Such claims should therefore only be used if the food contains prebiotic fibres with a scientifically proven effect and when such a claim is accompanied by a specific health claim. In practice, this is not yet the case and such claims are still very common on the market. Only a few (prebiotic) fibres have received favourable opinions from the EFSA. One example is the already-mentioned oat and barley grain fibres - whose bulking effects are determined by the higher extent of fermentation and thus an increase in the bacterial mass in faeces. On such a basis their ability to increase faecal bulk has been confirmed (EFSA 2011u) (Table 9). Several fibres were evaluated for other functions, i.e. for their role in maintaining healthy gastro-intestinal function by increasing the number of bifidobacteria in the gut. In some cases the studies clearly demonstrated a significantly increased number of bifidobacteria in the gut; however, there was no direct evidence provided that changes in the number of bifidobacteria in the gut are beneficial for gut function (EFSA 2009a). For such claims the beneficial effect needs to be shown.

6.4 Other food or food constituents
6.4.1 Antioxidants

Numerous food constituents possess antioxidant activity, yet, in the health claim evaluation process, only a few of them were assessed with a favourable outcome. As in the prebiotics case, the claim *contains antioxidants* is considered a health benefit and should be accompanied by a specific health claim. However, apart from vitamins and trace elements (Table 7), until now only olive oil polyphenols have been recognised to possess antioxidant activity which is beneficial to human health, specifically in the protection of LDL particles from oxidative damage (Covas et al. 2006a; Covas et al. 2006b; EFSA 2011v; Marrugat et al. 2004; Weinbrenner et al. 2004). Hydroxytyrosol and its derivatives (e.g. oleuropein complex and tyrosol) are the key compounds with such activity, and to bear the claim olive oil it should contain enough of them to provide 5mg of these compounds daily. It was noted that the concentrations in some olive oils does not allow consumption of such an amount of polyphenols in the context of a balanced diet (EFSA 2011v). Many other known antioxidants (including flavonoids and flavonols, lycopene, lutein etc.) received unfavourable opinions, mostly because of poor characterisation, non-specific health effects or poor evidence for such effects. In a recent draft of the *Guidance on the scientific requirements for health claims related to antioxidants, oxidative damage and cardiovascular health* the EFSA noted that it is not established that changes in the overall antioxidant capacity of plasma exert a beneficial physiological effect in humans. A beneficial physiological effect will therefore need to be proven for any specific antioxidant for a successful substantiation.

6.4.2 Probiotics

The claim *contains probiotics* is also considered a health benefit and should be accompanied by a specific health claim. However, the EFSA has not released a favourable opinion in relation to live organisms other than for live yoghurt cultures in yoghurt, which were shown to improve the digestion of lactose in yoghurt in individuals with lactose maldigestion (discussed below). The main reasons for the unfavourable opinion were that the microorganisms were not properly characterised (in either the health claim application or the supporting study of the claimed effect), or that there was poor evidence of the beneficial effect. However, many probiotics health claims were returned for re-evaluation and there has been a call to provide additional data for scientific evaluation. Discussion about the results of the re-evaluation has been very speculative, but it is clear that further research will be needed to support a beneficial physiological effect in humans in most (if not all) cases. The specific functions will need to be properly addressed.

6.4.3 Sport nutrition

A series of food constituents have been evaluated for sport related general function health claims. Caffeine was shown to contribute to increase in endurance performance, reduction in the rated perceived exertion during exercise and increased alertness (Table 10). However, it is noted that for children consumption of a dose of 5mg/kg body weight could result in transient behavioural changes, such as increased arousal, irritability, nervousness or anxiety. In relation to pregnancy and lactation, moderation of caffeine intake, from whatever source, is advisable (EFSA 2011i). The role of vitamin C in the maintenance of normal function of the immune system during and after intense physical exercise was also confirmed, but due to the high daily dosage (200mg vitamin C per day in addition to the usual diet) it remains a question if such a claim will be authorised. In some EU countries such a dosage of vitamin C is considered a medicinal use. Other health claims which received favourable opinions include protein and the maintenance of muscle mass, water and the maintenance of normal thermoregulation, and creatine and an increase in physical performance during short-term, high intensity, or repeated bouts of exercise. The proposed condition for use of these claims is presented in Table 10. The role of carbohydrate-electrolyte solutions in the enhancement of water absorption during exercise and in the maintenance of endurance performance was also evaluated. It was proposed that in order to bear the claim a carbohydrate-electrolyte solution should contain 80-350 kcal/L from carbohydrates, and at least 75 % of the energy should be derived from carbohydrates which induce a high glycaemic response, such as glucose, glucose polymers and sucrose. In addition, these beverages should contain between 460mg/L and 1150mg/L of sodium, and have an osmolality between 200-330mOsm/kg water (EFSA 2011j).

Food or ingredient	Function	Conditions of use	Reference
Vitamin C	Maintenance of normal function of the immune system during and after intense physical exercise	200mg vitamin C per day[1] (in addition to the usual diet)	(EFSA 2009{)
Protein	Maintenance of muscle mass	Source of protein	(EFSA 2010 |)
Caffeine	Increase in endurance performance	3mg/kg body weight (one hour prior to exercise)	(EFSA 2011h)
	Increase in endurance capacity		(EFSA 2011h)
	Reduction in the rated perceived exertion during exercise	4mg/kg body weight (one hour prior to exercise)	(EFSA 2011h)
	Increased alertness	75mg caffeine per serving (adults)	(EFSA 2011i)
Creatine	Increase in physical performance during short-term exercise	3g per day	(EFSA 2011m)
Water	Maintenance of normal thermoregulation	2.0 L per day	(EFSA 2011~)

Food or ingredient	Function	Conditions of use	Reference
Carbohydrate-electrolyte solutions	Enhancement of water absorption during exercise	Energy: 80-350kcal/L from carbohydrates and at least 75 % from sugars; Sodium: 460-1,150mg /L Osmolality: 200-330mOsm/kg	(EFSA 2011j)
	Maintenance of endurance performance		(EFSA 2011j)

Notes: [1]In some countries 200mg of vitamin C daily is considered a medicinal use.

Table 10. Conditions of use for sport related general function health claims as proposed by the EFSA

6.4.4 Weight management

For food producers weight-management products are appealing as these can be marketed successfully. Weight loss can be interpreted as the achievement of a normal body weight in previously overweight subjects. In this context, weight loss in overweight subjects without the achievement of a normal body weight is also considered beneficial to health (EFSA 2010t). However, only a few favourable opinions of the EFSA were published (EFSA 2010o; EFSA 2010t; EFSA 2011}). It was established that substituting two daily meals with meal replacements helps to lose weight in the context of energy restricted diets. In order to bear the claims, a food should contain a maximum of 250 kcal/serving and comply with specifications laid in legislation covering foods intended for use in energy-restricted diets for weight reduction (EFSA 2010t). Similarly, it was established that replacing the usual diet with a very low calorie diet (VLCD) helps to lose weight (EFSA 2011}). VLCDs are diets which contain energy levels between 450 and 800 kcal per day, and 100% of the recommended daily intakes for vitamins and minerals. They should contain not less than 50g of high-quality protein, should provide not less than 3g of linoleic acid and not less than 0.5g alpha-linolenic acid, with a linoleic acid/alpha-linolenic acid ratio between 5 and 15, and should provide not less than 50g of available carbohydrates (CODEX STAN 203-19956) (EFSA 2011}). VLCDs are typically used for 8-16 weeks.

6.4.5 Foods for individuals with symptomatic lactose maldigestion

Lactose maldigestion is a common condition characterised by intestinal lactase deficiency. It is most prevalent in Asian, African, Hispanic and Indian populations, but is also common in Europe. Most people with primary lactose maldigestion are usually able to tolerate small amounts of lactose (EFSA 2009o). Ingested lactose is hydrolysed by an enzyme of the microvillus membrane of the enterocytes, called lactase. It is split into the monosaccharides glucose and galactose, which are rapidly and completely absorbed within the small intestine. In persons with lactose maldigestion, undigested lactose reaches the colon where it is degraded to lactic acid, acetic acid, water and carbon dioxide by intestinal bacteria (EFSA 2011q). In some lactose maldigesters this can elicit symptoms of lactose intolerance, which may develop one to three hours after the ingestion of lactose. These symptoms include abdominal pain, bloating, flatulence and diarrhoea. It was established that consumption of foods with reduced amounts of lactose may help to decrease gastro-intestinal discomfort caused by lactose intake in lactose intolerant individuals (EFSA 2011q). However, it was not possible to propose a single condition of use because of the great variation in the individual tolerances of lactose intolerant individuals. Additionally, an improvement in lactose digestion may be of interest to lactose intolerant subjects (EFSA 2009o). A cause-and-effect relationship has been established between the externally administered lactase enzymes and breaking down lactose in individuals with symptomatic lactose maldigestion, which can alleviate lactose intolerance symptoms. The recommended dose was 4500 FCC (Food Chemicals Codex) units with each lactose-containing meal. It was noted that the dose may have to be adjusted to individual needs for lactase supplementation and consumption of lactose-containing products (EFSA 2009o). Live yoghurt cultures in yoghurt were also shown to improve the digestion of lactose in yoghurt in individuals with lactose maldigestion (EFSA 2010q). The effect has been confirmed in a number of human studies and is based on the ability of specific bacteria to produce active β-galactosidase enzymes in the digestive tract. It was proposed that in order to bear the claim, the yoghurt should contain at least 10^8 colony-forming units (CFU) live starter microorganisms (*Lactobacillus delbrueckii* subsp. *bulgaricus* and *Streptococcus thermophilus*) per gram (EFSA 2010q).

6.4.6 Chewing gums

It was established that there is a cause-and-effect relationship between the consumption of sugar-free chewing gum and plaque acid neutralisation, the maintenance of tooth mineralisation, and a reduction of oral dryness (EFSA 2009w). The solubility of tooth hydroxyapatite crystals drops with the lowering of pH and buffering of acids, and limiting the duration of periods of pH drop can prevent demineralisation and promote remineralisation of the hydroxyapatite crystals. Acid is produced in plaque through the fermentation of carbohydrates by acid-producing bacteria and studies have shown that chewing a sugar-free chewing gum enhances saliva flow and counteracts pH drops upon sugar-induced acid production. Chewing for at least 20 minutes after meals may be needed to obtain the beneficial effect. In the absence of fermentable carbohydrates, no clinically relevant reduction on plaque pH may be expected by the consumption of sugar-free chewing gum (EFSA 2009w). In a separate opinion it was established that sugar-free chewing gum with carbamide contributes to plaque acid neutralisation over and above the effect achieved with sugar-free chewing gums without carbamide (EFSA 2011z). At least 20mg carbamide should be added per piece to communicate such claims.

6.4.7 Borderline substances

In some cases the functionality of possible food ingredients enters the borderline between food and medicine. A series of such cases has also been revealed in the evaluation of general function health claims. The most obvious example is the evaluation of monacolin K from rice fermented with the red yeast *Monascus purpureus* (EFSA 2011t). Red yeast rice is a traditional Chinese food product which is still a dietary staple in many Asian countries. Depending on the *Monascus* strains used and the fermentation conditions, it may contain Monacolin K, in the form of hydroxy acid and lactone (also known as lovastatin), which received a favourable opinion for the maintenance of normal blood LDL-cholesterol concentrations (EFSA 2011t). On the EU market there are a number of lovastatin-containing medicinal products and it remains controversial if it is possible to set appropriate conditions of use that would put this product into a food category. In any case, red yeast rice has not been traditionally consumed as a food in Europe and must go through a novel food authorisation procedure. However, its use is already possible in some specific food categories (i.e. in food supplements) in some EU countries. Somewhat similar is the case of melatonin. While in a number of EU countries this hormone is considered a medicinal product, in some countries it is sold as a food supplement. In the health claim evaluation process it was confirmed that melatonin helps to reduce the time taken to fall asleep (EFSA 2011d) and contributes to the alleviation of subjective feelings of jet lag (EFSA 2010u). It is proposed that 1mg of melatonin is consumed close to bedtime (EFSA 2011d).

7. Nutrient profiles

To exclude the use of nutrition and health claims on foods with overall poor nutritional status nutrient profiles should be established, including the exemptions which food or certain categories of food must comply with in order to bear the claims (EC 2006). Unfortunately, this part of the legislation has not yet been implemented, even though the scientific criteria for this were prepared on time (EFSA 2008). The stakeholders have obviously been quite effective in lobbying against the setting of nutrient profiles and there is little evidence of progress in this area since 2009. Currently, it is not even clear if nutrition profiles will be implemented at all (Cappuccio and Pravst 2011). Nevertheless, the producers of functional food must be aware that the overall composition of a final product should provide health benefits. Particular care should be directed towards those nutrients with the greatest public health importance for EU populations, such as saturated fatty acids, sodium and sugar, intakes of which generally do not comply with nutrient intake recommendations in many Member States (EFSA 2008).

8. Quality of functional foods

The quality and safety of functional foods is entirely the responsibility of the producer but can be controlled by national authorities. In practice, such controls mainly focus on assuring adequate safety by controlling for contaminants and additives (Pravst and Žmitek 2011). Nutritional composition is usually not considered a health risk and is therefore less controlled. In fact, while labelling requirements have been in existence in many countries for more than a decade, analyses of many food constituents are still challenging. The EU legislation currently concentrates on regulating the use of vitamins and minerals, while the use of other substances with a nutritional or physiological effect is not regulated in detail. When discussing the safety and quality control assessment of foods containing particular ingredients with biological activity we must distinguish products on the basis of their active ingredients; i.e. chemically stable dietary minerals, less stable vitamins, chemical compounds other than vitamins and minerals, living microorganisms (i.e. probiotics) etc. (Pravst and Žmitek 2011). The appropriate content of these ingredients in final products must be achieved using suitable production standards (including quality control of both raw materials and the final product) and stability during the manufacturing process and shelf life. The low content of an ingredient in a final product is often connected with either improper manufacturing (inappropriate purity or insufficient ingredients used in the production, uncontrolled manufacturing conditions, and inappropriate formulation) or its decomposition during shelf life. In situations where decomposition occurs either during manufacturing or shelf life this not only misleads the consumer but might also create increased health risks due to the possibility of the uncontrolled formation of by-products. In contrast, when not enough ingredients are used during manufacturing the primary concern is about misleading the consumer. In some cases, such scenarios can also pose a risk to human health, i.e. in instances of adulteration.

A significant problem that arises in the evaluation of the quality of functional foods is that there are no generally accepted tolerances for the declaration of nutrients and other active ingredients in the EU (DG SANCO 2006), although guidelines on this issue have been accepted in some countries (Table 11). The task of setting tolerable margins was identified as a priority 10 years ago during the discussion that led to the adoption of Directive 2002/46/EC on food supplements, but this goal has not yet been achieved (Pravst and Žmitek 2011). Nevertheless, there is a general agreement that such tolerances should be defined at the Community level in order to avoid trade barriers and ensure consumer protection (DG SANCO 2006).

| | Tolerances for added nutrients [1] | | |
Country	Minerals	Water-soluble vitamins	Fat-soluble vitamins
Belgium	90% - 120%	90% - 120%	90% - 120%
Denmark	80% - 150%	80% - 150%	80% - 150%
France	80% - 200%	80% - 200%[2]	80% - 200%[2]
Italy	75% - 100%	80% - 130%[2]	80% - 130%[2]
Slovenia	80% - 150%	80% - 150%[2]	80% - 130%
Netherlands	80% - 150%	80% - 200%	80% - 200%
United Kingdom	50% - 200%	50% - 200%	70% - 130%

Notes: [1] if legislation prescribes minimum and maximum values for the addition of nutrients, the analysed amount must not exceed these limits; [2] with exceptions

Table 11. Tolerance values accepted or practiced in some EU member states (CIAA 2007; DG SANCO 2006; IVZ 2009).

9. Conclusions and future issues

In the last few years the regulation of nutrition and health claims has been one of the top food-related themes discussed in Europe. Regulation covering these areas was indeed required. Protecting consumers against misleading claims, along with the harmonisation of the European market, were the key issues that needed to be addressed. The regulation targets functional foods, a concept which emerged in Japan about 20 years ago to reduce the escalating health care costs with a category of foods offering potential health benefits, although from a different perspective. At the time, the USA and some EU-member states were also at the frontier of developments, but the European Union as a whole was lagging far behind. It was decided that the use of pre-approved evidence-based health claims on food labels would serve us best and in the ensuing time there has been a focus on creating a list of approved claims. In such a system, functional foods are basically defined by the limitations and the opportunities for the use of claims.

Essential nutrients are clearly the winner of the evaluation process. In cases where a well-established consensus among scientists exists on the biological role of a nutrient, the EFSA relied on that consensus and confirmed the cause-and-effect relationship without reviewing the primary scientific studies. In most cases, the proposed condition of use is to include at least 15% of the RDA of the nutrient per 100g of final product, to enable the use of health claims for such a nutrient. This will enable products which are a source of at least one such nutrient to communicate health claims, even in cases where there is no deficiency in the population. The consumer will recognise such a nutrient as a health added value and there are concerns that such claims might flood the market and enable consumers to be legally misled. While the authorisation of such health claims may pose a risk of misleading the consumer, there are also cases where concerns related to public health arise. Such an example is the claim concerning phosphorus and its role in the maintenance of normal bones. The intake of phosphorus easily exceeds the recommendations and a bigger intake might have adverse effects on bone health (Pravst 2011b). Therefore, both health and ethical concerns arise as to whether such claims should be allowed, even though science is not yet clear on this issue. A useful solution in such cases would be to authorise claims with more specific conditions of use.

Foods promoted with claims may be perceived by consumers as having a health advantage over other foods and this may encourage consumers to make choices which directly influence their total intake of individual nutrients in a way which would run counter to scientific advice. The regulation aims to avoid a situation where claims mask the overall nutritional status of a food product and confuse consumers when trying to make healthy choices in the context of a balanced diet with the introduction of nutrient profiles. However, these profiles have not yet been established and it is not even clear if they will be implemented at all. In a situation in which food producers have the power to stimulate the consumption of foods with a poor nutritional status we must count on their commitment to serving consumers (Pravst 2011a).

All of the above issues suggest that we are still far from the target. The scientific substantiation of health claims for non-essential ingredients is very important and substantial additional research will be needed to get new claims approved. A detailed examination of all the concerns raised by the EFSA in its published opinions, together with some additional advice about expectations related to the scientific substantiation of health claims, should result in the improved quality of clinical testing for bioactive components and functional foods.

10. Acknowledgments

I gratefully acknowledge Tobin Bales for providing help with the language. The work was financially supported by Ministry of Agriculture, Forestry and Food of the Republic of Slovenia, and Slovenian Research Agency and (Contract 1000-11-282007, Research project V7-1107).

11. References

AbuMweis, S.S., Jew, S. & Ames, N.P. (2010). Beta-glucan from barley and its lipid-lowering capacity: a meta-analysis of randomized, controlled trials. *Eur J Clin Nutr.* 64(12): 1472-1480.

Aggett, P.J., Antoine, J.M., Asp, N.G., Bellisle, F., Contor, L. et al. (2005). PASSCLAIM - Consensus on criteria. *European Journal of Nutrition.* 44 5-30.

Arvanitoyannis, I.S. & Houwelingen-Koukaliaroglou, M. (2005). Functional foods: A survey of health claims, pros and cons, and current legislation. *Crit. Rev. Food Sci. Nutr.* 45(5): 385-404.

Ashwell, M. (2002) *Concepts of Functional Foods.* ILSI - International LIfe Sciences Institute, Brussels.

Bech-Larsen, T. & Scholderer, J. (2007). Functional foods in Europe: consumer research, market experiences and regulatory aspects. *Trends in Food Science & Technology.* 18(4): 231-234.

Bitzios, M., Fraser, I. & Haddock-Fraser, J. (2011). Functional ingredients and food choice: Results from a dual-mode study employing means-end-chain analysis and a choice experiment. *Food Policy.* 36(5): 715-725.

Bjelakovic, G., Nikolova, D., Gluud, L.L., Simonetti, R.G. & Gluud, C. (2008). Antioxidant supplements for prevention of mortality in healthy participants and patients with various diseases. *Cochrane database of systematic reviews.*(2): CD007176.

Bonsmann, S.S.G., Celemin, L.F., Larranaga, A., Egger, S., Wills, J.M. et al. (2010). Penetration of nutrition information on food labels across the EU-27 plus Turkey. *European Journal of Clinical Nutrition.* 64(12): 1379-1385.

Burton, R.A. & Fincher, G.B. (2009). (1,3;1,4)-beta-D-Glucans in Cell Walls of the Poaceae, Lower Plants, and Fungi: A Tale of Two Linkages. *Molecular Plant.* 2(5): 873-882.

Cappuccio, F.P. & Pravst, I. (2011). Health claims on foods: Promoting healthy food choices or high salt intake? *British Journal of Nutrition.* In press (doi:10.1017/S0007114511002856)

Cazaubiel, M. & Bard, J.M. (2008). Use of biomarkers for the optimization of clinical trials in nutrition. *Agro Food Industry Hi-Tech.* 19(5): 22-24.

Celemin, L.F., Bonsmann, S.S.G., Carlsson, E., Larranaga, A. & Egger, S. (2011). Mapping public health nutrition awareness campaigns across Europe. *Agro Food Industry Hi-Tech.* 22(1): 38-40.

CIAA (2007) *CIAA manual on nutrition labelling (URL: http://www.ibec.ie/Sectors/FDII/FDII.nsf/vPages/Regulatory_Affairs~Consumer_Informati on~ciaa-nutrition-manual/$file/CIAA%20Nutrition%20Labelling%20Manual.doc, Accessed: June 2010).* CIAA, Brussels.

Colombo, M.L. (2010). Conventional and new foods: health and nutritional claims The new functional role of food. *Agro Food Industry Hi-Tech.* 21(1): 42-44.

Covas, M.I., de la Torre, K., Farre-Albaladejo, M., Kaikkonen, J., Fito, M. et al. (2006a). Postprandial LDL phenolic content and LDL oxidation are modulated by olive oil phenolic compounds in humans. *Free Radical Biology and Medicine.* 40(4): 608-616.

Covas, M.I., Nyyssonen, K., Poulsen, H.E., Kaikkonen, J., Zunft, H.J.F. et al. (2006b). The effect of polyphenols in olive oil on heart disease risk factors - A randomized trial. *Annals of Internal Medicine.* 145(5): 333-341.

D-A-CH (2002) *Reference Values for Nutrient Intake.* German Nutrition Society, Bonn.

Demonty, I., Ras, R.T., van der Knaap, H.C.M., Duchateau, G.S.M.J., Meijer, L. et al. (2008). Continuous dose-response relationship of the LDL-cholesterol lowering effect of phytosterol intake. *The Journal of Nutrition.* 139 271-284.

DG SANCO. 2006. Directive 90/496/EEC on Nutrition Labelling for Foodstuffs: Discussion Paper on Revision of Technical Issues (URL: http://ec.europa.eu/food/food/labellingnutrition/nutritionlabel/discussion_pap er_rev_tech_issues.pdf, Accessed: June 2010). Brussels, European Commission, DG SANCO.

Earl, S., Cole, Z.A., Holroyd, C., Cooper, C. & Harvey, N.C. (2010). Dietary management of osteoporosis throughout the life course. *Proceedings of the Nutrition Society.* 69(01): 25-33.

EC. 2006. Regulation (EC) No 1924/2006 on nutrition and health claims made on foods.

EFSA. (2005). Opinion of the Scientific Panel on Dietetic Products, Nutrition and Allergies on a request from the Commission related to an application concerning the use of betaine as a novel food in the EU. *The EFSA Journal.* 191 1-17. (doi:10.2903/j.efsa.2005.191)

EFSA. (2008). The setting of nutrient profiles for foods bearing nutrition and health claims pursuant to Article 4 of the Regulation (EC) No 1924/2006 - Scientific Opinion of the Panel on Dietetic Products, Nutrition and Allergies. *The EFSA Journal.* 191 1-17. (doi:10.2903/j.efsa.2005.191)

EFSA. (2009a). Bimuno™ and help to maintain a healthy gastro-intestinal function - Scientific substantiation of a health claim related to BimunoTM and help to maintain a healthy gastro-intestinal function pursuant to Article 13(5) of Regulation (EC) No 1924/2006. *The EFSA Journal.* 1107 1-10. (doi:10.2903/j.efsa.2009.1107)

EFSA. (2009b). Opinion on the substantiation of health claims related to alpha linolenic acid and maintenance of normal blood cholesterol concentrations (ID 493) and maintenance of normal blood pressure (ID 625) pursuant to Article 13(1) of Regulation (EC) No 1924/2006. *EFSA Journal.* 7(9): 1252. (doi:10.2903/j.efsa.2009.1252)

EFSA. (2009c). Scientific Opinion on substantiation of health claims related to thiamine and energy-yielding metabolism (ID 21, 24, 28), cardiac function (ID 20), function of the nervous system (ID 22, 27), maintenance of bone (ID 25), maintenance of teeth (ID 25), maintenance of hair (ID 25), maintenance of nails (ID 25), maintenance of skin (ID 25) pursuant to Article 13(1) of Regulation (EC) No 1924/2006. *EFSA Journal.* 7(9): 1222. (doi:10.2903/j.efsa.2009.1222)

EFSA. (2009d). Scientific Opinion on the substantiation of health claims related to beta glucans and maintenance of normal blood cholesterol concentrations (ID 754, 755, 757, 801, 1465, 2934) and maintenance or achievement of a normal body weight (ID 820, 823) pursuant to Article 13(1) of Regulation (EC) No 1924/2006. *EFSA Journal.* 7(9): 1524. (doi:10.2903/j.efsa.2009.1254)

EFSA. (2009e). Scientific Opinion on the substantiation of health claims related to biotin and energy-yielding metabolism (ID 114, 117), macronutrient metabolism (ID 113, 114, 117), maintenance of skin and mucous membranes (ID 115), maintenance of hair (ID 118, 2876) and function of the nervous system (ID 116) pursuant to Article 13(1) of Regulation (EC) No 1924/2006. *EFSA Journal.* 7(9): 1209. (doi:10.2903/j.efsa.2009.1209)

EFSA. (2009f). Scientific Opinion on the substantiation of health claims related to calcium and maintenance of bones and teeth (ID 224, 230, 231, 354, 3099), muscle function and neurotransmission (ID 226, 227, 230, 235), blood coagulation (ID 230, 236), energy-yielding metabolism (ID 234), function of digestive enzymes (ID 355), and maintenance of normal blood pressure (ID 225, 385, 1419) pursuant to Article 13(1) of Regulation (EC) No 1924/2006. *EFSA Journal.* 7(9): 1210. (doi:10.2903/j.efsa.2009.1210)

EFSA. (2009g). Scientific Opinion on the substantiation of health claims related to calcium and vitamin D and maintenance of bone (ID 350) pursuant to Article 13(1) of Regulation (EC) No 1924/2006. *EFSA Journal.* 7(9): 1272. (doi:10.2903/j.efsa.2009.1272)

EFSA. (2009h). Scientific Opinion on the substantiation of health claims related to copper and protection of DNA, proteins and lipids from oxidative damage (ID 263, 1726), function of the immune system (ID 264), maintenance of connective tissues (ID 265, 271, 1722), energy-yielding metabolism (ID 266), function of the nervous system (ID 267), maintenance of skin and hair pigmentation (ID 268, 1724), iron transport (ID 269, 270, 1727), cholesterol metabolism (ID 369), and glucose metabolism (ID 369) pursuant to Article 13(1) of Regulation (EC) No 1924/2006. *EFSA Journal.* 7(9): 1211. (doi:10.2903/j.efsa.2009.1211)

EFSA. (2009i). Scientific Opinion on the substantiation of health claims related to EPA, DHA, DPA and maintenance of normal blood pressure (ID 502), maintenance of normal HDL-cholesterol concentrations (ID 515), maintenance of normal (fasting) blood concentrations of triglycerides (ID 517), maintenance of normal LDL-cholesterol concentrations (ID 528, 698) and maintenance of joints (ID 503, 505, 507, 511, 518, 524, 526, 535, 537) pursuant to Article 13(1) of Regulation (EC) No 1924/2006. *EFSA Journal.* 7(9): 1263. (doi:10.2903/j.efsa.2009.1263)

EFSA. (2009j). Scientific Opinion on the substantiation of health claims related to fluoride and maintenance of tooth mineralisation (ID 275, 276) and maintenance of bone (ID 371) pursuant to Article 13(1) of Regulation (EC) No 1924/2006. *EFSA Journal.* 7(9): 1212. (doi:10.2903/j.efsa.2009.1212)

EFSA. (2009k). Scientific Opinion on the substantiation of health claims related to folate and blood formation (ID 79), homocysteine metabolism (ID 80), energy-yielding metabolism (ID 90), function of the immune system (ID 91), function of blood vessels (ID 94, 175, 192), cell division (ID 193), and maternal tissue growth during pregnancy (ID 2882) pursuant to Article 13(1) of Regulation (EC) No 1924/2006. *EFSA Journal.* 7(9): 1213. (doi:10.2903/j.efsa.2009.1213)

EFSA. (2009l). Scientific Opinion on the substantiation of health claims related to glucomannan and maintenance of normal blood cholesterol concentrations (ID 836, 1560) pursuant to Article 13(1) of Regulation (EC) No 1924/2006. *EFSA Journal.* 7(9): 1258. (doi:10.2903/j.efsa.2009.1258)

EFSA. (2009m). Scientific Opinion on the substantiation of health claims related to iodine and thyroid function and production of thyroid hormones (ID 274), energy-yielding metabolism (ID 274), maintenance of vision (ID 356), maintenance of hair (ID 370), maintenance of nails (ID 370), and maintenance of skin (ID 370) pursuant to Article 13(1) of Regulation (EC) No 1924/2006. *EFSA Journal.* 7(9): 1214. (doi:10.2903/j.efsa.2009.1214)

EFSA. (2009n). Scientific Opinion on the substantiation of health claims related to iron and formation of red blood cells and haemoglobin (ID 249, ID 1589), oxygen transport (ID 250, ID 254, ID 256), energy-yielding metabolism (ID 251, ID 1589), function of the immune system (ID 252, ID 259), cognitive function (ID 253) and cell division (ID 368) pursuant to Article 13(1) of Regulation (EC) No 1924/2006. *EFSA Journal.* 7(9): 1215. (doi:10.2903/j.efsa.2009.1215)

EFSA. (2009o). Scientific Opinion on the substantiation of health claims related to lactase enzyme and breaking down lactose (ID 1697, 1818) pursuant to Article 13(1) of Regulation (EC) No 1924/2006. *EFSA Journal.* 7(9): 1236. (doi:10.2903/j.efsa.2009.1236)

EFSA. (2009p). Scientific Opinion on the substantiation of health claims related to linoleic acid and maintenance of normal blood cholesterol concentrations (ID 489) pursuant to Article 13(1) of Regulation (EC) No 1924/2006. *EFSA Journal.* 7(9): 1276. (doi:10.2903/j.efsa.2009.1276)

EFSA. (2009q). Scientific Opinion on the substantiation of health claims related to magnesium and electrolyte balance (ID 238), energy-yielding metabolism (ID 240, 247, 248), neurotransmission and muscle contraction including heart muscle (ID 241, 242), cell division (ID 365), maintenance of bone (ID 239), maintenance of teeth (ID 239), blood coagulation (ID 357) and protein synthesis (ID 364) pursuant to Article 13(1) of Regulation (EC) No 1924/2006. *EFSA Journal.* 7(9): 1216. (doi:10.2903/j.efsa.2009.1216)

EFSA. (2009r). Scientific Opinion on the substantiation of health claims related to manganese and protection of DNA, proteins and lipids from oxidative damage (ID 309), maintenance of bone (ID 310), energy-yielding metabolism (ID 311), and cognitive function (ID 340) pursuant to Article 13(1) of Regulation (EC) No 1924/2006. *EFSA Journal.* 7(9): 1217. (doi:10.2903/j.efsa.2009.1217)

EFSA. (2009s). Scientific Opinion on the substantiation of health claims related to niacin and energy-yielding metabolism (ID 43, 49, 54), function of the nervous system (ID 44, 53), maintenance of the skin and mucous membranes (ID 45, 48, 50, 52), maintenance of normal LDL-cholesterol, HDL cholesterol and triglyceride concentrations (ID 46), maintenance of bone (ID 50), maintenance of teeth (ID 50), maintenance of hair (ID 50, 2875) and maintenance of nails (ID 50, 2875) pursuant to Article 13(1) of Regulation (EC) No 1924/2006. *EFSA Journal.* 7(9): 1224. (doi:10.2903/j.efsa.2009.1224)

EFSA. (2009t). Scientific Opinion on the substantiation of health claims related to pantothenic acid and energy-yielding metabolism (ID 56, 59, 60, 64, 171, 172, 208), mental performance (ID 57), maintenance of bone (ID 61), maintenance of teeth (ID 61), maintenance of hair (ID 61), maintenance of skin (ID 61), maintenance of nails (ID 61) and synthesis and metabolism of steroid hormones, vitamin D and some neurotransmitters (ID 181) pursuant to Article 13(1) of Regulation (EC) No 1924/2006. *EFSA Journal.* 7(9): 1218. (doi:10.2903/j.efsa.2009.1218)

EFSA. (2009u). Scientific Opinion on the substantiation of health claims related to phosphorus and function of cell membranes (ID 328), energy-yielding metabolism (ID 329, 373) and maintenance of bone and teeth (ID 324, 327) pursuant to Article 13(1) of Regulation (EC) No 1924/2006. *EFSA Journal.* 7(9): 1219. (doi:10.2903/j.efsa.2009.1219)

EFSA. (2009v). Scientific Opinion on the substantiation of health claims related to selenium and protection of DNA, proteins and lipids from oxidative damage (ID 277, 283, 286, 1289, 1290, 1291, 1293, 1751), function of the immune system (ID 278), thyroid function (ID 279, 282, 286, 1289, 1290, 1291, 1293), function of the heart and blood vessels (ID 280), prostate function (ID 284), cognitive function (ID 285) and spermatogenesis (ID 396) pursuant to Article 13(1) of Regulation (EC) No 1924/2006. *EFSA Journal.* 7(9): 1220. (doi:10.2903/j.efsa.2009.1220)

EFSA. (2009w). Scientific Opinion on the substantiation of health claims related to sugar free chewing gum and dental and oral health, including gum and tooth protection and strength (ID 1149), plaque acid neutralisation (ID 1150), maintenance of tooth mineralisation (ID 1151), reduction of oral dryness (ID 1240), and maintenance of the normal body weight (ID 1152) pursuant to Article 13(1) of Regulation (EC) No 1924/2006. *EFSA Journal*. 7(9): 1271. (doi:10.2903/j.efsa.2009.1271)

EFSA. (2009x). Scientific Opinion on the substantiation of health claims related to vitamin A and cell differentiation (ID 14), function of the immune system (ID 14), maintenance of skin and mucous membranes (ID 15, 17), maintenance of vision (ID 16), maintenance of bone (ID 13, 17), maintenance of teeth (ID 13, 17), maintenance of hair (ID 17), maintenance of nails (ID 17), metabolism of iron (ID 206), and protection of DNA, proteins and lipids from oxidative damage (ID 209) pursuant to Article 13(1) of Regulation (EC) No 1924/2006. *EFSA Journal*. 7(9): 1221. (doi:10.2903/j.efsa.2009.1221)

EFSA. (2009y). Scientific Opinion on the substantiation of health claims related to vitamin B12 and red blood cell formation (ID 92, 101), cell division (ID 93), energy-yielding metabolism (ID 99, 190) and function of the immune system (ID 107) pursuant to Article 13(1) of Regulation (EC) No 1924/2006. *EFSA Journal*. 7(9): 1223. (doi:10.2903/j.efsa.2009.1223)

EFSA. (2009z). Scientific Opinion on the substantiation of health claims related to vitamin B6 and protein and glycogen metabolism (ID 65, 70, 71), function of the nervous system (ID 66), red blood cell formation (ID 67, 72, 186), function of the immune system (ID 68), regulation of hormonal activity (ID 69) and mental performance (ID 185) pursuant to Article 13(1) of Regulation (EC) No 1924/2006. *EFSA Journal*. 7(9): 1225. (doi:10.2903/j.efsa.2009.1225)

EFSA. (2009{). Scientific Opinion on the substantiation of health claims related to vitamin C and protection of DNA, proteins and lipids from oxidative damage (ID 129, 138, 143, 148), antioxidant function of lutein (ID 146), maintenance of vision (ID 141, 142), collagen formation (ID 130, 131, 136, 137, 149), function of the nervous system (ID 133), function of the immune system (ID 134), function of the immune system during and after extreme physical exercise (ID 144), non-haem iron absorption (ID 132, 147), energy-yielding metabolism (ID 135), and relief in case of irritation in the upper respiratory tract (ID 1714, 1715) pursuant to Article 13(1) of Regulation (EC) No 1924/2006. *EFSA Journal*. 7(9): 1226. (doi:10.2903/j.efsa.2009.1226)

EFSA. (2009 |). Scientific Opinion on the substantiation of health claims related to vitamin D and maintenance of bone and teeth (ID 150, 151, 158), absorption and utilisation of calcium and phosphorus and maintenance of normal blood calcium concentrations (ID 152, 157), cell division (ID 153), and thyroid function (ID 156) pursuant to Article 13(1) of Regulation (EC) No 1924/2006. *EFSA Journal*. 7(9): 1227. (doi:10.2903/j.efsa.2009.1227)

EFSA. (2009}). Scientific Opinion on the substantiation of health claims related to vitamin K and maintenance of bone (ID 123, 127, 128, and 2879), blood coagulation (ID 124 and 126), and function of the heart and blood vessels (ID 124, 125 and 2880) pursuant to Article 13(1) of Regulation (EC) No 1924/2006. *EFSA Journal*. 7(9): 1228. (doi:10.2903/j.efsa.2009.1228)

EFSA. (2009~). Scientific Opinion on the substantiation of health claims related to zinc and function of the immune system (ID 291, 1757), DNA synthesis and cell division (ID 292, 1759), protection of DNA, proteins and lipids from oxidative damage (ID 294, 1758), maintenance of bone (ID 295, 1756), cognitive function (ID 296), fertility and reproduction (ID 297, 300), reproductive development (ID 298), muscle function (ID 299), metabolism of fatty acids (ID 302), maintenance of joints (ID 305), function of the heart and blood vessels (ID 306), prostate function (ID 307), thyroid function (ID 308), acid-base metabolism (ID 360), vitamin A metabolism (ID 361) and maintenance of vision (ID 361) pursuant to Article 13(1) of Regulation (EC) No 1924/2006. *EFSA Journal.* 7(9): 1229. (doi:10.2903/j.efsa.2009.1229)

EFSA. (2009). Water-soluble tomato concentrate (WSTC I and II) and platelet aggregation. *The EFSA Journal.* 1101 1-15. (doi:10.2903/j.efsa.2009.1101)

EFSA. (2010a). Scientific Opinion on Dietary Reference Values for carbohydrates and dietary fibre. *EFSA Journal.* 8(3): 1462. (doi:10.2903/j.efsa.2010.1462)

EFSA. (2010b). Scientific Opinion on Dietary Reference Values for fats, including saturated fatty acids, polyunsaturated fatty acids, monounsaturated fatty acids, *trans* fatty acids, and cholesterol. *EFSA Journal.* 8(3): 1461. (doi:10.2903/j.efsa.2010.1461)

EFSA. (2010c). Scientific Opinion on the substantiation of a health claim related to ethanol-water extract of *Caralluma fimbriata* (Slimaluma®) and helps to reduce waist circumference pursuant to Article 13(5) of Regulation (EC) No 1924/2006. *EFSA Journal.* 8(5): 1602. (doi:10.2903/j.efsa.2010.1602)

EFSA. (2010d). Scientific Opinion on the substantiation of health claims related to biotin and maintenance of normal skin and mucous membranes (ID 121), maintenance of normal hair (ID 121), maintenance of normal bone (ID 121), maintenance of normal teeth (ID 121), maintenance of normal nails (ID 121, 2877), reduction of tiredness and fatigue (ID 119), contribution to normal psychological functions (ID 120) and contribution to normal macronutrient metabolism (ID 4661) pursuant to Article 13(1) of Regulation (EC) No 1924/2006. *EFSA Journal.* 8(10): 1728. (doi:10.2903/j.efsa.2010.1728)

EFSA. (2010e). Scientific Opinion on the substantiation of health claims related to calcium and maintenance of normal bone and teeth (ID 2731, 3155, 4311, 4312, 4703), maintenance of normal hair and nails (ID 399, 3155), maintenance of normal blood LDL-cholesterol concentrations (ID 349, 1893), maintenance of normal blood HDL-cholesterol concentrations (ID 349, 1893), reduction in the severity of symptoms related to the premenstrual syndrome (ID 348, 1892), "cell membrane permeability" (ID 363), reduction of tiredness and fatigue (ID 232), contribution to normal psychological functions (ID 233), contribution to the maintenance or achievement of a normal body weight (ID 228, 229) and regulation of normal cell division and differentiation (ID 237) pursuant to Article 13(1) of Regulation (EC) No 1924/2006. *EFSA Journal.* 8(10): 1725. (doi:10.2903/j.efsa.2010.1725)

EFSA. (2010f). Scientific Opinion on the substantiation of health claims related to chromium and contribution to normal macronutrient metabolism (ID 260, 401, 4665, 4666, 4667), maintenance of normal blood glucose concentrations (ID 262, 4667), contribution to the maintenance or achievement of a normal body weight (ID 339, 4665, 4666), and reduction of tiredness and fatigue (ID 261) pursuant to Article 13(1) of Regulation (EC) No 1924/2006. *EFSA Journal.* 8(10): 1732. (doi:10.2903/j.efsa.2010.1732)

EFSA. (2010g). Scientific Opinion on the substantiation of health claims related to docosahexaenoic acid (DHA) and maintenance of normal (fasting) blood concentrations of triglycerides (ID 533, 691, 3150), protection of blood lipids from oxidative damage (ID 630), contribution to the maintenance or achievement of a normal body weight (ID 629), brain, eye and nerve development (ID 627, 689, 704, 742, 3148, 3151), maintenance of normal brain function (ID 565, 626, 631, 689, 690, 704, 742, 3148, 3151), maintenance of normal vision (ID 627, 632, 743, 3149) and maintenance of normal spermatozoa motility (ID 628) pursuant to Article 13(1) of Regulation (EC) No 1924/2006. *EFSA Journal.* 8(10): 1734. (doi:10.2903/j.efsa.2010.1734)

EFSA. (2010h). Scientific Opinion on the substantiation of health claims related to dried plums of 'prune' cultivars (Prunus domestica L.) and maintenance of normal bowel function (ID 1164) pursuant to Article 13(1) of Regulation (EC) No 1924/2006. *EFSA Journal.* 8(2): 1486. (doi:10.2903/j.efsa.2010.1486)

EFSA. (2010i). Scientific Opinion on the substantiation of health claims related to eicosapentaenoic acid (EPA), docosahexaenoic acid (DHA), docosapentaenoic acid (DPA) and maintenance of normal cardiac function (ID 504, 506, 516, 527, 538, 703, 1128, 1317, 1324, 1325), maintenance of normal blood glucose concentrations (ID 566), maintenance of normal blood pressure (ID 506, 516, 703, 1317, 1324), maintenance of normal blood HDL-cholesterol concentrations (ID 506), maintenance of normal (fasting) blood concentrations of triglycerides (ID 506, 527, 538, 1317, 1324, 1325), maintenance of normal blood LDL-cholesterol concentrations (ID 527, 538, 1317, 1325, 4689), protection of the skin from photo-oxidative (UV-induced) damage (ID 530), improved absorption of EPA and DHA (ID 522, 523), contribution to the normal function of the immune system by decreasing the levels of eicosanoids, arachidonic acid-derived mediators and pro-inflammatory cytokines (ID 520, 2914), and "immunomodulating agent" (4690) pursuant to Article 13(1) of Regulation (EC) No 1924/2006. *EFSA Journal.* 8(10): 1796. (doi:10.2903/j.efsa.2010.1796)

EFSA. (2010j). Scientific Opinion on the substantiation of health claims related to folate and contribution to normal psychological functions (ID 81, 85, 86, 88), maintenance of normal vision (ID 83, 87), reduction of tiredness and fatigue (ID 84), cell division (ID 195, 2881) and contribution to normal amino acid synthesis (ID 195, 2881) pursuant to Article 13(1) of Regulation (EC) No 1924/2006. *EFSA Journal.* 8(10): 1760. (doi:10.2903/j.efsa.2010.1760)

EFSA. (2010k). Scientific Opinion on the substantiation of health claims related to guar gum and maintenance of normal blood glucose concentrations (ID 794), increase in satiety (ID 795) and maintenance of normal blood cholesterol concentrations (ID 808) pursuant to Article 13(1) of Regulation (EC) No 1924/2006. *EFSA Journal.* 8(2): 1464. (doi:10.2903/j.efsa.2010.1464)

EFSA. (2010l). Scientific Opinion on the substantiation of health claims related to hydroxypropyl methylcellulose (HPMC) and maintenance of normal bowel function (ID 812), reduction of post-prandial glycaemic responses (ID 814), maintenance of normal blood cholesterol concentrations (ID 815) and increase in satiety leading to a reduction in energy intake (ID 2933) pursuant to Article 13(1) of Regulation (EC) No 1924/2006. *EFSA Journal.* 8(10): 1739.

(doi:10.2903/j.efsa.2010.1739)

EFSA. (2010m). Scientific Opinion on the substantiation of health claims related to iodine and contribution to normal cognitive and neurological function (ID 273), contribution to normal energy-yielding metabolism (ID 402), and contribution to normal thyroid function and production of thyroid hormones (ID 1237) pursuant to Article 13(1) of Regulation (EC) No 1924/2006. *EFSA Journal.* 8(10): 1800. (doi:10.2903/j.efsa.2010.1800)

EFSA. (2010n). Scientific Opinion on the substantiation of health claims related to iron and formation of red blood cells and haemoglobin (ID 374, 2889), oxygen transport (ID 255), contribution to normal energy-yielding metabolism (ID 255), reduction of tiredness and fatigue (ID 255, 374, 2889), biotransformation of xenobiotic substances (ID 258), and "activity of heart, liver and muscles" (ID 397) pursuant to Article 13(1) of Regulation (EC) No 1924/2006. *EFSA Journal.* 8(10): 1740. (doi:10.2903/j.efsa.2010.1740)

EFSA. (2010o). Scientific Opinion on the substantiation of health claims related to konjac mannan (glucomannan) and reduction of body weight (ID 854, 1556, 3725), reduction of post-prandial glycaemic responses (ID 1559), maintenance of normal blood glucose concentrations (ID 835, 3724), maintenance of normal (fasting) blood concentrations of triglycerides (ID 3217), maintenance of normal blood cholesterol concentrations (ID 3100, 3217), maintenance of normal bowel function (ID 834, 1557, 3901) and decreasing potentially pathogenic gastro-intestinal microorganisms (ID 1558) pursuant to Article 13(1) of Regulation (EC) No 1924/2006. *EFSA Journal.* 8(10): 1798. (doi:10.2903/j.efsa.2010.1798)

EFSA. (2010p). Scientific Opinion on the substantiation of health claims related to lactulose and decreasing potentially pathogenic gastro-intestinal microorganisms (ID 806) and reduction in intestinal transit time (ID 807) pursuant to Article 13(1) of Regulation (EC) No 1924/2006. *EFSA Journal.* 8(10): 1806. (doi:10.2903/j.efsa.2010.1806)

EFSA. (2010q). Scientific Opinion on the substantiation of health claims related to live yoghurt cultures and improved lactose digestion (ID 1143, 2976) pursuant to Article 13(1) of Regulation (EC) No 1924/2006. *EFSA Journal.* 8(10): 1763. (doi:10.2903/j.efsa.2010.1763)

EFSA. (2010r). Scientific Opinion on the substantiation of health claims related to magnesium and "hormonal health" (ID 243), reduction of tiredness and fatigue (ID 244), contribution to normal psychological functions (ID 245, 246), maintenance of normal blood glucose concentrations (ID 342), maintenance of normal blood pressure (ID 344, 366, 379), protection of DNA, proteins and lipids from oxidative damage (ID 351), maintenance of the normal function of the immune system (ID 352), maintenance of normal blood pressure during pregnancy (ID 367), resistance to mental stress (ID 375, 381), reduction of gastric acid levels (ID 376), maintenance of normal fat metabolism (ID 378) and maintenance of normal muscle contraction (ID 380, ID 3083) pursuant to Article 13(1) of Regulation (EC) No 1924/2006. *EFSA Journal.* 8(10): 1807. (doi:10.2903/j.efsa.2010.1807)

EFSA. (2010s). Scientific Opinion on the substantiation of health claims related to manganese and reduction of tiredness and fatigue (ID 312), contribution to normal formation of connective tissue (ID 404) and contribution to normal energy yielding metabolism

(ID 405) pursuant to Article 13(1) of Regulation (EC) No 1924/2006. *EFSA Journal.* 8(10): 1808. (doi:10.2903/j.efsa.2010.1808)

EFSA. (2010t). Scientific Opinion on the substantiation of health claims related to meal replacements for weight control (as defined in Directive 96/8/EC on energy restricted diets for weight loss) and reduction in body weight (ID 1417), and maintenance of body weight after weight loss (ID 1418) pursuant to Article 13(1) of Regulation (EC) No 1924/2006. *EFSA Journal.* 8(2): 1466. (doi:10.2903/j.efsa.2010.1466)

EFSA. (2010u). Scientific Opinion on the substantiation of health claims related to melatonin and alleviation of subjective feelings of jet lag (ID 1953), and reduction of sleep onset latency, and improvement of sleep quality (ID 1953) pursuant to Article 13(1) of Regulation (EC) No 1924/2006. *EFSA Journal.* 8(2): 1467. (doi:10.2903/j.efsa.2010.1467)

EFSA. (2010v). Scientific Opinion on the substantiation of health claims related to molybdenum and contribution to normal amino acid metabolism (ID 313) and protection of DNA, proteins and lipids from oxidative damage (ID 341) pursuant to Article 13(1) of Regulation (EC) No 1924/2006. *EFSA Journal.* 8(10): 1745. (doi:10.2903/j.efsa.2010.1745)

EFSA. (2010w). Scientific Opinion on the substantiation of health claims related to niacin and reduction of tiredness and fatigue (ID 47), contribution to normal energy-yielding metabolism (ID 51), contribution to normal psychological functions (ID 55), maintenance of normal blood flow (ID 211), and maintenance of normal skin and mucous membranes (ID 4700) pursuant to Article 13(1) of Regulation (EC) No 1924/2006. *EFSA Journal.* 8(10): 1757. (doi:10.2903/j.efsa.2010.1757)

EFSA. (2010x). Scientific Opinion on the substantiation of health claims related to pantothenic acid and mental performance (ID 58), reduction of tiredness and fatigue (ID 63), adrenal function (ID 204) and maintenance of normal skin (ID 2878) pursuant to Article 13(1) of Regulation (EC) No 1924/2006. *EFSA Journal.* 8(10): 1758. (doi:10.2903/j.efsa.2010.1758)

EFSA. (2010y). Scientific Opinion on the substantiation of health claims related to pectins and reduction of post-prandial glycaemic responses (ID 786), maintenance of normal blood cholesterol concentrations (ID 818) and increase in satiety leading to a reduction in energy intake (ID 4692) pursuant to Article 13(1) of Regulation (EC) No 1924/2006. *EFSA Journal.* 8(10): 1747. (doi:10.2903/j.efsa.2010.1747)

EFSA. (2010z). Scientific Opinion on the substantiation of health claims related to plant sterols and plant stanols and maintenance of normal blood cholesterol concentrations (ID 549, 550, 567, 713, 1234, 1235, 1466, 1634, 1984, 2909, 3140), and maintenance of normal prostate size and normal urination (ID 714, 1467, 1635) pursuant to Article 13(1) of Regulation (EC) No 1924/2006. *EFSA Journal.* 8(10): 1813. (doi:10.2903/j.efsa.2010.1813)

EFSA. (2010{). Scientific Opinion on the substantiation of health claims related to potassium and maintenance of normal muscular and neurological function (ID 320, 386) and maintenance of normal blood pressure (ID 321) pursuant to Article 13(1) of Regulation (EC) No 1924/2006. *EFSA Journal.* 8(10): 1469. (doi:10.2903/j.efsa.2010.1469)

EFSA. (2010|). Scientific Opinion on the substantiation of health claims related to protein and increase in satiety leading to a reduction in energy intake (ID 414, 616, 730), contribution to the maintenance or achievement of a normal body weight (ID 414, 616, 730), maintenance of normal bone (ID 416) and growth or maintenance of muscle mass (ID 415, 417, 593, 594, 595, 715) pursuant to Article 13(1) of Regulation (EC) No 1924/2006. *EFSA Journal.* 8(10): 1811. (doi:10.2903/j.efsa.2010.1811)

EFSA. (2010}). Scientific Opinion on the substantiation of health claims related to riboflavin (vitamin B2) and contribution to normal energy-yielding metabolism (ID 29, 35, 36, 42), contribution to normal metabolism of iron (ID 30, 37), maintenance of normal skin and mucous membranes (ID 31, 33), contribution to normal psychological functions (ID 32), maintenance of normal bone (ID 33), maintenance of normal teeth (ID 33), maintenance of normal hair (ID 33), maintenance of normal nails (ID 33), maintenance of normal vision (ID 39), maintenance of normal red blood cells (ID 40), reduction of tiredness and fatigue (ID 41), protection of DNA, proteins and lipids from oxidative damage (ID 207), and maintenance of the normal function of the nervous system (ID 213) pursuant to Article 13(1) of Regulation (EC) No 1924/2006. *EFSA Journal.* 8(10): 1814. (doi:10.2903/j.efsa.2010.1814)

EFSA. (2010~). Scientific Opinion on the substantiation of health claims related to selenium and maintenance of normal hair (ID 281), maintenance of normal nails (ID 281), protection against heavy metals (ID 383), maintenance of normal joints (ID 409), maintenance of normal thyroid function (ID 410, 1292), protection of DNA, proteins and lipids from oxidative damage (ID 410, 1292), and maintenance of the normal function of the immune system (ID 1750) pursuant to Article 13(1) of Regulation (EC) No 1924/2006. *EFSA Journal.* 8(10): 1727. (doi:10.2903/j.efsa.2010.1727)

EFSA. (2010). Scientific Opinion on the substantiation of health claims related to thiamin and reduction of tiredness and fatigue (ID 23) and contribution to normal psychological functions (ID 205) pursuant to Article 13(1) of Regulation (EC) No 1924/2006. *EFSA Journal.* 8(10): 1755. (doi:10.2903/j.efsa.2010.1755)

EFSA. (2010€). Scientific Opinion on the substantiation of health claims related to vitamin B12 and contribution to normal neurological and psychological functions (ID 95, 97, 98, 100, 102, 109), contribution to normal homocysteine metabolism (ID 96, 103, 106), maintenance of normal bone (ID 104), maintenance of normal teeth (ID 104), maintenance of normal hair (ID 104), maintenance of normal skin (ID 104), maintenance of normal nails (ID 104), reduction of tiredness and fatigue (ID 108), and cell division (ID 212) pursuant to Article 13(1) of Regulation (EC) No 1924/2006. *EFSA Journal.* 8(10): 1756. (doi:10.2903/j.efsa.2010.1756)

EFSA. (2010). Scientific Opinion on the substantiation of health claims related to vitamin B6 and contribution to normal homocysteine metabolism (ID 73, 76, 199), maintenance of normal bone (ID 74), maintenance of normal teeth (ID 74), maintenance of normal hair (ID 74), maintenance of normal skin (ID 74), maintenance of normal nails (ID 74), contribution to normal energy-yielding metabolism (ID 75, 214), contribution to normal psychological functions (ID 77), reduction of tiredness and fatigue (ID 78), and contribution to normal cysteine synthesis (ID 4283) pursuant to Article 13(1) of Regulation (EC) No 1924/2006. *EFSA Journal.* 8(10): 1759. (doi:10.2903/j.efsa.2010.1759)

EFSA. (2010,). Scientific Opinion on the substantiation of health claims related to vitamin C and reduction of tiredness and fatigue (ID 139, 2622), contribution to normal psychological functions (ID 140), regeneration of the reduced form of vitamin E (ID 202), contribution to normal energy-yielding metabolism (ID 2334, 3196), maintenance of the normal function of the immune system (ID 4321) and protection of DNA, proteins and lipids from oxidative damage (ID 3331) pursuant to Article 13(1) of Regulation (EC) No 1924/2006. *EFSA Journal.* 8(10): 1815. (doi:10.2903/j.efsa.2010.1815)

EFSA. (2010*f*). Scientific Opinion on the substantiation of health claims related to vitamin D and normal function of the immune system and inflammatory response (ID 154, 159), maintenance of normal muscle function (ID 155) and maintenance of normal cardiovascular function (ID 159) pursuant to Article 13(1) of Regulation (EC) No 1924/2006. *EFSA Journal.* 8(2): 1468. (doi:10.2903/j.efsa.2010.1468)

EFSA. (2010„). Scientific Opinion on the substantiation of health claims related to vitamin E and protection of DNA, proteins and lipids from oxidative damage (ID 160, 162, 1947), maintenance of the normal function of the immune system (ID 161, 163), maintenance of normal bone (ID 164), maintenance of normal teeth (ID 164), maintenance of normal hair (ID 164), maintenance of normal skin (ID 164), maintenance of normal nails (ID 164), maintenance of normal cardiac function (ID 166), maintenance of normal vision by protection of the lens of the eye (ID 167), contribution to normal cognitive function (ID 182, 183), regeneration of the reduced form of vitamin C (ID 203), maintenance of normal blood circulation (ID 216) and maintenance of normal a scalp (ID 2873) pursuant to Article 13(1) of Regulation (EC) No 1924/2006. *EFSA Journal.* 8(10): 1816. (doi:10.2903/j.efsa.2010.1816)

EFSA. (2010...). Scientific Opinion on the substantiation of health claims related to wheat bran fibre and increase in faecal bulk (ID 3066), reduction in intestinal transit time (ID 828, 839, 3067, 4699) and contribution to the maintenance or achievement of a normal body weight (ID 829) pursuant to Article 13(1) of Regulation (EC) No 1924/2006. *EFSA Journal.* 8(10): 1817. (doi:10.2903/j.efsa.2010.1817)

EFSA. (2010†). Scientific Opinion on the substantiation of health claims related to zinc and maintenance of normal skin (ID 293), DNA synthesis and cell division (ID 293), contribution to normal protein synthesis (ID 293, 4293), maintenance of normal serum testosterone concentrations (ID 301), "normal growth" (ID 303), reduction of tiredness and fatigue (ID 304), contribution to normal carbohydrate metabolism (ID 382), maintenance of normal hair (ID 412), maintenance of normal nails (ID 412) and contribution to normal macronutrient metabolism (ID 2890) pursuant to Article 13(1) of Regulation (EC) No 1924/2006. *EFSA Journal.* 8(10): 1819. (doi:10.2903/j.efsa.2010.1819)

EFSA. (2011a). General guidance for stakeholders on the evaluation of Article 13.1, 13.5 and 14 health claims. *EFSA Journal.* 9(4): 2135. (doi:10.2903/j.efsa.2011.2135)

EFSA. (2011b). Guidance on the scientific requirements for health claims related to gut and immune function. *EFSA Journal.* 9(4): 1984. (doi:10.2903/j.efsa.2011.1984)

EFSA. (2011c). Scientific and technical guidance for the preparation and presentation of an application for authorisation of a health claim (revision 1). *EFSA Journal.* 9(5): 2170. (doi:10.2903/j.efsa.2011.2170)

EFSA. (2011d). Scientific Opinion on the substantiation of a health claim related to melatonin and reduction of sleep onset latency (ID 1698, 1780, 4080) pursuant to Article 13(1) of Regulation (EC) No 1924/2006. *EFSA Journal*. 9(6): 2241. (doi:10.2903/j.efsa.2011.2241)

EFSA. (2011e). Scientific Opinion on the substantiation of health claims related to arabinoxylan produced from wheat endosperm and reduction of post-prandial glycaemic responses (ID 830) pursuant to Article 13(1) of Regulation (EC) No 1924/2006. *EFSA Journal*. 9(6): 2205. (doi:10.2903/j.efsa.2011.2205)

EFSA. (2011f). Scientific Opinion on the substantiation of health claims related to beta-glucans from oats and barley and maintenance of normal blood LDL-cholesterol concentrations (ID 1236, 1299), increase in satiety leading to a reduction in energy intake (ID 851, 852), reduction of post-prandial glycaemic responses (ID 821, 824), and "digestive function" (ID 850) pursuant to Article 13(1) of Regulation (EC) No 1924/2006. *EFSA Journal*. 9(6): 2207. (doi:10.2903/j.efsa.2011.2207)

EFSA. (2011g). Scientific Opinion on the substantiation of health claims related to betaine and contribution to normal homocysteine metabolism (ID 4325) pursuant to Article 13(1) of Regulation (EC) No 1924/2006. *EFSA Journal*. 9(4): 2052. (doi:10.2903/j.efsa.2011.2052)

EFSA. (2011h). Scientific Opinion on the substantiation of health claims related to caffeine and increase in physical performance during short-term high-intensity exercise (ID 737, 1486, 1489), increase in endurance performance (ID 737, 1486), increase in endurance capacity (ID 1488) and reduction in the rated perceived exertion/effort during exercise (ID 1488, 1490) pursuant to Article 13(1) of Regulation (EC) No 1924/2006. *EFSA Journal*. 9(4): 2053. (doi:10.2903/j.efsa.2011.2053)

EFSA. (2011i). Scientific Opinion on the substantiation of health claims related to caffeine and increased fat oxidation leading to a reduction in body fat mass (ID 735, 1484), increased energy expenditure leading to a reduction in body weight (ID 1487), increased alertness (ID 736, 1101, 1187, 1485, 1491, 2063, 2103) and increased attention (ID 736, 1485, 1491, 2375) pursuant to Article 13(1) of Regulation (EC) No 1924/2006. *EFSA Journal*. 9(4): 2054. (doi:10.2903/j.efsa.2011.2054)

EFSA. (2011j). Scientific Opinion on the substantiation of health claims related to carbohydrate-electrolyte solutions and reduction in rated perceived exertion/effort during exercise (ID 460, 466, 467, 468), enhancement of water absorption during exercise (ID 314, 315, 316, 317, 319, 322, 325, 332, 408, 465, 473, 1168, 1574, 1593, 1618, 4302, 4309), and maintenance of endurance performance (ID 466, 469) pursuant to Article 13(1) of Regulation (EC) No 1924/2006. *EFSA Journal*. 9(6): 2211. (doi:10.2903/j.efsa.2011.2211)

EFSA. (2011k). Scientific Opinion on the substantiation of health claims related to chitosan and reduction in body weight (ID 679, 1499), maintenance of normal blood LDL-cholesterol concentrations (ID 4663), reduction of intestinal transit time (ID 4664) and reduction of inflammation (ID 1985) pursuant to Article 13(1) of Regulation (EC) No 1924/2006. *EFSA Journal*. 9(6): 2214. (doi:10.2903/j.efsa.2011.2214)

EFSA. (2011l). Scientific Opinion on the substantiation of health claims related to choline and contribution to normal lipid metabolism (ID 3186), maintenance of normal liver function (ID 1501), contribution to normal homocysteine metabolism (ID 3090), maintenance of normal neurological function (ID 1502), contribution to normal cognitive function (ID 1502), and brain and neurological development (ID 1503) pursuant to Article 13(1) of Regulation (EC) No 1924/2006. *EFSA Journal.* 9(4): 2056. (doi:10.2903/j.efsa.2011.2056)

EFSA. (2011m). Scientific Opinion on the substantiation of health claims related to creatine and increase in physical performance during short-term, high intensity, repeated exercise bouts (ID 739, 1520, 1521, 1522, 1523, 1525, 1526, 1531, 1532, 1533, 1534, 1922, 1923, 1924), increase in endurance capacity (ID 1527, 1535), and increase in endurance performance (ID 1521, 1963) pursuant to Article 13(1) of Regulation (EC) No 1924/2006. *EFSA Journal.* 9(7): 2303. (doi:10.2903/j.efsa.2011.2303)

EFSA. (2011n). Scientific Opinion on the substantiation of health claims related to fats and "function of the cell membrane" (ID 622, 2900, 2911) and normal absorption of fat-soluble vitamins (ID 670, 2902) pursuant to Article 13(1) of Regulation (EC) No 1924/2006. *EFSA Journal.* 9(6): 2220. (doi:10.2903/j.efsa.2011.2220)

EFSA. (2011o). Scientific Opinion on the substantiation of health claims related to foods with reduced amounts of saturated fatty acids (SFAs) and maintenance of normal blood LDL cholesterol concentrations (ID 620, 671, 4332) pursuant to Article 13(1) of Regulation (EC) No 1924/2006. *EFSA Journal.* 9(4): 2062. (doi:10.2903/j.efsa.2011.2062)

EFSA. (2011p). Scientific Opinion on the substantiation of health claims related to foods with reduced amounts of sodium and maintenance of normal blood pressure (ID 336, 705, 1148, 1178, 1185, 1420) pursuant to Article 13(1) of Regulation (EC) No 1924/2006. *EFSA Journal.* 9(6): 2237. (doi:10.2903/j.efsa.2011.2237)

EFSA. (2011q). Scientific Opinion on the substantiation of health claims related to foods with reduced lactose content and decreasing gastro-intestinal discomfort caused by lactose intake in lactose intolerant individuals (ID 646, 1224, 1238, 1339) pursuant to Article 13(1) of Regulation (EC) No 1924/2006. *EFSA Journal.* 9(6): 2236. (doi:10.2903/j.efsa.2011.2236)

EFSA. (2011r). Scientific Opinion on the substantiation of health claims related to glycaemic carbohydrates and maintenance of normal brain function (ID 603, 653) pursuant to Article 13(1) of Regulation (EC) No 1924/2006. *EFSA Journal.* 9(6): 2226. (doi:10.2903/j.efsa.2011.2226)

EFSA. (2011s). Scientific Opinion on the substantiation of health claims related to intense sweeteners and contribution to the maintenance or achievement of a normal body weight (ID 1136, 1444, 4299), reduction of post-prandial glycaemic responses (ID 4298), maintenance of normal blood glucose concentrations (ID 1221, 4298), and maintenance of tooth mineralisation by decreasing tooth demineralisation (ID 1134, 1167, 1283) pursuant to Article 13(1) of Regulation (EC) No 1924/2006. *EFSA Journal.* 9(6): 2229. (doi:10.2903/j.efsa.2011.2229)

EFSA. (2011t). Scientific Opinion on the substantiation of health claims related to monacolin K from red yeast rice and maintenance of normal blood LDL cholesterol concentrations (ID 1648, 1700) pursuant to Article 13(1) of Regulation (EC) No 1924/2006. *EFSA Journal.* 9(7): 2304. (doi:10.2903/j.efsa.2011.2304)

EFSA. (2011u). Scientific Opinion on the substantiation of health claims related to oat and barley grain fibre and increase in faecal bulk (ID 819, 822) pursuant to Article 13(1) of Regulation (EC) No 1924/2006. *EFSA Journal.* 9(6): 2249. (doi:10.2903/j.efsa.2011.2249)

EFSA. (2011v). Scientific Opinion on the substantiation of health claims related to polyphenols in olive and protection of LDL particles from oxidative damage (ID 1333, 1638, 1639, 1696, 2865), maintenance of normal blood HDL cholesterol concentrations (ID 1639), maintenance of normal blood pressure (ID 3781), "anti-inflammatory properties" (ID 1882), "contributes to the upper respiratory tract health" (ID 3468), "can help to maintain a normal function of gastrointestinal tract" (3779), and "contributes to body defences against external agents" (ID 3467) pursuant to Article 13(1) of Regulation (EC) No 1924/2006. *EFSA Journal.* 9(4): 2033. (doi:10.2903/j.efsa.2011.2033)

EFSA. (2011w). Scientific Opinion on the substantiation of health claims related to resistant starch and reduction of post-prandial glycaemic responses (ID 681), "digestive health benefits" (ID 682) and "favours a normal colon metabolism" (ID 783) pursuant to Article 13(1) of Regulation (EC) No 1924/2006. *EFSA Journal.* 9(4): 2024. (doi:10.2903/j.efsa.2011.2024)

EFSA. (2011x). Scientific Opinion on the substantiation of health claims related to rye fibre and changes in bowel function (ID 825), reduction of post prandial glycaemic responses (ID 826) and maintenance of normal blood LDL-cholesterol concentrations (ID 827) pursuant to Article 13(1) of Regulation (EC) No 1924/2006. *EFSA Journal.* 9(6): 2258. (doi:10.2903/j.efsa.2011.2258)

EFSA. (2011y). Scientific Opinion on the substantiation of health claims related to sodium and maintenance of normal muscle function (ID 359) pursuant to Article 13(1) of Regulation (EC) No 1924/2006. *EFSA Journal.* 9(6): 2260. (doi:10.2903/j.efsa.2011.2260)

EFSA. (2011z). Scientific Opinion on the substantiation of health claims related to sugar-free chewing gum with carbamide and plaque acid neutralisation (ID 1153) pursuant to Article 13(1) of Regulation (EC) No 1924/2006. *EFSA Journal.* 9(4): 2071. (doi:10.2903/j.efsa.2011.2071)

EFSA. (2011{). Scientific Opinion on the substantiation of health claims related to the replacement of mixtures of saturated fatty acids (SFAs) as present in foods or diets with mixtures of monounsaturated fatty acids (MUFAs) and/or mixtures of polyunsaturated fatty acids (PUFAs), and maintenance of normal blood LDL cholesterol concentrations (ID 621, 1190, 1203, 2906, 2910, 3065) pursuant to Article 13(1) of Regulation (EC) No 1924/2006. *EFSA Journal.* 9(4): 2069. (doi:10.2903/j.efsa.2011.2069)

EFSA. (2011 |). Scientific Opinion on the substantiation of health claims related to the sugar replacers xylitol, sorbitol, mannitol, maltitol, lactitol, isomalt, erythritol, D-tagatose, isomaltulose, sucralose and polydextrose and maintenance of tooth mineralisation by decreasing tooth demineralisation (ID 463, 464, 563, 618, 647, 1182, 1591, 2907, 2921, 4300), and reduction of post-prandial glycaemic responses (ID 617, 619, 669, 1590, 1762, 2903, 2908, 2920) pursuant to Article 13(1) of Regulation (EC) No 1924/2006. *EFSA Journal.* 9(4): 2076. (doi:10.2903/j.efsa.2011.2076)

EFSA. (2011}). Scientific Opinion on the substantiation of health claims related to very low calorie diets (VLCDs) and reduction in body weight (ID 1410), reduction in the sense of hunger (ID 1411), reduction in body fat mass while maintaining lean body mass (ID 1412), reduction of post-prandial glycaemic responses (ID 1414), and maintenance of normal blood lipid profile (1421) pursuant to Article 13(1) of Regulation (EC) No 1924/2006. *EFSA Journal.* 9(6): 2271.

(doi:10.2903/j.efsa.2011.2271)

EFSA. (2011~). Scientific Opinion on the substantiation of health claims related to water and maintenance of normal physical and cognitive functions (ID 1102, 1209, 1294, 1331), maintenance of normal thermoregulation (ID 1208) and "basic requirement of all living things" (ID 1207) pursuant to Article 13(1) of Regulation (EC) No 1924/2006. *EFSA Journal.* 9(4): 2075. (doi:10.2903/j.efsa.2011.2075)

Geissler, C. and Powers, H. Geissler, C. and Powers, H. Eds.(2005) *Human Nutrition.* Elsevier, Edinburgh.

Gibson, G.R. & Roberfroid, M.B. (1995). Dietary Modulation of the Human Colonic Microbiota: Introducing the Concept of Prebiotics. *The Journal of Nutrition.* 125(6): 1401-1412.

Grossklaus, R. (2009). Codex recommendations on the scientific basis of health claims. *European Journal of Nutrition.* 48 S15-S22.

He, F.J. & MacGregor, G.A. (2003). How far should salt intake be reduced? Hypertension. *Hypertension.* 42 1093-1099.

He, F.J. & MacGregor, G.A. (2010). Reducing population salt intake worldwide: from evidence to implementation. *Prog Cardiovasc Dis.* 52 363-382.

Hoefkens, C., Verbeke, W. & Van Camp, J. (2011). European consumers' perceived importance of qualifying and disqualifying nutrients in food choices. *Food Quality and Preference.* 22(6): 550-558.

Howlett, J. (2008) *Functional Foods: From science to health and claims.* ILSI Europe, Brussels.

Ikeda, Y., Iki, M., Morita, A., Kajita, E., Kagamimori, S. et al. (2006). Intake of fermented soybeans, natto, is associated with reduced bone loss in postmenopausal women: Japanese population-based osteoporosis (JPOS) study. *J. Nutr.* 136(5): 1323-1328.

IVZ (2009) *Tolerance values for labeling of nutritional composition of foods (URL: http://ivz.arhiv.over.net/javne_datoteke/datoteke/2068-HVtolerance_27102009.doc, Accessed: June 2010).* Slovenian Institute of Public Health, Ljubljana.

Jadhav, S.J., Lutz, S.E., Ghorpade, V.M. & Salunkhe, D.K. (1998). Barley: Chemistry and value-added processing. *Critical Reviews in Food Science and Nutrition.* 38(2): 123-171.

Katsuyama, H., Ideguchi, S., Fukunaga, M., Fukunaga, T., Saijoh, K. et al. (2004). Promotion of bone formation by fermented soybean (Natto) intake in premenopausal women. *Journal of Nutritional Science and Vitaminology.* 50(2): 114-120.

Krutulyte, R., Grunert, K.G., Scholderer, J., Lahteenmaki, L., Hagemann, K.S. et al. (2011). Perceived fit of different combinations of carriers and functional ingredients and its effect on purchase intention. *Food Quality and Preference.* 22(1): 11-16.

Lalor, F., Kennedy, J., Flynn, M.A. & Wall, P.G. (2010). A study of nutrition and health claims? a snapshot of what's on the Irish market. *Public Health Nutrition.* 13(05): 704-711.

Ling, W.H. & Jones, P.J.H. (1995). Dietary phytosterols: A review of metabolism, benefits and side effects. *Life Sciences.* 57(3): 195-206.

Marrugat, J., Covas, M.I., Fito, M., Schroder, H., Miro-Casas, E. et al. (2004). Effects of differing phenolic content in dietary olive oils on lipids and LDL oxidation - A randomized controlled trial. *European Journal of Nutrition.* 43(3): 140-147.

Palacios, C. (2006). The role of nutrients in bone health, from A to Z. *Crit. Rev. Food Sci. Nutr.* 46(8): 621-628.

Pothoulaki, M. & Chryssochoidis, G. (2009). Health claims: Consumers' matters. *Journal of Functional Foods.* 1(2): 222-228.

Pravst, I. (2010). The evaluation of health claims in Europe: What have we learned? *AgroFOOD industry hi-tech.* 21(4): 4-6.

Pravst, I. (2011a). Health claims: Where are we now and where are we going? (editorial) *Agro Food Industry Hi-Tech.* 22(4): 2-3.

Pravst, I. (2011b). Risking public health by approving some health claims? – The case of phosphorus. *Food Policy.* 36: 725-727.

Pravst, I. & Žmitek, K. (2011). The coenzyme Q10 content of food supplements. *Journal fur Verbraucherschutz und Lebensmittelsicherheit-Journal of Consumer Protection and Food Safety.* 6: 457-463.

Pravst, I., Žmitek, K. & Žmitek, J. (2010). Coenzyme Q10 contents in foods and fortification strategies. *Crit. Rev. Food Sci. Nutr.* 50(4): 269-280.

Raspor, P. 2011. Definition of functional foods and safety aspects of functional nutrition, Functional Foods Seminar, ZKZP/GZS: Ljubljana, Slovenia.

Sioen, I., Matthys, C., Huybrechts, I., Van Camp, J. & De Henauw, S. (2011). Consumption of plant sterols in Belgium: consumption patterns of plant sterol-enriched foods in Flanders, Belgium. *British Journal of Nutrition.* 105(06): 911-918.

St Onge, M.P. & Jones, P.J.H. (2003). Phytosterols and human lipid metabolism: efficacy, safety, and novel foods. *Lipids.* 38(4): 367-375.

Strazzullo, P., D'Elia, L., Kandala, N.-B. & Cappuccio, F.P. (2009). Salt intake, stroke and cardiovascular disease: a meta-analysis of prospective studies. *Br Med J.* 339 b4567.

Talati, R., Baker, W.L., Pabilonia, M.S., White, C.M. & Coleman, C.I. (2009). The Effects of Barley-Derived Soluble Fiber on Serum Lipids. *Ann Fam Med.* 7(2): 157-163.

Turner, N.D. & Lupton, J.R. (2011). Dietary Fiber. *Advances in Nutrition: An International Review Journal.* 2(2): 151-152.

Urala, N. & Lahteenmaki, L. (2004). Attitudes behind consumers' willingness to use functional foods. *Food Quality and Preference.* 15(7-8): 793-803.

Verbeke, W. (2005). Consumer acceptance of functional foods: socio-demographic, cognitive and attitudinal determinants. *Food Quality and Preference.* 16(1): 45-57.

Verbeke, W. (2006). Functional foods: Consumer willingness to compromise on taste for health? *Food Quality and Preference.* 17(1-2): 126-131.

Verbeke, W. (2010). Consumer reactions to foods with nutrition and health claims. *Agro Food Industry Hi-Tech.* 21(6): 5-8.

Verbeke, W., Scholderer, J. & Lahteenmaki, L. (2009). Consumer appeal of nutrition and health claims in three existing product concepts. *Appetite.* 52(3): 684-692.

Vermeer, C., Shearer, M.J., Zittermann, A., Bolton-Smith, C., Szulc, P. et al. (2004). Beyond deficiency: Potential benefits of increased intakes of vitamin K for bone and vascular health. *Eur. J. Nutr.* 43(6): 325-335.

Weinbrenner, T., Fito, M., de la Torre, R., Saez, G.T., Rijken, P. et al. (2004). Olive oils high in phenolic compounds modulate oxidative/antioxidative status in men. Journal of Nutrition. 134(9): 2314-2321.

WHO (2003). Recomendations for preventing cardiovascular diseases. In: *WHO Technical Report Series.* pp. 81-94. World Health Organization, Geneva.

Wills, J.M., Schmidt, D.B., Pillo-Blocka, F. & Cairns, G. (2009). Exploring global consumer attitudes toward nutrition information on food labels. *Nutrition Reviews.* 67(5): S102-S106.

Zeisel, S.H. & Caudill, M.A. (2010). Choline. *Advances in Nutrition: An International Review Journal.* 1(1): 46-48.

10

The Industrial Meat Processing Enterprises in the Adaptation Process of Marketing Management of the European Market

Ladislav Mura

Dubnica Institute of Technology, Department of Specialised Subjects
Slovak Republic

1. Introduction

Coming on globalization, international economical interdependency, generate new business chances and move into territorially and qualitatively new dimensions of enterprising. In the current sharp competitive struggle the successful and quick application of results of the research and technological improvement together with innovation of products and services to forecast the needs of consumers are on the first place. Adaptation to the new conditions of the market is expected in the field of marketing management selection of adequate marketing and competitive strategy for building a strong market position and for perspective additional development of business. [6]

Agro-food market conditions have dynamically changed in the last 20 years. The biggest influence on the market had the ownership transformation, later the penetrating of foreign investors into the several branches of food industry. [11]

Formation of business environment was markedly influenced by the accession of theSlovak Republic into the EU, together with the need of adaptation to new conditions of the united market. Enterprises have adjusted to competitive environment, they changed the intracompany management, and implemented innovations.

Especially for the branches of food industry it was important to modernize the technological accessories, to fulfill the challenging quality and hygienic standards asked by the European norms. Aforesaid aspect was obligatory to put into the marketing management adaptation of concrete businesses. [1] New conditions, which are characterized by a highly developed market - where the supply overtops the demand - relative consumers saturation of their basic wants, called for adaptation of supplying companies to consumers needs. The marketing management is one of the basic expectations of effective and successful company operations. The managers of Slovak companies are aware of the demanding market background. [7]

2. Materials and methodology

Based on exploration of marketing management adaptation and on analyzed sample of meat processing plants in the period of changing market conditions, in the horizon of the years 2002 - 2006, is the main aim of the article to define the determining factors of successful marketing management adaptation to the conditions of the common EU market. Partial aims are as follows: to analyze the marketing management of the examined enterprises before the enter of the Slovak Republic into the EU, to analyze the marketing management of the examined enterprises after the Slovakia´s entry into the EU, to characterize the business activities of meat processing plants on the domestic and foreign markets, to prepare the SWOT analysis and to specify the factors of the successful marketing management of analyzed meat processing plants.

Objects of research are the biggest and the most important plants of meat processing industry of Slovakia. The influential sample is created by the enterprises producing 74,8% of the Slovak meat processing companies production.

To fulfill the stated aim a primary research within the meat processing companies was needed. Basic information and data were obtained by a questionnaire, by managed interviews with top management and by panel discussion in order to define the key factors of successful adaptation of the marketing management. To keep the sensitive data we will use the following identifications: "comp.1", "comp.2", etc. in our article. Additional sources of information were secondary sources such as the Slovak agricultural and food industry reports (so-called "Green report"), analytical works of the Slovak payment agency, analytical and internal materials of the Ministry of Agriculture of the Slovak Republic, and outputs of research works in the given topic.

To process the primary and secondary information logical methods, selected mathematical and statistical methods, methods of descriptive statistics, SWOT analysis were used. Relations between particular characters were quantified and examined on the significance level of $\alpha=0,05$ by the chi quadrate test and the correlation coefficients. Interpretations of the outcomes are done by the "p-value".

3. Theoretical scopes

Conditions of business activities in the sector of agro-food complex are markedly influenced by turbulent background of the agro-market, changing rules of financial supports, social and economical spheres of the life in Slovakia. In the current period, just the smaller parts of the managements of agro-food industry apply marketing practices. [8] With a consecutive enter of international companies into the Slovak food processing enterprises, the increasing need of marketing management in relation to successfulness on the market can be observed. A marketing approach to the business management is the eminent condition for successful business activities.

The success of an enterprise depends on numbers of factors and conditions, among which the qualitative marketing management takes the first place in dynamically changing conditions. Marketing management is a systematic and goal-seeking activity, aimed at the maximum utilization of abilities and properties of the enterprise, with the goal of stable status on the market and competitive advantage besides meeting the customers´ needs. In today´s globalization conditions and marketing structures integration, the territorial expansion of marketing management becomes a scope on the target markets. [8]

Based on identified demands and requests, a company creates the most adequate marketing strategy for placement on the given market. A selected marketing strategy is declared by a marketing mix. Practically, the successful marketing mix application depends on the three conditions [4]:

- tools of marketing mix must chronologically form a constant and harmonic unit,
- facilities of marketing mix tools have to reflex of eventual market development and company situation
- intensity of the usage of several tools of marketing mix must be sufficient

Seriousness of the listed operations consists in the moving conditions in time and in their correlations.

The latest trends in marketing shows, that marketing is an integrated complex of actions focused on customers and the market. Marketing steps must be at the same time re-bounded with other processes in the company and to be an integral part of the management.

Marketing management is a continuous process of analysis, planning, implementation and control. Its sense is the creation and maintenance of long-standing relationships with target customers and consumers, which help companies to reach the given targets. [2] The aim of marketing management is to identify consumers´ needs and wishes , to create a vision of innovative products and to set up the company processes in the way to be able to present for competition a product of higher quality and efficiency.

Coming on with globalization of the world economics and the integration processes causes that enterprises operate their businesses more and more in the international environment, rank into the foreign markets to reach a better valorization of the company capital. [5] The companies are under an extreme competitive pressure.

4. Result and discussion

Integration of the Slovak Republic into the market structures of the EU besides the positive sides was also taken negatively by the business sector. This is valid for most of the agro-food companies. On one hand, for Slovak companies it was a chance to join the united market of the EU, on the other hand, it was a must to fulfill the demanding conditions of hygienic, qualitative and veterinary norms. The high level of these parameters caused the downfall of many companies, which were not able to adapt to the changed conditions. The fulfillment of particular norms and standards necessitated serious investments into the technological facilities and the reconstruction of existing producing companies.

Adaptation of meat processing companies to the new legal and market conditions was the first and basic premise of the successful business activities on the internal EU market. This form of adjustment, so called compulsory adaptation, allows the concrete enterprises to practice their business activities in the field of agro-food business.

The European Union represents an internationally, extra-nationally and nationally marked business environment. Some of the legal, political, economical and technological factors of a marketing mix of the EU members do not have clear international or national character, but rather a combination of national and extra-national norms, rules and politics. The objects of company marketing management of integrated economics are first of all the high quality production orientation, its ability to compete and the ability to achieve a place on the united market. Together with the strategic marketing management they are focused on the product differentiation and increasing the surplus value of products. [12] This field of marketing management can be denominated as a voluntary form of adaptation for the new conditions with the aim to achieve a competitive advance and a bigger market share. The choice which marketing strategy will used to reach the marked target is on the concrete company.

The objects of our analysis were nine biggest animal firms producing plants in Slovakia whose production covers 74,8% of the market demand. Data obtained by research are considered as a case study. The focus is on selected marketing management and economical indicators. The secondary data were obtained from Statistical office of the Slovak Republic. The table 1 presents the biggest meat processing factories in the years 2002 - 2006 in Slovakia based on data of the Statistical office of the Slovak Republic. [10]

Nr.	2002	2003	2004	2005	2006
1.	Tauris, a. s.	Tauris, a. s.	Tauris, a. s.	Tauris, a. s.	Tauris, a. s.
2.	Hrádok Mäsokombinát, s.r.o.	Hrádok Mäsokombinát, s.r.o.	Hrádok Mäsokombinát s.r.o.	Mecom, a. s.	Mecom, a. s.
3.	Mecom, a. s.	Mecom, a. s.	Mecom, a. s.	Hrádok Mäsokombinát, s.r.o.	THP, a. s.
4.	THP, a. s.	THP, a. s.	THP, a. s.	PM Zbrojníky, a. s.	Hrádok Mäsokombinát, s.r.o.
5.	Hyza, a. s.	PM Zbrojníky, a. s.	PM Zbrojníky, a. s.	THP, a. s.	PM Zbrojníky, a. s.
6.	Hydina ZK, a. s.	Tauris Danubius, a. s.	Tauris Danubius, a. s.	Tauris Danubius, a. s.	Hyza, a. s.
7.	Tauris Danubius, a. s.	Hyza, a. s.	Hyza, a. s.	Hyza, a. s.	Tauris Danubius, a. s.
8.	Hydina, a. s.	Hydina ZK, a. s.	Hydina ZK, a. s.	Hydina ZK, a. s.	Hydina ZK, a. s.

Table 1. The biggest meat-processing companies in Slovakia. Source: own processing

As the table shows, the biggest and at the same time the most successful business entity in the branch of meat-processing in 2002 - 2006 was the company Tauris, a.s. It holds its leading position for a long term thanks to business strategy, innovations and the effective marketing management. The second and third most important subjects – Hrádok Mäsokombinát, a. s. and Mecom, a. s. hold their position stable. The companies THP, a. s. and Hyza, a. s. started to fall behind the leading processing companies and as separate subjects they could not stand the competing struggle what culminated in a fusion to a single company.

For the identification of strengths and opportunities, the weaknesses and threats there was prepared a SWOT analysis in the sample of analyzed meat-processing companies. A summary review about the situation in the particular companies, overall information about the products, delivered certificates, turnover development, strengths and opportunities, the specification of weaknesses and the forecast of threats from the point of view of the company documents are given in the table 2.

The business entities which took part in the research reacted the changes in their macro-background and stepwise adapted to the new social and economical conditions. Examined enterprises not only implemented, but also permanently kept the production process in accordance to the strict conditions of critical control points of HACCP (Hazard Analysis Critical Control Point). Besides of this system they had implemented a system of quality management based on norms of ISO 9001, or ISO. 9002. The research showed the time slip in the implementation of systems of quality management at the particular companies which caused a competitive advance for some of them. Company 1 and company 3 which were flexible with the implementation of systems of the quality management based on norms of ISO became the leaders on the market with an adequate development of their turnover.

Adaptation to the strong technological norms asked for high investments to purchase and implement new technologies and technological procedures for meat processing to meat products. The investment activities were insured by companies by credit lines what on the other hand caused a high finance and capital indebtedness (Company 2 and Company 5). The opportunity to obtain the indebted companies motivated financial groups to enter the agro-food industry. Financial and investment groups as Penta Investments, a. s. and Eco Invest, a. s. are such examples.

	Company 1	Company 2	Company 3	Company 4
Product portfolio	132 types of products	252 types of products	224 types of products	110 types of products
System of quality management and the date of implementation	ISO 9001 – 2003 SK 15 – 1996 SK 5 – 2001 SK 618 – 2004	ISO 9001 – 2003 SK 6061 – 2004	ISO 9001 – 1998 ISO 9002 – 1998 BRC Food – 2003 IFS – 2003 SK 63 – 2003	ISO 9001 – 2002 ISO 14001 – 2006 SK 61 – 2003
HACCP	from the year 1997	from the year 1998	from the year 2004	from the year 2001
The turnover evolution during the analyzed period	107,5 mil. € - 107,9 mil. € ↑	75,6 mil. € - 63,0 mil. € ↓	54,8 mil. € - 81,3 mil. € ↑	70,3 mil. € - 38,6 mil. € ↓
Strengths	modern technology, product innovations, capable human resources, long term relationships with chains	modern technology, capable human resources, original recipes, good geographical location in relation to foreign markets, own abattoir	experienced management, strategic marketing, investments to the technologies, extension of producing capacities, acquisitions of small companies	capable human resources, own cannery
Weaknesses	Capital indebtedness, improper capital structure	tight range in chains, promotion	duplicity of management, improper organization and management structure	stray marketing, absence of management skills
Opportunities	penetration to the foreign markets, EU founds, increasing of assurance of consumers	penetration to the foreign markets, reinforcement of promotional activities, wider product portfolio in the chains	penetration to the foreign markets, innovation of product portfolio	penetration to the foreign markets, concentration of the branch, acquisition of companies out of EU norms
Threats	overhead costs, price pressure of chains - product quality fall down, competitive faith	overhead costs, the real income fall down of consumers	overhead costs, competitive faith	overhead costs, price pressure of chains - product quality fall down,
Promotion	active promotion by participation on home and foreign exhibitions	active promotion by participation on home and foreign exhibitions	active promotion by participation on home and foreign exhibitions	active promotion by participation on home and foreign exhibitions

Table 2. Some aspects of marketing mix and SWOT analysis. Source: own processing

	Company 5	Company 6	Company 7	Company 8	Company 9
Product portfolio	100 types of products	190 types of products	40 types of products	slaughter	40 types of products
System of quality management and the date of implementation	ISO 9001 – 2000 SK 64 – 2004	ISO 9001 – 2004 SK 3092 – 2004	ISO 9002 – 1996 SK 15 – 1996	SK 26 – 1997	ISO 9001 – 2004 SK 630 ES – 2004 SK 3031 – 2004
HACCP	From the year. 1998	From the year 2004	From the year 1995	From the year 1999	From the year 2000
The turnover evolution during the analyzed period	256 mil. € - 430 mil. € ↑	95 mil. € - 161 mil. € ↑	373 mil. € - 336 mil. € ↓	-	140 – 484 mil. € ↑
Strengths	quality of products, own retailing, capital interconnection whit the company 1	Flexibility, adaptability and giving on regional changes, available prices, tailor-made approach	product and packing innovations, capital interconnection whit the company 1	available prices, modern technology, bio-meat production	original recipes, flexibility, capable human resources, own abattoir,
Weaknesses	absence of marketing activities	only local entreprene-urship, weak marketing policy	high operating costs (special financial costs), absence of export	only jointing meat, fluctuation of human resources	absence of export, narrow range of goods, weak marketing policy
Opportunities	capital interconnection whit company 1, penetration to the foreign markets, new hypermarkets	penetration to the foreign markets,	penetration to the foreign markets, innovation of product portfolio	promotion, penetration to the foreign markets	penetration to the foreign markets, promotion
Threats	absolute submission of business policy to company 1, competitors fight	competitors struggle	operating costs, competitors fight, force of trade string	force of trade string, operating costs	operating costs, force of trade string, competitors fight
Propagation	Passive propagation	Passive propagation	active promotion by participation on home and foreign exhibitions	Passive propagation	Passive propagation

Table 3. Some aspects of marketing mix and SWOT analysis in other companies. Source: own processing

The tables 2 and 3 give a synthetic overview about the situation in particular businesses and give an integrated information about the products, quality certificates, sales dynamics, strengths and weaknesses of companies inclusive the identification of opportunities and threats from the point of view of a concrete company.

The synthesis of determinated facts of the SWOT analysis identifies the absence of strategic management elements in the selected group of companies. In one case there is an absence of the strategy of human resources development with limited motivation, two fifths of businesses do not use actively the possibilities of the communication mix and none of the companies entered an international market during the analyzed period. The potential threats are the epidemic diseases of animals (BSE, KMO), low net margin in the meat production, the enforcement of the law, limited defense of the home market. As substandard factors remain the increasing prices of energies and of raw materials and the decreased consumer´s acceptance. The managements of enterprises agree that under the impact of the entrance of Slovakia into the EU domestic and external competition sharpened and the companies do not feel an adequate defense of their home market against external suppliers. Discriminating practices of foreign commercial chains express themselves by abusing their dominant position and by increasing power against processers.

Strengths are positive factors influencing the future successfulness what concerns the analyzed sample of enterprises: the implemented system of quality management, modern technological equipment's, qualitative and innovated products, capital cohesion of companies with basic industry. With the implementation of quality systems we can see the time difference in achieving it, which is followed by a competitive advantage for companies with an earlier certification. On the other hand, the purchase of modern technologies caused indebtedness of some subjects and consequently a takeover by stronger subjects.

Opportunities are new distribution channels (purchasing alliances, hypermarkets), shopping practices of consumers (packed meat, meat semi-products), internalization of business, penetrating into the foreign markets, and reinforcement of marketing activities.

Sharpened competitive struggle on the market expresses itself in the dynamics of sales of analyzed companies. We were interested in the development of total sales in the period 2002 - 2006. Based on our detections, the growth achieved six subjects (66,66%), stagnation two companies (22,22%) and decrease - one company (11,11%).

Through built-up questionnaire we detected changes in the amount of sales while the companies were selling goods under a private brand name. We tested the following hypotheses:

H_0: We expect that there is no dependence between the share of sales on the foreign market and sales under the private brand names.

H_1: We expect that there is a dependence between the share of sales on the foreign market and sales under the private brand names.

The hypothesis was examined with the chi-square test of independence. The strength of dependence was determined by the Persons contingency coefficient C. The results are shown in the table 4.

Test description	Test	Statistics	P-value
test of independence	chi-square	18,7682	0.0009
strength of dependence	contingency coefficient		0.7145

Table 4. Testing of independence. Source: SAS software

Calculated value of test criteria χ^2 is bigger than the critical value of χ_{tab}. We refuse the hypothesis H_0 and accept the hypothesis H_1 based on which there exists a dependency between listed qualitative attributes. According to the values of Persons contingency coefficient we can declare a very strong dependency. Production of goods under private brand names as well the expansion in the foreign markets show a strong adaptability for the intensive competitive environment of the company and show utilization of potential chances of achieving a position on the market.

In the next part of the research we focused on the store types which take the biggest part on the sales of goods during the analyzed period. The conclusions are in the graph 1.

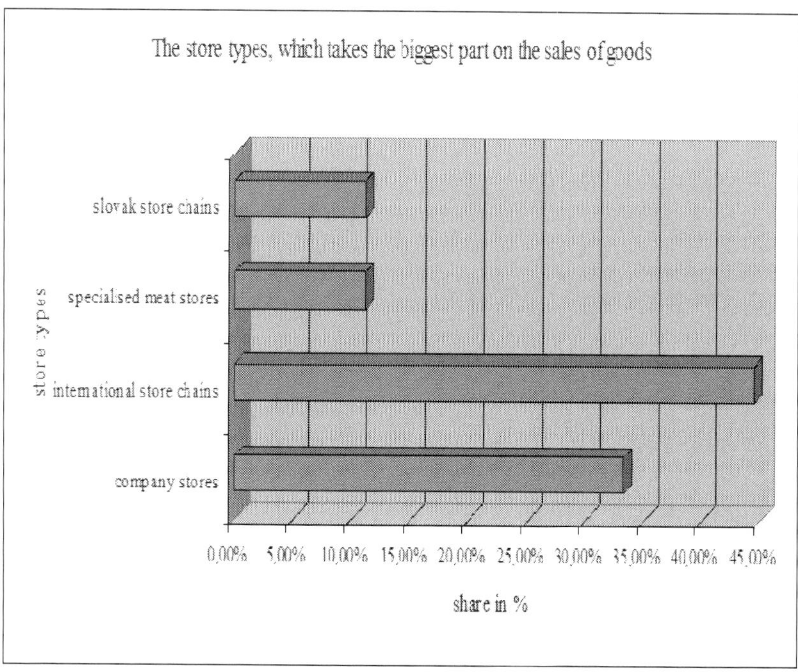

Fig. 1. The store types, which takes the biggest part on the sales of goods. Source: own processing

As the graph 1 shows, the biggest part of production is sold through international chains – four companies with share of 44,44%. Thanks to mentioned store types the production of Slovak meat processing companies is taken onto the foreign markets too. This fact helps enterprises to ensure the sales of bigger part of the production, to increase the profit and to build up the status on the target markets. By 11% less, the Slovak chains participate on sales of production - in case of three analyzed companies. The same share on sales - 11,11% have specialized meat stores and the company store. It is true in one case in company 1. In spite of the low share on the sales channels the specialized meat stores and the company stores have, they provide to customers the full range together with products with a higher added value.

The Slovak entry into the EU meant for Slovak companies a possibility to achieve the foreign markets of the EU members too, but on the other hand the competition accelerated. We have tested if there is a dependency between increased domestic competition after the Slovakia´s entry into the EU and if there are larger requests to adapt to the quality requirements and to the adherence of quality standards. We have verified the hypothesis:

H_0: We expect that there is no dependence between increased domestic competition after the EU entry and the larger requests to adapt to the quality requirements and to the adherence of quality standards.

H_1: We expect that there is a dependence between increased domestic competition after the EU entry and the larger requests to adapt to the quality requirements and to the adherence of quality standards.

The hypothesis was examined with chi-square test of independence. In regarding to achieved values the strength of dependence was not determined. The results are given in the table 5.

Test description	Test	Statistics	P-value
test of independence	chi-square	1,1688	0,2796
strength of dependence	contingency coefficient	-	-

Table 5. Testing of independence. Source: SAS software

Calculated value of test criteria χ^2 is lower than the critical value of $\chi_{tab.}$ We accept the hypothesis H_0 based on which there does not exist a dependency between listed qualitative attributes. We refuse the hypothesis H_1. It means that companies were not motivated to take a bigger attention to ensure the quality of their production after the entry to the EU. In the future perspectives we recommend companies to revaluate the possible impact of sharpened competition on their businesses and to actively participate on the building of their position on markets.

In the frame of marketing management analysis in the examined group of processing companies we asked the enterprises if there is an influence of various marketing steps to increase the sales. The scale of answers was from 1 to 10. The lowest influence has 1, the biggest 10. The businesses rated the influence of concrete marketing actions differently. To analyze the results we used the methods of descriptive statistics - median. The found status is illustrated in the graph 2.

The marketing step "price, price proceeding, price reduction" has clearly achieved the highest score on the ranking scale. Based on the opinion of analyzed companies, this has the biggest influence. The second most eminent marketing step is the B2B marketing which includes good relationships with consumers , contacts, business presentations. A close result was achieved within the factor "production under private brand names". It helps companies to rank into the different market segments. According to experience of managements within effects less influencing the sales increase there are factors such as promotion, in-store testing and product innovations. Different usage of marketing actions and tools in practice causes the differentiation of producers´ offers.

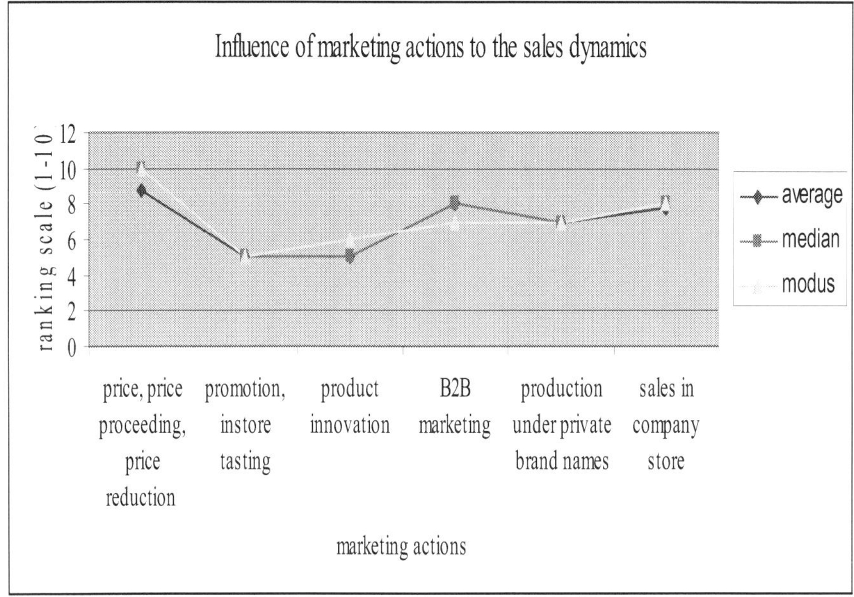

Fig. 2. Influence of marketing actions to the sales dynamics. Source: own processing

5. Conclusion

The branch of meat processing is a significant part of the food industry branch. It cannot be separated as it is an integrated part of food industry. The research revealed weak and limited connections between agricultural production and capacities of processing companies and sales to the stores network. Based on the achieved information and the carried out research we recommend the management of meat processing companies to focus on the following fields which we consider as a market expansion:
- production of meat products with pro-biotics,
- production of meat products with reduced content of fat,
- product portfolio diversification
- internalization of business activities, especially a penetration into the markets of V4 countries and the Russian federation,
- to administer the innovation strategies of products,
- active development of PR possibilities,
- to stabilize business relationships by informal meetings and common undertakings,
- to develop the B2B marketing,
- to renew human resources with language skills and business experience on foreign markets.

The target market identification and the identification of several market segments helps to fulfill not only the current aims but also potential expansive objectives. Selected aspects of marketing management in the business practice of meat processing companies and perspective fields for expansion determinate the winning factors of successful marketing management adaptation for the new EU conditions. Based on Agricultural payment agency [10] it is possible to expect in the future a sharp competitive struggle on the domestic markets of the EU and the countries with open economics such as in Slovakia which must adapt to the import of cheap meat and meat products for example from Poland or Romania.

6. Acknowledgment

This scientific work has been supported by the internal research project of Dubnica Institute of Technology: „Internationalization of small and medium enterprises in chosen region".

7. References

Horská, E.; Oremus, P. (2008). Territorial Approach to International Marketing Channel and Value Added: Case of Agribusiness. In: *Impacts of Globalization on Agribusiness: Trends and Policies*, IV. International Conference on Applied Business Research ICABR, ACCRA. Ghana, Publisher: Mendel University in Brno, 2008, ISBN 978-80-7375-154-8

Kotler, P.; Armstrong, G. (2004). *Marketing*. Praha: Grada Publishing, 2004, pp. 43 – 47, ISBN 80-247-0513-3

Kozelová, D.; Mura, L. et al. (2011). Organic products , consumer behavior on market and European organic product market situation. In: *Potravinárstvo* Vol. V, No. 3/2011, pp. 20-25, ISSN 1338-0230 02.08.2011 Available from: http://www.potravinarstvo.com/journal1/index.php/potravinarstvo/article/view/96/131

Kretter, A. et. al. (2010). *Marketing*. Nitra: SPU, 2010, 288 s., ISBN 978-80-552-0355-3

Malá, E. (2009). Language and intercultural dimension in the process of internationalisation of higher education. In: *Ianua ad Linguas Hominesque Reserata II*. Paris: INALCO 2009, p 100 – 123. ISBN 978-2-91525596-9

Mura, L. (2011). The network approach of internationalization – study case of SME segment. In: *Scientific Papers of the University of Pardubice – Series D*, No. 19, Vol. XVI č. 1/2011, pp. 155-161, ISSN 1211-555X 27.06.2011 Available from: http://www.upce.cz/fes/veda-vyzkum/fakultni-casopisy/scipap/posledni-obsah.pdf

Mura, L.; Gašparíková, V. (2010). Penetration of small and medium sized food companies on foreign markets. In: *Acta Universitatis Agriculturae et Silviculturae Mendelianae Brunensis*, MZLU Brno, LVIII, 2010, 3, pp. 157 – 164, ISSN 1211-8516 27.06.2011 Available from: http://www.mendelu.cz/dok_server/slozka.pl?id=45392;download=63053

Oremus, P. (2008). Adaptácia marketingového manažmentu podnikov potravinárskeho priemyslu na vnútornom trhu Európskej únie. Nitra: DDP, 2008, 150 p.

Šimo, D. (2006). Agrárny marketing. Nitra: SPU, 300 p., ISBN 80-8069-726-4

ŠÚ SR. Štatistiky. Poľnohospodárstvo. 14.06.2011 Available from http://portal.statistics.sk/showdoc.do?docid=11005

Šúbertová, E. (2010). The structural changes and results of agricultural companies in the Slovak republic. In: Megatrend Review, Vol. 7, 2/2010, pp. 53-62, ISSN 1820-4570

Zelená správa. (2011). Správa o poľnohospodárstve a potravinárstve. 12.06.2011 Availed from: http://www.land.gov.sk/sk/?start&navID=121

Zelené štatistiky. (2011). Pôdohospodárska platobná agentúra pri Ministerstve pôdohospodárstva, životného prostredia a regionálneho rozvoja SR. 12.06.2011 Availed

11

Facilitating Innovations in a Mature Industry-Learnings from the Skane Food Innovation Network

Håkan Jönsson[1], Hans Knutsson[1] and Carl-Otto Frykfors[2]
[1]Lund University,
[2]Linköping University
Sweden

1. Introduction

The food industry is today one of the most globalized industries and subjected to growing strains. Globalization and the shift towards innovation-driven knowledge-based competitiveness between firms and regions have rendered traditional national and regional policies and concerted efforts for enhancing industrial economy and prosperity less effective. Baldwin (2006) refers to globalization as "the great unbundling". It is not just a matter of slicing up of value chains and relocating various stages of production to more comparably advantageous regions. Innovation and intrinsic knowledge creation are important parts of the renewal of national and regional policies and efforts to compete internationally.

The food industry, though, is a mature industry. Such tend to be regarded less innovative than emerging Science & Technology-based industries. A possible reason is that they tend to be governed by social and technological regimes (Winter, 1983) related to experience and tacit knowledge built up during the long history of the sector. The food industry is not only mature, but is further characterized by complex and long value chains – shaped and formed over time.

This chapter was written from a regional perspective. In the region of Skane, located in the south of Sweden, a strong food industry has developed over centuries. The potential of Skane as a food region has been acknowledged and supported by changes in the national Swedish R&D and industrial growth policies. The concept of sectorial innovation system and a triple helix approach is now used for enhancing the knowledge-based transition in the Swedish system (Frykfors & Klofsten, 2011).

The major challenge in Skane was that the different actors in the food industry had different ultimate goals. Another challenge was that no single actor owned the overall strategic problem of enhancing the innovative dynamic capabilities (Teece 2007) of the local food industry forming an innovative cluster. It was necessary to reach an agreement about how to govern this joint strategic problem.

The Skane Food Innovation Network (SFIN) was established at the time of Sweden joining the European Community. As the international competition grew more apparent to companies in the Skane region, companies, universities and public authorities saw the need to establish a model for facilitating innovations in the food sector and to create a joint ownership of strategic questions. Since 1994, SFIN is gradually shaping the concept of innovation community. It creates dedicated sub-communities for specific areas and the innovative capability in this mature industry is gradually evolving.

This chapter describes and analyses the policy-making impact on the transition of a locally mature food industry of the Swedish province Skane into an innovative and ever more competitive international food region. The theoretical frame of reference is based upon the concept of social and technological regimes. These regimes, as will be shown, differed between major stakeholders in the region. This called for bridging activities. Several of these activities have been orchestrated by Skane Food Innovation Network (SFIN) and have helped forming new cognitive maps, in turn helping new innovation communities to form and navigate in new market spaces (Frykfors & Jönsson, 2010).

The aim of this chapter, thus, is to present and discuss the learnings of the development of the Skane Food Innovation Network. This relates directly to the ongoing discussion about how stakeholders, both private and public, may facilitate the development of a regional innovation system.

2. Background

2.1 The Skane food industry

Skane is located in Scandinavia, in the south of Sweden and has approximately 1.2 million inhabitants. It is the center of Sweden's food industry. The region has a high density in food-related activities: within the 11 000 square kilometers, all sectors of the food business area are found, covering the total chain from farm to fork. Primary production, the food processing industry, packaging, production machinery, distribution, warehousing and quality control can be found in the cluster. Competence in product and process development, both in industry and academia, is abundant and the marketing and R&D functions of large companies are well developed. Retailing and food distribution companies are also present in the cluster (Lagnevik 2006).

In 2007, the industries forming the core of the Skane food industry employed approximately 25,000 people (Henning et al 2010). Definitions of the boundaries of the food industry, which companies should be included etc, are debated. While discussing the dimensions of the food industry, it has been argued that it not only includes the traditional areas of the food industry, but also industries and disciplines that have a strong link to or act in symbiosis with the food industry (Oresund food 2011:40). If the Skane food industry is broadly defined, including related businesses, an employment at about 100,000 people can be accounted for. This is high in relation to the population in the region (Wastenson et al, 1999). Moreover, Skane accounted for about 21 per cent of the total number of employees in the Swedish food sector's core industries. As Skane has about 12 per cent of the total number of economically active individuals in Sweden, this indicates a strong regional position of the food industry (Henning et al 2010).

2.2 A changing foodscape

During most part of the 20th century, the Swedish food market was protected from international competition. The idea of national self-subsistence guided Swedish food policy. International competitiveness was not a major issue, since surplus production was limited.

During the last decades though, the Swedish food industry has been rapidly changing from a sheltered national industry into an industry exposed to strong international competition. This has occurred gradually in three steps (Lagnevik 2006).

In 1986 the Swedish government declared that the food sector gradually should be exposed to international competition. This induced a change in the Swedish agricultural sector and many Swedish agricultural companies began to adapt to the new working conditions. The second step and a major change in the competitive situation occurred when Sweden joined the European Community on January 1st, 1995. By the entry, trade barriers for finished food products were completely removed. The third step occurred on May 1st, 2004, as the new member states joined the European Community. The Swedish food industry and agriculture is now exposed to fierce international competition. The closest neighbor states, e.g. Lithuania and Poland, produce agricultural bulk products at a cost that cannot be met by Swedish farms and agricultural companies. In addition, the food industry has experienced revolutions in the IT and the Biotech sectors. These technological revolutions have radically changed the working environment for food companies and increased consumer interest for organic and local food, food safety and healthy eating. All in all, this means that the whole context in the food sector has changed dramatically the last 25 years.

It is not only the competitive situation that has changed, but also how the consumers view food quality and food safety. A term developed to capture the multidimensional aspects of food, trying to give equal value to the material, social and mental aspects of food is "foodscape". The term has been used by, among others, Rick Dolphijn (2005) and Pauline Adema (2009). Building on Arjun Appadurai's (1990) influential use of different "scapes" in order to understand the processes of deterritorialization in a global world economy, we use the term 'foodscape' to try to capture the complex intertwining of people, food products, places, emotions etc. that happens in food-related situations, resulting in communal identities as well as economic, physical and social structures linked to food. This broader view on the changing conditions for the food actors is necessary in order to understand the context within which the activities of SFIN have taken place.

2.3 The triple helix setting

The companies in the Skane food industry are a mixture of larger national and multinational companies (such as IKEA Food services, Nestle, Findus and Atria), and SMEs, from specialized food manufacturers to small innovative research-based companies, such as Oatly. Food-related education and research is to be found at all four universities in the region, especially at Lund University (Scandinavia's largest establishment for higher education and research with over 40,000 students, founded in 1666), but also at the Swedish University of Agricultural Science, Alnarp, at the Malmö University, and at Kristianstad University College. Research in the region covers all kinds of scientific knowledge in the food chain from farming to consumer studies as well as scientific knowledge in industries related to and supporting the food chain. The Skane region also has a strong position in research and development in the companies of the food branch. Many Swedish companies have located their R&D centers in this region, as have packaging, processing and distribution plants.). Among the food related industries, a special note should be made about the packaging industry as it represents a large sector in the region with internationally successful actors such as Tetra Pak.

State, regional and municipal authorities support development in the food industry. In particular, the Region of Skane invests actively in the development of food-related initiatives and pursues a vision of becoming the food center of northern Europe in 2025. The state-governed county council supports local companies and initiatives consciously and forcefully.

2.4 Features of the Swedish innovation system and innovation & research policies

The arena for the Swedish food industry has changed quickly and radically in just a few years. Skane, being the center of Sweden's food industry and closest to the international market, has faced an urgent need for increasing the innovative capabilities of the food industry. A brief look at the Swedish innovation policy is needed in order to describe the framework in which SFIN operates.

Understanding the rationalities of the current Swedish use of the triple helix approach requires recognition of the main structure of the Swedish economy. It reveals several distinctive features grounded in *la longue durée* of niche management, e.g., industrial specialization and creative use of innovative technology procurement. This has orchestrated the main features of the innovation and research policy system and its growth model for industrial development and prosperity (Freeman, 1987, Lundvall, 1992).

The concept of "development pairs" (Fridlund, 1999) describes the long-term development cooperation and close partner relationship formed between specific state client and private firms in order to exploit and develop new technology. Such relationships often played the role of driving force and a catalyst for growth of industrial development blocks (Dahmén, 1987), which later matured into defined innovation system or what Carlsson & Stankiewicz (1995) called "technology systems". The evolution of the Swedish innovation system offers a number of examples of niche management, e.g the state power authority which historically enjoyed great freedom from political interventions and the Swedish telecom industry, and different key companies or group of key firms (Fridlund, 1999, Laestadius & Berggren, 2000).

However, the application of development pairs has exposed some negative structural features of the concept. The most striking feature is the bifocal policy structure that is split between two subsystems – one that supports innovation and business development, and another for academic education and research. Another shortcoming in the approach is a diminutive structure of intermediaries and research institutes. A third characteristic feature is that the regional governance level is by-passed and weakened (Pierre & Peters, 2005). The regional politico-administrative level has primarily been pre-occupied with financial support and the governance of common healthcare and communication infrastructure, while policies for research, technology and innovation (RTI) always have been a state affair. As Sweden became member of the EU in 1995, the country faced new regulations for compliance with EG legislation of procurement and trade. As a result, development pairs being the significant part of the Swedish innovation system were no longer viable. This called for a change in national innovation and research policies.

2.5 The dawn of the Skane model of food innovation

Influenced by the EU, the regional level in Sweden has to be more influential in issues of innovation and development policies. The shift in command, underlined by ongoing globalization of markets and of growing multinational companies, has made national innovation and growth policies less effective and more difficult to manage solely from a national level. In 2003, VINNOVA, the state agency for innovation and systems-oriented research, launched a regionally oriented program for research, technology and deployment/demonstration (RTD) in line with this shift. The program is based on the concept of sectoral innovation systems and the triple helix approach. The aim is to promote upgrading and renewal of local innovation and R&D capabilities and skill building in certain important growth areas with strong regional profiles. The focus on enhancing the strength of regional constituents and supporting structures of the sector innovation system involving regional authorities has materialized in the Winn-Growth program. This program, though, calls for the development of new types of governance and practices.

One initiative within the Winn-Growth program was granted to the Skane Food Innovation Network, an innovation intermediary, a regional "facilitator" composed of companies, universities and regional authorities. The idea was to increase the return on investment and added value in the regional food industry by strategically enhancing the R&D and innovation capacity in the food sector. A previous innovation strategy for the food sector of Skane, based on increasing internationalization and more knowledge-driven innovations, had been outlined prior to the Winn-Growth program. This had been done together with the establishment of an academic functional food research program (Nilsson 2008). Each of the regional Winn-Growth programs suggested by Vinnova run for a 10-year period and are subdivided in approximately 3-year stages.

3. The Skane food innovation network

The changing competitive conditions that occurred during the 1990's had spurred thoughts among leading industrialists, in the regional government and among researchers that measures were needed to be taken if the food industry in the region should stay competitive. In 1994, the three parties consequently created a joint network organization, Triple Helix style, and named it 'The Skane Food Innovation Network'. The objective was to facilitate the exchange of research and development between universities and the food industry. In the organization, leading actors from the region did investigations and started joint actions in areas related to competence and competitiveness. The insight that major changes were ahead made leading actors work hard together to improve the future for the food industry. In this context, it is worth noting that the original initiative came from the food industry. This has created a triple helix organization in which the industry has had a leading role for more than one decade (Asheim & Coenen 2005, Coenen 2006). From the outset, it was decided that the network CEO should have an industrial background and that the chairman of the board should be the County Governor.

The urgent need for a transition away from traditional bulk production was the major incentive for creating the organization. New knowledge and collaborations with other actors were needed in order to increase the production of specialized and highly processed products. This made some of the major players in the food industry dedicated to the organization. At the time, functional foods and convenience foods seemed to have a very promising future. Both segments demanded academic research and, although to a lesser extent, regional collaboration.

The Skane Food Innovation Network was organized as a flexible network organization, with a small, operative executive board. Through the co-operation over the years, it came to be seen as a constructive force in the regional development from both the industry and the regional authorities. In spite of the small size of the organization, when competing for the

"Winn-Growth programme" it was seen as logical that SFIN should represent the joint innovation efforts in the region.

3.1 From project management to innovation communities

The Winn-Growth program multiplied SFIN's resources in a single blow. As any organization experiencing sudden growth, measures needed to be taken in order to fulfill the new goals. Over the years, continuous reconsiderations have been made. These can be divided into four phases: the starting phase, the foresight phase, the governance phase, and the community phase. The different phases were all induced by discussions about strategic issues and are therefore described in some detail. The story is based on interviews with key actors, both during the foresight process in 2006-2007 (Jönsson & Sarv 2008) and on follow-up interviews in 2009-2010. Two of the authors (Jönsson and Knutsson) became members of the management group in 2009. The material is thus complemented by inside information, in an anthropological tradition described as participant observation (Davies 2008).

Phase 1 – Making things start

The SFIN Winn-Growth initiative's point of departure, outlined by professor Magnus Lagnevik at Lund International Food Studies, was that many innovations are born at interfaces. The primary aim was therefore to encourage interaction at interfaces between different areas of scientific knowledge, between different technologies, between academic research and commercial enterprises and between private and public organizations. Knowledge integration (Grant, 1996) between different knowledge areas should produce new products, services and concepts. The Triple Helix dimension was the second cornerstone of the SFIN organization, reflected in the board already from the start, with the county governor as chairman and representatives from the food business and academia in the board and management group.

In order to get things started quickly, without having to start by building a central organization, the vast majority of the funding was canalized through existing organizations. Four key development areas were defined; Functional foods, International Marketing, Food Service, and Innovation. The former three were canalized through The Functional Food research centre, Lund International Food studies and Food Centre Lund, all academic research centers within Lund University. The latter was assigned to Ideon Agro Food, a foundation working mainly on strengthening the links between academia and business, aiming to support innovation projects and technological development in the food industry.

The key development area during the first years was Functional Foods. The area was at the time believed to have an enormous growth potential, and the need for knowledge integration between all sides of the triple helix was obvious. On the research side, nutrition, medicine, food technology, food engineering, marketing and consumer behavior were all relevant but shared neither traditions nor agendas. Food companies, food ingredient companies and packaging companies could join from the corporate side, all with a more congruent agenda. The county government, responsible for the healthcare in the region, could enter the cooperation with a lot of knowledge, but also as a major receiver and user of the innovation results.

During the first three-year period, the program was primarily engaged in stimulating the innovation system by providing new competence and deepening knowledge about the needs, prerequisites and opportunities for renewal of the food industry. Several PhD projects were initiated, primarily in the functional foods and marketing areas. The research projects were quite successful, as were many of the smaller innovation projects. Yet still, doubts were raised when the first three years of the program was about to be concluded.

Within the literature on clusters and cluster management, questions regarding the governance of clusters emerge (Nilsson 2008). Is it at all possible to govern a cluster, based upon intricate networks of actors, imprinted with different ways of thinking and acting? The Skane food innovation network had expectations from both its members and its funders (primarily Vinnova) to govern the innovation efforts in the Skane food industry. These expectations created an urgent need for rethinking the initial approach, based on the idea of outsourcing dedicated projects to organizations beyond the control of the SFIN organization. There were good reasons for the reconsideration: there was a wish to keep administration at a minimum and a clear need to position SFIN distinctly in relation to other facilitating organizations active in the industry. After the first three years, the problem of organizing the innovation efforts became obvious. It was probably efficient to get research and R&D project conducted by delegating them to specialized organizations and institutions. However, with a central organization without the resources to fully coordinate the outcomes of the separate projects, the Winn-Growth project appeared to be any other funder of research and innovation projects, with the strategic challenge and direction being left out. Another problem was the decreasing engagement from the business. The first years saw massive investment in research, which takes time to deliver useful results, especially since many of them were four-year PhD projects. The fact that the hype around functional foods cooled down these years made the connections between business and academia weaker. Concerns were raised about the dedication of resources from the industry. A long-term strategic discussion was initiated, both from Vinnova and from SFIN. What was the role of SFIN, really?

Phase 2 – Foresight as pit stop

Each of the regional Winn-Growth programs granted by VINNOVA ran for a 10-year period and was subdivided in approximately 3-year stages. Prior to each stage, an update of the program content was undertaken, involving the reconsideration of a relevant set of regional stakeholders and measures. In the intersection between the first and second stage of the SFIN program, a foresight process was suggested in order to navigate in a changing "foodscape" since both key stakeholders and markets were in transition. The foresight process should also strengthen the Innovation Network's governance of the "regional food cluster" of the sectoral innovation system, in accordance with the experiences mentioned above. The foresight exercise was part of an EU-funded four party foresight development project titled Foresight Lab (www.innovating-regions.org). The main goal for the "systemic" foresight was not to do a full-scale foresight, but to strengthen the innovation community in order to make the food cluster more competitive when future opportunities emerged.

The foresight identified three major challenges for the governance of the food cluster. The first and foremost was to make actors in the triple helix–perspective more profoundly committed to the idea of building a regional-based food innovation community. The regional perspective was not self-evident to many of the actors – they were regularly working in organizations where either a larger geographical area (national and/or global for the major companies) or a smaller area (sub-regional or local for the micro businesses) was being in focus for the business strategy and development activities.

The second was the heterogeneous character of the food innovation system. There were many small businesses throughout the value chain, from farmers over food producers to grocery stores, restaurants and public eating-places (retail and wholesale being somewhat more centralized). Some actors did not feel connected to other levels in the food chain, while others felt a need to establish collaborations, but uncomfortable with the process. At this point in the process it became obvious that the Swedish food system (as many other mature industrial sectors) had established strong regimes on each level of the food chain, with little or no understanding of the actors on other levels.

The third challenge for the foresight activities rose from the different social and technological regimes of the actors. This called for bridge-building activities that could translate and transform the disparate cognitive maps in order to build an innovation community that could navigate together in the new market. All actors were imprinted with values emergent from a time when the Swedish food system was based on national self-subsistence, with limited market competition. This was a major impediment for cross-sectoral innovation, since the new situation demanded new ways of collaboration and consistent interpretation of the foodscape. Although difficult, the timing was right. Since the regimes already were in transition, all actors were aware of the need for new working methods and collaborations.

The foresight confirmed several strengths of the Skane food industry, such as the presence of all vital parts of the supply chain, a strong, academic research tradition, and a well developed, common sense approach for food pleasure at the regional community levels. But findings also pointed especially to some weaknesses and threats in terms of "gaps". Hence, different needs and measures for strengthening the "bridges" across knowledge areas, long and short term plans and actions, big and small businesses, parties in the supply chain, research and business and among different levels of the innovation system (Jönsson & Sarv 2008).

Four potential sub-communities were identified from the mapping as being of particular interest to SFIN. In line with the systemic idea of developing the communities from a learning partition perspective, with knowledge and governance partition support, four systemic meetings were launched, with open invitations to all relevant actors. After the meetings, an ex post analysis of the foresight exercise and its outcomes was made. The focus was upon governance and how the organization could be developed in order to take a strategic leadership of a heterogeneous food innovation and production system. Governance-related services were hereby identified as particularly interesting services in the upcoming foresight work. Consequently, four focus areas were proposed: 1) *launch new services* (such as systemic meetings for sub communities in the food innovation system and dedicated knowledge partition services for hot areas), 2) *develop innovation community and foresight governance*, 3) *increase communication and bridge building* by continuously publish innovation community progress on an interactive basis, 4) *investigate in innovation system research* (research that not only formulates bases for food innovations, but also contributes to the development of the innovation system as a whole).

For all focus areas, specific activities were proposed. The analysis finally concluded that the different suggestions reinforced one another, and should therefore be developed concurrently, on an experimental and learning-oriented basis. Both "direct innovations" and cross discipline research projects were noted in the results in addition to the new knowledge and experiences regarding foresight work. In particular, the linking of the food innovation system with the health innovation system and the improved integration of retailers into the food innovation network was very promising (Jönsson & Sarv 2008). The foresight became a catalyst in the development process that Skane Food Innovation Network initiated in 2006. In 2010, most of the proposals of the Foresight's concluding report have been implemented.

Phase 3 – Establishing governance and credibility

Parallel and in continuous dialogue with the foresight process, organizational changes within SFIN took place. A new CEO, Lotta Törner, was recruited in 2006, and a re-organization of the board with the new county governor as chairman took place. In the dialogue with the major players from the business side, it was obvious that, from their perspective, there had been too much emphasis on long term academic research. Given the question what SFIN as an organization could do, the answer was primarily to enhance the attractiveness of the food business. For a long time, the industry had faced problems with recruiting highly educated younger people. A stronger focus on supporting the latter phases of the innovation process, to get new innovative products, concepts and services on the market, was further proposed.

The period led to a greater focus on meeting places. A number of meeting places were developed. Here, representatives of various interests and competencies could exchange and discuss ideas and develop creative solutions and business ideas. The existing meeting places were upgraded. The annual "Network day" were turned into a meeting place for most of the food sector in the region by inviting internationally renowned speakers and awarding research prizes and scholarships of a combined value of almost 100 000 euros in collaboration with a large foundation in the food and health area.

Furthermore, dedicated sub networks were established. A network of CEO's was formed, in which the most prominent CEOs of the food business now meet on a regular basis to exchange ideas and to discuss present and future challenges. This was a direct result of a foresight activity, where it was identified that the CEO's needed a special network in order to be committed to development projects outside their own companies. A Research Network was created as a meeting place for interdisciplinary contacts for food scientists from all faculties and universities in the region. The network is organizing seminars and workshops on "hot areas". A Retailer Network was organized in order to get the retailers more involved in the Innovation community. This was arranged as a part of a platform for innovative market places in order to integrate the retail side in the food innovation community and to promote innovative market solutions. The platform Future Meal Service, focusing on meals in the public sector was another initiative during this period. An Entrepreneur Council, where entrepreneurs can present their ideas and get professional advice and seed money in order to make their business ideas successful was further established.

The constant need for upgrading the creative capability of the innovation system was recognised as an important factor for the future, not least since the Winn-Growth program ends in 2013. The establishment of foresight as a continuous process was decided as a strategy in this work. Systemic meetings are being used whenever interesting projects materialize. A new arena for developing contacts between the packaging industry and the food industry in order to develop "Innovative food in innovative packages" is one example of the results generated from the systemic meetings. As part of the foresight process, there has been an increased focus on international bench learning.

A lot of focus was given to the Student Recruitment Program, reflecting the business representatives wish for activities dedicated at enhancing the attractiveness for the food industry among younger people. A starting point was to establish an Advisory Board with students. Ten students from different educational programs gathered together with the assignment to give the Network and the individual companies valuable input on how the best brains can be recruited to the food sector in the future. In order to improve contacts between companies and educational programs, the students on the Advisory Board arranged field trips to companies and lectures by CEOs at universities and colleges. In collaboration with Skane Food Innovation Network's communication manager they have designed a web-based "career" site. The advisory board's activity led to the establishment of a joint trainee program, where five companies assigned in total eight trainees for a period of sixteen months. The trainee program offered a mix of company-based work and joint activities. The establishment of a joint trainee program showed the high degree of credibility as a neutral arena as the SFIN had established from the companies perspective.

Finally, attempts were made to use diversity as a tool for adding new perspectives to the innovation climate. Traditionally, the food industry has been dominated by male, ethnic Swedes. This means that both women, and the fast growing immigrant population, have been under-represented in key positions in the food industry, even though there is a lot of competence and entrepreneurship in both groups. The research project Power over Food, that started in 2009 is addressing the question "How can better knowledge of gender and equal opportunity issues in organizations and companies create more and better food innovations?" There are suspicions that many capable female researchers and entrepreneurs terminate their careers in the food field. That implies a risk that many ideas for potential innovations come to nothing. The project wants to find out if this loss is related to attitudes about gender and equal opportunity, and then come up with possible changes that can encourage more people to remain in the business. The development of "Etnos", a project devoted to producers of ethnic foods in Malmö funded by the Skane Food Innovation Network and the Skane County Administrative Board considerably increased the diversity in the program. This ethnic network consists of food entrepreneurs with background in other countries than Sweden.

Phase 4 – Expanding the innovation community

After reassuring commitment from the main initiators of the network, SFIN in 2009 took several steps in order to strengthen the governing capacity and expanding the innovation community. The management group was expanded and re-organized into strategic areas. These were "Strategy & Cooperation", "Jobs & Careers", "Tomorrow's Meal Services", "Innovation & Entrepreneurship", and "Taste of Skane" (formed in 2010).

The distinct areas of responsibility demand a deliberate cross-fertilization and commitment from the different areas. The heterogeneous food innovation system requires "multilingual" skills. This was explicitly searched for by the CEO when recruiting the members of a new, larger management group. Apart from having knowledge and authority in the various areas, the members were constantly asked to make sure that the integration and cross-fertilization between the different areas was functioning.

Strategy & Cooperation was dealing with the overarching questions for the whole SFIN network. *Jobs & Careers* were targeted at the attractiveness of the food business, with the advisory board, trainee program and establishing connections between gifted students at universities and university colleges and the food industry. *Tomorrow's Meal Services* is focusing on meals in the public sector. Special emphasis is given to education and innovative purchasing procedures. The existing procedures don't promote new meal solutions or food products and the average education level in this part of the industry doesn't suffice in the contacts with private suppliers and competitors. In this platform the collaboration between all parts of the triple helix is most prominent, based on the idea of getting better meal solutions. Food for elderly people and hospital meals are top priorities within the area. *Innovation & Entrepreneurship* is the area in which the development and marketing of new products and processes are supported, either by dedicated small funding or by competent coaching from experienced business people connected to SFIN.

The initiative for SFIN had come from the traditional large-scale food industry, with the regional authorities and academia as main supporters. However, the changes in the foodscape, especially at the demand side of the food system, had created new opportunities and challenges for the food sector. While the food industry faced decreasing margins and fierce competition, the gastronomic side of the food business started to boom during the 1990s. From being a nation in the culinary outskirt, Sweden and its neighbouring Scandinavian nations had a rapid development in the fine dining sector. Swedish chefs started to win prestigious culinary awards and the restaurants with Michelin stars had an impressive development. Sourdough bakeries made people lining up for buying crispy bread at prices that must have seemed like a wet dream for the large bakeries delivering bread to discount stores and super markets. The media's infotainment programs made new generations discover the pleasures of cooking, starting to demand high-class products, new vegetables and spices, and rare cuts of meat from the local supermarket. And the on-going debate on food safety and food quality led to a growing appreciation for food experiences based on authentic, traditional and local values. Small-scale food manufacturers got a revival, often integrating their production with culinary tourism.

Restaurants and small-scale food manufacturers were already connected in networks, based on mutual commercial interests. But the connections with the large-scale food and retail systems were poor. The region hosted a separate development organisation for culinary tourism, restaurants and food manufacturers. However, a mutual interest had been starting to grow between the separated food domains. The food industry saw that large numbers of consumers were willing to pay considerably higher prices for authentic high-class taste experiences, an added value that not even the hi-tech products of convenience and functional foods had been able to accomplish. The small food manufacturers had limited access to necessary retail and distribution channels in order to expand and/or make their sales profitable. Here they saw a potential in collaborating with the larger players.

In 2010, the regional authorities decided to let their separate development organisation for culinary tourism and small-scale food manufacturers be handed over to SFIN. The former platform Innovative Market Places was integrated as a part of the platform *Taste of Skane*, directed at the concrete local development of on one side the small-scale food production and distribution, and on the other side the local culinary tourism. From now on, the whole food business, from fine dining to bulk production, was hosted by one single organisation in order to facilitate cross-fertilization between small-scale and large-scale food activities.

If the different regimes of the food industry, the academia and the regional authorities had been somewhat difficult to bridge, this was nothing compared to the separate traditions of food service and restaurants, and small-scale and large-scale food producers. Some of the differences were hidden, but some were clearly outspoken. Many of the small food manufacturers had started their business in direct opposition to the food industry's way of growing and processing food. Although in essence being committed to collaboration, the mistrust is not easy to overcome. But some development projects have been mutually attractive. A successful pilot project has concerned the development of a joint retail brand for local food. The retailers' network had already been discussing the need for an easy way for the consumer to buy local premium food in ordinary supermarkets. Despite consumers' positive attitudes, local food (i.e. local provenance) had not been commonly promoted in the Swedish supermarkets. Local food with a strong local profile had mainly reached the consumers through alternative channels, such as the Farmers' Market (Ekelund & Fernqvist 2007). Since Swedish food retailing is highly concentrated to four actors, accounting for 76 per cent of all food retail sales (Market Magasin, 2008), it seemed necessary for food products promoted as local, to be sold through the supermarkets in order to grow beyond being pure niche products. By relocating the food products in the store, and developing information material and a joint brand, the small-scale food products are now easy to locate within the stores of the participating retailers. The brand "Taste of Skane – carefully selected local food" remedies the producers' concern that their products were insufficiently exposed in ordinary supermarkets. So was their concern about low profitability, since the retailers agreed upon having somewhat lower margins on the selected products. It shows the potential of collaboration if there is both a short-term gain for the users, and a larger good (regional food) that could engage the actors apart from strict business considerations.

A parallel development is to be found in the SFIN "Chefs' network". It turned out that some of the more prominent chefs in the region were interested to contribute with the experience and knowledge to the public sector. Since this particular part of the food industry, municipal and county foodservices, is a low-status, low-wage part of an already meagre sector, the initiative was welcomed. One of the most renowned chefs in the region started up a small chefs' network with only two chefs from the public sector and two from the private sector. These four people had the assignment of identifying urgent development areas where the cross-fertilization between sectors could be meaningful. A new dimension was introduced to the SFIN agenda, industry development hand in hand with corporate social responsibility. Gradually, the chefs' network has turned into a "meal network", where a much wider array of competencies and backgrounds now join forces in order to revitalise the entire meal situations in various areas within the public sector. It is behavioural consultants, architects, chefs, nutritionists, etc. Different projects have up until today covered elderly care, education in food and meal knowledge, health care and disabled people. They all share the characteristics of not being commercially "hot" and the innovative solutions are all needed in the entire sector. The pilot projects are high-risk and designed with replication criteria and business potential, in case the projects turn out well. The common denominator for the meal network is to redefine the meal situations for those who do not choose for themselves what and when to eat. Although the agenda is clear, the bridging of different areas is not easy. Being public operations, these are all relying on political decisions. The public foodservice has been neglected for a long time and is devoid of any national, regional and even local co-ordination and suffers from poor funding and low status. However, the regime of public foodservice is increasingly scrutinized, media interest grows and points out the social as well as nutritional importance of public meals. The sector has a potential to raise the bar for product and process innovation within the entire Swedish food industry. Thus, the natural extension of the meal network within the area of Tomorrow's Meal Services is the establishment of a politicians' network, which is currently underway. The public authorities may turn out to be the most vital part of the triple helix structure underpinning the SFIN operations – not only as a primary funding source, but also as an increasingly demanding end-user in the Swedish food industry.

Both the retailers' network and the meal network initiate and run risky pilot projects with a business potential for involved entrepreneurs and other incumbents. Pilot projects are carefully selected and consciously and strategically granted. In this way, relatively small initial pilot project fundings may reproduce themselves in the industry, on sound commercial conditions.

4. Summary on findings

The story of Skane Food Innovation Network and its efforts in stimulating innovation in the regional food industry boils down to four main topics: the organization has been developed, managerial crossroads have been designed, stakeholder interests have been aligned, and SFIN has found a viable modus operandi in its risky pilot projects with reproduction and diffusion potential.

4.1 Organizational development
The SFIN organization has evolved, in seven years, from strict and arms-length R&D funding into a multilevel and multidisciplinary innovation community.
An important insight was the need for dedicated sub-communities, i.e. "focal networks", with hands-on activities, in turn co-ordinated in the overarching food innovation network. This combination creates opportunities for direct innovation in the sub-communities as well as creating a breeding ground for cross-fertilization between the sub-communities.

4.2 Managerial crossroads
The SFIN story reveals how the Triple Helix approach has been gradually operationalized through board representation and a "multilingual" management group. The role of the management team has been sharpened and requires both depth and breadth from the area managers. The managers need to be skilled intermediaries with an ability to identify and translate differences across differing regimes. Cultural differences between large and small, public and private, primary production and retailing, etc, influence community building activities. At the same time, cultural impediments may well hold the potential innovation (Jönsson, 2008).

4.3 Alignment of stakeholder interests
SFIN has gone through a drastic change from being preoccupied with the funding of R&D projects without any joint strategy, into engaging in cross-sectorial collaborations. A recent study of the Skane food industry concluded that SFIN have had a significant impact on reducing the fragmentation in the industry (Henning et al, 2010). Gradually, joint arenas become legitimate and the strategic and long-term nature of the SFIN operations spread among different actors. This makes it less threatening to engage in open innovation-kind of collaboration, as long as it resides under the "neutral" SFIN label.

4.4 Emergence of a modus operandi
In seven years, SFIN has turned into a network of cross-functional networks defining and funding risky pilot projects, commercially viable and with diffusion potential. The diffusion process is backed up by support activities such as marketing and communication, but the business potential is for the entrepreneurs to realize. In this way, funding exploits the

incentives of entrepreneurs, at the same time receiving a "reality-check" from the level of interest expressed by entrepreneurs in the first place. Furthermore, small funding may be leveraged into, at best, increased economic activity and societal improvement.

5. Discussion and outlook

The Skane Food Innovation Network operates on many fronts to boost the impact of innovation in the food industry. The Network also tries to create the conditions for innovation by enhancing the attractiveness of the food business. It has been a tricky road, with successful paths as well as some dead ends. Although promising, the work of SFIN is only at the beginning of the transition processes in the Skane food sector. The changing foodscape, with completely new situations for both farmers, industries and retailers, calls upon new and innovative solutions. Most of the mature industries tend to have the same problems, such as well-established regimes, created in very different markets. These regimes are de facto impediments to innovation. In this section, we will discuss the learnings of the work of SFIN with facilitating innovations in a mature industry, focusing on the use of a multilevel triple helix approach, the importance of analyzing and bridging regimes and the importance of an end user perspective as a guiding principle.

5.1 The multilevel triple helix network approach

The food innovation system is heterogeneous, with a mix of large national and inter-nordic producers and retailers and many small local businesses all through the value chains. It spans from farmers over food producers to grocery stores, restaurants and public eating-places. This characteristic was initially a major problem when trying to build an overarching innovation community and common ground for concerted efforts. Some actors did not feel connected to other levels of a presumptive cluster in the food chain, while others felt a need to establish collaborations but felt insecure about the process. This mirrored the fact that the Swedish food system, like many other mature industrial sectors, had established strong regimes on each level of the food producer-consumer chains, with little or no understanding of the actors on other levels. All actors were imprinted with values emergent from a time when the Swedish food system was based on national self-subsistence and strong public regulations, resulting in limited market competition (Beckemann 2011). This was a major impediment for cross-sectoral innovation, since the new situation demanded new ways of collaboration.

The triple helix approach is concerned with joint efforts and coordination, in our case between universities, business firms, local production units of large multinational companies and regional authorities. Different stakeholders have their individual rationales and logic. In order to be successful, the triple helix approach should rest on mutual trust and goal congruence between incumbents. Bridging regimes is like an evolution, an emerging understanding of the common good of a change of strategy and behavior.

5.2 The importance of analysing and bridging regimes

The fact that processes underlying innovation and industrial and economic transformation are governed by social and technological regimes have been acknowledged by, among others, Cooke, (2005), Geels, (2008), Bergek et al, (2007), Klein Wollthuis et al, (2005) and Malerba and Orsenigo, (1997). Winter (1983) defines regimes in a sector as a specific set of not only regulative institutions and norms but regimes also regulate codified formal as well as tacit informal habits and routines related to common collective and individual practices and beliefs. These practices and beliefs shape and coordinate actions between various groups, individuals, and organizations in the sector. The notion that technological regimes and their production of knowledge are shaped by historical and cultural factors have, from a different starting point, been repeatedly argued in the tradition of cultural analysis, which have become used more often in both product development and marketing in the latest decades (Pink, 2005. Sunderland & Denny, 2005. Kedia & Van Willigen, 2005). The tradition of cultural analysis further stresses that technological systems do not function independently from the human actors within the system. The actors are seen as embedded in social groups with cultural requisites, such as traditions, norms and beliefs (Law, 1999).

Breschi and Malerba (1997) concluded in their studies of sector characteristics of national innovation systems that technological regimes are defined by the level and type of opportunity and appropriate condition to innovate. This is bonded to the history, nature and the cumulativeness of knowledge as well as to the means of communication and transmission of knowledge within the sectoral systems of production. Following Levinthal (1991) and Scott (1995) regimes have three dimensions: i) cognitive rules, related to belief systems, ii) normative rules expressed in missions, goals, and identity, and iii) strategies and strategic orientations towards the surrounding external socio- technical and politico–economic environment. Regimes are closely related to the concept of institution. An institution could be defined as "patterns of routinized behaviour" (Hodgson, 1988) and may be analysed on a number of different levels.

In the networks of SFIN, it is a number of sub-systems who engage and challenge current regimes. The result is ideas, tested in "risky pilot projects" with diffusion and profitability potential. The networks may be understood as "liquid environments", where different knowledge, experience and values meet (Johnson, 2010). Such liquid environments define the so-called "adjacent possible" (ibid.). This means that the configuration of single networks and the links between the individual networks constitute the limits to what the network may produce. i.e., configurations are imperative. The regimes are influenced by the experience, thoughts and ideas, values and objectives of each and every individual within the different networks. SFIN organizes the network of networks and initiate pilot projects. On this level, cognitive rules are tested and different belief-systems are bridged in the work of individuals.

Leaving the individuals' level, the next level may be approached from a business angle. Business firms are normally run with a profit incentive. Innovations aim at creating new or better value to customers, leading to sales and profits. Following Christensen (1997), the average company inherently faces difficulties innovating a thriving business. Organizational routines and activities are shaped for efficient use of resources. Business innovation implies the change of product offerings, markets or resource use and the re-shaping of the "theory of the business" (Drucker, 1994). In terms of regimes, innovation by definition alters the business regime in one way or another, disruptively or incrementally (Christensen, ibid.). Govindarajan and Trimble (2010) are pre-occupied with "solving the execution challenge", focusing the way that an on-going business may handle challenging ideas and taking them to market. A new idea could form a spin-off initiative and be the start of an entirely new company. However, firms also need to innovate their current businesses, why it is necessary to establish a formalised co-operation between the existing business and the innovation initiative. What Govindarajan and Trimble (ibid.) suggest is a gradual and well-managed integration of old and new regimes in terms of both social and technological challenges. SFIN relieves the established firms from the direct disturbances of challenging ideas, working as an outside test-lab without worrying the on-going business. Still, the CEO network and the Entrepreneur Council bridge the gap between established practices and innovations developed in the pilot projects. Both social and technological regimes are bridged by way of the SFIN networks.

Different regimes that have their own specific cultures have been developed during the long history of the national food system in Sweden. By opening up arenas for individuals to meet, to identify and test new ideas, SFIN helps established firms to engage in innovation without compromising their running business. At the same time, entrepreneurs are invited to a vibrant group of people all joined in the common interest of developing the food industry by "open innovation".

5.3 The importance of an end user perspective

The end-user perspective is notable in the literature on open innovation (Chesbrough, 2006. von Hippel, 2005. Wallin, 2006), which stresses the importance of integrating the demand side in the study with the development of innovation systems. The action taken in the extended network all followed one of the major recommendation from the foresight process: take the end user perspective as a starting point. The user side was also used as a start for community building activities, whether it was the student recruitment program, the local food or the food for elderly activities.

Bringing the end-users helped to synchronize the agendas of the different actors in a multilevel food triple helix space and a multilevel foodscape. The end user perspective made the participating actors really feel that although they couldn't solve the problem by themselves, they all had important contributions to make. We conclude that the user side cannot be reduced to the result of the innovation system or the triple helix actors' achievements in a conventional way, since the user side interacts with every level and affects the outcomes from an early stage. The end-users have of course always been the important landing point for innovation work, the place where the success of the attempts to innovate is determined. But our point is that they may also be the best starting point, since it is the only level to which all actors of the triple helix can relate.

5.4 A new innovation systems model

The experience this far shows that SFIN engage a wide range of stakeholders in its different networks. Small-scale food producers, public servants, small service businesses, large retailers, politicians, entrepreneurs, large international food-related companies, researchers etc join the different networks of SFIN. Individuals meet in focal networks, form pilot projects which drive economic development in the industry from within. The former Swedish innovation system using "development pairs" in order to direct – "top-down" – the formation of an entire industry through a single company is gradually supplanted by the bottom-up network model strengthening the inherent innovative capabilities of a wide range of small and medium-sized firms, as well as larger corporations and public organizations. We suggest that the SFIN triple helix-based network form of organization holds several strengths. It is dynamic in its formation, it is resilient to temporary failure and it is cost-efficient in its selection and execution phases – it uses entrepreneurial incentives and helps isolating innovation initiatives from on-going business in established firms. Although it remains to be tested, this could be considered an efficient way of bridging strong regimes of a stifled and mature industry. It could be the Swedish food industry, but the mechanisms controlling the network of networks may well be transferred to other industries sharing these characteristics.

6. Concluding remarks

We would argue that the related work methodology of SFIN may be part of a transition from the prevalent Swedish innovation and development mode and work as a model for facilitating innovations in mature industries. The combination of an overarching innovation network responsible for issues of governance, in combination with dedicated sub-communities implementing hands-on activities and projects was a major step forward from the original SFIN organization, which was based on a traditional way or organizing innovation facilitators in Sweden. We would like to call the refined methodology a *Régime-bridging strategy*, with a *multilevel triple helix approach* and an *end user perspective* as fundamental cornerstones.

7. Notes on contributors

Håkan Jönsson, associate professor in European Ethnology is researcher and lecturer at the Department of Arts And Cultural Sciences at Lund University, where he teaches at the Master of Applied Cultural Analysis program (www.maca.ac). He is also head of operations in the Skane Food Innovation Network, responsible for the area of small scale food manufacturers and culinary tourism.

Hans Knutsson is assistant professor at the School of Economics and Management, Lund University. He teaches accounting, management control, and strategy and focuses his research on public management and cluster development. He is head of operations in the Skane Food Innovation Network, running the area Foodservice of Tomorrow.

Carl-Otto Frykfors is affiliated to the Department of Management and Engineering at Linköping University and prior senior program manager at VINNOVA, The principal governmental agency for knowledge driven industrial renewal and innovation. He was further director of The Dahmén Institute in charge for evaluation of Foresight activities related to an European Foresight project between regions in Sweden, Italy, Germany and Poland.

8. References

Adema, P. 2009. Garlic Capital of the World: Gilroy, Garlic, and the Making of a Festive Foodscape. Jackson: University Press of Mississippi.

Appadurai, A. 1990. 'Disjuncture and Difference in the Global Cultural Economy' Public Culture 2: 1990, pp 1-24.

Asheim, B. T. & Coenen, L 2005 Knowledge bases and regional innovation systems: Comparing Nordic Clusters in Research Policy 34 pp 1173-1190

Baldwin,R., 2006. Globalisation: the great unbundling(s). Report Prime Minster's Office, Economic Council of Finland.

Beckemann, M. 2011. The Potential for Innovation in the Swedish Food Sector. Lund: Lund University Press (diss.)

Bergek, A., S. Jacobsson, M. Hekkert, and K. Smith. 2007. Functionality of innovation systems as a rationale for, and guide to innovation policy. In: Innovation policy, theory and practice: An International handbook, ed. R. Smits, S. Kuhlmann and S. Shapira. Cheltenham, UK: Edward Elgar.

Breschi, F., and F. Malerba. 1997. Sectoral innovation system: Technological regimes, Schumpeterian dynamics and spatial boundary. In System of innovation, technologies, institutions and organizations, ed. C. Edqvist. London: Pinter.

Carlsson, B. and Stankiewicz, R. (1995): 'On the nature, function and composition of technological systems'. In: Carlsson, B. (ed.): *Technological Systems and Economic Performance: The Case of Factory Automation*. Dordrecht: Kluwer Academic Publishers. pp. 21-56.

Chesbrough, H.W, 2006. in Open Innovation: Researching a new Paradigm, eds Chesbrough; Vanhwerbeke; West, Oxford University Press.

Christensen, C.M. 1997 The innovator's dilemma. When new technologies cause great firms to fail. Harvard Business School Press, Boston.

Coenen, L 2006 Faraway, so close! The changing geographies of regional innovation. CIRCLE. Lund University, Lund.

Cooke, P. 2005 Regionally asymmetric knowledge capabilities and open innovation: Exploring Globalisation 2 -A new model of industry organization. Research policy 34(8):1128-49

Davies, C.A 2008: Reflexive Ethnography. New York: Routledge

Dahmén, E. 1987. 'Development Blocks' in Industrial Economics'. *Scandinavian Economic History Review*.

Dolphijn, R., 2005. Foodscapes. Towards a Deleuzian Ethics of Consumption. Delft. Eburon Publishers.

Drucker, P.F. 1994, The Theory of the Business, Harvard Business Review, September, pp. 95-105.

Ekelund, L., Fernqvist, F 2007 'Organic Apple Culture in Sweden'. The European Journal of Plant Science and Biotechnology

Freeman, C. 1987 Technology Policy and Economic Performance: lesson learned from Japan. London, Printer Publisher.

Fridlund, M, 1999. Switching Relations and Trajectories: The Development Procurement of the Swedish AXE Switching Technology, in: Charles Edquist, Leif Hommen & Lena Tsipouri, eds., Public Technology Procurement and Innovation, Economics of Science, Technology and Innovation v. 16. Norwell, Mass.: Kluwer Academic Publishers, 143–165.

Frykfors, C-O., Klofsten, M. 2011. Emergence of the Swedish Innovation System and the Support for Regional Entrepreneurship: A Socio-Economic Perspective, in Science and Technology Based Regional Entrepreneurship: Global Experience in Policy and Program Development, ed. Mian, S.A. Edward Elgar.

Frykfors, C-O, Jönsson, H; 2010 'Reframing the multilevel triple helix in a regional innovation system: a case of systemic foresight and regimes in renewal of Skåne's food industry' Technology Analysis & Strategic Management 2010, 22, 819-829

Geels, F. W. 2004. From sectoral systems of innovation to socio-technical system. Insights about dynamics and change from sociology and institutional theory. Research Policy 33 (6-7) pp 897-92.

Govindarajan, V., & Trimble, C. 2010 The other side of innovation. Solving the execution challenge. Harvard Business Press, Boston.

Grant, R. M 1996, 'Toward a knowledge based theory of the firm', Strategic Management Journal, 17, Special Issue Winter, 93–109.

Henning, M, Moodysson, J, Nilsson, M 2010. Innovation and regional transformation From clusters to new combinations. Malmö: Region Skåne

Von Hippel, E 2005 Democratizing Innovation. MIT Press Chambridge

Hodgson, G. 1988 Economics and Institutions, Polity Press.

Johnson, S. 2010 Where good ideas come from. The natural history of innovation, Penguin Group, New York.

Jönsson, H 2008. The Cultural Analyst – an Innovative Intermediary?' // ETN 2/2008. Lund: Etnologiska Institutionen.

Jönsson, H & Sarv, H 2008. On the Art of Shaping Futures. Lund, Skånes Livsmedelsakademi.

Kedia, S.; Van Willigen, J. 2005. eds. Applied Anthropology. Domains of Application. Westport, CT: Greenwood.

Klein Woolthuis, R.; Lankhuizenb, M.; Gilsing, V, 2005. A system failure framework for innovation policy design in Technovation 25 (6), pp 609-19.

Laestadius, S., Berggren, C., 2000. The embeddedness of industrial clusters : the strength of the path in the Nordic telecom system. Stockholm: Royal Institute of Technology.

Lagnevik, M 2006. 'Food innovation at interfaces: experience from the Öresund region', in Hulsink, W.; Dons, H. (Eds.) Pathways to High-Tech Valleys and Research Triangles. Wageningen: Wageningen UR Frontis Series, Volume 24

Law, J. 1999, ed. Actor Network Theory and After. Oxford, Blackwell.

Levinthal, D.A. 1991. Organizational adaption and environmental selections-interrelated processes of change. Organization Science 2: 140–45.

Lundvall, B-Å. 1992, National Innovation System : Towards a Theory of Innovation and Interactive Learing, London: Printer Publisher.

Malerba,F.; Orsenigo, L. 1997 Technological Regimes and Sectoral Patterns of Innovative Activities. Industrial and Corporate Change 6: 83-119.

Market Magasin 2008

Nilsson, M. 2008. A tale of two clusters. Sharing resources to compete. Lund. Lund Business Press.

Oresund Food. 2011 Redefining the Food sector. Copenhagen: Oresund Food

Pierre, J.; Peters, B.G. 2005. Governing Complex Societies: Trajectories and Scenarios. Basingstoke, Palgrave.

Pink, Sarah, 2005. ed. Applications of Anthropology. Professional Anthropology in the Twenty-first Century. Oxford, Berghahn Books.

Scott, W.R. 1995 Institutions and Organizations, Sage, Sunderland, P. L.; Denny, R. M., 2007. Doing Anthropology in Consumer Research. Walnut Creek: Left Coast Press.

Teece, D.J. 2007 Explicating Dynamic Capabilities: The Nature and Microfoundations of (Sustainable) Enterprise Performance, Strategic Management Journal 28, 1319- 1350.

Wallin, J. 2006. Business Orchestration: Strategic leadership in the Era of Digital Convergence. John Wiley & Sons Ltd.

Wastenson, L., T. Germundsson, P. Schlyter, the Swedish Society for Anthropology and Geography, Statistics Sweden, Lund University. Department of Social and Economic Geography, National Atlas of Sweden and National Land Survey of Sweden 1999. Sveriges nationalatlas (National Atlas of Sweden). Vällingby, Gävle, National Atlas of Sweden (SNA); Publisher and distributor of maps.

12

Organic Food Preference: An Empirical Study on the Profile and Loyalty of Organic Food Customers

Pelin Özgen
Atılım University
Turkey

1. Introduction

Eating habits in the world have shown some phases over time. In the last century, the sole aim was to feed oneself, however, later in time, food industry has been affected from industrialization trend, and agriculture turned into a sector in which large-scale food products are produced and consumed primarily based on their being cost effective. In parallel with this trend, one of the major challenges in agriculture is to increase efficiency. As a result of this challenge, two types of food production have aroused- one being the *conventional method* and the other method utilizing *genetic engineering* which emerged in the last decade.

Conventional method is the oldest and the most widely used technique. According to Knorr and Watkins (1984) conventional agriculture is defined as "capital-intensive, large-scale, highly mechanized agriculture with monocultures of crops and extensive use of artificial fertilizers, herbicides and pesticides, with intensive animal husbandry". However, heavy reliance on synthetic chemical fertilizers and pesticides is said to have serious impacts on public health and the environment (Pimentel et al. 2005). Therefore, as people are becoming more environmental conscious, this method is being questioned and tried to be developed during the last decades.

The second technique, called as genetically modified foods (GM foods or GMO foods), were first put on the market in the early 1990s (wikipedia.org). These food products are derived from genetically modified organisms, (GMOs), which are obtained by using advanced techniques of genetic engineering. Currently, genetic modification is mostly applied to soybean, corn, canola, cotton seed and sugar beet and the application area is observed to expand everyday.

Naturally, genetically modified foods are not without advantages and disadvantages. The biggest advantage of using genetically modified foods is the ability to grow faster and bigger crops. In addition to that, weaknesses against certain types of disease and insects might be eliminated by genetic modification (hubpages.com(1)). Moreover, higher crop yields are thought to make food prices decrease and therefore lead to less starvation in the world. Besides these advantages, GDOs have also disadvantages, such as their tendency to make harm to other organisms (such as in the case of monarch butterflies which are poisoned by GMO corns), possible damages to environment in the long run, possible health problems in humans and unforeseen risks and dangers due to the complexity of nature (hubpages.com(2)).

Apart from the technological developments and the increased need for food, beginning with the 1970s, consciousness about health and environmental issues has aroused. This awakening led to many changes in both production and consumption patterns of food. In parallel with these, environmental friendly agriculture, which is also called as ecological or organic agriculture, started to be employed and supported by the governments.

Detailed information about organic foods, discussion on customer loyalty in food sector and empirical study are given next.

2. Organic foods

The first formal organization to promote and regulate organic agriculture is the International Federation of Organic Agriculture Movement (IFOAM), which was established in 1972. According to IFOAM, organic agriculture is "a production system that sustains the health of soil, ecosystem and people. It relies on ecological processes, biodiversity and cycles adapted to local conditions, rather than the use of inputs with adverse effects." In other words, organic foods are foods which are produced by using organic farming techniques, in which use of synthetic fertilizers, pesticides, fungicides, growth regulators and livestock feed additives and antibiotics as well as genetic modification are strictly prohibited (Lohr, 2001). Because of the naturalness of the production, organic foods are said be superior over food products that are produced with other techniques. Therefore, due to this perceived superiority, there is an increased attention towards organic practices.

According to a study conducted by Hau and Joaris (2000), certified organic products make up about 2% of the world food market. Despite its small market share, organic food products is the fastest growing segment of the food industry especially in the developed countries such as USA, Japan and EU countries (Raynolds, 2004). In the USA, the demand for organic food market was reported to increase at a rate of 18.5% (Klonsky, 2007), and in France the demand increase rate was about 10% annually (Monier, et.al, 2009), which is about ten times the rate of demand for total food products. This also stands as an evidence that the organic food market is growing very rapidly and special attention should be given for this segment in the food industry. Consequently, some studies (for example Vindigni et. al, 2002, Thompson, 1998, Makatouni, 2002, Lohr, 2002, Davies et.al, 1995) have been addressed to this issue in developed countries, yet there exist very few studies addressed in developing countries such as Turkey.

According to Turkish Statistical Institute (TUIK), about 43 million people live in Turkey in 20-60 age group, who decide personally on their food and may be considered as possible organic food consumers. This large number is appealing for marketers of several products, including organic food. However, previous research reveals that organic food production in Turkey was started to be employed not until 1985 (Karakoc and Baykam, 2009)- about 15 years later than it was started to be encouraged in the international markets with the establishment of International Federation of Organic Agriculture Movement (IFOAM) in 1972 (www.ifoam.org). Official encouragement of organic food production in Turkey was started only a decade ago, with the establishment of Association of Ecological Agriculture Organization (ETO) (Yanmaz, 2005). Currently, in Europe, about 6% of the agricultural fields are allocated to organic farming, whereas in Turkey only 1% of the area is reserved for the same purpose (Deniz, 2007) and organic food products are still considered to be new products for Turkish consumers. Despite the short history of organic food in Turkey, currently about 250 different organic products are produced and almost all certified products are exported to developed countries, which are European Union countries, USA and Japan in particular. Also, it is reported that Turkey holds the market leader position in dry and dried organic fruits (www.eto.org.tr). However, it should not be overlooked that, logistics and certification process causes international trade to hold more problems than marketing domestically, especially in food products where freshness, food standards and reliability are the major concerns. However, despite its difficulties, as stated above, almost all of the organic food is produced for international markets and it is clear that the demand for organic food in domestic market needs to be promoted more. Considering that Turkey has a large population, who are getting more conscious in organic food production and consumption, reaching domestic customers might be more rewarding for businesses.

3. Customer loyalty in the food sector

Customer loyalty is another issue that is examined in this research. According to Jacoby and Kryner (1974), customer loyalty is defined as customer's repeat purchase which is resulted from a series of psychological processes. However, it should be noted that, if people only focus on the repeated purchase, than it might be misleading, and this repeated buying should not be treated as customer loyalty. As Dick and Basu (1994) point out, even a relatively important repeat purchase may not reflect true loyalty, but may merely be the result of situational conditions. Therefore, many studies (for example, Jacoby & Chestnut, 1978; Kahn & Meyer, 1991; Dick & Basu, 1994) are available in the literature suggesting that loyalty should be divided into 2 types of loyalty: behavioral loyalty and attitudinal loyalty. In order to make a satisfying and comprehensive definition, both attitudinal and behavioral components should be present (Kim et.al, 1994). In parallel with this, Samuelson and Sandvick (1997) state that, behavioral approach to loyalty is still valid as a component of loyalty, however, attitudinal approaches to loyalty should also be present to supplement the behavioral approach.

Dick and Basu (1994) have developed a Loyalty Model, in which loyalty is shown to have different levels, affected from different attitudinal levels. As seen in the figure below (Figure 1), true loyalty can only exist when there is a highly positive relative attitude accompanied with a behavioral measure, which is called here as repeated patronage.

		Repeated Patronage	
		High	Low
Relative	High	True Loyalty	Latent Loyalty
Attitude	Low	Spurious Loyalty	No Loyalty

Fig. 1. Dick and Basu's Loyalty Model (Garland and Gendall, 2004)

Looking at the financial perspective of a firm, one of the most important roles of marketing is to increase the market for a product and to create continous cash flows for the company. According to many researchers, (such as Gupta and Zeithaml (2006), Rust et.al (2000) Srinivasan et.al, (2005), Baloglu (2002)) creating loyal customers to the firm is the first step and is very essential for increasing the market for a product. Keeping in mind that the market for organic food products needs to be enlarged, customer loyalty concept should also be examined in depth. Therefore, results and implications of the empirical study will be discussed in the following sections of this study.

4. Methodology

Survey method is used in order to gather data in this research. The questionnaire was formed based on previous research (such as Sarikaya, 2007; Monier, 2009; McIver, 2004, etc.) in addition to questions formed by the researcher, parallel with the aim of the study. Since the questionnaire is newly formed, it should be tested for internal consistency before employing it as the actual data gathering instrument (Tabacknick and Fidell, 2001). Therefore the questionnaire was applied to a test group of 37 in order to see the internal consistency coefficient, known as Cronbach's Alfa. After running the test, the Cronbach's Alfa was found to be .72. Since the acceptable level of Cronbach Alfa is .70 for internal consistency (Nunnally and Bernstein, 1974), the created questionnaire was found to be suitable to be used in the current research and it is applied for the actual study.

The questionnaire consists of five parts. In the first part, attitudes towards organic food products and buying patterns are investigated with 22 questions. Based on author's personal experience and observation, there is a debate on the fact that people's choice of organic food show difference according to for whom they are buying the food. In other words, when people are shopping for their children, it is seen that tendency towards buying organic alternative increases. Whether this observation is true for Turkish customers is also tried to be answered with the help of questions in the first part.

In the second part, the accessibility of organic food products and the place where respondents get their organic foods are asked. In the third part, respondents are asked to rank the reasons for choosing organic foods, where in the fourth part, they are kindly asked to reveal their opinions about what needs to be developed in the organic food sector.

The final part of the questionnaire is devoted to demographic questions, consisting of educational level, marital status, family size, gender, monthly income and age.

All of the questions in the questionnaire are formed as structured and pre-coded questions; therefore reluctance towards participating in the study is minimized due to minimized effort required from respondents.

The questionnaire was applied on June 2011 in Turkey's two largest cities- Ankara and Istanbul, due to convenience and purposive reasons. Respondents were chosen via mall-intercept method and before applying the questionnaire, a filtering question whether or not they purchase organic products was asked and it is made sure that the sample group consists of only organic customers.

A number of 138 questionnaires were filled, however, after preliminary screening, only 122 of them were found to be useful. The data gathered was analyzed with respect to descriptive and univariate analysis such as frequency tables, t-test and ANOVA by using SPSS 15.0 software package.

5. Empirical results / findings

The current study was conducted in Ankara and Istanbul in June 2011. Respondents were selected via mall-intercept method among organic food purchasers, with respect to availability. According to the results of the study, majority of the respondents are found out to be female, have university degree and are married with children. Considering that the respondents are chosen among organic food buyers, these demographic findings might be considered as the general profile of organic food buyers segment. The characteristics of the respondents are given in the tables below.

Gender	Frequency	%	Cum. %
Male	35	29	29
Female	87	71	100
Total	122	100	

Table 1. Gender distribution of respondents

As seen from the table above, 71% of the respondents are female. This result might imply that, female are the major customers of organic food products.

Age	Frequency	%	Cum. %
< 25	8	6.6	6.6
26-30	16	13.1	19.7
31-35	38	31.1	50.8
36-40	28	23.0	73.8
> 41	32	26.2	100
Total	122	100	

Table 2. Age distribution of respondents

When the age distribution is examined, it is seen that about 30% of the respondents are between 31 and 35, and the smallest group with respect to age is composed of people who are younger than 25. This result might be interpreted as younger people are less likely to buy organic food products. However, this interpretation should be approached with caution, because the sample is not taken via random sampling method and therefore is susceptible to sampling errors (Malhotra, 2011).

Marital Status	Frequency	%	Cum. %
Single	38	31.1	31.1
Married	84	68.9	100
Total	122	100	

Table 3. Marital Status of the respondents

According to the results, 84 of the respondents are married and remaining 38 of respondents are single. The married group consists of about 70% of the total respondents. In order to increase the expressiveness of this result, family size is also measured, as shown below.

Family Size	Frequency	%	Cum. %
Spouse and myself	20	16.4	15.6
Spouse, myself, and kid(s) younger than 3 years old	46	37.7	54.1
Spouse, myself, and kid(s) older than 3 years old	41	33.6	87.7
Myself and my parents	5	4.1	91.8
Only myself	10	8.2	100
Total	122	100	

Table 4. The family size of the respondents

As shown in the table above (Table 4), 71.3% of the respondents are living with spouse and kid(s). The smallest group with respect to family size is the group who are living with his/her parents, composing only 4% of the respondents. This result may lead to a stereotyping that, there is an increased tendency towards buying organic food products if a person has a child. Whether this stereotype is true or not, in other words, whether there is a significant difference between buying food for kids and buying for adults will be tested as an hypothesis in the following parts of this study.

Education Level	Frequency	%	Cum. %
< High school	4	3.3	3
High school	22	18	21.3
University	67	54.9	76.2
Master's degree	20	16.4	92.6
PhD/Doctorate	9	7.4	100
Total	122	100	

Table 5. Education level of respondents

According to the results, it is seen that 97% of the respondents have at least high school degree. In addition to this, it is seen that 67 of the people have a university diploma, which corresponds to 55% of the respondents. Moreover, 29 of the respondents have graduate degree, corresponding to 23%. By looking at this education data, it can be concluded that, organic foods are preferred by mostly educated people.

Monthly Income	Frequency	%	Cum. %
1001- 2000 TL	20	16.4	16.4
2001- 3000 TL	39	32.0	48.4
3001- 4000 TL	38	31.1	79.5
> 4000 TL	25	20.5	100
Total	122	100	

Table 6. Monthly Income of respondents

As seen in the table above (Table 6), 48% of the respondents earn less than 3000 TL a month and 52% have monthly income higher than 3000 TL. According to TUIK (2011), the average monthly income for household in Turkey is 1750 TL for the year 2010. This result might be interpreted as that, the shoppers for organic food products have a higher income level than Turkish households.

5.1 Attitudes towards organic foods and organic food preference

Respondent's attitudes towards organic food products are measured with Likert- scaled items in the questionnaire. The items are scaled from 1 to 5, with "1" being "strongly disagree" and "5" being strongly agree" to given statements.

By looking at the results, it is seen that, the average score for the statement "organic foods are more delicious than traditional food products" is 4.26, implying that respondents strongly agree that there is a difference in taste in favor of organically produced food products. Similarly, the statement of "organic foods are more nutritious" has received 3.8 points and the statement "organic food products are healthier than traditional food products" have received an average of 4.12. These results show that, there is a general strong belief towards organic food's being healthier.

When they are asked for their buying behavior, it is seen that respondents are not reluctant to pay a premium price in order to buy an organic food product. Currently, organic foods are about twice more expensive than regular food products in Turkish market. Even though this 100% premium price is paid, either willingly or unwillingly, 79 of the respondents (64.7%) believe that, in order to increase the demand for organic food products, the price of the organic food products should be decreased. In addition to price, respondents state that, the barriers to purchase organic foods include availability and organic food product range. These factors should also be developed, if organic food market is wanted to expand.

In an attempt to find out the reasons why people do not complain much about premium price they pay for organic foods, an open ended question is placed in the questionnaire. The most frequent answer is related with health concerns. 83 of the respondents (68%) believe that, organic foods are worth a price premium due to being healthier and have more nutritive value. Other reasons for household's willingness to pay more are listed as quality, certification and environmental concerns, where environmental concerns being seen as the least important reason. This finding contradicts with the research made by Bellows et.al (2008), in which environmental concerns are seen as more important factor for demand for organic products.

As stated above, customer loyalty is very essential in order to expand a market of a product. Since nothing can be improved unless measured, two questions were placed in the questionnaire in order to find out loyalty levels of the respondents towards organic food products. In the first question, which was asked in order to identify the behavioral component of customer loyalty, respondents are asked "if they prefer to buy the organic alternative if there existed one". This question has received an average score of 3.74, implying that there is a tendency to buy the organic alternative. The second question about loyalty is asked to test for attitudinal loyalty. In this question respondents are asked "if they recommend organic foods to other people". This question has received an average score of 4.2, which may be translated as that they make recommendations to other people which leads to a result that they have attitudinal loyalty. By comparing these two loyalty scores, it might be concluded that respondents are both behaviorally and attitudinally loyal to organic food products. This result may imply that organic food product market might continue to expand in the following years as well.

5.2 Changes in behavior according to whom the food is bought for

Based on author's personal experience and observation, it is seen that people become selective in shopping if they are buying goods for other people, especially if they are buying foods for children. Eventhough less attention is paid to health concerns while buying for himself/herself, the picture changes when it comes to shop for the children. Considering that there is a positive attitude towards organic food products, it is expected that people would buy organic food products for their children more often than they buy for themselves. In order to answer the question whether this observation is statistically provable for Turkish customers, questions in the first part of the questionnaire should be analyzed.

As shown above in Table 4, 70% of the respondents are living with spouse and kid(s), leading to a thought that there is an increased tendency towards buying organic food products if a person has a child. However, if this stereotyping is statistically significant or not should be tested.

A one sample t-test is conducted to see if there is significant change between the intention to buy organic foods for respondent himself or for his children. Firstly, the general tendency towards organic food buying is tested. The sample mean for "tendency to buy organic food products for children" is 4.19, which is found to be significantly different from the mean for "tendency to buy organic food products for himself", which has an average of 3.15 (t(121)=1.82, p=.04) at 95% confidence level.

After t-test is employed for testing the difference in general buying tendencies, individual tests, with respect to some product groups are conducted. These product groups include fresh vegetables, fresh fruits, milk and dairy products and dried fruits. However, no significant change is observed in the tendency to buy organic alternative in any of the individual product groups cited above. This result shows that despite the observed significant difference in tendency to buy organic food for children and for adults, there is no significant relation between tendency to buy organic foods for these two groups, with respect to special product groups.

6. Conclusion

Nowadays the society is mainly concerned with topics such as global warming, ecological impacts, health issues and better nutrition. In the area of food production, organic production techniques are seen as the best alternative for these issues. In parallel with the increased consciousness on health and environmental issues, the demand for organic food products is rapidly increasing worldwide. It is reported that annual growth rate in demand for organic foods is about 10 times the rate of demand for total food products (Monier et.al.2009). Despite the increasing domestic demand in Turkey, it is seen that majority of the produced organic food is exported to European countries, USA and to Japan. In spite of the high potential for organic food production in Turkey, it is seen that only about 1% of the total agricultural area is devoted to organic farming. Considering that demand is increasing everyday, Turkey should act intelligently to utilize its potential to become a major local and international organic food supplier. Yet, certification process, issues in labeling and logistics of the organic foods constitute important barriers to exporting. Therefore, in order to increase the production of organic products, it is firstly essential to expand the domestic market for organic products.

For the purpose of increasing the domestic demand, the general attitude of Turkish consumers towards organic foods, the profile of organic buyers and customer loyalty in organic food products market is tried to be investigated in this study. A survey is applied to 122 respondents, of which the majority of the respondents are highly educated females, who are married and have children. In addition to that, respondents have about 3000 TL monthly income, which is seen to be higher than Turkey average income. This profile is parallel with other studies (such as Sarıkaya, 2007, Monier et.al, 2009, Yanmaz, 2005) concerning the buyers for organic food products.

According to the results, there is a strong belief that organic foods are more delicious than other foods, and they are believed to have more nutritious value. In an overall assessment, organic products are preferred over conventionally produced or genetically modified food, especially if people are buying for their children. The fact that preference for organic foods differs according to whom the food is bought for is also tested and is proved to have a statistical significance. However, no significant change is observed in the tendency to buy organic alternative for specific product groups. The reason for not observing a significant difference might be due to small sample size and signals that tests should be repeated in further studies. In addition to this, in general, it can be said that organic food products are preferred mainly due to health concerns. Nevertheless, trust is an important issue in customer's minds and it is believed that strict controls and procedures in both production labeling should be implemented.

Another topic investigated in this research is about customer loyalty. According to the results, it is seen that here is a high loyalty among organic customers both in attitudinal and behavioral dimensions. This result is especially important for organic food producers, because high attitudinal loyalty is considered as a signal that consumers are willing to buy the organic alternative if there exists one and they recommend organic products to their families and friends. Moreover, it is seen that people are not satisfied with the currently available product range. Therefore, one can conclude that organic demand is congruent and the market for organic food products are expected to expand provided that the industry and the retailers ensure regular and easy supply with a high product variability.

To sum it up, according to the results of the empirical study, it is seen that the domestic market for organic food products is eager for new products and there is a strong loyalty among organic customers towards these products. In order to utilize this market potential, availability should be increased via utilizing supermarkets and alternative marketing channels more effectively. Currently, marketing and distribution for organic foods are relatively inefficient due to small volumes, which leads to high costs. Provided that marketing channels are better organized for organic foods, then the prices will eventually decrease, which will lead to an increase in demand. Moreover, there is still a lack of information about organic food products in the domestic market. Both the producer and customers should be better informed. Considering the demographic profile of the organic customers, marketing of the organic food products should be targeted mostly to educated women who have children. Finally, the variability in the organic foods should be increased since current customers for organic products are eager to buy organic alternative if possible. Provided that these actions are taken, then it should be no surprise to see Turkey as the leader in organic food production and consumption.

7. Limitations and suggestions for further study

As in every study, this study has also its limitations. One of major limitations of the study is due to application of survey method. There might be some errors due to factors such as social desirability or affect of the interviewer. It is believed that more reliable results could be obtained if survey method could have been backed up with other research methods such as observation or even experimentation. Especially, market basket analysis is expected to lead to interesting findings in consumer behavior in food sector.
Even though 122 is a satisfying sample size, the results of the findings could be more reliable and maybe the statistical associations which could not be observed could be observed if the sample size was larger.
Therefore, in further studies, it is recommended to study with a larger sample size and with other techniques in order to increase the generalizability and reliability of the results.

8. References

Association of Ecological Agriculture http://www.eto.org.tr
Baloglu, S. (2002) Dimensions of Customer Loyalty: Seperating Friends From Well Wishers. *Cornell Hotel and Restaurant Administration Quarterly*, 43, (1), 47-59
Bellows, Anne C.; Onyango, Benjamin; Diamond, Adam; and Hallman, William K. (2008) Understanding Consumer Interest in Organics: Production Values vs. Purchasing Behavior. *Journal of Agricultural & Food Industrial Organization*. 6, 1, pp.1-31
Davies, A., Titterington, A.J. & Cochrane, C. (1995). Who buys organic food?: A profile of the purchasers of organic food in Northern Ireland, *British Food Journal*, 97, 10, pp. 17 – 23
Deniz, N. (2007) Turkish Export potential. Paper presented in 1st Organic Agriculture Congress, Bahcesehir University, October 19-20 2007 Istanbul, Turkey
Gupta, S.& Zeithaml, V. (2006). Customer Metrics and Their Impact on Financial Performance. *Marketing Science*, Nov-Dec. 718-739.
Hau and Joaris, (2000) Organic farming, EU report, The European Commission in Vindigni, G.,Janssen, M.A., Jager, W. (2002). Organic Food Consumption. A multi-theoretical framework of consumer decision making. *British Food Journal*, 104, 8, pp. 624-642

Hubpages.com (1): http://benjimester.hubpages.com/hub/Genetically-Modified-foods-Pros-and-Cons

Hubpages.com (2): http://hubpages.com/hub/GMO-advantages-and-disadvantages

International Federation of Organic Agriculture Movement (IFOAM) www.ifoam.org

Jacoby, J. & Chestnut, R.W. (1978). *Brand Loyalty: Measurement and Management.* Wiley, ISBN:0471028452 New York

Jacoby, J.& Kryner, D.B. (1973). Brand Loyalty vs. Repeat Purchasing Behaviour. *Journal of Marketing Research,* Vol. 10, pp. 1-9

Karakoc, U. & Baykam, B.G. (2009). Türkiye'de Organik Tarım Gelisiyor. Betam, Arastirma Notu, 09-35. Retrieved on April 6, 2011 from www.betam.bahcesehir.edu.tr

Klonsky, K. (2007) Organic Agriculture and the US Farm Bill. University of Carolina, Agricultural Issues Center As cited in Monier, S., Hassan, D., Nichèle, V., Simioni, M. (2009). Organic Food Consumption Patterns, *Journal of Agricultural & Food Industrial Organization,* 7, 2. Article 12.

Knorr, D., Watkins, T.R (1984) *Alterations in Food Production. New York* As cited in Beus, C.E., Dunlap, R.E. (1990), Conventional versus Alternative Agriculture: The Paradigmatic Roots of the Debate, *Rural Sociology,* 55, 4, pp.590-616

Lohr, L. (2001) Factors Affecting International Demand and Trade in Organic Food Products. As cited in Regmi, A. (Editor) *Changing Structure of Global Food Consumption and Trade.* Market and Trade Economics Division, Economic Research Service / WRS-01-1

Makatouni, A. (2002) What Motivates Consumers to Buy Organic Food in the UK?: Results from a Qualitative Study", *British Food Journal,* 104, 3/4/5, pp. 345 – 352

Malhotra, N. (2011) *Marketing Research: An Applied Orientation,* Pearson International Press.

McIver, H. (2004). Organic hip: Popular Picks at Health Food Stores. *Better Nutrition,* 66, 2

Monier, S., Hassan, D., Nichèle, V. & Simioni, M. (2009) "Organic Food Consumption Patterns" *Journal of Agricultural & Food Industrial Organization,* 7, 2. Article 12.

Nunnally, J.C.& Berstein, I.H. (1994). *Psychometric Theory.* 3rd Ed., New York: McGraw-Hill.

Pimentel, D., Hepperly, P., Hanson, J., Douds, D. & Seidel, R. (2005). *"Environmental, Energetic, and Economic Comparisons of Organic and Conventional Farming Systems",* Bioscience, 55(July), 7, pp.573-582

Raynolds, L.T. (2004). The Globalization of Organic Agro-Food Networks. *World Development* 32, 5, pp.725-743

Rust, R.T., Zeithaml,V.A. & Lemon, K.N. (2000). *Driving Customer Equity.*The Free Press, ISBN: 0684864665 New York

Sarıkaya, N. (2007) Organik Ürun Tüketimini Etkileyen Faktörler ve Tutumlar Üzerine bir Saha Çalısması (A fieldwork on Factors Affecting the Consumption and Attitudes Towards Organic Foods). *Kocaeli Universitesi Sosyal Bilimler Enstitusu Dergisi,* 14,2, pp. 110- 125.

Srinivasan, V., Park, C.S. & Chang, D.R. (2005). An Approach to the Measurement, Analysis and Prediction of Brand Equity and Its Sources. *Management Science,* 51, 9,pp. 1433-1448

Thompson, G.D.(1998). Consumer Demand for Organic Foods: What We Know and What We Need to Know. *American Journal of Agricultural Economics,* 80, pp. 1113-1118

Turkish Statistical Institute (TUIK) www.tuik.gov.tr

Vindigni, G., Janssen, M.A., Jager, W. (2002). Organic Food Consumption. A multi-theoretical Framework of Consumer Decision Making. *British Food Journal*, 104, 8, pp. 624-642

Genetically Modified Food. Retrieved from
http://en.wikipedia.org/wiki/ Genetically_modified_food on 22.07.2011

Yanmaz, R. (2005) Organik Ürünlerin Pazarlanması ve Ticareti. (Marketing and Commercial Trade of Organic Products) *Symposium on Food Safety and Reliability*, Ankara-Turkey Conference Proceedings, Full Text, pp. 349-365

Part 3

Health Aspects

13

Yeast, the Man's Best Friend

Joana Tulha, Joana Carvalho, Rui Armada,
Fábio Faria-Oliveira, Cândida Lucas, Célia Pais,
Judite Almeida and Célia Ferreira
University of Minho/CBMA (Centre of Molecular and Environmental Biology)
Portugal

1. Introduction

In most cultures, bread making depends on a fermentation step. The flour leavening ability was, at first, most probably dependent on spontaneous fermentation. It became a controlled process by the maintenance of fresh innocula from one preparation to the next and this kind of environmental constraints eventually generated a particular type of yeast and bacteria biodiversity, adapted to ferment a certain type of flour mixture, conferring specific organoleptic characteristics to the dough. Nowadays, although some types of bread are still prepared using dough carried over from previous makings as a starter, the baking industry generally uses commercially available strains of *Saccharomyces cerevisiae* for bread making. Although the flour types, geographical origin and mixtures introduce organoleptic diversity in bread, the leavening is a crucial step in order to achieve the traditional specific flavours and textures of each population and region. The trend of globalization of baker's yeast market decreased worldwide bread diversity and the cultural values associated, simultaneously increasing the dependence of local producers on world-scale yeast producers. Sustainability demands assessing yeast biodiversity and devising simple and cheap methods for maintaining dough and multiply yeast. The following sub-chapters address these possibilities.

1.1 Biodiversity at the bakery

The better example of traditional practises still available is the use of sourdough, an extremely diverse fermented product widely used for the production of bread and sweet leavened baked goods. The production of sourdough bread can be traced back to ancient times (Rothe et al., 1973). The products are characterized by their unique flavour, enhanced shelf life and nutritional value, and favourable technological properties (Hämmes & Gänzle, 1998; Salovaara, 1998). Traditionally, sourdoughs have been used to produce many types of bread with rye, maize or wheat flours. This variety underlies the generalized utilization of the term sourdough as a synonym of leavening bread found in the literature. Although the primary purpose of the sourdough is leavening by the yeasts, a simultaneous souring action also takes place due to the activity of the lactic acid bacteria (LAB) present, resulting in bread with a good grain, an elastic crumb and, usually, the characteristic sensory quality of sourdough bread (Gobbetti et al., 1994a).

1.2 Sourdoughs: A yeast and bacteria synergistic ecosystem

Three types of sourdough have been defined based on common principles used in artisanal and industrial processes (Bocker et al., 1995). One type is produced with traditional techniques and is characterized by continuous propagation to keep the microorganisms in an active state, as indicated by high leavening ability. Examples of baked goods so obtained are San Francisco sourdough, French bread, the Italian panettone, and three-stage sourdough rye bread. The industrialization of the baking process for rye bread led to the development of another type of sourdoughs, which serve mainly as dough acidifiers. These sourdoughs are fermented for long periods (up to 5 days) at temperatures of 30° C, and high dough yields permit pumping of the dough. The microorganisms are commonly in the late stationary phase and therefore exhibit restricted metabolic activity only. Another type of sourdoughs is dried doughs, which are used as acidifier supplements and aroma carriers. These two last types of dough require the addition of baker's yeast for leavening (Meroth et al., 2003a, 2003b). Although the fermentation process runs under nonaseptic conditions, microbial associations present may last for years, as shown for certain industrial sourdough processes (de Vuyst & Vancanneyt, 2007).

Hundreds of different types of traditional sourdough breads exist in Europe, in particular in Italy. They differ in the type of flour, other ingredients, and the applied technology and fermentation process. Because of their artisan and region-dependent handling, sourdoughs are an important source of diverse LAB species and strains that are metabolically active or can be reactivated upon addition of flour and water. Some of these strains play a crucial role during the sourdough fermentation process and are or can be used as sourdough starters. Sourdough is a unique food ecosystem in that it (i) selects for LAB strains that are adapted to their environment, and (ii) harbours LAB species specific for sourdough (Dal Bello et al., 2005; de Vuyst & Neysens, 2005; Gobbetti et al., 2005).

Sourdoughs are stable ecosystems obtained using daily propagations of LAB and yeasts. Dough acidification is due mostly to homofermentative and heterofermentative LAB species, mainly belonging to the *Lactobacillus* genus (Hämmes & Vogel, 1995), while yeasts are primarily responsible for the leavening action through the production of carbon dioxide and consequent increase in dough volume.

The microorganisms present in sourdough usually originate from flour, dough ingredients or the environment. The variety and number of species in the dough are influenced by several endogenous and exogenous factors, such as type of flour, temperature and time of fermentation, redox potential, and length of time that the "starter dough" has been maintained (Hämmes & Gänzle, 1998). In fact, strong effects are exerted by process parameters such as dough yield, amount and composition of the starter, number of propagation steps, and fermentation time. The impact of these parameters during continuous propagation of sourdough causes the selection of a characteristic microflora consisting of LAB and usually yeasts (Gobbetti et al., 1994b).

Extensive research efforts have been directed towards the study of the species diversity and identification of lactic acid bacteria involved in sourdough fermentation processes. Lactobacilli, obligatory homofermentative and facultative or obligatory heterofermentative, are the typical sourdough LAB. These usually belong to the genus *Lactobacillus*, but occasionally *Leuconostoc* spp., *Weissella* spp., *Pediococcus* spp., and *Enterococcus* spp. have been found. In general, heterofermentative *Lactobacillus* species dominate the sourdough microbiota (de Vuyst & Neysens, 2005). *Lactobacillus sanfranciscensis* (Trûper & De' Clari, 1997), *Lb. plantarum* and *Lb. brevis* are the most frequently isolated lactobacilli.

Recent biodiversity studies of particular sourdough ecosystems throughout Europe resulted in the description of new LAB species. During the last years, several new LAB species have been isolated from traditional sourdoughs that were continuously propagated by back-slopping (repeated cyclic re-inoculation) at ambient temperature: *Lb. mindensis, Lb. spicheri, Lb. rossiae, Lb. zymae, Lb. acidifarinae, Lb. hammesii,* and *Lb. nantensis.* Some of these species have been described on one single isolate only. The distribution of the taxa of LAB is highly variable from one sourdough ecosystem to another. Therefore, it is difficult to define correlations between population composition and both the type of sourdough or the geographic location (Gobbetti, 1998).

Adaptations of certain LAB to a sourdough environment include (i) a unique central metabolism and/or transport of specific carbohydrates such as maltose and fructose, being maltose the most abundant fermentable carbohydrate and fructose being an important alternative electron acceptor; (ii) an activated proteolytic activity and/or arginine deiminase pathway; (iii) particular stress responses; and (iv) production of antimicrobial compounds. For instance, dough acidification is a prerequisite for rye baking to inhibit the flour α-amylase (de Vuyst & Vancanneyt, 2007).

Although LAB initially isolated from sourdough are not necessarily unique for sourdough ecosystems, some correlations can be seen between specific LAB species and the type of sourdough, and sometimes the origin of the sourdough. In practice, sourdoughs are either continuously propagated by using a piece of dough from the preceding fermentation process, or produced by using once a week a commercial starter followed by back-slopping for several days. Therefore, large differences can often be seen in species composition within and among sourdough types (de Vuyst &Vancanneyt, 2007).

This is well illustrated in a more recent study of spontaneously fermented wheat sourdoughs from two regions of Greece, where a total of 136 lactic acid bacteria strains were isolated. *Lactobacillus sanfranciscensis* were dominant in the sourdoughs from Thessaly and *Lb. plantarum* sub spp. *plantarum* in the sourdoughs from Peloponnesus. The latter was accompanied by *Pediococcus pentosaceus* as secondary microbiota. In this case, none of the lactic acid bacteria strains isolated produced antimicrobial compounds (Paramithiotis et al., 2010).

1.3 The particular roles of yeasts in sourdoughs

Several of the most important functions in bread making are fulfilled by yeast. They contribute to leavening and produce metabolites such as alcohols, esters, and carbonyl compounds, which are involved in the development of the characteristic bread flavour (Corsetti et al., 1998; Damiani et al., 1996; B. Hansen & Á. Hansen, 1994; Martinez-Anaya et al., 1990a, 1990b). Furthermore, the enzymatic activities of yeasts by enzymes such as proteases, lecithinases, lipases α-glucosidase, β-fructosidase, and invertase have an influence on the dough stickiness and rheology as well as on the flavour, crust colour, crumb texture, and firmness of the bread (Antuna & Martinez-Anaya, 1993; Collar et al., 1998; Meroth et al., 2002).

S. cerevisiae is the species most frequently found in sourdoughs but several other yeast species may be present in these ecosystems. In early studies the amount of *S. cerevisiae* may have been overestimated due to the lack of reliable systems for identifying and classifying yeasts from this habitat (Vogel, 1997). To study the sourdough yeast microbiota traditional cultivation methods in combination with phenotypic (physiological and biochemical) and/or genotypic (randomly amplified polymorphic DNA [RAPD]-PCR and restriction fragment length polymorphism [RFLP] analysis) identification methods have commonly been used (Corsetti et al., 2001; Galli et al., 1988; Mäntynen et al., 1999; Paramithiotis et al., 2000; Rocha & Malcata, 1999). These studies focused mainly on the characterization of ripe doughs and revealed the presence of 23 yeast species belonging especially to the genera *Saccharomyces* and *Candida* (Brandt, 2001; Ottogalli et al., 1996; Rossi, 1996). In particular *S. exiguus* (imperfect state *Torulopsisholmii* or *Candida holmii*, physiologically similar to *C. milleri*), and *C. krusei*, *Pichia norvegensis* and *P. anomala* are yeasts associated with LAB in sourdoughs. The LAB/yeast ratio in sourdoughs is generally 100:1 (Gobbetti et al., 1994a). Like other fermented foods produced by mixed microflora, the organoleptic, health and nutritional properties of baked sourdough goods depend on the cooperative activity of LAB and yeasts (Gobbetti, 1998). No data are available on the competitiveness of yeasts; thus, the effects of ecological factors and process conditions on the development of yeast biota during sourdough fermentation processes are virtually unknown (Meroth et al., 2003a). In a simulation of the complex natural sourdough ecosystem, the competition for substrates was studied in model co-cultures showing that yeasts only partially compete with the LAB for the nitrogen sources present and synthesize and excrete essential and stimulatory amino acids which enhance the cell yield of the LAB (Gobetti et al., 1994a). These findings contribute to the interpretation of some of the complex interactions which occur during sourdough leavening and which are difficult to understand because of the extensive proteolytic activity that occurs in sourdough (Spicher & Nierle, 1984).

Several recent studies have given emphasis to the yeast microbiota associated with spontaneous sourdough fermentations (Paramithiotis et al., 2010; Valmorri et al., 2010; Vrancken et al., 2010). A total of 167 yeast and 136 lactic acid bacteria strains were isolated from spontaneously fermented wheat sourdoughs from two regions of Greece, namely Thessaly and Peloponnesus. Identification of the isolates exhibited dominance of *Torulaspora delbrueckii* with sporadic presence of *S. cerevisiae* (Paramithiotis et al., 2010). In another study, conducted in 20 sourdoughs collected from central Italy, PCR-RFLP analysis identified 85% of the isolates as *S. cerevisiae*, with the other dominant species being *C. milleri* (11%), *C. krusei* (2.5%), and *T. delbrueckii* (1%). RAPD-PCR analysis performed with primers M13 and LA1, highlighted intraspecific polymorphism among the *S. cerevisiae* strains. The diversity of the sourdoughs from the Abruzzo region is reflected in the chemical composition, yeast species, and strain polymorphism (Valmorri et al., 2010). In contrast with the Greek study, the high presence of *S. cerevisiae* had already been reported in Italian sourdoughs by other authors (Corsetti et al., 2001; Gobbetti et al., 1994b; Iacumin et al., 2009; Pulvirenti et al., 2004; Succi et al., 2003). The current general opinion is that cross contamination from bakery equipment and working environment by baker's yeast is commonly associated with the presence of *S. cerevisiae* in sourdoughs. The other species detected (*Candida milleri*, *C. krusei* and *Torulaspora delbrueckii*) are typically associated with sourdoughs (Corsetti et al., 2001; Garofalo et al., 2008; Gobbetti et al., 1994b; Halm et al., 1993; Iacumin et al., 2009; Obiri-Danso, 1994; Ottogalli et al., 1996; Pulvirenti et al., 2004; Rossi, 1996; Succi et al., 2003; Sugihara et al., 1971; Vernocchi et al., 2004a, 2004b).

In rural areas in the north of Portugal a corn and rye bread is still prepared using a piece of dough usually kept in cool places, covered with a layer of salt. Prior to bread making this piece of dough is mixed with fresh flour and water and, when fully developed, serves as the inoculum for the bread dough. This starter dough is a natural biological system characterized by the presence of yeast and lactic acid bacteria living in complex associations in a system somewhat similar to that existing in sourdough. In a survey carried in 33 dough samples from farms mainly located in the north of Portugal, 73 yeast isolates were obtained belonging to eight different species. The predominant species was *S. cerevisiae* but other yeasts also occurred frequently, among which *Issachenkia orientalis*, *Pichia membranaefaciens* and *Torulaspora delbrueckii* were the most abundant, being present in about 40% of the doughs examined. Only six of the doughs contained a single yeast species. Associations of two species were found in 48% of the bread doughs, 30% presented three different species and the remainder consisted of a mixture of four yeast species. Associations of *S. cerevisiae* and *T. delbrueckii, I. orientalis* and/ or *P. membranaefaciens* were the most frequent. All mixed populations included at least one fermentative species with the exception of the association between *P. anomala, P. membranaefaciens*and *I. orientalis*, which was found in one of the doughs (Almeida & Pais, 1996a). Apparently this dough is somehow similar to the San Francisco sour dough in which maltose-negative *S. exiguus* is predominantly found and the fermentation may be carried out by lactic acid bacteria (Sugihara et al., 1971). In another Portuguese study in which, besides sourdough, maize and rye flour were examined the most frequently isolated yeasts were *S. cerevisiae* and *C. pelliculosa* (Rocha & Malcata, 1999).
In conclusion, yeasts and lactic acid bacteria (LAB) are often encountered together in the fermentation of wheat and rye sourdough breads. To optimize control of the fermentation, there has been an increased interest in understanding the interactions that occur between the LAB and yeasts in the complex biological ecosystem of sourdough.

2. Sustainability: The old made new

Sustainability aims are all about using simple ideas, mixing with old procedures and new materials, adding inventive solutions, generating innovation. In the baking market, in particular in the baking industry, there is considerable space for improvement. The present procedure of bread making in developed countries consists of using block or granular baker's yeast identically produced all around the world. Consequently, the above mentioned old procedure of using the previous leaven to the next leavening step was lost as baking became progressively an industrialized process. Producing and conserving large amounts of yeast requires energy wasting biotechnological plants and expensive technical support, favouring the standardization and centralization of production. Additionally, the conservation processes involving freezing temperatures compromise *S. cerevisiae* viability but also its desirable leavening ability and organoleptic properties.

2.1 Frozen yeast and frozen dough
2.1.1 Yeast response to cold
All living organisms, from prokaryotes to plants and higher eukaryotes are exposed to environmental changes. Cellular organisms require specific conditions for optimal growth and function. Growth is considered optimal when it allows fast multiplication of the cells, and the preservation of a favourable cell/organism internal composition, *i.e.,* homeostasis.

Therefore, any circumstance that provokes unbalance in a previous homeostatic condition may generally be considered stressful, as is the case of sudden changes in the external environment. These generally cause disturbances in the metabolism/regulation of the cells, tissues or organs, eventually disrupting their functions and preventing growth. Cellular organisms have to face this constant challenge and, therefore, rapidly adapt to the surroundings, adjusting their internal milieu to operate under the new situation. For this purpose, uncountable strategies have been developed to sustain the homeostasis. Whereas, multicellular organisms can make use of specialized organs and tissues to provide a relatively stable and homogenous internal environment, unicellular organisms have built up independent mechanisms in order to adjust to drastic environmental changes. Several approaches have been described for the most diverse microbes, from bacteria to fungi, involving responses at the level of gene expression as well as metabolism adaptation by faster processes like protein processing, targeting and inactivation, or iRNA interference (K.R. Hansen et al., 2005), just to mention the more general processes.

Yeasts, in particular, in their natural habitat can be found living in numerous, miscellaneous and changeable environments, since they can live as saprophytes on, either plants, or animals. As examples we can name fruits and flowers, humans, animals, etc. Likewise, in their substrates, yeasts are also exposed to highly variable milieus. On such diverse ambiences, it can be expected that yeasts regularly withstand fluctuations in the types and quantities of available nutrients, acidity and osmolarity, as well as temperature of their environment. In fact, the most limiting factors cells have to cope are the low water activity (a_w), *i.e.* availability of water, and temperature. Being yeast unicellular organisms, cell wall and plasma membrane are the first barriers to defeat environment and its alterations. Both changes on the water content and temperature lead to physical and functional modifications on plasma membrane, altering its permeability that are on the basis of cell lyses and ultimately cell death (D'Amico et al., 2006; Simonin et al., 2007).

Actually, the variations on the permeability of plasma membrane, attributed to transitions of the phospholipid phase in the membranes (Laroche & Gervais, 2003), are associated to loss of viability during dehydration/rehydration stress (Laroche & Gervais, 2003; Simonin et al., 2007). In a physical perspective, membrane phospholipid bilayers, under an optimum temperature level and favourable availability of water, are supposedly in a fluid lamellar liquid-crystalline phase. When temperature levels drop or under any other cause of dehydration, such organization suffers alterations as the hydrophilic polar head groups of phospholipids compulsorily gather. This phenomenon leads to the loss liquid-crystalline regular phase and conversion into a gel phase and consequent reduction of membrane fluidity (D'Amico et al., 2006; Simonin et al., 2007; Aguilera et al., 2007).

Still, a decline in temperature has other effects besides the reduction in membrane fluidity. Aside with the alterations on plasma membrane permeability (primarily but on the other physiological membranes as well) and hence changes on the transport of nutrients and waste products occurs the formation of intracellular ice crystals, which damage all cellular organelles and importantly reduces the a_w, under near-freeze temperatures. Furthermore, it has been evoked that temperature downshifts can cause profound alterations on protein biosynthesis, alterations in molecular topology or modifications in enzyme kinetics (Aguilera et al., 2007). Other crucial biological activities involving nucleic acids, such as DNA replication, transcription and translation can also suffer from exposure to low temperatures. This happens through the formation and stabilization of RNA and RNA secondary or super-coiled structures (D'Amico et al., 2006; Simonin et al., 2007; Aguilera et al., 2007). In turn, the stabilization of secondary structures of RNAs takes place, for instance, at the level of the inhibition of the expression of several genes that would be unfavourable for cell growth at low temperatures (Phadtare & Severinov, 2010). The latter occurs since the transcription of the mentioned genes is impaired, as well as the RNA degradation becomes ineffective (Phadtare & Severinov, 2010).

In yeast, as in most organisms, the adaptive response to temperature downshifts, commonly referred to as the cold-shock response comprises orchestrated adjustments on the lipid composition of membranes and on the transcriptional and translational machinery, including protein folding. These adjustments are mostly elicited by a drastic variation in the gene expression program (Aguilera et al., 2007; Simonin et al., 2007). Still, some authors name cold-shock response to temperature falls in the region of 18- 10°C and near-freezing response to downshifts below 10°C. In fact, yeast cells appear to initiate quite different responses to one or another situation, which can be rationalized since yeast can actively grow at 10–18 °C, but growth tends to stop at lower temperatures (Al-Fageeh & Smales, 2006). To withdraw misinterpretations we will focus mainly in low/near-freezing temperatures, which is also a cold response. Some works on genome-wide expression analysis have explored the genetic response of *S. cerevisiae* to temperature downshifts (L. Zhang et al., 2001; Rodrigues-Vargas et al., 2002; Sahara et al., 2002; Murata et al., 2006). In *S. cerevisiae* exposed to low temperature, 4°C, together with the enhanced expression of the general stress response genes, other groups of genes were induced as well. These include genes involved in trehalose and glycogen synthesis (*TPS1, GDB1, GAC1*, etc.), which may suggest that biosynthesis and accumulation of those reserve carbohydrates are necessary for cold tolerance and energy preservation. Genes implicated on phospholipids biosynthesis (*INO1, OPI3*, etc.), seripauperin proteins (*PAU1, PAU2, PAU4, PAU5, PAU6* and *PAU7*) and cold shock proteins (*TIP1, TIR1*, etc.) displayed as well increased expression, which is consistent with membrane maintenance and increased permeability of the cell wall. Conversely, the observed induction of Heat Shock genes (*HSP12, HSP104, SSA4*, etc.) can possibly be linked with the demand of enzyme activity revitalization, and the induction of glutathione related genes (*TTR1, GTT1, GPX1*, etc.) required for the detoxification of active oxygen species. On the other hand, it is also described the down-regulation of some genes, like the ones associated with protein synthesis (*RPL3, RPS3*, etc.), reflecting the reduction of cell growth, which in turn may be a sign of a preparation for the following adjustment to the novel conditions (Fig. 1). A rationalization of all the data from genome-wide expression analysis and also from the numerous works on yeast cold response developed on the last years, led to the idea that there are two separated responses to temperature downshifts (Aguilera et al., 2007; Al-Fageeh & Smales, 2006). One is a general response, which involves certain clusters of genes. These include members of the *DAN/TIR* family encoding putative cell-wall mannoproteins, temperature shock inducible genes (*TIR1/SRP1, TIR2* and *TIR4*) and seripauperins family, which have some phospholipids interacting activity. The other is a time dependent separated response, meaning that the transcriptional profile changes are divided in a time succession (Aguilera et al., 2007; Al-Fageeh & Smales, 2006). For instance, within the first two hours would be observed an over-expression of genes involved in phospholipid synthesis (like *INO1, OPI3*, etc.), in fatty-acid desaturation (*OLE1*), genes related to transcription, including RNA helicases, polymerase subunits and processing proteins, and also some ribosomal protein genes.

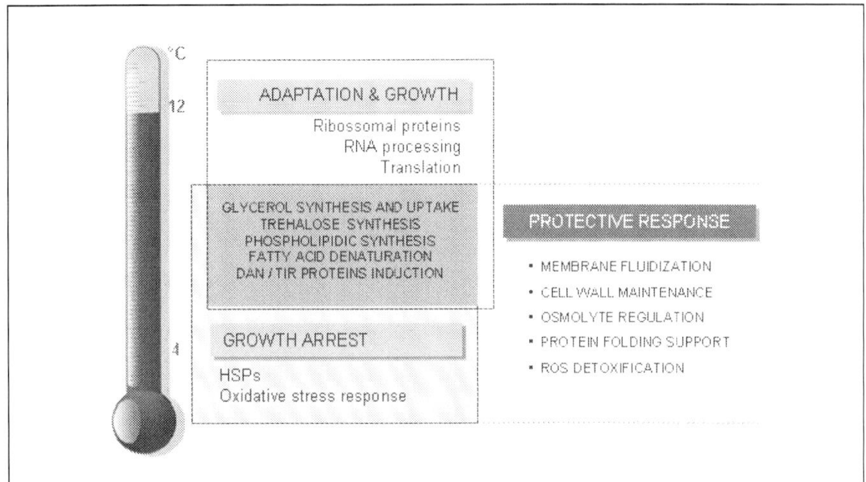

Fig. 1. *S. cerevisiae* major response to a temperature downshift (Adapted from Aguilera et al., 2007).

Whereas, in a second stage the latter genes (transcription related ones) are silenced and is promoted the induction of another set of genes, such as some of the heat shock protein (*HSP*) genes, also of genes associated with the accumulation of glycogen (*GLG1, GSY1, GLC3, GAC1, GPH1* and *GDB1*) and trehalose (*TPS1, TPS2* and *TSL1*), of genes in charge of the detoxification of reactive oxygen species (ROS) and defence against oxidative stress (including catalase, *CTT1*; glutaredoxin, *TTR1*; thioredoxin, *PRX1* and glutathione transferase, *GTT2*) (Aguilera et al., 2007; Al-Fageeh & Smales, 2006).

2.1.2 Improving baker's yeast frozen dough performance

Preservation by low temperatures is widely accepted as a suitable method for long-term storage of various types of cells. Specially, freezing has become an important mean of preservation and storage of strains used for many types of industrial and food processing, such as those used in the production of wine, cheese and bread. Bread, in particular, is a central dietary product in most countries of the world, and presently frozen dough technology is extensively used in the baking industry. Yet, the loss of leavening ability, and organoleptic properties, but mainly the loss of viability of the yeasts after thawing the frozen dough is a problem that persists nowadays.

In-depth knowledge concerning yeast genetics, physiology, and biochemistry as well as engineering and fermentation technologies has accumulated over the time, and naturally, there have been several attempts to improve freeze-thaw stress tolerance in *S. cerevisiae*. A recent work described that genes associated with the homeostasis of metal ions were upregulated after freezing/thawing process and that mutants in some of these genes, as *MAC1* and *CTR1* (involved in copper homeostasis), exhibited freeze-thaw sensitivity (Takahashi et al., 2009). Furthermore, the researchers showed that cell viability after freezing/thawing process was considerably improved by supplementing the broth with copper ions. Those results suggest that insufficiency of copper ion homeostasis may be one of the causes of freeze-thaw injury; yet, these ions toxicity does not allow their easy incorporation in food products. A very promising study reported an improved freeze-resistant industrial strain, in which the aquaporin was overexpressed (Tanghe et al., 2002). Nonetheless, this enhancement was not attained in larger dough preparations (under industrial conditions), wherein freezing rate is not that rapid (Tanghe et al., 2004). Another recent approach addressed the impact of unsaturated fatty acids on freeze-thaw tolerance by assaying the overexpression of two different desaturases (*FAD2-1* and *FAD2-3*) from sunflower in *S. cerevisiae*. This resulted into increased membrane fluidity and freezing tolerance (Rodriguez-Vargas et al., 2007). Also the heterologous expression of antifreeze proteins (antifreeze peptide GS-5 from the polar fish grubby sculpin (*Myxocephalusaenaeus*)) was tested in an industrial yeast strain, leading to both improved viability and enhanced gas production in the frozen dough (Panadero et al., 2005). A very current study confirmed the role of hydrophilins in yeast dehydration stress tolerance yeast cells, since overexpression of *YJL144W* and *YMR175W* (*SIP18*) become yeast more tolerant to desiccation and to freezing (Dang & Hincha, 2011). An alternative work, showed improved freezing resistance by expressing of *AZI1* (Azelaic acid induced 1) from *Arabidopsis thaliana* in *S. cerevisiae* (Xu et al., 2011). Other approaches devoid of genetic engineering were also taken. Cells were cultured in diverse conditions, including media with high concentration of trehalose or glycerol (Hirasawa et al., 2001; Izawa et al., 2004a); with poly-γ-glutamate (Yokoigawa et al., 2006), and with soy peptides (Izawa et al., 2007) acquiring improved tolerance to freeze–thaw stress and also retaining high leavening ability.

The benefits of cryoprotectants, substances that promote the excretion of water, thus decreasing the formation of ice crystals that happens during the freezing process, were also addressed. These include Me_2SO (Momose et al., 2010); proline (Terao et al., 2003; Kaino et al., 2008) and charged aminoacids as arginine and glutamate (Shima et al., 2003); trehalose (Kandror et al., 2004) as well as glycerol (Izawa et al., 2004a, 2004b; Tulha et al., 2010). A comparative analysis of yeast transcriptional responses to Me_2SO and trehalose revealed that exposure to cryoprotectants prior to freezing not only reduce the freeze-thaw damage, but also provide various process to the recovery from freeze-thaw injury (Momose et al., 2010). Yet, the use of Me_2SO in food preparation is not possible due to its toxicity. Intracellular proline accumulation was found to enhance freeze-thaw tolerance, thus several engineering strains emerged, overexpressing glutamyl metabolic related enzymes *PRO1* and *PRO2* or specific alleles (Terao et al., 2003), and self-cloned strains in which *PRO1* specific alleles combined with disruption of proline oxidase *PUT1* (Kaino et al., 2008). Moreover, it was shown that an arginase mutant (disrupted on *CAR1* gene) accumulates high levels of arginine and/or glutamate (depending on the cultivation conditions), with increased viability and leavening ability during the freeze-thaw process (Shima et al., 2003).

Trehalose and glycerol are not only cryoprotectants but also confer resistance to osmotic stress. A correlation between the intracellular trehalose content and freeze–thaw stress tolerance in *S cerevisiae* was described (Kandror et al., 2004). The same correlation has been made for glycerol (Izawa et al., 2004a, 2004b; Tulha et al., 2010). Furthermore, it has been reported that, beyond the cryoprotection, an increased level of intracellular glycerol has several benefits for the shelf life of wet yeast products and for the leavening activity (Myers et al., 1998; Hirasawa & Yokoigawa 2001; Izawa et al., 2004a) and no effect on final bread quality in terms of flavour, colour, and texture (Myers et al., 1998).

2.1.3 Role of glycerol for the baker's yeast frozen dough

S. cerevisiae accumulates intracellular glycerol as an osmolyte under osmotic stress but also under temperature (high and low) stress through the high osmolarity glycerol signaling pathway (HOG pathway) (Siderius et al., 2000; Hayashi & Maeda, 2006; Ferreira & Lucas, 2007; Tulha et al., 2010). Moreover, it was reported that a pre-treatment of yeast cells with osmotic stress was an effective way to acquire freeze tolerance, probably due to the intracellular glycerol accumulation attained. Some engineering approaches were performed in order to increase the intracellular glycerol accumulation in baker's yeast. For instance, Izawa and co-authors (Izawa et al., 2004a) showed that the quadruple mutant on the glycerol dehydrogenase genes ($ara1\Delta gcy1\Delta gre3\Delta ypr1\Delta$), responsible for the alternative pathway of glycerol dissimilation (Fig. 2) has an increased level of intracellular glycerol with concomitant freeze-thaw stress resistance. Similarly, the overexpression of the isogenes *GPD1* and *GPD2* that encode for glycerol-3-phosphate dehydrogenase (Fig. 2) (Ansell et al., 1997) also lead to an increase in intracellular glycerol levels (Michnick et al., 1997; Remize et al., 1999) and probably improved freeze-thaw tolerance. One of the most promising genetic modifications was the deletion of *FPS1* encoding the yeast glycerol channel. Fps1p channel opens/closes, regulating extrusion and retention of massive amounts of glycerol in response to osmotic hyper- or hipo-osmotic shock (Luyten et al., 1995; Tamás et al., 1999). The engineered cells deleted on *FPS1* showed an increased intracellular glycerol accumulation accompanied by higher survival after 7 days at -20° C (Izawa et al., 2004b). Yet, the dynamics of the channel under this type of stress remains unexplored. The mentioned study was considered quite innovative, it was even suggested the possibility that the *fps1Δ* mutant strain could be applicable to frozen dough technology. This because the *fps1Δ* mutant strain displayed the higher intracellular glycerol content attained so far, and (similarly to the previous engineered strains) avoided the exogenous supply of glycerol into the culture medium, which was at the time too expensive for using at an industrial scale. Our group has recently described a simple recipe with high biotechnological potential (Tulha et al., 2010), which also avoids the use of transgenic strains. We found that yeast cells grown on glycerol based medium and subjected to freeze-thaw stress displayed an extremely high expression of the glycerol/H$^+$ symporter, Stl1p (Ferreira et al., 2005), also visible at activity level. This permease plays an important role on the fast accumulation of glycerol; under those conditions, the strains accumulated more than 400 mM glycerol (whereas the mutant *stl1Δ* presented less than 1 mM) and survived 25-50% more. Therefore, any *S. cerevisiae* strain already in use can become more resistant to cold/freeze-thaw stress just by simply adding glycerol (presently a cheap substrate) to the broth. Moreover, as mentioned above glycerol also improves the leavening activity and has no effect on final bread quality (Izawa et al., 2004a; Myers et al., 1998).

3. Low-cost yeasts, a new possibility

The industrial production of baker's yeast is carried out in large fermentors with working volumes up to 200.000 l, using cane or sugar beet molasses as carbon source. These are rich but expensive substrates. Quite the opposite, glycerol, once a high value product, is fast becoming a waste product due to worldwide large surplus from biofuels industry, with disposal costs associated (Yazdani & Gonzales, 2007). Glycerol represents approximately 10% of the fatty acid/biodiesel conversion yield. Due to its chemical versatility, glycerol has countless applications, yet, new applications have to be found to cope with the amounts presently produced. This underlies the global interest for glycerol, which became an attractive cheap substrate for microbial fermentation processes (Chatzifragkou et al., 2011).

3.1 Metabolism of glycerol in yeasts

A significant number of bacteria are able to grow anaerobically on glycerol (da Silva et al., 2009). In the case of yeasts, most of the known species can grow on glycerol (Barnett et al., 2000) but this is achieved under aerobic conditions. *S. cerevisiae* is a poor glycerol consumer, presenting only residual growth on synthetic mineral medium with glycerol as sole carbon and energy source. In order to obtain significant growth on this medium a starter of 0.2% (w/v) glucose is needed (Sutherland et al., 1997). Yet, glycerol is a very important metabolite in yeasts, including *S. cerevisiae*. Importantly, its pathway is central for bulk cell redox balance, because it couples the cytosolic potential with mitochondria's. Furthermore, glycerol is the only osmolyte known to yeasts, in which accumulation, cells depend for survival under high sugar, high salt (Hohmann, 2009), high and low temperature (Siderius et al., 2000), anaerobiosis and oxidative stress (Påhlman et al., 2001).

Recently, it was suggested that *S. cerevisiae* glycerol poor consumption yields could be due to a limited availability of energy for gluconeogenesis, and biomass synthesis (X. Zhang et al., 2010). Nevertheless, the weak growth performances have long been attributed to a redox unbalance caused by the intersection of glycerol pathway with glycolysis at the level of glycerol-P shuttle (Fig. 2) (Larsson et al., 1998). Fermenting cultures of *S. cerevisiae* produce glycerol to reoxidize the excess NADH generated during biosynthesis of aminoacids and organic acids, since mitochondrial activity is limited by oxygen availability, and ethanol production is a redox neutral process (van Dijken & Scheffers, 1986) (Fig. 2). This is the reason why glycerol is a major by-product in ethanol and wine production processes. Consistently, the mutant defective in the above mentioned isogenes encoding the glycerol 3-P dehydrogenases (*Δgpd1Δgpd2*) is not able to grow anaerobically (Ansell et al., 1997; Påhlman et al., 2001). This ability was partially restored supplementing the medium with acetic acid as electron acceptor (Guadalupe Medina et al., 2010).

S. cerevisiae takes up glycerol through the two transport systems above mentioned, the Fps1 channel and the Stl1 glycerol/H+ symporter (Ferreira et al., 2005). Fps1p is expressed constitutively (Luyten et al., 1995; Tamás et al., 1999), while *STL1* is complexly regulated by a number of conditions (Ferreira et al., 2005; Rep et al., 2000). It is derepressed by starvation, and inducible by transition from fermentative to respiratory metabolism, as happens during diauxic shift at the end of exponential growth on rich carbon sources. Additionally, it is also the most expressed gene under hyper-osmotic stress (Rep et al., 2000), highly expressed at high temperature, overcoming glucose repression (Ferreira & Lucas, 2007) and under low-near-freeze temperatures (Tulha et al., 2010). In yeasts glycerol can be consumed through two alternative pathways (Fig. 2), the most important of which involving the glycerol 3-P shuttle above mentioned, directing glycerol to dihydroxyacetone-P through respiration and mitochondria. According to very disperse literature, other yeasts, better glycerol consumers than *S. cerevisiae*, appear to have equivalent pathways, though they should differ substantially in the underlying regulation to justify the better performance. At the level of transcription, significant ability to consume glycerol depends on the constitutive expression of active transport (Lages et al., 1999). Possibly, unlike in *S. cerevisiae* where *GUT1* is under glucose repression (Rønnow & Kielland-Brandt, 1993; Grauslund et al., 1999), glycerol consumption enzymes could be identically regulated. This should be in accordance with the yeasts respiratory/fermentative ability. Related or not, *S. cerevisiae* respiratory chain differs from a series of other yeasts classified as respiratory, which are resistant to cyanide (CRR - Cyanide resistant respiration) (Veiga et al., 2003). Cyanide acts at the level of Cytochrome Oxidase complexes. CRR owes its resistance to an alternative oxidase (AOX) that short-circuits the main respiratory chain, driving electrons directly from ubiquinone to oxygen, by-passing complex III and IV. Although exhaustive data are not available, CRR appears to occur quite frequently in yeasts that are Crabtree[1] negative or simply incapable of aerobic fermentation (Veiga et al., 2003), all of which are good glycerol consumers (Lages et al., 1999; Barnett et al., 2000). Interestingly, CRR may not be constant, occurring only under specific physiological conditions like diauxic shift, in *P. membranifaciens* and *Y. lipolytica*, or early exponential phase, in *D. hanseni* (Veiga et al., 2003). In *S. cerevisiae*, both conditions highly and transiently induce the glycerol transporter *STL1* expression (Ferreira et al., 2005; Rep et al., 2000, Lucas C. unpublished results).

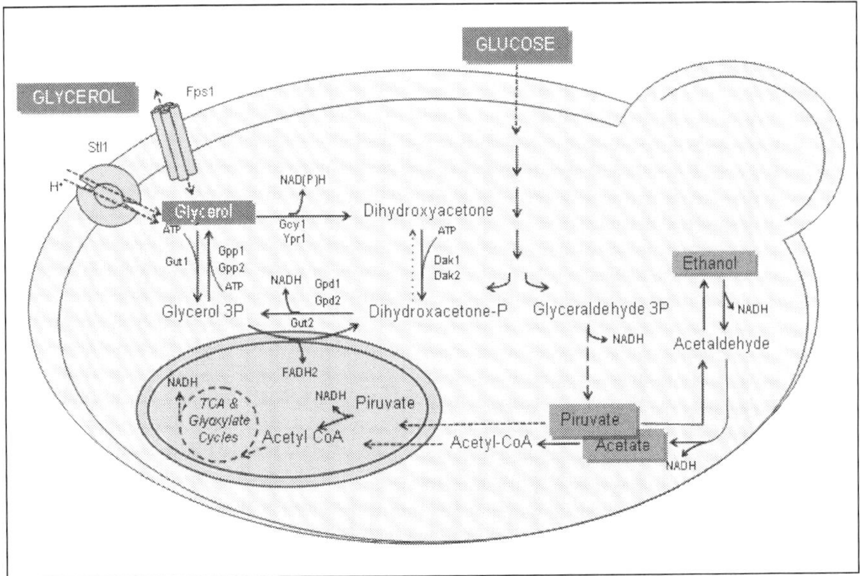

Fig. 2. Glycerol transport and metabolism in *S. cerevisiae* and coupling to main metabolic pathways.

Baker's yeast is a Crabtree[1] positive yeast. Fermentation begins instantly when a glucose pulse is added to glucose-limited, aerobically grown cells. Crabtree effect has been seldom addressed in the last two decades, although it is still a recognized important variable in industrial processes (Ochoa-Estopier et al., 2011). The molecular regulation and main players of this process remain obscure. A relation of Crabtree effect with respiration was discarded (van Urk et al., 1990). Instead, the piruvate decarboxilase levels were found to be 6 times higher in the Crabtree positive yeasts *S. cerevisiae, T. glabrata* (today *C. glabrata*) and *S. pombe*. This presented an increased glucose consumption rate that the authors attributed

---

[1] Crabtree effect is the phenomenon whereby *S. cerevisiae* produces ethanol aerobically in the presence of high external glucose concentrations. Instead, Crabtree negative yeasts instead produce biomass via TCA. In *S. cerevisiae*, high concentrations of glucose accelerate glycolysis, producing appreciable amounts of ATP through substrate-level phosphorylation. This reduces the need of oxidative phosphorylation done by the TCA cycle via the electron transport chain, inhibits respiration and ATP synthesis, and therefore decreases oxygen consumption.

to glucose uptake (van Urk et al., 1990), that did not correspond to equivalent growth improvement, but instead to ethanol production through fermentation. Concurrently, growth on glycerol is supposedly entirely oxidative (Gancedo et al., 1968; Flores et al., 2000), which underlies the good and bad performance of respectively Crabtree negative and positive yeasts. In order to turn glycerol broths commercially attractive for *S. cerevisiae*-based biotechnology, in particular baker's yeast cultivation, several approaches were assayed.

One of the most straightforward strategies is metabolic engineering, obtained through genetic manipulation (Randez-Gil et al., 1999). However, this needs precise knowledge on the strain/species genome, available molecular tools (mutants and vectors to the least), and deep knowledge of the metabolic process involved, which are not always available. Additionally, cellular processes are hardly under the control of a single gene and simply regulated. Because of this, available molecular and informatics tools are combined for engineering industrial strains of interest (Patnaik, 2008). In the particular case of baker's yeast, the industrial strains are mostly aneuploids and homothallic, impairing easy genetic improvement (Randez-Gil et al., 1999). In view of the Crabtree effect regulation complexity, and these genetic characteristics, the improvement of baker's yeast glycerol consumption can hardly be possible by genetic engineering. All this, and the general skepticism of consumers towards the use of genetically modified organisms in the food industry, led to the search of alternative strategies for the baking industry.

3.2 Improbable hybrids

The traditional way of producing new strains is by the generation of hybrids through mating. This approach allows the indirect *in vivo* genetic recombination and the propagation of phenotypes of interest. It can be achieved through intra- or inter-specific hybridization. The most resourceful way is the intra-specific recombination of strains with desirable phenotypes. To achieve this, it is necessary to induce sporulation of the target diploid strains, usually by nitrogen starvation. The haploid ascospores are then isolated and their mating type determined, followed by the mating of ascospores from opposite mating type, and the formation of a new heterozygous diploid. Several wine and baker's yeast strains available commercially are the result of such hybridization (Higgins et al., 2001; Pretorius & Bauer, 2002; Marullo et al., 2006).

These strategies demand for a deep knowledge of the phenotypes and the underlying metabolic and molecular processes. As an example, Higgins and collaborators (Higgins et al., 2001) generated a *S. cerevisiae* strain able to combine efficient maltose metabolism, indispensable for fermentative ability of unsugared dough's, with hyperosmotic resistance for optimization of growth on sugared dough's. Loading *S. cerevisiae* with glycerol has been shown to improve the fermentation of sweet doughs (Myers et al., 1998), therefore the selection for osmotolerant phenotype. On the other hand, unlagged growth on maltose is due to the constitutive derepression of maltase and maltose permease (Higgins et al., 1999), as well as of invertase (Myers et al., 1997), but this was previously reported to negatively influence the leavening of sweet doughs (Oda et al., 1990). This difficulty was overcome by the use of massive random mating upon sporulation enrichment, yielding approximately 10% of interesting isolates for further detailed screening (Higgins et al., 2001).

In the particular case of baker's yeast, the sporulation ability of industrial strains is extremely reduced and most strains are homothallic yielding random-mating spores. This is due to their frequent aneuploidy and the consequent heterogenous coupling of their chromosomes during meiosis. This raises the need of using *assexual* approaches, as spheroplast fusion or cell-spore mating (Sauer, 2001), as well as other mass mating strategies that may circumvent isolated spores inability to mate (Higghins et al., 2001). In spheroplast fusion, after appropriate cell wall digestion, it is possible to force the fusion of cells with different levels of ploidy. These are though in many cases phenotypically and reproductively unstable non-resilient multinuclear cells unfit for industry.

3.3 Evolutionary engineering

The alternative solution to extensive and expensive genetic manipulation is evolutionary engineering (Chatterjee & Yuan; 2006, Fong, 2010). This strategy allows the improvement of complex phenotypes of interest, for example stress resistance combined with carbon source utilization. The methodology is based in the combination of confined environmental selection and natural variability. It was first used in a work of Butler and co-workers (Butler et al., 1996), who selected different genetic strains of *Streptomyces griseus* under selective conditions. Evolutionary engineering aims the creation of an improved strain based in selection of behavioral differences between individual cells within a population. For this reason, the generation of genetic variability is vital to this approach, accelerating the adaptive confined evolution based on spontaneous mutations which demands extremely prolonged cultivation under selective conditions (Aguillera et al., 2010; Faria-Oliveira F., Ferreira C. & Lucas C. unpublished results).

One of the simplest ways of generating variability within a population is the introduction of random genetic mutations. Within a population, there is naturally occurring mutagenesis, either through local changes in the genome or larger modifications like DNA rearrangements and horizontal transfers (Sauer, 2001). Nevertheless, spontaneous mutations occur at very low rate, mainly due to the DNA proof-reading mechanism of the organisms and high fidelity of the DNA polymerases. However, it is known that under adverse conditions the mutation rate is enhanced. This feature is crucial to increase the genetic variability within the population to a level propitious to adaptation to challenging environmental constraints. This selection through survival is the basic principle behind the evolutionary engineering.

Several methodologies are available for the generation of variability, namely physical or chemical mutagenesis, sporulation followed by mating, spheroplast fusion, whole genome shuffling, and so on (Fong, 2010; Petri & Schmidt-Dannert, 2004). Mutagenesis is the most common practice, being technically simple and applicable to most organisms (Fong, 2010). The most common mutagens are either chemicals, like ethyl methane sulfonate (EMS), ethidium bromide (EB), or radiation, namely ultraviolet (UV). These mutagens are rather unspecific, and for this reason are widely used (Sauer, 2001). The main drawback of such approaches is the low rate of useful mutations, and the high rate of lethal and neutral mutations. Most chemicals, like EB, introduce preferentially alterations to the DNA like nucleotide exchanges or frame shifts, but other like EMS can induce deletions (Nair & Zhao, 2010). These are responsible for important DNA rearrangements and severe phenotypic alterations. Yet, some chemicals have affinity for certain genome sub-regions, and its utilization in sequential rounds of mutation/selection can be rather reductive. Physical mutagens, namely UV radiation and X-rays, are more prone to chromosomal structural changes and nucleotide frame shifts.

The simplest method to ameliorate a baker's yeast strain relies, as mentioned above, on spontaneous mutations and prolonged cultivations. This strategy has the advantage of doing without the manipulation of dangerous chemicals or radiation, and the disadvantage of the long time needed to obtain results. These batch or fed-batch cultivations have to cover more than 100 generations (Aguilera et al., 2010; Merico et al., 2011; Ochoa-Estopier et al., 2011; Faria-Oliveira F., Ferreira C. & Lucas C. unpublished results) and can last for several months depending on the severity of the environmental constraints. This procedure was applied with success to transform *S. cerevisiae* into a good lactose consumer (Guimarães et al., 2008), to improve freeze and salt tolerance (Aguilera et al., 2010), and turning baker's yeast into an efficient consumer of glycerol as sole carbon and energy source to industrially acceptable biomass yields (Faria-Oliveira F., Ferreira C. & Lucas C. unpublished results). This was obtained through the use of a simple evolution strategy consisting of sequential aerobic batch cultivations on synthetic glycerol-based media for 150 generations, followed by several cycles of cultivation on rich media for ensuring phenotypic/mutation stability. The resulting strains were able to grow up to 4 g biomass dry weight in 2% (w/v) raw biodiesel centrifuged glycerol[2] which corresponds to a biomass yield of 0,4 g.g^{-1} glycerol. Although yields were not significantly different from cultures obtained on reagent grade glycerol, specific growth rates were 3 times higher in raw glycerol (μ_g 0.12 h^{-1}) and lag phase was reduced to a minimum of 2 h. Merico and collaborators (Merico et al., 2011) describe the selection and characterization of an identically evolved *S. cerevisiae* strain which, additionally, also exhibited a high resistance to freeze and thaw stress after prolonged storage at -20° C. Screening capacity can nowadays be expanded by high throughput techniques, but more importantly, as exemplified above, there has to be prior extensive knowledge to be able to design clever phenotype selection platforms.

Alternatively, yeast species other than *S. cerevisiae*, displaying good characteristics for the baking industry can be used. For example, the above mentioned *T. delbrueckii* strains isolated from corn and rye traditional bread doughs display dough-raising capacities and yield production similar to the ones found in commercial baker's yeasts (Almeida & Pais, 1996a), and maintain approximately the same leavening ability during storage of frozen doughs for 30 days, showing a very high tolerance to freezing (Almeida & Pais, 1996b). Furthermore, in one of these strains no loss of cell viability was observed after 120 days of freezing at -20°C, whereas a loss of 80% was observed in a commercial baker's yeast after 15 days (Alves-Araújo et al., 2004). These characteristics make them candidates of great potential value to the baking industry, mainly to be used in frozen dough products.

4. Conclusion

Nowadays, the man made baker's yeast strains, as well as their associated technological process particularities are fast disappearing due to globalization of yeast and dough industry. Nevertheless, sustainability demands contradict globalization trends, asking for solutions empowering local populations with tools that decrease their dependence on global markets. Yeasts, as always, have a role in this desired change of paradigm.

---

[2] Raw glycerol was diluted with water (1:3) and the pH adjusted to 4.5 with HCl. This was centrifuged at 5000 rpm for 15 min at 4°C. Fat separates from glycerol forming an upper layer which is sucked with a vacuum pump to liberate the cleaner glycerol fraction. The pH adjustment increases the separation efficiency.

Going after the regional tastes for unique types of bread and other bakery products could improve the revenue for the local economical players. This can be achieved through the reintroduction of lost biodiversity in the leavening processes. The industrial production of such yeasts and bacteria can be done using unconventional substrates like biodiesel-derived glycerol, since most bacteria and yeast can consume this substrate naturally, in opposition to the traditional baker's yeast strains. Yet still, if necessary, baker's yeast can be improved for producing interesting biomass yields at the expense of glycerol by clever and simple accelerated and confined evolution strategies. Finally, the glycerol-based broths can by themselves improve the shelf-life span of doughs and leavens.

Biotechnological processes need improvement in order to meet sustainability objectives. No simple unique solution exists. Instead, sustainability can be achieved by increasing diversity of processes, tools and products, for which clever simplicity-generating solutions can be devised.

5. Acknowledgment

Authors would like to acknowledge Hugh S. Johnson for the several critical readings of the manuscript regarding proper English usage. Fábio Faria-Oliveira is supported by a PhD grant from FCT-SFRH/BD/45368/2008. This work was financed by FEDER through COMPETE Programme (Programa Operacional Factores de Competitividade) and national funds from FCT (Fundação para a Ciência e a Tecnologia) project PEst-C/BIA/UI4050/2011.

6. References

Aguilera, J., Andreu, P., Randez-Gil, F. & Prieto, J.A. (2010). Adaptive evolution of baker's yeast in a dough-like environment enhances freeze and salinity tolerance. *Microb Biotechnol*, Vol.3, No.2, pp. 210-221, ISSN 1751-7915

Aguilera, J., Randez-Gil, F. & Prieto, J.A. (2007). Cold response in *Saccharomyces cerevisiae*: new functions for old mechanisms. *FEMS Microbiology Reviews*, Vol.31, No.3, pp. 327-341, ISSN 1574-6976

Al-Fageeh, M.B. & Smales, C.M. (2006). Control and regulation of the cellular responses to cold shock: the responses in yeast and mammalian systems. *Biochem J*, Vol.397, No.2, pp. 247-259

Almeida, M.J. & Pais, C. (1996b). Leavening ability and freeze tolerance of yeasts isolated from traditional corn and rye bread doughs. *Appl Environ Microbiol*, Vol.62, No.12, pp. 4401-4404, ISSN 0099-2240

Almeida, M.J. & Pais, C.S. (1996a). Characterization of the yeast population from traditional corn and rye bread doughs. *Letters in Applied Microbiology*, Vol.23, No.3, pp. 154-158, ISSN 1472-765X

Alves-Araújo, C., Almeida, M.J., Sousa, M.J. & Leão, C. (2004). Freeze tolerance of the yeast *Torulaspora delbrueckii*: cellular and biochemical basis. *FEMS Microbiol Lett*, Vol.240, No.1, pp. 7-14, ISSN 0378-1097

Ansell, R., Granath, K., Hohmann, S., Thevelein, J.M. & Adler, L. (1997). The two isoenzymes for yeast NAD[+]-dependent glycerol 3-phosphate dehydrogenase encoded by *GPD1* and *GPD2* have distinct roles in osmoadaptation and redox regulation. *EMBO J*, Vol.16, No.9, pp. 2179-2187, ISSN 0261-4189

Antuna, B. & Martinez-Anaya, M.A. (1993). Sugar uptake and involved enzymatic activities by yeasts and lactic acid bacteria: their relationship with bread making quality. *Int J Food Microbiol*, Vol.18, No.3, pp. 191-200, ISSN 0168-1605

Barnett, J., Payne R.W. & Yarrow D. (Ed(s).) (2000). *Yeasts: Characteristics and identification* (3rd), Cambridge University Press, ISBN 978-052-1573-96-2, Cambridge

Bocker, G., Stolz, P. & Hammes, P. (1995). Neue Erkenntnisse zum Okosystem Sauerteig und zur Physiologie der sauerteigtypischen Stamme *Lactobacillus sanfrancisco* und *Lactobacillus pontis*. *Getreide Mehl Brot*, Vol.49, pp. 370–374

Brandt, M.J. (2001). Mikrobiologische Wechselwirkungen von technologischer Bedeutung. Ph.D. thesis. University of Hohenheim, Stuttgart, Germany.

Butler, P.R., Brown, M. & Oliver, S.G. (1996). Improvement of antibiotic titers from *Streptomyces* bacteria by interactive continuous selection. *Biotechnol Bioeng*, Vol.49, No.2, pp. 185-196, ISSN 0006-3592

Chatterjee, R. & Yuan, L. (2006). Directed evolution of metabolic pathways. *Trends Biotechnol*, Vol.24, No.1, pp. 28-38, ISSN 0167-7799

Chatzifragkou, A., Makri, A., Belka, A., Bellou, S., Mavrou, M., Mastoridou, M., Mystrioti, P., Onjaro, G., Aggelis, G. & Papanikolaou, S. (2011). Biotechnological conversions of biodiesel derived waste glycerol by yeast and fungal species. *Energy*, Vol.36, No.2, pp. 1097-1108, ISSN 0360-5442

Collar, C., Andreu, P. & Martínez-Anaya, M.A. (1998). Interactive effects of flour, starter and enzyme on bread dough machinability. *Zeitschrift für Lebensmitteluntersuchung und - Forschung A*, Vol.207, No.2, pp. 133-139, ISSN 1431-4630

Corsetti, A., Gobbetti, M., Balestrieri, F., Paoletti, F., Russi, L. & Rossi, J. (1998). Sourdough Lactic Acid Bacteria Effects on Bread Firmness and Stalin. *Journal of Food Science*, Vol.63, No.2, pp. 347-351, ISSN 1750-3841

Corsetti, A., Lavermicocca, P., Morea, M., Baruzzi, F., Tosti, N. & Gobbetti, M. (2001). Phenotypic and molecular identification and clustering of lactic acid bacteria and yeasts from wheat (species *Triticum durum* and *Triticum aestivum*) sourdoughs of Southern Italy. *Int J Food Microbiol*, Vol.64, No.1-2, pp. 95-104, ISSN 0168-1605

da Silva, G.P., Mack, M. & Contiero, J. (2009) Glycerol: A promising and abundant carbon source for industrial microbiology. *Biotechnology Advances*, Vol.27, No.1, pp. 30-39, ISSN 0734-9750

Dal Bello, F., Walter, J., Roos, S., Jonsson, H. & Hertel, C. (2005). Inducible Gene Expression in *Lactobacillus reuteri* LTH5531 during Type II Sourdough Fermentation. *Appl. Environ. Microbiol.*, Vol.71, No.10, pp. 5873-5878

Damiani, P., Gobbetti, M., Cossignani, L., Corsetti, A., Simonetti, M.S. & Rossi, J. (1996). The sourdough microflora. Characterization of hetero- and homofermentative Lactic Acid Bacteria, yeasts and their interactions on the basis of the volatile compounds produced. *Lebensmittel-Wissenschaft und-Technologie*, Vol.29, No.1-2, pp. 63-70, ISSN 0023-6438

D'Amico, S., Collins, T., Marx, J.C., Feller, G. & Gerday, C. (2006). Psychrophilic microorganisms: challenges for life. *EMBO Rep*, Vol.7, No.4, pp. 385-389, ISSN 1469-221X

Dang, N.X. & Hincha, D.K. (2011). Identification of two hydrophilins that contribute to the desiccation and freezing tolerance of yeast (*Saccharomyces cerevisiae*) cells. *Cryobiology*, Vol.62, No.3, pp. 188-193, ISSN 0011-2240 de Vuyst, L. & Neysens, P. (2005). The sourdough microflora: biodiversity and metabolic interactions. *Trends in Food Science & Technology*, Vol.16, No.1-3, pp. 43-56, ISSN 0924-2244

de Vuyst, L. & Vancanneyt, M. (2007). Biodiversity and identification of sourdough lactic acid bacteria. *Food Microbiology*, Vol.24, No.2, pp. 120-127, ISSN 0740-0020

Ferreira, C. & Lucas, C. (2007). Glucose repression over *Saccharomyces cerevisiae* glycerol/H+ symporter gene *STL1* is overcome by high temperature. *FEBS Lett*, Vol.581, No.9, pp. 1923-1927, ISSN 0014-5793

Ferreira, C., van Voorst, F., Martins, A., Neves, L., Oliveira, R., Kielland-Brandt, M.C., Lucas, C. & Brandt, A. (2005). A member of the sugar transporter family, Stl1p is the glycerol/H+ symporter in *Saccharomyces cerevisiae*. *Mol Biol Cell*, Vol.16, No.4, pp. 2068-2076, ISSN 1059-1524

Flores, C.L., Rodriguez, C., Petit, T. & Gancedo, C. (2000). Carbohydrate and energy-yielding metabolism in non-conventional yeasts. *FEMS Microbiol Rev*, Vol.24, No.4, pp. 507-529, ISSN 0168-6445

Fong, S.S. (2010). Evolutionary engineering of industrially important microbial phenotypes, In: *The metabolic pathway engineering handbook: tools and applications*, C.D. Smolke, pp. 1/1-1/13, CRC Press, ISBN 978-142-0077-65-0, New York

Galli, A., Franzetti, L. & Fortina. M.G. (1988). Isolation and identification of sour dough microflora. *Microbiol. Aliment. Nutr.*, Vol.6, pp. 345-351

Gancedo, C., Gancedo, J.M. & Sols, A. (1968). Glycerol metabolism in yeasts. Pathways of utilization and production. *Eur J Biochem*, Vol. 5, No.2, pp. 165-172, ISSN 0014-2956

Garofalo, C., Silvestri, G., Aquilanti, L. & Clementi, F. (2008). PCR-DGGE analysis of lactic acid bacteria and yeast dynamics during the production processes of three varieties of Panettone. *J Appl Microbiol*, Vol.105, No.1, pp. 243-254, ISSN 1365-2672

Gobbetti, M. (1998). The sourdough microflora: Interactions of lactic acid bacteria and yeasts. *Trends in Food Science & Technology*, Vol.9, No.7, pp. 267-274, ISSN 0924-2244

Gobbetti, M., Corsetti, A. & Rossi, J. (1994a). The sourdough microflora. Interactions between lactic acid bacteria and yeasts: metabolism of amino acids. *World Journal of Microbiology and Biotechnology*, Vol.10, pp. 275-279

Gobbetti, M., Corsetti, A., Rossi, J., La Rosa, F. & de Vincenzi, S. (1994b). Identification and clustering of lactic acid bacteria and yeasts from wheat sourdoughs of central Italy. *Italian J Food Sci*, Vol.6, pp. 85-94

Gobbetti, M., De Angelis, M., Corsetti, A. & Di Camargo, R. (2005). Biochemistry and physiology of sourdough lactic acid bacteria. *Trends in Food Science & Technology*, Vol.16, pp. 57-69

Grauslund, M., Lopes, J.M. & Rønnow, B. (1999). Expression of *GUT1*, which encodes glycerol kinase in *Saccharomyces cerevisiae*, is controlled by the positive regulators Adr1p, Ino2p and Ino4p and the negative regulator Opi1p in a carbon source-dependent fashion. *Nucleic Acids Res*, Vol.27, No.22, pp. 4391-4398, ISSN 1362-4962

Guadalupe Medina, V., Almering, M.J., van Maris, A.J. & Pronk, J.T. (2010). Elimination of glycerol production in anaerobic cultures of *Saccharomyces cerevisiae* engineered for use of acetic acid as electron acceptor. *Appl Environ Microbiol*, Vol.76, No.1, pp. 190-195, ISSN 0099-2240

Guimarães, P.M., Francois, J., Parrou, J.L., Teixeira, J.A. & Domingues, L. (2008). Adaptive evolution of a lactose-consuming *Saccharomyces cerevisiae* recombinant. *Appl Environ Microbiol*, Vol.74, No.6, pp. 1748-1756, ISSN 1098-5336

Halm, M., Lillie, A., Sorensen, A.K. & Jakobsen, M. (1993). Microbiological and aromatic characteristics of fermented maize doughs for kenkey production in Ghana. *Int J Food Microbiol*, Vol.19, No.2, pp. 135-43, ISSN 0168-1605

Hämmes, W.P. & Gänzle, M.G. (1998). Sourdough breads and related products, In: *Microbiology of Fermented Foods*, B.J.B. Wood, W.H. Holzapfel, pp. 199-216, Blackie Academic and Professional, ISBN 978-075-1402-16-2, London

Hämmes, W.P. & Vogel, R.F. (1995). The genus *Lactobacillus*, In: *The genera of lactic acid bacteria*, B.J.B. Wood, W.H. Holzapfel, pp. 19-54, Blackie Academic and Professional, ISBN 0-7514-0215X, London

Hansen, B. & Hansen, Å. (1994). Volatile compounds in wheat sourdoughs produced by lactic acid bacteria and sourdough yeasts. *Zeitschrift für Lebensmitteluntersuchung und - Forschung A*, Vol.198, No.3, pp. 202-209, ISSN 1431-4630

Hansen, K.R., Burns, G., Mata, J., Volpe, T.A., Martienssen, R.A., Bahler, J. & Thon, G. (2005). Global effects on gene expression in fission yeast by silencing and RNA interference machineries. *Mol Cell Biol*, Vol.25, No.2, pp. 590-601, ISSN 0270-7306

Hayashi, M. & Maeda, T. (2006). Activation of the HOG Pathway upon cold stress in *Saccharomyces cerevisiae*. *Journal of Biochemistry*, Vol.139, No.4, pp. 797-803

Higgins, V.J., Bell, P.J., Dawes, I.W. & Attfield, P.V. (2001). Generation of a novel *Saccharomyces cerevisiae* strain that exhibits strong maltose utilization and hyperosmotic resistance using nonrecombinant techniques. *Appl Environ Microbiol*, Vol.67, No.9, pp. 4346-4348

Higgins, V.J., Braidwood, M., Bissinger, P., Dawes, I.W. & Attfield, P.V. (1999). Leu343Phe substitution in the Malx3 protein of *Saccharomyces cerevisiae* increases the constitutivity and glucose insensitivity of *MAL* gene expression. *Curr Genet*, Vol.35, No.5, pp. 491-498, ISSN 0172-8083

Hirasawa, R. & Yokoigawa, K. (2001). Leavening ability of baker's yeast exposed to hyperosmotic media. *FEMS Microbiol Lett*, Vol.194, No.2, pp. 159-162, ISSN 0378-1097

Hirasawa, R., Yokoigawa, K., Isobe, Y. & Kawai, H. (2001). Improving the freeze tolerance of bakers' yeast by loading with trehalose. *Biosci Biotechnol Biochem*, Vol.65, No.3, pp. 522-526, ISSN 0916-8451

Hohmann, S. (2009). Control of high osmolarity signalling in the yeast *Saccharomyces cerevisiae*. *FEBS Letters*, Vol.583, No.24, pp. 4025-4029, ISSN 0014-5793

Iacumin, L., Cecchini, F., Manzano, M., Osualdini, M., Boscolo, D., Orlic, S. & Comi, G. (2009). Description of the microflora of sourdoughs by culture-dependent and culture-independent methods. *Food Microbiol*, Vol.26, No.2, pp. 128-135, ISSN 1095-9998

Izawa, S., Ikeda, K., Maeta, K. & Inoue, Y. (2004b). Deficiency in the glycerol channel Fps1p confers increased freeze tolerance to yeast cells: application of the *fps1Δ* mutant to frozen dough technology. *Appl Microbiol Biotechnol*, Vol.66, No.3, pp. 303-305, ISSN 0175-7598

Izawa, S., Ikeda, K., Takahashi, N. & Inoue, Y. (2007). Improvement of tolerance to freeze-thaw stress of baker's yeast by cultivation with soy peptides. *Appl Microbiol Biotechnol*, Vol.75, No.3, pp. 533-537, ISSN 0175-7598

Izawa, S., Sato, M., Yokoigawa, K. & Inoue, Y. (2004a). Intracellular glycerol influences resistance to freeze stress in *Saccharomyces cerevisiae*: analysis of a quadruple mutant in glycerol dehydrogenase genes and glycerol-enriched cells. *Appl Microbiol Biotechnol*, Vol.66, No.1, pp. 108-114, ISSN 0175-7598

Kaino, T., Tateiwa, T., Mizukami-Murata, S., Shima, J. & Takagi, H. (2008). Self-cloning baker's yeasts that accumulate proline enhance freeze tolerance in doughs. *Appl Environ Microbiol*, Vol.74, No.18, pp. 5845-5849, ISSN 1098-5336

Kandror, O., Bretschneider, N., Kreydin, E., Cavalieri, D. & Goldberg, A.L. (2004). Yeast adapt to near-freezing temperatures by *STRE*/Msn2,4-dependent induction of trehalose synthesis and certain molecular chaperones. *Molecular Cell*, Vol.13, No.6, pp. 771-781, ISSN 1097-2765

Lages, F., Silva-Graça, M. & Lucas, C. (1999). Active glycerol uptake is a mechanism underlying halotolerance in yeasts: a study of 42 species. *Microbiology*, Vol.145, No.9, pp. 2577-2585, ISSN 1350-0872

Laroche, C. & Gervais, P. (2003). Achievement of rapid osmotic dehydration at specific temperatures could maintain high *Saccharomyces cerevisiae* viability. *Appl Microbiol Biotechnol*, Vol.60, No.6, pp. 743-747, ISSN 0175-7598

Larsson, C., Pahlman, I.L., Ansell, R., Rigoulet, M., Adler, L. & Gustafsson, L. (1998). The importance of the glycerol 3-phosphate shuttle during aerobic growth of *Saccharomyces cerevisiae*. *Yeast*, Vol.14, No.4, pp. 347-357, ISSN 0749-503X

Luyten, K., Albertyn, J., Skibbe, W.F., Prior, B.A., Ramos, J., Thevelein, J.M. & Hohmann, S. (1995). Fps1, a yeast member of the MIP family of channel proteins, is a facilitator for glycerol uptake and efflux and is inactive under osmotic stress. *EMBO J*, Vol.14, No.7, pp. 1360-1371, ISSN 0261-4189

Mäntynen, V.H., Korhola, M., Gudmundsson, H., Turakainen, H., Alfredsson, G.A., Salovaara, H. & Lindstrom, K. (1999). A polyphasic study on the taxonomic position of industrial sour dough yeasts. *Syst Appl Microbiol*, Vol.22, No.1, pp. 87-96, ISSN 0723-2020

Martinez-Anaya, M.A., Pitarch, B., Bayarri, P. & Barber, C.B. (1990a). Microflora of the sourdoughs of wheat flour bread. Interactions between yeasts and lactic acid bacteria in wheat doughs and their effects on bread quality. *Cereal chem*, Vol.67, pp. 85-91

Martinez-Anaya, M.A., Torner, J.M. & de Barber, C.B. (1990b). Microflora of the sour dough of wheat flour bread. *Zeitschrift für Lebensmitteluntersuchung und -Forschung A*, Vol.190, No.2, pp. 126-131, ISSN 1431-4630

Marullo, P., Bely, M., Masneuf-Pomarede, I., Pons, M., Aigle, M. & Dubourdieu, D. (2006). Breeding strategies for combining fermentative qualities and reducing off-flavor production in a wine yeast model. *FEMS Yeast Res*, Vol.6, No.2, pp. 268-279, ISSN 1567-1356

Merico, A., Ragni, E., Galafassi, S., Popolo, L. & Compagno, C. (2011). Generation of an evolved *Saccharomyces cerevisiae* strain with a high freeze tolerance and an improved ability to grow on glycerol. *J Ind Microbiol Biotechnol*, Vol.38, No.8, pp. 1037-1044, ISSN 1476-5535

Meroth, C.B., Brandt, M.J. & Hamme, W.P. (2002). Die hefe macht's. *Brot Backwaren*, Vol.4, pp. 33-34

Meroth, C.B., Hammes, W.P. & Hertel, C. (2003a). Identification and population dynamics of yeasts in sourdough fermentation processes by PCR-denaturing gradient gel electrophoresis. *Appl Environ Microbiol*, Vol.69, No.12, pp. 7453-7461

Meroth, C.B., Walter, J., Hertel, C., Brandt, M.J. & Hammes, W.P. (2003b). Monitoring the bacterial population dynamics in sourdough fermentation processes by using PCR-denaturing gradient gel electrophoresis. *Appl Environ Microbiol*, Vol.69, No.1, pp. 475-482

Michnick, S., Roustan, J.L., Remize, F., Barre, P. & Dequin, S. (1997). Modulation of glycerol and ethanol yields during alcoholic fermentation in *Saccharomyces cerevisiae* strains overexpressed or disrupted for *GPD1* encoding glycerol 3-phosphate dehydrogenase. *Yeast*, Vol.13, No.9, pp. 783-793, ISSN 0749-503X

Momose, Y., Matsumoto, R., Maruyama, A. & Yamaoka, M. (2010). Comparative analysis of transcriptional responses to the cryoprotectants, dimethyl sulfoxide and trehalose, which confer tolerance to freeze-thaw stress in *Saccharomyces cerevisiae*. *Cryobiology*, Vol.60, No.3, pp. 245-261, ISSN 1090-2392

Murata, Y., Homma, T., Kitagawa, E., Momose, Y., Sato, M.S., Odani, M., Shimizu, H., Hasegawa-Mizusawa, M., Matsumoto, R., Mizukami, S., Fujita, K., Parveen, M., Komatsu, Y. & Iwahashi, H. (2006). Genome-wide expression analysis of yeast response during exposure to 4 degrees C. *Extremophiles*, Vol.10, No.2, pp. 117-128, ISSN 1431-0651

Myers, D.K., Joseph, V.M., Pehm, S., Galvagno, M. & Attfield, P.V. (1998). Loading of *Saccharomyces cerevisiae* with glycerol leads to enhanced fermentation in sweet bread doughs. *Food Microbiology*, Vol.15, No.1, pp. 51-58, ISSN 0740-0020

Myers, D.K., Lawlor, D.T. & Attfield, P.V. (1997). Influence of invertase activity and glycerol synthesis and retention on fermentation of media with a high sugar concentration by *Saccharomyces cerevisiae*. *Appl Environ Microbiol*, Vol.63, No.1, pp. 145-150

Nair, N.U. & Zhao, H. (2010). The metabolic pathway engineering handbook: tools and applications, In: *The metabolic pathway engineering handbook: tools and applications*, C.D. Smolke, pp. 2/1-2/37, CRC Press, ISBN 978-142-0077-65-0, New York

Obiri-Danso, K. (1994). Microbiological studies on corn dough ferment. *Cereal chem*, Vol.7, pp. 186-188

Ochoa-Estopier, A., Lesage, J., Gorret, N. & Guillouet, S.E. (2011). Kinetic analysis of a *Saccharomyces cerevisiae* strain adapted for improved growth on glycerol: Implications for the development of yeast bioprocesses on glycerol. *Bioresour Technol*, Vol.102, No.2, pp. 1521-1527, ISSN 1873-2976

Oda, Y. & Ouchi, K. (1990). Effect of invertase activity on the leavening ability of yeast in sweet dough. *Food Microbiology*, Vol.7, No.3, pp. 241-248, ISSN 0740-0020

Ottogalli, G., Galli, A. & Foschino, R. (1996). Italian bakery products obtained with sour dough: characterization of the tipical microflora. *Adv Food Sci*, Vol.18, pp. 131-144

Påhlman, I.L., Gustafsson, L., Rigoulet, M. & Larsson, C. (2001). Cytosolic redox metabolism in aerobic chemostat cultures of *Saccharomyces cerevisiae*. *Yeast*, Vol.18, No.7, pp. 611-620, ISSN 0749-503X

Panadero, J., Randez-Gil, F. & Prieto, J.A. (2005). Validation of a flour-free model dough system for throughput studies of baker's yeast. *Appl Environ Microbiol*, Vol.71, No.3, pp. 1142-1147, ISSN 0099-2240

Paramithiotis, S., Muller, M.R.A., Ehrmann, M.A., Tsakalidou, E., Seiler, H., Vogel, R. & Kalantzopoulos, G. (2000). Polyphasic identification of wild yeast strains isolated from Greek sourdoughs. *Syst Appl Microbiol*, Vol.23, No.1, pp. 156-164, ISSN 0723-2020

Paramithiotis, S., Tsiasiotou, S. & Drosinos, E. (2010). Comparative study of spontaneously fermented sourdoughs originating from two regions of Greece: Peloponnesus and Thessaly. *European Food Research and Technology*, Vol.231, No.6, pp. 883-890, ISSN 1438-2377

Patnaik, R. (2008). Engineering complex phenotypes in industrial strains. *Biotechnol Prog*, Vol.24, No.1, pp. 38-47, ISSN 8756-7938

Petri, R. & Schmidt-Dannert, C. (2004). Dealing with complexity: evolutionary engineering and genome shuffling. *Curr Opin Biotechnol*, Vol.15, No.4, pp. 298-304, ISSN 0958-1669

Phadtare, S. & Severinov, K. (2010). RNA remodeling and gene regulation by cold shock proteins. *RNA Biol*, Vol.7, No.6, pp. 788-795, ISSN 1555-8584

Pretorius, I.S. & Bauer, F.F. (2002). Meeting the consumer challenge through genetically customized wine-yeast strains. *Trends Biotechnol*, Vol.20, No.10, pp. 426-432, ISSN 0167-7799

Pulvirenti, A., Solieri, L., Gullo, M., de Vero, L. & Giudici, P. (2004). Occurrence and dominance of yeast species in sourdough. *Lett Appl Microbiol*, Vol.38, No.2, pp. 113-117, ISSN 0266-8254

Randez-Gil, F., Sanz, P. & Prieto, J.A. (1999). Engineering baker's yeast: room for improvement. *Trends Biotechnol*, Vol.17, No.6, pp. 237-244, ISSN 0167-7799

Remize, F., Roustan, J.L., Sablayrolles, J.M., Barre, P. & Dequin, S. (1999). Glycerol overproduction by engineered *Saccharomyces cerevisiae* wine yeast strains leads to substantial changes in by-product formation and to a stimulation of fermentation rate in stationary phase. *Appl Environ Microbiol*, Vol.65, No.1, pp. 143-149, ISSN 1098-5336

Rep, M., Krantz, M., Thevelein, J.M. & Hohmann, S. (2000). The transcriptional response of *Saccharomyces cerevisiae* to osmotic shock. *Journal of Biological Chemistry*, Vol.275, No.12, pp. 8290-8300

Rocha, J.M. & Malcata, F.X. (1999). On the microbiological profile of traditional Portuguese sourdough. *J Food Prot*, Vol.62, No.12, pp. 1416-1429, ISSN 0362-028X

Rodriguez-Vargas, S., Estruch, F. & Randez-Gil, F. (2002). Gene expression analysis of cold and freeze stress in Baker's yeast. *Appl Environ Microbiol*, Vol.68, No.6, pp. 3024-3030

Rodriguez-Vargas, S., Sanchez-Garcia, A., Martinez-Rivas, J.M., Prieto, J.A. & Randez-Gil, F. (2007). Fluidization of membrane lipids enhances the tolerance of *Saccharomyces cerevisiae* to freezing and salt stress. *Appl Environ Microbiol*, Vol.73, No.1, pp. 110-116

Rønnow, B. & Kielland-Brandt, M.C. (1993). GUT2, a gene for mitochondrial glycerol 3-phosphate dehydrogenase of *Saccharomyces cerevisiae*. *Yeast*, Vol.9, No.10, pp. 1121-1130, ISSN 0749-503X

Rossi, J. (1996). The yeasts in sourdough. *Adv Food Sci*, Vol.18, pp. 201-211

Rothe, M., Schneeweiss, R. & Ehrlich. R. (1973). Zur historischen entwicklung von getreideverarbeitung und getredeverzehr. *Ernahrungsfirschung*, Vol.18, pp. 249-283

Sahara, T., Goda, T. & Ohgiya, S. (2002). Comprehensive expression analysis of time-dependent genetic responses in yeast cells to low temperature. *J Biol Chem*, Vol.277, No.51, pp. 50015-50021, ISSN 0021-9258

Salovaara, H. (1998). Lactic acid bacteria in cereal based products, In: *Lactic Acid Bacteria - Technology and Health Effects*, Salminen, S., von Wright, A., pp. 115-338, Marcel Dekker, New York

Sauer, U. (2001). Evolutionary engineering of industrially important microbial phenotypes. *Adv Biochem Eng Biotechnol*, Vol.73, pp. 129-169, ISSN 0724-6145

Shima, J., Sakata-Tsuda, Y., Suzuki, Y., Nakajima, R., Watanabe, H., Kawamoto, S. & Takano, H. (2003). Disruption of the *CAR1* gene encoding arginase enhances freeze tolerance of the commercial baker's yeast *Saccharomyces cerevisiae*. *Appl Environ Microbiol*, Vol.69, No.1, pp. 715-718

Siderius, M., van Wuytswinkel, O., Reijenga, K.A., Kelders, M. & Mager, W.H. (2000). The control of intracellular glycerol in *Saccharomyces cerevisiae* influences osmotic stress response and resistance to increased temperature. *Mol Microbiol*, Vol.36, No.6, pp. 1381-1390, ISSN 0950-382X

Simonin, H., Beney, L. & Gervais, P. (2007). Sequence of occurring damages in yeast plasma membrane during dehydration and rehydration: mechanisms of cell death. *Biochim Biophys Acta*, Vol.1768, No.6, pp. 1600-1610, ISSN 0006-3002

Spicher, G. & Nierle, W. (1984). The microflora of sourdough. The influence of yeast on the proteolysis during sourdough fermentation. *Zeitschrifl flit Lebensmittel-Untersuchung und-Forschung*, Vol.179, pp. 109-112

Succi, M., Reale, A., Andrighetto, C., Lombardi, A., Sorrentino, E. & Coppola, R. (2003). Presence of yeasts in southern Italian sourdoughs from *Triticum aestivum* flour. *FEMS Microbiol Lett*, Vol.225, No.1, pp. 143-148, ISSN 0378-1097

Sugihara, T.F., Kline, L. & Miller, M.W. (1971). Microorganisms of the San Francisco sour dough bread process. Yeasts responsible for the leavening action. *Appl Microbiol*, Vol.21, No.3, pp. 456-458, ISSN 0003-6919

Sutherland, F., Lages, F., Lucas, C., Luyten, K., Albertyn, J., Hohmann, S., Prior, B. & Kilian, S. (1997). Characteristics of Fps1-dependent and -independent glycerol transport in *Saccharomyces cerevisiae*. *J. Bacteriol.*, Vol.179, No.24, pp. 7790-7795

Takahashi, S., Ando, A., Takagi, H. & Shima, J. (2009). Insufficiency of copper ion homeostasis causes freeze-thaw injury of yeast cells as revealed by indirect gene expression analysis. *Appl Environ Microbiol*, Vol.75, No.21, pp. 6706-6711, ISSN 1098-5336

Tamás, M.J., Luyten, K., Sutherland, F.C., Hernandez, A., Albertyn, J., Valadi, H., Li, H., Prior, B.A., Kilian, S.G., Ramos, J., Gustafsson, L., Thevelein, J.M. & Hohmann, S. (1999). Fps1p controls the accumulation and release of the compatible solute glycerol in yeast osmoregulation. *Mol Microbiol*, Vol.31, No.4, pp. 1087-1104, ISSN 0950-382X

Tanghe, A., van Dijck, P., Colavizza, D. & Thevelein, J.M. (2004). Aquaporin-mediated improvement of freeze tolerance of *Saccharomyces cerevisiae* is restricted to rapid freezing conditions. *Appl Environ Microbiol*, Vol.70, No.6, pp. 3377-3382

Tanghe, A., van Dijck, P., Dumortier, F., Teunissen, A., Hohmann, S. & Thevelein, J.M. (2002). Aquaporin expression correlates with freeze tolerance in baker's yeast, and overexpression improves freeze tolerance in industrial strains. *Appl Environ Microbiol*, Vol.68, No.12, pp. 5981-5989

Terao, Y., Nakamori, S. & Takagi, H. (2003). Gene dosage effect of L-proline biosynthetic enzymes on L-proline accumulation and freeze tolerance in *Saccharomyces cerevisiae*. *Appl Environ Microbiol*, Vol.69, No.11, pp. 6527-6532

Trüper, H.G. & De' Clari, L. (1997). Taxonomic note: Necessary correction of specific epithets formed as substantives (Nouns) "in Apposition". *International Journal of Systematic Bacteriology*, Vol.47, No.3, pp. 908-909

Tulha, J., Lima, A., Lucas, C. & Ferreira, C. (2010). *Saccharomyces cerevisiae* glycerol/H$^+$ symporter Stl1p is essential for cold/near-freeze and freeze stress adaptation. A simple recipe with high biotechnological potential is given. *Microb Cell Fact*, Vol.9, No.82, ISSN 1475-2859

Valmorri, S., Tofalo, R., Settanni, L., Corsetti, A. & Suzzi, G. (2010). Yeast microbiota associated with spontaneous sourdough fermentations in the production of traditional wheat sourdough breads of the Abruzzo region (Italy). *Antonie Van Leeuwenhoek*, Vol.97, No.2, pp. 119-129, ISSN 1572-9699

van Dijken, J.P. & Scheffers, W.A. (1986). Redox balances in the metabolism of sugars by yeasts. *FEMS Microbiology Letters*, Vol.32, No.3-4, pp. 199-224, ISSN 0378-1097

van Urk H., Voll W.S.L., Scheffers W.A. & van Dijken J. (1990). Transient-state analysis of metabolic fluxes in Crabtree-positive and Crabtree-negative yeasts. *Appl. Environ. Microbiol*, Vol.56, No.1, pp. 281-287

Veiga, A., Arrabaca, J.D. & Loureiro-Dias, M.C. (2003). Stress situations induce cyanide-resistant respiration in spoilage yeasts. *J Appl Microbiol*, Vol.95, No.2, pp. 364-371, ISSN 1364-5072

Vernocchi, P., Valmorri, S., Dalai, I., Torriani, S., Gianotti, A., Suzzi, G., Guerzoni, M.E., Mastrocola, D. & Gardini, F. (2004a). Characterization of the yeast population involved in the production of a typical Italian bread. *Journal of Food Science*, Vol.69, No.7, pp. 182-186, ISSN 1750-3841

Vernocchi, P., Valmorri, S., Gatto, V., Torriani, S., Gianotti, A., Suzzi, G., Guerzoni, M.E. & Gardini, F. (2004b). A survey on yeast microbiota associated with an Italian traditional sweet-leavened baked good fermentation. *Food Research International*, Vol.37, No.5, pp. 469-476, ISSN 0963-9969

Vogel, R. (1997). Microbial Ecology of Cereal Fermentations. *Food Tecnhol. Biotechnol.*, Vol.35, pp. 51-54, ISSN 1330-9862

Vrancken, G., de Vuyst, L., van der Meulen, R., Huys, G., Vandamme, P. & Daniel, H.M. (2010). Yeast species composition differs between artisan bakery and spontaneous laboratory sourdoughs. *FEMS Yeast Res*, Vol.10, No.4, pp. 471-481, ISSN 1567-1364

Xu, Z.-Y., Zhang, X., Schläppi, M. & Xu, Z.-Q. (2011). Cold-inducible expression of *AZI1* and its function in improvement of freezing tolerance of *Arabidopsis thaliana* and *Saccharomyces cerevisiae*. *Journal of Plant Physiology*, Vol.168, No.13, pp. 1576-1587, ISSN 0176-1617

Yazdani, S.S. & Gonzalez, R. (2007). Anaerobic fermentation of glycerol: a path to economic viability for the biofuels industry. *Curr Opin Biotechnol*, Vol.18, No.3, pp. 213-219, ISSN 0958-1669

Yokoigawa, K., Sato, M. & Soda, K. (2006). Simple improvement in freeze-tolerance of bakers' yeast with poly-γ-glutamate. *J Biosci Bioeng*, Vol.102, No.3, pp. 215-219, ISSN 1389-1723

Zhang, L., Ohta, A., Horiuchi, H., Takagi, M. & Imai, R. (2001). Multiple mechanisms regulate expression of low temperature responsive (*LOT*) genes in *Saccharomyces cerevisiae*. *Biochem Biophys Res Commun*, Vol.283, No.2, pp. 531-535, ISSN 0006-291X

Zhang, X., Shanmugam, K.T. & Ingram, L.O. (2010). Fermentation of glycerol to succinate by metabolically engineered strains of *Escherichia coli*. *Appl. Environ. Microbiol.*, Vol.76, No.8, pp. 2397–2401

14

Trends in Functional Food Against Obesity

José C.E. Serrano, Anna Cassanyé and Manuel Portero-Otin
Department of Experimental Medicine, University of Lleida
Spain

1. Introduction

The increasingly accepted notion of the relationship between diet and health has opened new perspectives on the effects of food ingredients on physiological functions and health. Among the nutritional complications, increased incidence of obesity and its associated medical complications is creating a pressure from consumers towards the food industry which may provide an opportunity for the development of functional foods designed for the prevention and/or treatment of these pathologies.

Obesity is a multifactor disease where several factors may influence its onset, which includes the contributions of inherited, metabolic, behavioural, environmental, cultural, and socioeconomic factors as it is shown in Figure 1. Most of these factors may play together in different grades of contribution, which may differ between patients, and may influence treatment objectives in each individual.

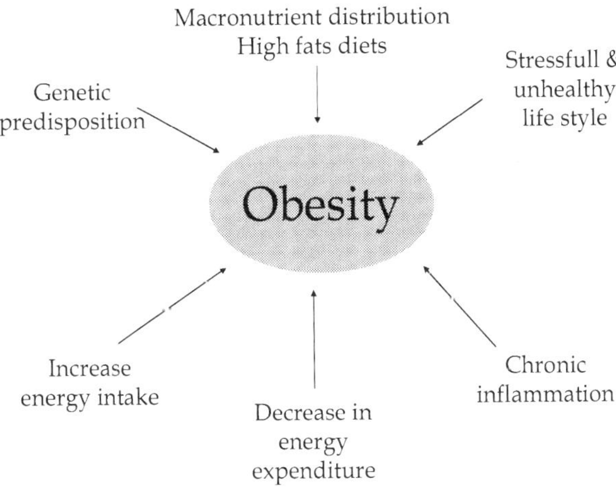

Fig. 1. Obesity, a multifactorial disease.

Moreover, overweight and obesity may raise the risk of other related pathologies like high blood pressure, high blood cholesterol, heart disease, stroke, diabetes, certain types of cancer, arthritis, and breathing problems. As weight increases, so does the prevalence of health risks. The health outcomes related to these diseases, however, may be improved through weight loss or, at a minimum, no further weight gain. The main goal of any nutritional intervention is to individually determine the principal factors that may contribute with individual obesity predisposition and find specific tools to counteract each factor. Food Industry may play an important role providing enough tools, functional foods, for the prevention and treatment of obesity

In simplified terms, overweight and obesity can be defined as an imbalance where the amount of energy intake exceeds the amount of energy expended. Treatment and prevention of obesity requires changes in one or two of the components of this simplified equation. In this sense, the development of functional foods should be aimed to decrease the amount of energy intake (by lowering the energy density of foods or reducing the food intake) or increasing caloric expenditure through the stimulation of thermogenesis and/or modifying the distribution and use of nutrients as energy fuel between tissues, discouraging fat deposition. A summary of possible strategic features for the development of functional foods against obesity is shown in Figure 2.

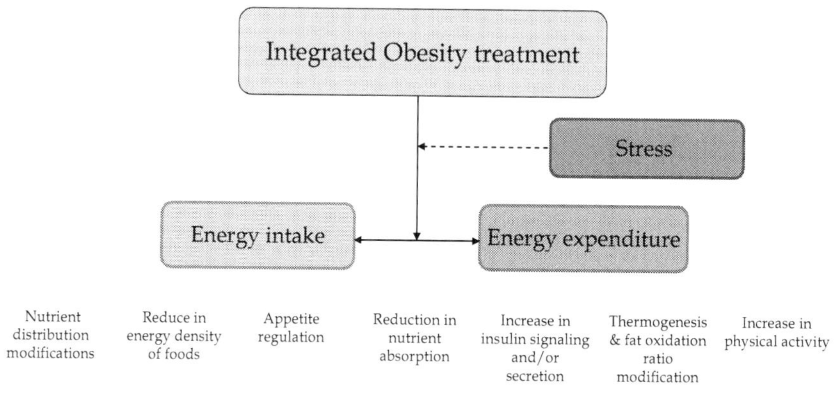

Fig. 2. Main strategic features for the development of functional foods against obesity.

Since several decades ago, in the fight against obesity, food manufactures had offer, a variety of food products named in the beginnings as "dietetic products" based principally in substitutions of sugars and fat by non-nutritive sweeteners and fat replacers respectively. Nowadays, the research and development of new products, should offer the market, food products indicated especially for obese people that besides their low caloric content, can offers the possibility to influence the energy metabolism as well as in the physiological sensation of satiety. Currently, there is a wide variety of products in the market with a low energy density, while the supply of products with bioactive ingredients that decrease appetite, increase caloric expenditure and/or affect the distribution of body fat is scare and in some cases of doubtful effectiveness.

In this context, the new European regulation regarding food labelling, may encourage the food industry to carry out more investment in research and in the determination of the effectiveness of the functional products launched to the market at different levels (biochemical, molecular, genomic and psychological). In this way, the confidence and scientifically contrasted effectiveness of these products as preventive and therapy tools against obesity, may contribute to reduce the incidence of obesity in the whole population. The first step in the development of functional foods is the identification of functional factor, condition or compound that produces a specific effect, which is effective as an adjunct in the treatment of obesity. The new European regulation demands that the effectiveness of such functional foods should be properly established with sufficient scientific evidence, including intervention studies in human populations. It is also desirable to establish the possible interactions of the functional ingredient within the body at different levels (genomic, molecular, cellular and psychological). On the other hand, is very important and necessary to investigate the functional ingredients incorporated into food as such, taking into account possible interactions between "functional ingredient" and other food matrix components, its dose, culinary preparation processes and the usual form of consumption. The key points in the evaluation of functional foods would be the safety and efficacy, thereby avoiding misleading advertising to the consumer.

In this sense, it is necessary to establish specific biomarkers (e.g. body mass index, blood cholesterol levels, percentage of body fat) the effectiveness of consumption of functional foods designed against obesity. However, there is still no consensus on the specific relevance and applicability of each of these biomarkers in the context of obesity, so that there is unanimity on tests of functional assessment for all food companies to launch a new functional food to the market.

The objective of this chapter is to describe possible guidelines for the development of functional foods based on the scientific evidence of the actions of several bioactive compounds and nutritional/technological modifications of foods to be used for the prevention and/or treatment of obesity. It describes possible actions that could be undertaken from different levels, starting with the technological modification of food with the aim to produce a satiety feeling towards the incorporation of functional ingredients that may modify energy intake and expenditure.

2. Energy balance, factors that influence energy intake & expenditure

As mentioned before, the strategy for functional food development should be based in the reduction of energy intake and/or in the increase in energy expenditure. The reduction in energy intake can be obtained by increasing satiety feeling either by the activation of satiety centres or by the modification of hunger feeling delaying it's unset. Optimally, a good satiety functional food must be satisfying and have a reduce energy content. The increase in energy expenditure can be regulated by the modification in the metabolic rate via an increase in thermogenesis or by the control in hormonal energetic metabolisms like insulin sensitivity. Additionally it should be also taken into consideration other factors like inflammation, psychological and physiological stress and life styles that may influence the response and adaptation to energy intake and expenditure.

This section includes a brief description of physiological mechanisms that may influence energy balance and possible strategies for counteracting the effect of each mechanism in obesity, as a tool for the manufacture of functional foods.

2.1 Satiety control, a tool for energy intake reduction

Human behaviour towards food can be defined as a physiological and psychological process that may be influenced by genetic and environmental factors in which the individual is involved. The physiological regulation of the act of eating (hunger and satiety sensations) is a complex interaction between peripheral signals and central nervous system interpretation of these signals, to which must be added physio-psychological variables, such as differences in taste perception and the strictly psychological variables likely influenced by the individual's surrounding environment.

From a physiological point of view the satiety and hunger regulation has been described using two paradigms: the glucostatic hypothesis (Mayer & Thomas, 1967) and the lipostatic model (Kennedy, 1953). The glucostatic hypothesis is based on the assumption that small changes in plasma glucose levels induced signal initiation and termination of eating. However, this model does not take into account how the body regulates the long-term storage and use of energy. The lipostatic model hypothesizes that there are peripheral signals that gives information about the amount of fat or stored energy and therefore the amount of energy needed to maintain a good energy balance. This hypothesis has been supported by the discovery of leptin, an adipokine that is released by adipose tissue in proportion to its fat content. However, since there are no significant fluctuations throughout the day on the composition of body fat and thus leptin, this model may not explain the dynamic behaviour and varying feelings of satiety-hungry induced throughout the day.

The interpretation of these signals is done by the central nervous systems. Recently it has been reported that short- and long-term satiety and hunger feelings may be regulated by several neural circuits at the ventromedial, dorsomedial and paraventricular hypothalamic nuclei for satiety sensations and at the lateral hypothalamus for hunger sensations. Although the hypothalamus is an important centre in the energy balance regulation, there are other brain regions such as the medulla oblongata and cortical and striatal structures, essential for the eating behaviour modulation. For example, some neural circuits of the medulla oblongata seem to have an important role in autonomic eating regulation, limiting the quantity of ingested food through the satiety responses regulation. Whereas, other parts of the brain, like the nucleus accumbens and ventral tegmental area, where dopamine, opioids and cannabinoids signals are integrated, regulated the motivation to eat, the rewards and the acts before eating. In this context, although hunger is connected to the biological needs, there are also psychological factors involved in the food intake regulation. Learning and emotions play a powerful role in determining what to eat, when to eat, and even how much to eat. In this context, the psychological desire of eating and its complicated mechanisms of influence in satiety interpretation difficult the design of functional foods for this purpose.

Although, there are many peripheral signals that can contribute to feeding behaviour and body weight regulation and can be modified by food and food ingredients. It is important to recognize that short-term and long-term food intake and energy balance are regulated through distinct, but interacting, mechanisms. Figure 3, shows a brief review of nowadays known possible satiety signals that may influence eating behaviour which included, beside short and long-term signals, individual social behaviour and other metabolism compounds that may influence satiety feeling.

Short-term regulation of food intake results from an integrated response from neural and humoral signals that originate mainly at the brain, gastrointestinal tract and adipose tissue. Ingested food evokes satiety in the gastrointestinal tract primarily by two distinct ways, i.e. by mechanical stimulation and therefore stimulation of the nerve endings; and by the release of satiety peptides. The scheme is more complicated as both ways seems to be intimately related, since many of the intestinal peptides released may inhibit also gastric emptying thus enhancing gastric mechanoreceptor stimulation.

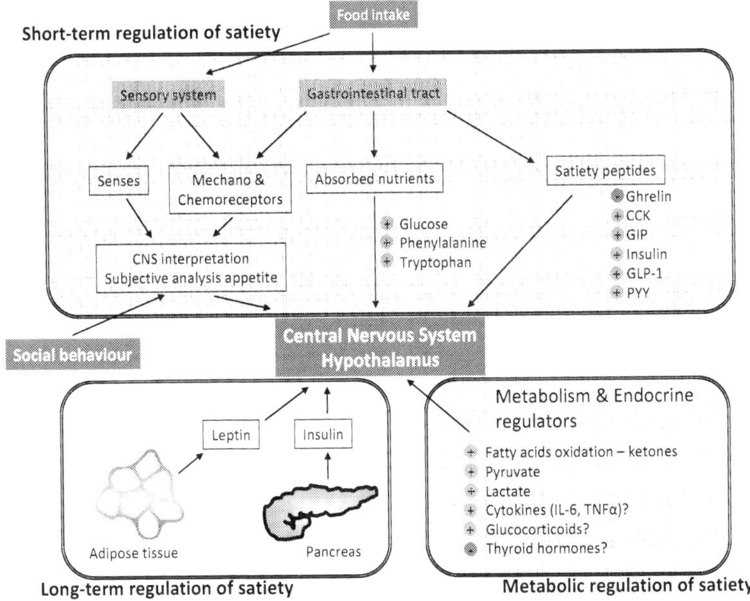

Fig. 3. Short- and long-term regulation of satiety. Principal signals implied in energy regulation and satiety.

The postprandial satiety consequences of food intake are determined both by the specific chemical composition and the characteristics physical properties of the food. Accordingly, different foods, despite their equal energy content, can differ in their capacity to affect postprandial metabolism, especially secretion of gastrointestinal peptides, thereby regulating energy homeostasis. A classical example is fibre were several differences in chemical structures and characteristic physical properties can be observed. For example, bulk/volume, viscosity, water-holding capacity, adsorption/binding, or fermentability may determine the subsequent physiological behaviour of fibre eventhough it is ingested in the same quantity. Table 1 and below is included a brief description of the principal satiety signals derived from the gastrointestinal tract and its possible effects in food intake behaviour.

Peptide	Organ of synthesis	Receptor related with satiety signals	Effects on food intake
CCK	Proximal intestine I-cells	CCK1R	Decrease
GLP-1	Distal intestine L-cells	GLP1R	Decrease
PYY$_{3-36}$	Distal intestine L-cells	Y2R	Decrease
PP	Pancreatic F cells	Y4R, Y5R	Decrease
Amylin	Pancreatic β-cells	CTRs, RAMPs	Decrease
Gastric leptin	Stomach P-cells	Leptin receptor	Decrease
Ghrelin	Gastric X/A-cells	Ghrelin receptor	Increase

Table 1. Gastrointestinal satiety peptides that may regulate food intake.

Ghrelin. Ghrelin is the only mammalian substance that has been shown to increase appetite and food intake when delivered to humans. Circulating ghrelin levels typically rise just before and fall shortly after a meal, thus playing a role in meal time hunger and meal initiation. The postprandial ghrelin response is affected principally by the caloric content of meals. Thus, high energy-rich meals suppress ghrelin more than lower ones. In humans, across the range of intakes of 220 to 1000 kcal, the lowest point of postprandial ghrelin was found to decrease by about 2.4% for every 100 kcal increase of energy intake (Callahan et al., 2004). Recent findings suggest that postprandial suppression of ghrelin is not mediated by nutrients in the stomach or duodenum but rather from post-ingestive increases in lower intestinal osmolarity (via enteric nervous signaling) as well as from insulin surges (Cummings, 2006). In contrast, the short-term parenteral administration of glucose and insulin in physiological doses may not suppress ghrelin levels (Gruendel et al., 2007). Moreover, ghrelin concentration is not affected by stomach distension, since the administration of water did not influence its concentrations (Shiiya et al., 2002). In relation to the possible modification of its plasmatic leveles, it is known that increased fibre content of the meal has shown both to decrease postprandial ghrelin concentrations as well as to inhibit the decrease.

Cholecystokinin (CCK). The inhibitory effect of CCK on food intake has been confirmed in numerous species, including humans. It is however short-lived, lasting less than 30 min. Therefore, CCK may inhibit food intake within the meal by reducing meal size and duration but does not affect the onset of a next meal. Thus it may have an important role in the causal chain leading to satiation or meal termination. Gastric distension augments the anoretic effects of CCK in humans. However, other mechanisms including activation of duodenal chemosensitive fibres and activation of CCK receptor 1 in the pyloric sphincter thay may slow down gastric emptying may be implicated.

Glucose-dependent insulinotropic polypeptide (GIP). GIP is released in response to the presence of nutrients in the intestinal lumen. The major stimuli for GIP release are dietary fat and carbohydrates. Protein seems to have no effect, although some evidence exists indicating that the intraduodenal administration of aminoacids can stimulate GIP release.

Glucagon-like peptide 1 (GLP-1). GLP-1 is an incretin hormone released in response to food intake. GLP-1 is typically very low in the fasting state, but rise quickly after food intake, especially after carbohydrate intake. The rise of GLP-1 has been correlated with increased satiety and less hunger. GLP-1 is thought to play an important role in the "ileal break", a mechanism that regulates the flow of nutrients from the stomach into the small intestine. It is also suggested that portal GLP-1 might influence the production of ghrelin (Lippl et al., 2004) and the increase in β-cell mass in the pancreas, thus improving the insulin production.

Peptide tyrosine-tyrosine (PYY). PYY is a member of the pancreatic polypeptide fold family including neuropeptide Y (NPY) and pancreatic polypeptide (PP). PYY mediates ileal and colonic breaks, mechanisms that ultimately slow gastric emptying and promote digestive activities to increase nutrient absorption. Plasma concentration of PYY increases after meals consistent with a meal-related signal of energy homeostasis. Nutrients stimulate PYY release within 30 minutes of ingestion, reaching usually a maximum within 60 min. The release is directly proportional to caloric intake; however meal composition may affect postprandial PYY release. In humans, infusion of PYY_{3-36} (active form) comparable to those after a meal result in decreased energy intake at subsequent meals compared with a control group (Batterham et al., 2002)

Amylin. Amylin is co-secreted together with insulin from pancreatic β-cells. It is a 37 amino acid peptide with anorexigenic effects that have shown to reduce meal size as well as the number of meals. The inhibitory effect of amylin on food intake is thought to be due to the inhibition of gastric emptying.

Pancreatic polypeptide (PP). PP is secreted by F-cells on the endocrine pancreas comprising approximately less than 5% of islet volume. The main function of PP is thought to be the inhibition of exocrine pancreas. Its secretion is controlled by the parasympathetic nervous system and shows a biphasic manner in proportion to food intake.

Leptin. Leptin is a peptide hormone which is released from white adipose tissue and acts in the hypothalamus to promote weight loss, both by reducing appetite and food intake and by increasing energy expenditure. Circulating leptin concentrations are highly positively correlated with body mass index. Despite this strong association, leptin levels shows large individual variation for a given degree of adiposity, indicating the likely effect of variables other than adipose mass, such as genetic and environmental factors. Additionally, food consumption stimulates leptin secretion after a meal and high carbohydrate meals results in greater leptin responses. Although leptin does not seem to play an important role in the short-term regulation of food intake, when subjects are in energy balance, plasma leptin is negatively correlated with appetite and food intake when energy balance is disturbed. Leptin therefore seems to have a role in the regulation of food intake when energy stores changes.

Insulin. Insulin is the major endocrine and metabolic polypeptide hormone secreted by β-cells of the endocrine pancreas and one of the key adiposity signals in the brain influencing energy homeostasis. Plasma insulin concentrations are in direct proportions to changes in adipose mass. Insulin concentrations are increased at positive energy balance and decreased at the times of negative energy balance. Additionally, plasma insulin concentrations are largely determined by peripheral insulin sensitivity which is related to the amount and distribution of body fat in insulin-resistant patients. Insulin thus provides information to the central nervous system about the size and distribution of the adipose mass to regulate metabolic homeostasis.

Besides the described satiety peptides and hormones, other absorbed food derived compounds, metabolites and hormones may also serve as satiety signals for the central nervous system. For example, aminoacids such as phenylalanine and tryptophan that are precursors to monoamine neurotransmitters suppress food intake in humans. The ratio of plasma tryptophan to other amino acids may influence brain serotonin levels, which are known to have inhibitory influence on food intake. Some authors have suggested that the oral administration of 5-hydroxytryptophan may reduce food intake, as well as a reduction in carbohydrate intake and a higher satiety feeling in obese subjects (Cangiano et al., 1992).

Other metabolisms by-products like ketones (from fatty acids), which are metabolic substrate for the central nervous system, are known to inhibit feeding. Moreover, lactate and pyruvate have been reported to induce satiety effects in animal models (Nagase et al., 1996). And finally, endocrine regulators of food intake like cytokines (IL-6 and TNFα), glucocorticoids and thyroid hormones have also been described to regulate satiety and hunger feelings. However the mechanisms of how they may influence feeding behaviour, possibly related with energy expenditure administration, are not well understood.

Food formulation for the induction of the release of satiety peptides or the decrease of ghrelin could be basically obtained by the incorporation of functional ingredients that may increase satiety peptides release in the gut, as well as by rheological modifications with the objective of producing sensory stimulation in the first phases of digestion and also by a higher gastric distension and emptying rate time. The possible strategies are described below.

2.1.1 Energy reduced satisfying foods, combined strategy to reduce energy intake

The combination of energy reduced foods with functional ingredients that stimulate satiety peptides release may be an interesting and ambitious approach. It should be taken in consideration that the release of satiety peptides is normally proportional to the energy intake and energy density of food. In this sense, it seems difficult to design energy reduced food with also the ability to stimulate satiety. However, some approximations to this goal can be developed.

There are some evidences that macronutrients may regulate the secretion of satiety peptides. For example, fats and proteins may increase the release of CCK and reduce ghrelin concentrations by its effect in gastric emptying delay. Other authors suggest also some effects in GLP-1, GIP and PYY, however the exact mechanisms of action is still unknown.

In the other hand, carbohydrates, especially viscous and fermentable fibre (with a reduce energy content), have been widely investigated because of its satiety effects in humans. Figure 4 briefly schematises the possible mechanisms of action of fibres in the induction of satiety.

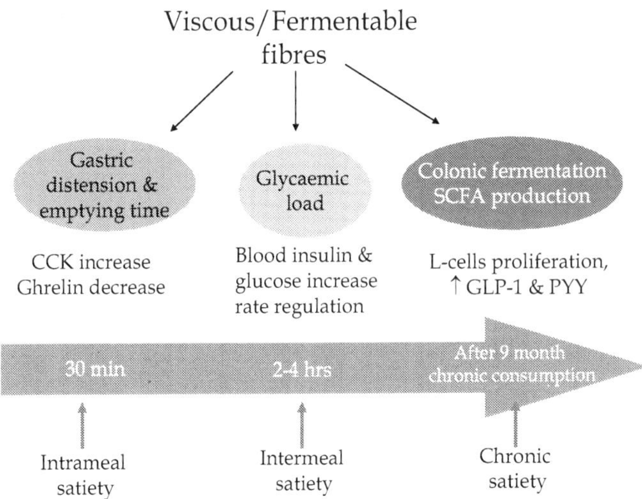

Fig. 4. Mechanisms of action of fibre in satiety induction. SCFA: short-chain fatty acids.

Viscous fibres have been related to satiety because of its effects in gastric distension and emptying rates which may increase CCK and reduce ghrelin secretion. Later, in the small intestine, viscous fibre may regulate bolus transit time, which may affect the total glycemic load, regulating glucose and insulin levels, hence inducing direct satiety stimulation in central nervous system satiety centre. And finally, fibre colonic fermentation by products like short-chain fatty acids are recognized to induce L-cells proliferation in the colon, which are the main cells where GLP-1 and PYY is produced. Thus, chronic fibre intake may chronically increase the amount of these satiety peptides in plasma. Notwithstanding, several authors suggest that the amount of fibre that may exert satiety feeling in humans should be higher than 8 g/day, which difficult its incorporation in technologically modified foods. Moreover, it should be taken into consideration that viscous fibre may produce acute satiety feeling (intrameal and intermeal satiety); whereas fermentable fibres may produce chronic satiety feeling, however after a chronic consumption higher than 9 months.

Other bioactive compounds like polyphenols have been described also to induce a higher rate of secretion of satiety peptides like GLP-1 possibly because of its effects in the inhibition of glucose uptake by enterocytes (McCarty, 2005). In this context, the addition of antioxidant dietary fibre (fibres with associated antioxidant compounds) to food formulation may help to increase satiety feeling of foods.

In brief, it can be said that glucose and fat are the main inductors of satiety peptides. However, it is widely accepted that the increase in glucose and fat uptake are strongly related with obesity. The next step should be the reduction of energy density of formulated foods.

Currently there are a variety of food ingredients that can be used with the purpose to replace fat and sugar in foods formulations (Table 2 & 3) with the intention to reduce the total calorie intake. For example, the average consumption of simple sugars in the Spanish diet by 2005 (MAPA, 2005) was around 113 g/person/day an estimated 18% of total calories (2424 kcal/person/day), exceeding by 8% the recommended dietary intake of simple sugars. Dairy products, beverages, cakes, pastries, etc.., target products for the replacement of simple sugars with non-nutritive sweeteners, contributed to 68% and 62% of the total intake of simple sugars and sucrose in the Spanish diet respectively (Figure 5). Its replacement with non-nutritive sweeteners could result in a reduction in energy intake around 310 kcal/day. However, the abuse in the intake of non-nutritive sweeteners is also criticized by the scientific community, since some of them are banned in other countries, such as the use of cyclamate by the Food and Drug Administration in the United States.

Generic name (commercial name)	Sweetener capacity (compare with sucrose)
Potassium acesulfame (Sunnett)	180-200
Alitame	2000
Aspartame (Equal, Nutrasweet)	180
Cyclamates	30-50
Glycyrrhizin	30-50
Lo han guo	30
Neotame	8000-13000
Perillartine	2000
Saccharine (Sweet'n low)	300
Stevioside	40-300
Sucralose (Splenda)	600

Table 2. Available non-nutritive sweeteners in the market.

Generic name	Commercial name	Caloric value
Based on proteins		
Whey protein	Simplesse, Dairy-Lo	1-2 kcal/g
Based on carbohydrates		
Cellulose	Avicel cellulose gel, Methocel, Solka-Floc	0 kcal/g
Dextrins	Amylum, N-Oil	4 kcal/g
Gums	Kelcogel, Keltrol	0 kcal/g
Inulin	Raftiline, Fruitafit, Fibruline	1-1.2 kcal/g
Maltodextrins	Crystal Lean, Lorelite, Lycadex, Maltrin	4 kcal/g
Polydextrose	Litesse, Sta-Lite	1 kcal/g
Polyols	Several brands	1.6-3.0 kcal/g
Fat analogous		
Olestra	Olean	0 kcal/g
Sorbestrin		1.5 kcal/g
Salatrim	Benefat	5 kcal/g
Emulsifiers	Dur-Lo	9 kcal/g

Table 3. Fat replaces available in the market.

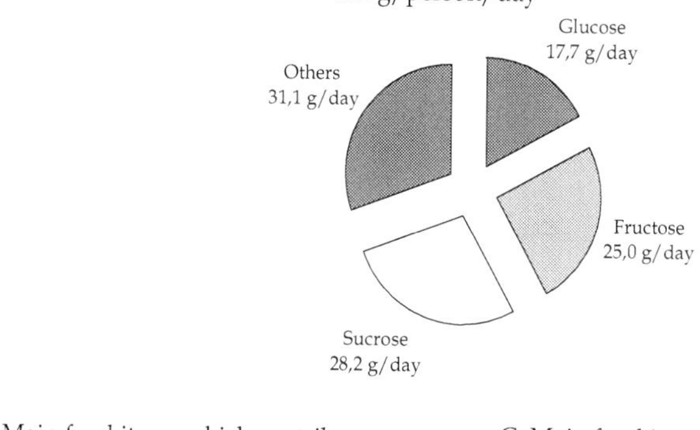

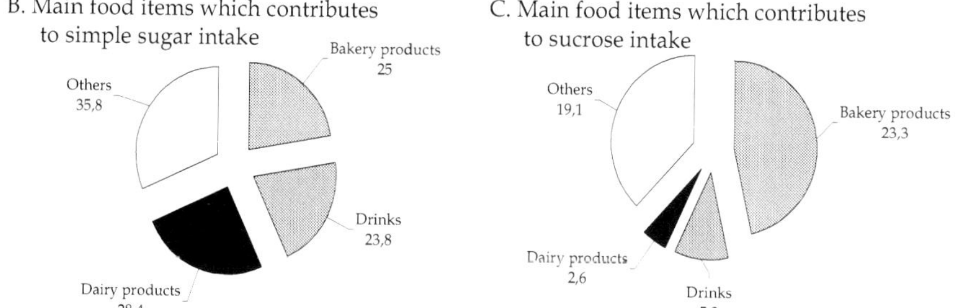

Fig. 5. Simple sugars intake and principal food items that contributes to its intake in the Spanish diet.

With respect to fat substitutes, several fat replacers have been developed to assimilate the sensory properties of fat. It can be divided basically into 3 groups; 1) based on proteins, used mainly in dairy products; 2) based on carbohydrates, used in dressings, meat products, etc. and 3) based on modified fat, that may contains organoleptic characteristics similar to fats but with a lower caloric content. Among them, Olestra approved in 1996 by the United States to replace fats and oils, is a sucrose polyester containing between 6 and 8 fatty acids per molecule, that has organoleptic properties similar to those of typical fat, but without the capacity of being hydrolyzed by lipases; and Salatrim, a mixture of long- (mainly stearic acid) and short-chain fatty acids (acetic, propionic and butyric) esterified with glycerol that greatly reduce the caloric intake per gram of fat. The main drawback of the use of fat-based substitutes is the possible decrease in the absorption of fat soluble vitamins, which must be taken into account in the formulation or modification of food that are sources of fat-soluble vitamins in the diet.

Naturally, foods with lower energy density and satiety properties, such as fruits and vegetables, compared to energy-dense foods, are low in fat and high in water and/or dietary fibre content, since these compounds may add weight and volume to foods without increasing its caloric content. Therefore, increasing the water content through wetting agents and/or increasing the content of dietary fibre could be a possible approach.

Possibly the best option is to reduce the energy density of foods by the incorporation of fibre (with an average energy value of 1-2 kcal/g) to the formulations, which also includes the possibility to increase the water content of food products. Physiologically, fibre may decrease the absorption of macronutrients like fat and carbohydrates reducing its energy value. On the other hand, some fibres obtained from fruits and vegetables may also provide functional antioxidant compounds valuables for obesity treatments.

It should be noted, however, that an increased intake of foods with low energy density is not enough to lose weight, except when they move to higher-energy density. The choice of food can often be influenced also by their quantity, volume or weight, in this sense changes in the appearance of food can be another simple strategy to develop other types of functional foods.

2.1.2 Rheological modifications of foods to induce satiety

The main objective in developing this type of functional foods is to induce satiety by modifying the perception of consumers toward food. That is, sensory (sight, smell and taste mainly) stimulations that able to production of an "apparent" fullness feeling in the early stages of feeding. For example, some studies suggest that the volume of food ingested psychologically affect hunger, satiety and the amount of food that individuals want to eat (Rolls et al., 2000). In this sense, an example could be the development of products with a high volume and low density in both weight and energy (for example by cereal extrusion or air emulsions desserts like "mousse").

Satiety may also be induced through the stimulation of retro-nasal aroma by food (Ruijschop & Burgering, 2007). There are some indications that not all types of foods produce the same quality (flavour) or quantity (intensity) of sensory stimulation. The physical structure largely seems to be responsible for stimulating aroma. For example, solids tend to produce a greater sense of satiety than liquid foods, probably because of the increased contact time of food in the oral cavity and therefore the greater sensory stimulation. It is also suggested the addition of encapsulated flavours that may extend specialized sensory flavour in the mouth in foods with a low energy density.

Rheological modifications for satiety peptide release are mainly based in the induction of a higher gastric distension and a reduction in gastric emptying which may promote the release of CCK and the decrease of ghrelin levels. Viscous fibres are the main nutrient associated with the ability to slow gastric emptying as discussed above.

2.1.3 Digestive enzyme inhibition and gut energy absorption inhibition

Another possible strategy is to limit the absorption of nutrients in the intestinal tract, by limiting the action of digestive enzymes and/or interacting with them to physically interrupt its absorption in the intestinal tract. The most common example is the use of fibre, which modulates the intestinal transit time resulting in increased satiety, reduced physical accessibility of nutrients for being absorbed and a reduced the calorie intake of the formulation.

An interesting example of functional fibre is chitosan (product obtained from chitin, located in the shells of shellfish), which is a positively charged polymer that could bind to negatively charged fat molecules in the intestinal lumen (mainly free fatty acids) therefore inhibiting its absorption. On the other hand there are other bioactive compounds with the ability to inhibit the activity of digestive enzymes, the most common digestive enzyme inhibitor found in foods are condensed tannins, which have the ability to precipitate proteins (including enzymes), reducing its action.

In connection with the absorption of carbohydrates, some researchers have shown that certain polyphenols (for example from tea) have the ability to inhibit *in vitro* the translocation of glucose transporter, GLUT2, in the intestinal epithelial cells, thus inhibiting the absorption of glucose (Kwon et al., 2007). This same effect has been demonstrated *in vivo* by oral sucrose tolerance test in the presence of epigallocatechin gallate, observing a decrease in blood glucose values (Serrano et al., 2009).

2.2 Thermogenesis and possible modifications in energy expenditure

Small differences in energy expenditure might have long-term effects on body weight. The human energy budget is usually divided into four major components (Figure 6), which, together, constitute the total energy expenditure: 1) basal metabolic rate; 2) diet-induced thermogenesis; 3) physical activity and 4) adaptative thermogenesis.

One of the suggested metabolic factors involved in the development of obesity is adaptive thermogenesis. It is defined as the regulated production of heat in response to environmental temperature or diet. It can be seen as a mechanism for dissipation, in a regulated manner, of the food energy as heat instead of its accumulation in fat. Skeletal muscle is potentially one of the largest contributors to adaptative thermogeneis in humans. For example, an adrenaline infusion may stimulate muscles to consume over a 90% more oxygen, thus increasing energy expenditure.

The possible mechanism of action of the increase in heat production may be due to the higher stimulation/synthesis of uncoupling proteins at the mitochondria. The uncoupling protein can dissipate a proportion of metabolic energy through uncoupled metabolism that creates heat rather than the generation of ATP. Consequently, these have been the topic of extensive research as possible targets to fight against obesity. The favoured route has been to develop specific β3-adrenergic compounds that might stimulate UCP1 without the undesirable β1 and β2 effects on heart rate and blood pressure.

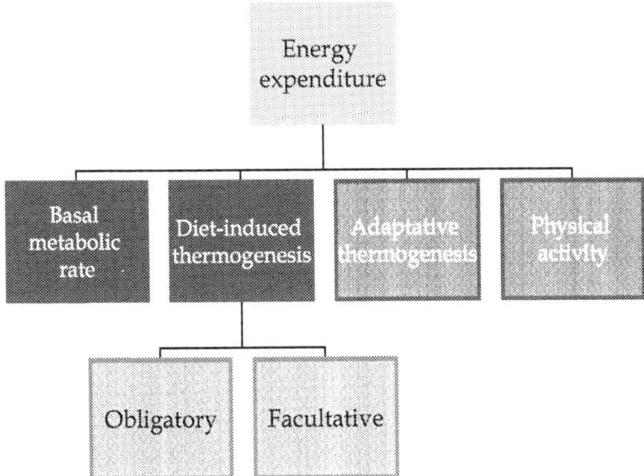

Fig. 6. Principal components of energy expenditure in humans.

In relation to nutrition, it has been shown that carbohydrates induced an increased adrenaline concentration, resulting in increased muscle thermogenesis (Astrup et al., 1986). Other authors have reported a variety of compounds that can alter energy expenditure. For example, in rodents, acute treatment with retinoic acid increases the thermogenic capacity in brown adipose tissue, which further in time may induce a significant decrease in body weight and adiposity (Bonet et al., 2000). Green tea extracts have been also described to stimulate thermogenesis in brown adipose tissue, mainly due to the interaction between its high content of catechins and caffeine that may stimulate the noradrenaline released by the sympathetic nervous system. Overall, the action of caffeine and catechins may prolong the stimulatory effects of noradrenaline on energy metabolism and lipid.

In another context, it is interesting to observe that after overfeeding, the same amount of excess energy intake and nutrients does not always invoke the same body weight gain in all people because of differences in diet-induced thermogenesis (energy expenditure in response to food intake).

Diet-induced thermogenesis can be divided in two categories: obligatory and facultative thermogenesis. The obligatory part consists of all process related to the digestion, absorption and processing of food. Stimulation of adenosine triphosphate (ATP) hydrolysis during intestinal absorption, initial metabolic steps and nutrient storage, are responsible for this food thermic effect. For example, measured thermic effects of nutrient digestion are 0-3% for fat, 5-10% for carbohydrates and 20-30% for proteins. While, the facultative component enables wasting of energy after a high caloric meal and prevents the storage of energy.

In relation to facultative thermogenesis, it has been observed that the thermic effect of food is reduced in obese and insulin-resistant patients, possibly because of the effect of the autonomic nervous system activation by certain nutrients and the insulin secretion stimulation. Insulin resistances may induce a decrease turnover of ATP by muscle tissue, therefore reducing its metabolic rate. It has also been suggested that insulin, via unidentified receptors, most probably located in the central nervous system, may stimulate muscle sympathetic nerve activity and facultative thermogenesis, and therefore its probably reduce in obese insulin resistance patients. Functional foods against insulin resistance may also help to reduce obesity by increasing insulin sensibility thus optimizing the use of energy. This aspect is described deeply in the next section.

And finally, protein turnover defined as degradation of proteins into amino acids and resynthesis of new proteins could be responsible for large part of the energy expenditure, around 15-20% of basal metabolic rate. Most tissues exhibit protein turnover, specifically the skeletal muscle tissue, liver, skin and small intestine. In humans, it has been shown that after carbohydrate overfeeding, protein turnover is increased by 12%, may be because of the reduction in protein intake.

2.3 Insulin metabolism and the regulation of adiposity

The causal links between obesity and insulin resistance are complex and controversial. Weight gain from overfeeding induces insulin resistance, whereas weight loss by caloric restriction reverse insulin resistance. The main concern should be to reduce the insulin secretion, associated to an increase in fat deposition in the adipocyte; as well as to increase its sensibility which is associated to an increase in metabolic rate.

For food formulation it is important to know that increased plasma glucose levels increase the secretion of insulin from pancreatic β-cells, although other substrates such as free fatty acids, ketone bodies, and certain amino acids can also directly stimulate insulin's release or augment glucose's ability to trigger insulin release.

It has been suggested that a low-glycemic diet can help control obesity because of the ability to reduce the rate of glucose uptake in the small intestine and therefore the insulin secretion, as well as to increase satiety value of food and appetite regulation (Brand-Miller et al., 2002; Roberts, 2003). While, high glycemic index foods, by its higher stimulation in insulin secretion, may promote a higher postprandial oxidation of glucose at the expense of fat oxidation, which can lead to increased body weight gain.

The glycemic index of a food is directly proportional to the degree of intestinal absorption of carbohydrates, in this sense the incorporation of factors that replace simple carbohydrates or decrease the absorption of carbohydrates such as fibre, fat or by inhibiting the action of digestive enzymes can reduce the glycemic index of food.

However, not all strategies to reduce the glycemic index of foods are widely accepted by the scientific community. For example, replacing simple sugars like sucrose with fructose, as a sweetener with low glycemic index, has been associated with increased body adiposity (Bray et al., 2004). After the ingestion of fructose, insulin is not increased, leptin is reduced, and ghrelin is not inhibited. Because these hormones play important roles in regulating food intake, the combined effects of excessive fructose intake could result in a lower induction of satiety and increase in total intake. Moreover, these foods should be low in saturated fatty acids, especially *trans* fatty acids, which are associated with a higher induction in adiposity. In this context, the addition of conjugated linoleic acid appears to have an effect based on the inhibition of the activity of lipoprotein lipase, which may reduce the uptake of lipids by the adipocyte.

3. Functional diets based on traditional foods

The use of appropriate combinations of nutrients that affect different processes of energy intake and expenditure could be the best strategy to tackle obesity control, usually a multi-causal problem. A higher effectiveness can be obtained from the combination of functional ingredients that inhibit the appetite sensations, the bioavailability of macronutrients and induce a thermogenic response to an individual. An interesting approach could be the formulation of "Ready Meal" type product.

A ready meal is a type of convenience food that consists of a pre-packaged meal that needs little preparation. In 2009 the global market for ready meals was worth $71.6bn. In the Western Europe and the US the ready meals markets are fairly mature and well-established with moderate growth rates. The only pre-requisite for such diets is that they should be nutritionally appropriate and balanced with a low caloric content that allows consumers to replace any meal time with some of these diets on the market.

4. Conclusions

To create and consume a satisfying diets and foods with high intrameal and intermeal satiety feeling that may help to fight against obesity, ideally should consist of low-energy dense foods with high palatability; however, such foods do not commonly exist. Moreover, individual genetic variation may influence the effectiveness of such preparations against obesity.

Notwithstanding, the diversity of strategies to combat the obesity problem makes possible to develop a variety of food products that may satisfy individual requirements. However, before its market launch, the metabolic effects should be widely contrasted as well as it should be defined the concrete profile of the people who can benefit from the consumption of the formulated functional food.

5. References

Astrup, A., Bulow, J., Christensen, N.J., Madsen, J. & Quaade, F. (1986). Facultative thermogenesis induced by carbohydrate: a skeletal muscle component mediated by epinephrine. American Journal of Physiology, Vol. 250, pp. E226-E229.

Batterham, R.L., Cowley, M.A., Small, C.J., Herzog, H., Cohen, M.A. & Deakin, C.L. (2002). Gut hormone PYY(3-36) physiologically inhibits food intake. Nature, Vol. 418, pp. 415-421.

Bonet, M.L., Oliver, J., Pico, C., Felipe, F., Ribot, J., Cinit, S. & Palou, A. (2000). Opposite effects of feeding a vitamin A-deficient diet and retinoic acid treatment on brown adipose tissue uncoupling protein 1 (UCP-1), UCP2 and leptin expression. Journal of Endocrinology, Vol. 166, pp. 511-517.

Brand-Miller, J.C., Holt, S.H., Pawlak, D.B. & McMillan J. (2002). Glycemic index and obesity. American Journal of Clinical Nutrition, Vol. 76, pp. 281S-285S.

Bray, G.A., Nielsen, S.J. & Popkin, B.M. (2004) Consumption of high fructose corn syrup in beverages may play a role in the epidemic of obesity. American Journal of Clinical Nutrition, Vol. 79, pp. 537-543.

Callahan, H.S., Cummings, D.E., Pepe, M.S., Breen, P.A., Matthys, C.C. & Weigle, D.S. (2004). Postprandial suppression of plasma ghrelin level is proportional to ingested caloric load but does not predict intermeal interval in humans. Journal of Clinical Endocrinology & Metabolism, Vol. 89, pp. 1319-1324.

Cangiano, C., Ceci, F., Cascino, A., Del Ben, M. & Laviano, A. (1992). Eating behaviour and adherence of dietary prescriptions in obese adults subjects with 5-hydroxytryptophan. American Journal of Clinical Nutrition, Vol. 56, pp. 863-867.

Cummings, D.E. (2006). Ghrelin and the short- and long-term regulation of appetite and body weight. Physiology & Behaviour, Vol. 89, pp. 71-84.

Gruendel, S., Otto, B., Garcia, A.L., Wagner, K., Mueller, C., Weickert, M.O., Heldwein, W. & Koebnick, C. (2007). Carob pulp preparation rich in insoluble dietary fibre and polyphenols increases plasma glucose and serum insulin response in combination with a glucose load in humans. British Journal of Nutrition, Vol. 98, pp. 101-105.

Kennedy, G.C. (1953). The role of depot fat in the hypothalamic control of food intake in the rat. Proceedings of the Royal Society of London B, Biological Sciences, Vol. 140, pp. 578-596.

Kwon, O., Eck, P., Chen, S., Corpe, C.P., Lee, J.H., Kruhlak, M. & Levine, M. (2007). Inhibition of the intestinal glucose transporter GLUT 2 by flavonoids. FASEB Journal, Vol. 21, pp. 366-377.

Lippl, F., Kircher, F., Erdmann, J., Allescher, H.D., Schusdziarra V. (2004). Effect of GIP, GLP-1, insulin and gastrin on ghrelin release in the isolated rat stomach. Regulatory Peptides, Vol. 119, pp. 93-98.

MAPA Ministerio de Agricultura Pesca y Alimentación. (2006). La Alimentación en España. Secretaría General de Agricultura y Alimentación, Madrid, Spain.

Mayer, J. & Thomas, D.W. (1967). Regulation of food intake and obesity. Science, Vol. 156, pp. 328.

McCarty, M.F. (2005). A chlorogenic acid-induced increase in GLP-1 production may mediate the impact of heavy coffee consumption on diabetes risk. Medical Hypotheses, Vol. 64, pp. 848-853.

Nagase, H., Bray, G.A. & York, D.A. (1996). Effects of pyruvate and lactate on food intake in rat strains sensitive and resistant to dietary obesity. Physiology & Behaviour, Vol. 59, pp. 555-560.

Roberts, S.B. (2003). Glycemic index and satiety. Nutrition in Clinical Care, Vol. 6, pp. 20-26.

Rolls, B.J., Bell, E. A. & Waugh, B.A. (2000). Increasing the volume of a food by incorporating air affects satiety in men. American Journal of Clinical Nutrition, Vol. 72, pp. 361-368.

Ruijschop, R.M. & Burgering, M.J.M. (2007). Aroma induced satiation possibilities to manage weight through aroma in food products. Agro Food Industry Hi-Tech, Vol. 18, pp. 37-39.

Serrano, J., Boada, J., Gonzalo, H., Bellmunt, M.J., Delgado, M.A., Espinel, A.E., Pamplona, R. & Portero-Otin, M. (2009). The mechanism of action of polyhenols supplementation in oxidative stress may not be related to their antioxidant properties. Acta Physiologica, Vol. 195, pp. 61.

Shiiya, T., Nakazato, M., Mizuta, M., Date, Y., Mondal, M.S., Tanaka, M., Nozoe, S., Hosoda, H., Kangawa, K. & Matsukura, S. (2002). Plasma ghrelin levels in lean and obese humans and the effect of glucose on ghrelin secretion. Journal of Clinical Endocrinology Metabolism, Vol. 87, pp. 240-244.

15

Improving Nutrition Through the Design of Food Matrices

Rommy N. Zúñiga[1] and Elizabeth Troncoso[2]
[1]Center for Research and Development CIEN Austral, Puerto Montt
[2]Pontificia Universidad Católica de Chile, Santiago,
Chile

1. Introduction

Increasing epidemiological evidence has linked the prevalence of diseases, such as obesity, cardiovascular disease, hypertension, type II diabetes mellitus, and even cancer, to dietary factors. Besides, overweight and obesity are major risk factors to the prevalence of the mentioned diet-related diseases. The obesity epidemic around the world has been attributed to energy imbalance, mainly because of increased food consumption and/or sedentary lifestyle. Changes in eating behavior and the massive amount of high-calorie foods readily available has allowed obesity to reach epidemic proportions around the world. The logical strategy to attack the obesity problem is lowering the total energy intake along with a reduction in fat and sugar/digestible carbohydrate intake, which can have a substantial impact on body weight (American Dietetic Association [ADA], 2005). In past years, the food industry developed "light" products by diminishing or replacing the amount of fat and/or sugar from high-calorie products. However, the replacement of fat and sugar decreases the palatability of foods and for consumers foods must be simultaneously safe, healthy, delicious and convenient (German & Watzke, 2004). Hence, the challenge for the food industry is much more complex than simply providing healthy foods; to provide healthy and delicious foods is the real challenge.

The control of digestion and the release of nutrients from the food matrix is an alternative approach to attack the obesity problem and other diet-related diseases. The concept of a food matrix points to the fact that nutrients are contained into a larger continuous medium that may be of cellular origin (*i.e.*, fruits and vegetables) or a structure produced by processing, where nutrients interact at different length scales with the components and structures of the medium (Aguilera & Stanley, 1999). Research on human digestion has often been undertaken with a view to changing the rates of digestion and delivery sites of macronutrients that might affect satiety and thus caloric intake. Several aspects of human eating behavior and food digestion may be relevant for identifying effective measures to treat or prevent diet-related diseases. Although most of people in developed and developing countries eat an unbalanced diet, an increasing part of the consumers are progressively more aware of the relationship between diet and health. Thus, the demand for functional food products that address specific health benefits is growing steadily (Palzer, 2009). The food industry is currently responding by reformulating its products, especially looking at salt, sugar and fat content with a particular emphasis on healthier fat compositions (Lundin et al., 2008). New designed foods with lower amounts of fat, controlled release of bioactives, in-body self-assembly structures or slowly digestible starches are already being developed by the food industry. The structure of these products must be modified accordingly to equalize the physical (*e.g.*, rheology) and sensory (*e.g.*, taste and release of aromas) properties of the original food. Therefore, designed food structures are required to diminish the sensorial impacts of such product modifications.

The food industry has assembled considerable information about composition, biochemistry, structure, and physical properties of foods. The principles of process engineering of biomaterials and the fundamental role of food structure in understanding the behavior of foods during processing are well established (Aguilera, 2005; German & Watzke, 2004). This knowledge is the basis for the food industry to formulate, process, preserve and distribute foods. Although the total amount of a nutrient in natural and formulated foods may be obtained from composition tables, its bioavailability (*i.e.*, the rate and extent to which a nutrient contained in a food is absorbed and become available at the site of action) depend on many factors, for instance, food microstructure, processing conditions, presence of other components, among others. However, there is still a lack of information about the performance of foods inside our body, which has limited the capacity of the industry to create products with tailored nutritional properties. Therefore, this chapter is an attempt to relate how the changes induced in food matrices affect their physicochemical properties and macronutrient (protein, fat, and carbohydrate) bioavailability, improving the nutritional performance of tailored foods. This chapter initially reviews the current situation about consumer trends and health. Next, the structures of the main macronutrients are briefly revised and the steps of the digestion process are explained. Finally, examples of structured foods to improve nutrition are presented and discussed.

2. Why improving the nutritional performance of foods? Past and present situation

Health and wellbeing are the major drivers for the food industry today. Scientific evidence that the quantity, composition and microstructure of the food ingested affects health is growing steadily (Norton et al., 2007; Parada & Aguilera, 2007). Although food digestibility and nutrient bioavailability had not been taken into account in the food design until now; a better understanding of the relationship between food properties, digestion and absorption would help in the rational design of foods with enhanced nutritional properties. This section deals with modifications in consumer perception about foods and the increasing need for foods for health and wellbeing.

2.1 Changes in human eating behavior

As a consequence of changes in lifestyle, diet is increasingly affecting health. A fast-paced lifestyle has left less time to cook, thus the consumption of fast food meals (mostly cheap foods available in large quantities) is augmenting constantly. In developed and developing countries families are eating out of home more often, portion sizes of foods consumed are getting larger at the same time body weights of people continue to increase. Taking as example the United States, the incidence of obesity increased continuously over the past decades (Fig. 1). Over-consumption of high–energy-dense (kcal/g) foods and beverages, and the increased portion sizes, may contribute to positive energy balance and lead to increasing incidence and prevalence of overweight and obesity. High-fat and sugar foods are problematic for the regulation of energy intake because they are high in energy density and very palatable. In addition, a decrease in physical activity due to the increasingly sedentary nature of many forms of work, changing modes of transportation, and increasing urbanization has contributed to the energy imbalance.

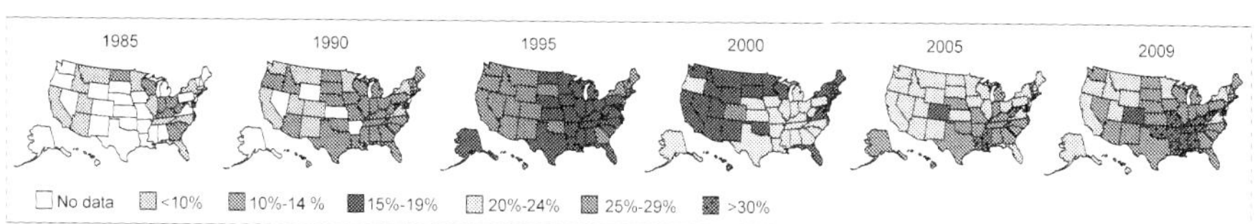

Fig. 1. Evolution of percentage of United States population by state, with a body mass index above 30, regarded as obese. Source: Centers for Disease Control and Prevention.

Nowadays, eating is driven by necessity and by pleasure. From the beginning of the humanity foods had been consumed to obtain energy and essential nutrients for living, but in modern societies foods are also consumed to have the pleasuring feeling of food flavors and textures into the mouth. Therefore, we eat because we have to and because we like to. As societies increase their purchase level, they begin to recognize the pleasurable aspects of eating, as comfort or reward and to satisfy, delight or stimulate the senses. The modern food industry has recognized this opportunity and much effort has been investing in the development of foods with enhanced sensory properties, transforming raw materials in palatable structures to gain acceptability from the consumer. Currently, we are surrounded by a huge variety of tempting high-calorie foods and cutting out or drastically restricting this kind of foods is simply not sustainable, probably because tastier foods are more rewarding and initiates some kind of seeking behavior.

The worldwide increase in the above mentioned diet-related problems are likely to change eating habits, processing technologies, and products. Hence, the food chain is facing a major challenge and now it is the consumer who indicate to producers what they want to eat. This global tendency is reshaping the industry into one that provides, in addition to safe and high-quality foods, products that contribute to the health and wellness of consumers. On this regard, the major food companies have understood that health and wellbeing are the major drivers for the current food industry, which is well represented in their logos, "*Good food, good life*" from Nestlé, "*Feel good, look good and get more out life*" from Unilever and "*The Beverage Institute for health and wellness*" from The Coca-Cola Company. The above are some examples of the messages that these companies want to give to the current health-conscious consumer.

2.2 Food-related diseases focused on obesity

Obesity, a preventable disease, with its co-morbidities such as the metabolic syndrome and cardiovascular diseases, are the major medical problems of the last decades. The health consequences and compromised quality of life associated with obesity provide major incentives to reduce the continuing obesity epidemic. The large numbers of children entering adulthood overweight together with weight gain in adulthood, produce an enormous burden in terms of human suffering, lost productivity and health care expenditures. The solution to overweight and obesity problem is body weight maintenance after body weight loss. This seems simple, but the required conditions are difficult to achieve for many individuals. Fortunately, the evidence has shown that many of the health risks associated with obesity can be reversed with weight loss.

Once considered a problem only in developed countries, obesity is now dramatically increasing in developing countries. According to the World Health Organization (WHO, 2011), some important key facts about obesity are:

- Worldwide obesity has more than doubled since 1980.
- 1.5 billion adults were overweight in 2008.
- 43 million children under the age of five were overweight in 2010.
- 2.8 million adults die each year as a result of being overweight or obese.
- 44% of the diabetes burden, 23% of the ischemic heart disease burden and between 7% and 41% of certain cancer burdens are attributable to overweight and obesity.
- 2.3 billion adults will be overweight and more than 700 million obese by 2015.

According to Wansink (2007), over 85% of the population with weight problems has consumed an average excess of only 25 kcal/day over a prolonged period of time. A sustained reduction in daily calorie consumption could prevent or reduce this long term weight gain in a large proportion of the population, and thus to reduce the incidence of obesity and associated health problems. Fatty foods have high-energy density and palatability but exert a relatively weak effect on satiation (compared calorie per calorie with protein and carbohydrate foods), which may encourage calorie over-consumption (Marciani et al., 2007). Lowering energy density of foods can decrease energy intake independent of macronutrient content and palatability.

The human mechanism of appetite regulation is highly complex involving neurophysiological interactions between the gut and the brain. The stomach is able to signal by distension of its wall how full it is and, therefore, how much more should be eaten, and duodenal receptors sensitive to the nutrient content of the chyme (*i.e.*, semifluid mass of partly digested food expelled by the stomach into the duodenum) also signal satiety through the secretion of gut hormones. Besides, neurological pathways including the hypothalamus, where two major neuronal populations stimulate or inhibit food intake, and the brain steam are involved (Marciani et al., 2001a; Suzuki et al., 2010). Understanding the neurophysiological mechanism of appetite regulation could help to food technologists in developing more satiating foods.

The energy density of foods has been demonstrated to have a robust and substantial effect on both satiety and satiation. Satiety refers to the effects of a food or meal after eating has ended, whereas satiation (sensation of fullness) refers to the process involved in the termination of a meal (ADA, 2005). Satiation has been found to be independent of the administered macronutrient (fat, protein or carbohydrate) for isocaloric liquid foods, but it was linearly related to the meal volumes, suggesting that stomach distension is a key factor in the sensation of fullness (Goetze et al., 2007).

Enhancing the satiety level of foods while keeping a low energy density may restrict the daily food intake and the desire of overeating, which it can be a strategy for preventing over-consumption and energy imbalances. Reduced-energy foods should preserve the sensory properties of the original foods to play a potential role in helping against obesity. Designed foods with tailored mechanical properties of the material, low caloric density and designed flavor properties may help in developing new foods for a sustainable reduction in energy intake, thus helping us of fighting against obesity.

2.3 Evidence relating the impact of food structure on nutrition

Physicochemical and sensory properties of manufactured foods depend largely on the food matrix structure. Currently, there is an emerging interest in the impact of food structure on digestion behavior and its relationship to human nutrition (Lundin et al., 2008). New interest has arisen regarding the function that food structure may play once foods are inside the body and, consequently, in our nutrition, health and wellness. Attention is further supported by the increased belief that foods and not nutrients are the fundamental unit of nutrition (Jacobs & Tapsell, 2007). The last assumption is based on recent scientific data demonstrating that in the case of certain nutrients the state of the food matrix of natural or processed foods may favor or hinder their nutritional response *in vivo*.

In recent years, there has been an upsurge in efforts to understand how food structure influences the rates of macronutrients digestion. This research is being undertaken with a view to developing novel foods that regulate calorie intake, provide increased satiety responses, provide controlled lipid digestion and/or deliver bioactive molecules (Singh & Sarkar, 2011). Foods are consumed to maintain human biological processes and the food matrix influences these processes (Jacobs & Tapsell, 2007). It has been shown that disruption of the natural matrix may influence the release, transformation and subsequent absorption of certain nutrients in the digestive tract (Parada & Aguilera, 2007).

Food processing modifies physical and chemical properties of food and thus may influence the release and uptake of nutrients from the food matrix. In this complex scenario, food scientists have proposed to develop novel foods to control the impact of physical properties and food microstructure on the digestion behavior and its relationship to human nutrition, because in many cases the interactions between individual macronutrients control the rate of digestive processes, conditioning the absorption of nutrients. Thus, the release of nutrients from the food matrix as well as the interactions between food components and restructuring phenomena during transit in the digestive system becomes far more important than the original contents of nutrients (Troncoso & Aguilera, 2009).

2.4 Modifying the food matrix to minimize or maximize nutrients bioavailability

According to Aguilera (2005) *"the creation of new products to satisfy expanding consumer's demands during this century will be based largely on interventions at the microscopic level"*. Thus, to make this next generation of designed foods, a combination of understanding of material chemistry and material science is needed, together with an understanding of how the processing affects food structure, chemistry and attractiveness (Norton et al., 2006).

The food industry is extremely Innovative in terms of new products, but is highly traditional in term of processes. For a particular product type, a limited range of unit operations have been employed for some considerable time and the same process lines are used to make a range of different product structures. The current challenge is to develop novel functional structures through innovative processing or new units operations. It is conceivable that enhanced nutritional properties of foods may be achieved by proper assembly of hierarchical structures from the microscopic level up to the macroscale (*i.e.*, bottom up approach). These new fabrication techniques will require understanding and precise control of assembly processes at all scales.

Nutrient bioavailability is gaining considerable attention in food technology. While one of the ongoing concerns of the food industry has always been to produce and provide the consumer with safe foods, the nutritional and caloric composition is now becoming equally important. For this reason, during the last years considerable research has been conducted to modify food matrices in basis on their physicochemical properties and effects on food digestion. For instance, the direct effect of physical properties (*e.g.*, microstructure, particle size and physical state) of foods has been evaluated by its influence on nutrient absorption. So, different intentional modifications could be induced by food technologists to design and fabricate foods with controlled bioavailability.

Foods can be viewed as delivery systems of macro and micronutrients to improve nutrition. Delivery systems to encapsulate, protect and deliver bioactive components are widely used by the pharmaceutical industry to carry these active agents to specific locations within the gastrointestinal tract (GIT) and release them at controlled rates. Using this knowledge the food industry, through the design of food matrices, is developing similar systems to encapsulate, protect and deliver food components. However, the effectiveness of the encapsulation process relies on the preservation of the bioavailability of the encapsulated component and the release of it in the correct portion within the GIT. As mentioned before, little it is known about the influence of food structure and breakdown on nutrient release in the GIT. This point is the primary importance because only what is released can be bioavailable for absorption (Parada & Aguilera, 2007).

Structuring matrices for nutrient delivery is a subject of enormous interest and several structuring food biopolymers are under study. To develop structured foods and develop a strategy for controlled release of food nutrients at desired sites in the GIT, it is essential to understand the kinetics of food disintegration and predict its digestion and subsequent metabolism. Biochemical, physiological, and physicochemical parameters that influence these processes need to be understood. This knowledge will benefit the food-processing industry in developing proper food structures for health purposes. The possibility of predicting the release of nutrients from food matrices under simulated GIT conditions is the upmost relevance in order to be able to define relationships between food matrix-nutrient, as well as for looking at the interactions of ingredients with the enzymes involved in the digestive process. Although *in vivo* methods provide direct data of bioavailability, ethical restrictions and complex protocols limit this type of studies when humans are used in biological research (Parada & Aguilera, 2007). Therefore, the need for validated *in vitro* methods is urgent in order to evaluate and compare the effect of the microstructure over the amount and the rate of nutrients release in the GIT.

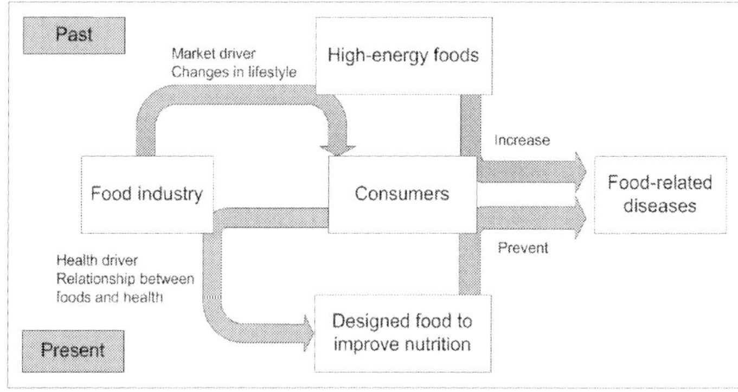

Fig. 2. Past and present scenarios of the relationship between consumers and food industry.

Summarizing, the terms "quality" and "health" are two main drivers of the modern food industry and the food microstructure influences both. Figure 2 gathers the principal subjects discussed in this section. To accomplish the ultimate goal of designing healthy foods with enhanced quality, knowledge of the overall food structure including the main building blocks of foods (proteins, carbohydrates and lipids) and a vast comprehension of the human digestion process are urgently needed to properly address the design of the future foods, which it will be discussed in the next sections.

3. Understanding the main food components and their structures

Foods consist of a complex group of components that differ in chemical composition and physical structure. Foods are largely composed by polymers (*i.e.*, proteins, carbohydrates, and lipids), micronutrients and water. Altogether, these components are arranged in different food matrices as the result of the multiple interactions that polymers can display under different conditions in an aqueous medium, as well as of the abundance of each respective substance. Proteins, carbohydrates and lipids are simple macromolecules made up of even simpler repeating units that play an important role in human nutrition. Each of these polymers has well-defined structures in the foods that contain them. This section discusses about the structure of proteins, carbohydrates, and lipids in foods and its impact on nutrient bioavailability.

3.1 Proteins

Protein is the most effective macronutrient found in foods providing satiety, on an energy equivalent basis, and protein hydrolysates comprising short chain peptides have been shown to induce release of satiety hormones, such as cholecystokinin (CCK), as part of this process (Lundin et al., 2008; Mackie & Macierzanka, 2010).

Proteins are certainly the most complex macromolecule encountered in foods. This complexity is evident in their secondary structure, which is intrinsically related to the sequence of amino acids (AAs) along the backbone. It is commonly recognized that 20 AAs form the building blocks of most proteins, being linked by peptide (amide) bonds. As a consequence, various proteins with different AA sequences will also differ in structure, modifying their physical properties.

Four levels of hierarchical organization are used to describe protein structure. The primary structure of a protein refers to the order of AA sequence in the polypeptide chain. The secondary structure describes the regular local conformations of the polypeptide backbone. The description of the overall three-dimensional folding of the protein is referred to as its tertiary structure. It involves the folding pattern of the polypeptide backbone including the secondary structures, arrangements of motifs into domains, and conformations of the side chains. Ultimately, the quaternary structure refers to the fourth-dimensional level of structure of protein complexes that may arise from the association of identical or heterogeneous polypeptide chains (Li-Chan, 2004). Changes in the secondary, tertiary, and quaternary structures without cleavage of backbone peptide bonds constitute "denaturation". In practical terms, denaturation means the unfolding of the native structure (Aguilera & Stanley, 1999). Protein denaturation usually has a negative connotation, since it is associated with loss of functionality in foods. However, it is often a prerequisite for improved digestibility, biological availability, and improved performance (*e.g.*, foam formation, emulsification or gelation). Denaturation not only affects the physical state of proteins but also their susceptibility to pepsin and trypsin/chymotrypsin mixture during digestion (Troncoso & Aguilera, 2009).

The most important factor in proteolysis is the accessibility of the enzyme to substrate and this involves both exposure of the cleavage site and local flexibility of the molecule and its side chains (Mackie & Macierzanka, 2010). In addition, hydrolysis of food proteins by digestive enzymes generates bioactive peptides that may exert a number of physiological effects *in vivo*, for example on the gastrointestinal, cardiovascular, endocrine, immune and nervous systems. On the other hand, some proteolysis-resistant proteins may produce allergenic reactions. Different factors regulate the allergenic potential (*i.e.*, defined as immunoglobulin E-mediated hypersensitivity reactions) of protein foods, such as type and composition of AAs, physicochemical characteristics of proteins, and relative abundance of the allergen in the food.

It is known that β-lactoglobulin (β-lg), a recognized allergenic protein, is hardly hydrolyzed by enzymes such as pepsin and chymotrypsin, but not by trypsin. The relative resistance to proteolysis is generally explained by the compact globular tertiary structure of the protein at low pH (<pH 3), which protects most of the peptide bonds susceptible to enzyme action. Physical and/or chemical denaturation of β-lg generally leads to a higher rate of hydrolysis by gastrointestinal enzymes. Stănciuc et al. (2008) showed that thermal treatment caused partial unfolding of β-lg and consequently an increased rate of hydrolysis by both trypsin and chymotrypsin at 37 °C. This phenomenon was attributed to the accessibility of the specific peptide bonds to the enzymes being enhanced. After heating at 78°C, a decrease in hydrolysis after prolonged heating time was thought to be due to aggregation phenomena, which could have hidden some susceptible bonds. High pressure treatment (HP) of β-lg between 400 and 800 MPa promoted digestion by pepsin, because HP treatment caused significant unfolding of the protein molecule (Chicón et al., 2008; Zeece et al., 2008).

Aggregation (cross-linking) of proteins results in the formation of high-molecular-weight polymers. These modifications are expected to affect the kinetics of protein digestion due to the reduced availability of reactive sites to proteolysis. Depending on the pH, temperature and presence of ions, proteins can be transformed into fibrils, spherical micro-gels, micro-particles, fractal aggregates and gels (Schmitt et al., 2007). Such controlled aggregation allows the management of molecular interactions to achieve the desired protein structures. Enzymatically cross-linked β-casein decreased their digestibility in comparison with native β-casein (Monogioudi et al., 2011). Heat treatment of milk (sterilization) increases protein resistance to the *in vitro* digestion in comparison with unheated milk, probably because of the structural changes caused by denaturation and aggregation of whey proteins and/or casein with whey proteins through disulfide bounds (Almaas et al., 2006; Dupont et al., 2010). In addition, structuring β-lg into fibrils (formed by heating solutions at 80°C and pH 2.0 for 20 h) induces a more complete digestion by pepsin (Bateman et al., 2010).

Hence, by using carefully controlled processing parameters (mainly temperature, pH, ionic strength, and pressure), protein structures can be designed to release selected bioactive peptides, decrease allergenicity of protein and also induce a higher sensation of satiety during protein digestion.

3.2 Lipids

It is usually considered that, in the normal diet, some 25 to 30% of the total calories are conveniently supplied as lipid. Certain lipids are needed for good health (*e.g.*, essential fatty acids and fat-soluble vitamins). Nevertheless, the over-consumption of some dietary lipids (*e.g.*, cholesterol, saturated fats, and *trans*-fatty acids) increases the prevalence of some public health problems, including cardiovascular disease and obesity (Simopoulus, 1999). In this scenario, an improved knowledge of lipid structure could address the rational design of dietary lipids to enhance or retard lipid digestion, aspect having strong nutritional and health impact.

Lipids are a chemically heterogeneous group of compounds characterized by their insolubility in water, but soluble in organic solvents. In general, lipids are conformed by fats and oils. The basic unit of natural fats and oils is the triglyceride (TG) molecule; however, others molecules, such as monoglycerides (MGs), diglycerides (DGs), and phospholipids (PLs), can be structuring dietary lipids too. The TG molecule consists of a glycerol backbone to which are acylated three fatty acids (FAs). Two FAs are at the ends of the glycerol molecule (sn-1 and sn-3 positions) pointing in one direction, while the third FA molecule in the middle (sn-2 position) is pointing in the opposite direction. The dietary FAs molecules that compose the TG structure may vary in the number of carbon atoms and the presence of double bounds from saturated FA to unsaturated FA. The type and position of the FA chains on the glycerol backbone affect the structure of the TG molecule and, consequently, its bioavailability. TGs containing short- and medium-chain FAs have higher rate of FA release during lipolysis than for long-chain FAs, and is faster for FAs at the sn-1 and sn-3 positions than in sn-2 position (Fave et al., 2004). On the other hand, a typical TG molecule can exhibit three physical states: crystal, bulk, and interface. These physical states will determine the properties and characteristics of the physical state of the lipid phases in foods, which vary from liquid to crystalline phases. The lipid phases in most foods are usually liquid at body temperature (~37°C), but in some foods they may be either fully or partially crystalline. The crystallinity of the lipid phase may alter the ability of enzymes to digest the emulsified lipids. For example, the release rate of a lipophilic drug can be decreased with increasing crystallinity degree of lipid droplets (Olbrich et al., 2002). Hence, according to the above mentioned aspects, composition, structural organization and physical state of lipid phases will control lipid digestion. In addition, lipids have the property of forming, in the presence of water, different self-assembled structures. These mesophase structures include monolayers, micelles, reverse micelles, bilayers, and hexagonal phases. The aqueous medium behavior of lipids is determined by intrinsic parameters such as lipid polarity, length and branching of acylated fatty acids, presence of double bounds, among others (Ulrich, 2002). Mesophase structures of lipids have different sizes and configurations and, for this reason, they are finding many applications in food, pharmaceutical, and cosmetics industries. Nowadays, bilayer vesicles (*i.e.*, often called liposomes) have received widespread attention because of their ability to entrap functional components. This characteristic allows use of these structures as drug-delivery vehicles for controlling the release of incorporated agents.

The lipid molecules in foods may be organized into a number of different forms, including as bulk, structural, emulsified, colloidal, or interfacial structures (McClements, 2005). However, invariably, all ingested lipids ends up as an emulsion either by gastric emulsification or prior to ingestion during manufacturing process. The structure of these emulsions is determined by the nature of the lipid phase, the aqueous phase and the interface (Lundin et al., 2008). Particularly the properties of the interface modulate lipid digestion. The interface of emulsified lipids is determined by the physicochemical properties of lipid, such as lipid droplet size (which determines the interfacial area of lipid droplets), structure of lipid droplet, and the molecular structure of the TGs that constitute the lipid droplet (Armand et al., 1999). Several works have investigated the impact of emulsion structure on lipid digestion. For example, it has been demonstrated that a lower initial fat droplet size facilitates fat digestion by lipase (Armand et al., 1999). Additionally, the composition of the emulsion interface can limit the lipase activity. Interfaces composed by phospholipid limit fat digestion in the absence of bile salts because there are few possible points where lipase can access the emulsified substrate (Wickham et al., 1998).

In principle, rational design of lipid structures may be a useful tool for food processors to control lipid digestibility and bioavailability. Nevertheless, there is clearly a need for further research to establish the key physicochemical factors that impact the performance of food lipids within the GIT.

3.3 Starch

Carbohydrates constitute the most heterogeneous group among the major food elements, ranging widely in size, shape, and function. Polysaccharides such as starch, cellulose, hemicellulose, pectic substances and plant gums provide textural attributes such as crispness, hardness, and mouthfeel to many foods. Many of them can form gels that will constitute their structure and also enhance viscosity of solutions owing to their high molecular weight (Aguilera & Stanley, 1999).

Although carbohydrates are not essential in the diet, they generally make up ~40–45% of the total daily caloric intake of humans, with plant starches generally comprising 50–60% of the carbohydrate calories consumed (Goodman, 2010). Hence, starch becomes the most important source of energy in the human diet. Due to its functionality, starch is used in a wide range of foods for a variety of purposes including thickening, gelling, adding stability and replacing expensive ingredients. There are a number of different structural scales to be considered in starch, ranging through the distribution of individual starch branches, through the overall branching structure of the starch molecules in a granule, to the macroscopic structure of the grain (Dona et al., 2010).

Starch is formed by two polymers of glucose, amylose and amylopectin. Starch is a plant storage polysaccharide synthetized in the form of granules (normally 1-100 μm in diameter) in which molecules are organized into a radially anisotropic, semi-crystalline unit. Starch molecule is composed of the straight-chain glucose polymer amylose (with α-1,4 glycosidic linkages) and the branched glucose polymer amylopectin (with α-1,6 glycosidic bonds) (Pérez et al., 2009). Normal starches, such as maize, rice, wheat and potato, contain 70–80% amylopectin and 20–30% amylose (Jane, 2009). Amylopectin, the major component in starch, strongly influences its physicochemical and functional properties.

There are several levels of structural complexity in starch granules. The first level is the "cluster arrangement", in which most starch granules are made up of alternating amorphous and crystalline shells. This structural periodicity is due to regions of ordered, tightly packed parallel glucan chains (crystalline zones) alternate with less ordered regions corresponding to branch points (amorphous zones). Thus, the starch granule appears to be formed by alternating concentric rings (growth rings).

Starch granules are insoluble in cold water and are densely packed in its native form. In the raw form, the native granule is generally indigestible for humans due to this semi-crystalline structure. The addition of water and the application of heat to native granules is essential to transform them into foods with pleasing textural attributes. Gelatinization and retrogradation are the main microstructural changes related to starch digestion. Gelatinization is the collapse (disruption) of the molecular order within the starch granule manifested in irreversible changes in properties such as granular swelling, native crystallite melting and starch solubilization (Mason, 2009). When granules of starch are heated in the presence of water, the amorphous regions that pervade the whole granule swell (up to 50 times), forming a continuous gel phase. As the temperature exceeds a value between 50 and 80°C (depending on the crystallinity degree), the crystalline structure of the matrix is broken down. As gelatinization proceeds, the granule network is destroyed and amylose diffuses into the aqueous medium, increasing its viscosity. Further heating and/or shear disrupts the granule and a starch paste consisting of a continuous phase of amylose and amylopectin is formed (Aguilera & Stanley, 1999; Jane, 2009). The total amount of bioaccessible starch is the principal factor affected by the gelatinization process, thus a gelatinized granule is more digestible than a raw granule because the digestive enzymes can attack more easily the active sites. In fully isolated amylose or amylopectin molecules the digestion rate would be basically the same and occur relatively fast (less than 10 min). Then, the change in the total amount of digestible starch in a food can be explained because in most foods starch is present as a combination of raw or partly gelatinized granules (more resistant to digestion) and gelatinized granules (more digestible).

Retrogradation is the phenomenon occurring when the molecules comprising gelatinized starch begin to re-associate in an ordered structure (Mason, 2009). During aging, starch molecules can re-associate into crystalline segments (retrograde), losing water from the structure and undergoing an incomplete re-crystallization. Amylose and amylopectin have different behaviors, for example, amylose has a high tendency to retrograde and produce tough gels, whereas amylopectin, in an aqueous dispersion, is more stable and produces soft gels (Jane, 2009). Re-crystallization can lower the digestibility of starch because the resulting structure is less accessible to enzymes (Parada & Aguilera, 2011a).

Starch retrogradation and water-limited gelatinization should present an opportunity to redesign starchy foods aiming at reducing the glycemic response (GR), defined as the concentration of glucose in the blood after ingestion. An understanding of the internal organization of starch granules is crucial for food scientists and engineers to comprehend the functionalities and improve the nutritional properties of designed starch products.

4. Digestion: A step-by-step process controlled by different food structures

The manner in which a food is structured impacts its breakdown during digestion and consequently certain nutrional functions such as nutrient release and bioaccessibility (Parada & Aguilera, 2007). This requires understanding what happens inside the gut when a food is ingested and how the GIT behaves as the interface between the body and the food supply (Norton et al., 2007). The GIT is a highly specialized system that allows humans to consume diverse food matrices in discrete meals to meet nutrient needs. The main organs of the GIT include the mouth, the stomach and the small intestine. Figure 3 summarizes some components and phenomena occurring in the GIT during food digestion.

The digestive system is connected to the vascular, lymphatic and nervous systems to facilitate regulation of the digestive response, delivery of absorbed compounds to organs in the body and the regulation of food intake (Schneeman, 2002). In the mouth the food is broken down by the chewing action, mixed with saliva, and undergoes a temperature change as heat flow occurs (heating or cooling to body temperature). The foods are processed to the stage at which a bolus can be formed and swallowed, passing through the esophagus to the stomach (Lucas et al., 2002). The motility of the stomach (*i.e.*, with a maximum contraction force ranging from 0.1 to 1 N) continues the process of mixing food with the digestive secretions (Marciani et al., 2001b), now including gastric juice which contains acid and some digestive enzymes (*e.g.*, gastric lipase and pepsin), inducing further reduction of particle size, remixing and phase separation (*e.g.*, oil and aqueous phases). Typically, there is an appreciable increase in the pH of the stomach contents after ingestion of a food, followed by a gradual decrease to around pH 2 over the next hour or so. After being ingested a food component may remain in the stomach for a period ranging from a few minutes to a few hours depending on its quantity, physical state, structure and location (Weisbrodt, 2001). In the stomach occurs a complex dynamic step affecting the kinetics and pattern of subsequent digestion in the small intestine. The action of the stomach continues to break down the food into smaller particles prior to passage to the small intestine (Hoebler et al., 2002). Once the chyme is in the small intestine, peristaltic motor activity propels it along the length of the intestine and segmentation allows mixing with digestive juices, which include pancreatic enzymes, bile salts (BSs) and sloughed intestinal cells. Digestion of macronutrients, which began in the mouth, continues in the small intestine primarily through the action of enzymes. Each of the macronutrients has a unique set of enzymes that break the macromolecules into sub-units that can be taken up by the absorptive systems in the intestinal cells, allowing the subsequent transport of nutrients to the systemic circulation (Salminem et al., 1998). In this section we give a brief overview of the basic physiological events that occur during the digestion of the main macronutrients (proteins, lipids and carbohydrates) found in natural and processed foods.

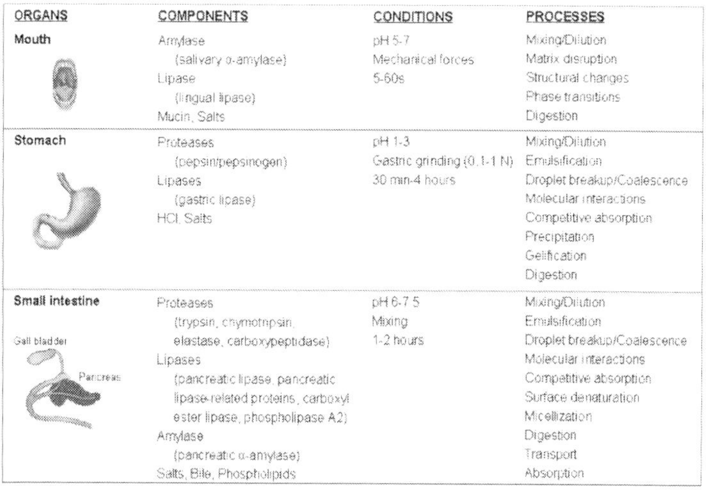

ORGANS	COMPONENTS	CONDITIONS	PROCESSES
Mouth	Amylase (salivary α-amylase) Lipase (lingual lipase) Mucin, Salts	pH 5-7 Mechanical forces 5-60s	Mixing/Dilution Matrix disruption Structural changes Phase transitions Digestion
Stomach	Proteases (pepsin/pepsinogen) Lipases (gastric lipase) HCl, Salts	pH 1-3 Gastric grinding (0.1-1 N) 30 min-4 hours	Mixing/Dilution Emulsification Droplet breakup/Coalescence Molecular interactions Competitive absorption Precipitation Gelification Digestion
Small intestine Gall bladder Pancreas	Proteases (trypsin, chymotrypsin, elastase, carboxypeptidase) Lipases (pancreatic lipase, pancreatic lipase-related proteins, carboxyl ester lipase, phospholipase A2) Amylase (pancreatic α-amylase) Salts, Bile, Phospholipids	pH 6-7.5 Mixing 1-2 hours	Mixing/Dilution Emulsification Droplet breakup/Coalescence Molecular interactions Competitive absorption Surface denaturation Micellization Digestion Transport Absorption

Fig. 3. Summary of the major components and processes involved in the digestion of foods.

4.1 Protein digestion

Proteins typically make up about 10% of caloric intake in a normal diet, being a dietary component essential for nutritional homeostasis in humans. In general, ingested protein undergoes a complex series of degradative processes promoted by hydrolytic enzymes (*i.e.*, proteases) originating in the stomach, pancreas, and small intestine. The product of this proteolytic activity is a mixture of AAs and small peptides that are absorbed by the small intestinal enterocytes (Erickson & Sim, 1990). Consequently the nutritional value of protein, also known as protein quality, is related to its AA content, to its digestibility (which can be regulated by the food processing) and to the subsequent physiological utilization of specific AAs after digestion and absorption (Friedman, 1996).

Protein digestion begins in the stomach by the action of gastric proteases. When the bolus enters the gastric lumen is not only exposed to hydrochloric acid and salts but also to different pepsins, the major gastric proteases. An acidic milieu is required for the proteolytic activity of pepsins, with an optimum activity between pH 1.8 and 3.2 (Untersmayr & Jensen-Jarolim, 2008). The action of the gastric proteases results in a mixture of large polypeptides, smaller oligopeptides, and some free AAs. These hydrolytic products regulate diverse gastric functions that are under hormonal control, such as secretion of acid and pepsinogen and rate of gastric emptying (Erickson & Sim, 1990). Subsequently, the remaining proteins and polypeptides present in the chyme are released into the small intestine, where they are exposed to a variety of proteases and peptidases (see Figure 3) synthesized and released by the pancreas (*i.e.*, the major source of proteases in the digestive system), and by the specific peptidases of the brush border of the intestinal mucosa (Whitcomb & Lowe, 2007). The single AAs, di-, and tri-peptides resulting from the intestinal digestion are taken up by enterocytes (where small peptides of up to 3 AAs are split into AAs by cytosolic peptidases) and then are used as nutrients for the human body (Untersmayr & Jensen-Jarolim, 2008).

Nowadays, the fate of dietary proteins during gastrointestinal digestion has become of particular interest because of the potential role that digestion may play in determining the allergenic potential of foods. In this context, the fundamental role of the stomach as the primary organ of protein digestion is very well recognized, leading to the classification of proteins as digestion-resistant or digestion-labile proteins (Moreno, 2007). Resistance of proteins to pepsin digestion has been proposed as a marker for potential allergenicity because it does appear to be a characteristic shared by many food allergens (*e.g.*, β-lg, α-lactalbumin and casein in cow's milk, parvalbumin in fish, ovomucoid and ovalbumin in egg, tropomyosin in shellfish and seafood, and Ara h 1 and Ara h 3 in peanut) (Mills & Breiteneder, 2005). Hence, there are groups of proteins, such as storage proteins or structural proteins, more stable to proteolysis in the GIT. Consequently, it has been postulated that for a food protein sensitizing an individual, it must have properties which preserve its structure from degradation in the GIT (such as resistance to low pH, bile salts, and proteolysis), thus allowing enough allergen to survive in a sufficiently intact or immunologically active form to be taken up by the gut and sensitize the mucosal immune system (Moreno, 2007). In this scenario, further investigation is needed to reveal the mechanisms controlling the allergenic potential of foods, as well as the influence of the food matrix on the gastrointestinal digestion and absorption of protein allergens.

4.2 Lipid digestion

Lipids are a major source of calories (9 kcal/g) in our daily diets and food emulsions (*e.g.*, mayonnaise, sauces, dressings) a major carrier of fat calories. The composition, structure and properties of fatty foods change appreciably during digestion. The structural organization and properties of the lipids within the bolus depend on their initial structural organization within the food, as well as the duration and intensity of the mastication process. In most cases, the lipids in the bolus are present as oil droplets stabilized at their interfaces by particles, proteins or surfactants, which may vary in size from around a micrometer (for some food emulsions) to more than a millimeter (for some bulk fats). These emulsified oil droplets may have been present in the original food or they may have been formed within the mouth due to breakdown of a bulk fat phase and the interaction with proteins (Hernell et al., 1990). In general, there is still a relatively poor understanding of the physicochemical and structural changes that occur within the mouth when fatty foods are consumed, although considerable progress has been made for some lipid foods such as emulsions and gels (Malone et al., 2003; Vingerhoeds et al., 2005).

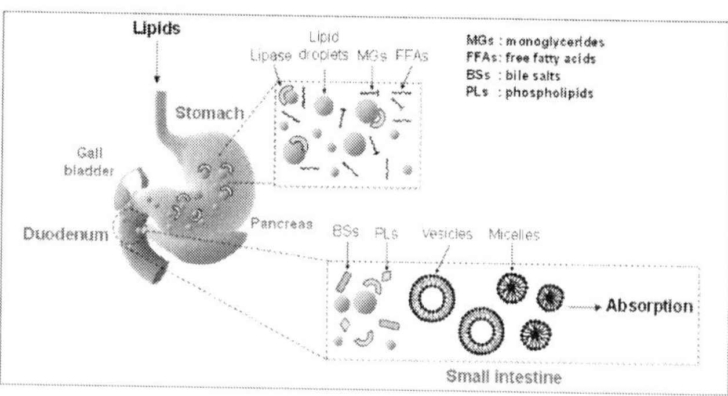

Fig. 4. Scheme of the digestive process of lipid foods.

After the food is swallowed it rapidly passes down the esophagus and enters the stomach, where it is mixed with acidic digestive juices containing gastric enzymes, minerals and various surface active compounds, and is also subjected to mechanical agitation due to movements of the stomach. During gastric digestion occurs breakdown of the food matrix structure, lipid droplet coalescence/disruption processes, and changes in the interfacial composition of the lipid phase due to adsorption/desorption of surface active substances (*e.g.*, proteins) to the surfaces of lipid droplets. Emulsion stability in the acidic gastric environment can readily be manipulated by altering the nature and state of the emulsifier agents, as revealed by non-invasive MRI studies in humans (Marciani et al., 2004). To be effective, the gastric lipase has to reach the surface of the lipid droplets in order to hydrolyze the emulsified TGs to DGs, MGs, and free fatty acids (FFAs). Typically, lipid hydrolysis stops when 10-30% of the FFAs have been released from the TGs (Armand, 2007). The emulsified lipids within the chyme are transferred from the stomach to the duodenum where they are mixed with sodium bicarbonate, BSs, phospholipids (PLs) and enzymes. The sodium bicarbonate secreted into the small intestine causes the pH to increase from highly acidic in the stomach to neutrality (pH 6.0-7.5) in the duodenum, where the pancreatic enzymes work most efficiently (Bauer et al., 2005). BSs and PLs are surface-active agents that can facilitate emulsification of the lipids by adsorbing to the droplet surfaces (Porter et al., 2007). The combined action of BSs and pancreatic juice brings about a marked change in the physicochemical form of the luminal lipid emulsion. In addition, the chyme is subjected to shear flow patterns in the small intestine that promote mixing and further emulsification. The lipid hydrolysis continues within the duodenum through the actions of lipases originating from the pancreas (Armand, 2007). A pancreatic lipase/colipase complex has to bind to the surface of the lipid droplet to hydrolyze the TGs to DGs, MGs, and FFAs, hence any compound covering the surface must be previously liberated (Jurado et al., 2006). Finally, lipid digestion continues in the small intestine with desorption and dispersion of insoluble lipid into an absorbable form. The digested lipids are solubilized in the lumen of the small intestine into at least two types of nanostructures: bile salt micelles and unilamellar vesicles. These assemblies are subsequently absorbed into the enterocyte's brush border membrane lining the surface of the small intestine. Thus, the absorption of digested fat and fat-soluble molecules that occurs in the small intestine is usually >90% (relatively high efficiency)

(German & Dillard, 2006). Figure 4 shows a highly schematic diagram of lipid digestion in the GIT where it is possible to observe that the dynamics of lipid digestion leads to the breakdown of complex structures, which disassemble during transit in the GIT.

4.3 Carbohydrate digestion

Starch, sucrose and lactose are the most important digestible carbohydrates in the human diet. However, from a nutritional point of view, the starch is the main carbohydrate derived from foods contributing significantly to the exogenous supply of glucose and total food energy intake (Slaughter et al., 2001). Starch is stored in plants as partially crystalline granules that contain two distinct polysaccharide fractions - amylose and amylopectin (*i.e.*, both polysaccharides composed of glucose molecules linked by digestible glycoside bonds). Upon cooking in the presence of water starch granules swell and partly disintegrate, facilitating the action of degrading enzymes that progressively transform them into maltose and glucose. The main characteristic of starch, compared with simple carbohydrates, is its slow digestion in the small intestine, which it produces a moderate GR. Additionally, food microstructure also affects the kinetics of starch hydrolysis and, consequently, the GR of starchy products, which it can be considered a reflection of the final nutritive effect of the food (Parada & Aguilera, 2007). In this way, the effect of the microstructure on digestibility of starchy foods may be manifested in two ways: the degree of starch gelatinization and the effect of the food matrix (non-starchy). These microstructural factors change from one food to another, modifying the digestibility of starch and its glycemic response (Fernandes et al., 2005). In humans there are several enzymes that hydrolyze starch. In the mouth, saliva contains α-amylase, enzyme that hydrolyses accessible starch, but since this starch remains in the mouth for a short period, the level of digestion is small. Once the food is partly digested in the mouth it is transported to the stomach, where there is no starch digestion but breakdown of food matrix will facilitate the access of hydrolytic enzymes to active sites for starch degradation in the small intestine. In the small intestine, the food receives pancreatic juice that contains pancreatic α-amylase, which hydrolyzes glycoside bonds producing glucose (absorbed in the intestinal epithelium), oligosaccharides and dextrins (Tester et al., 2004). However, it is known that the extent of starch digestion within the small intestine is variable and that a substantial amount of starch escapes digestion in the small intestine and enters the colon. The reasons for the incomplete digestion of starch may be separated into intrinsic factors (*i.e.*, physical form, presence of amylose-lipid complexes, and α-amylase inhibitors) and extrinsic factors (*e.g.*, chewing and transit through the bowel) (Cummings & Englyst, 1995).

Commonly, the GR is a way to know the bioavailability of starch and the nutritive effect of starchy foods. This metabolic response after food ingestion is not the same for different foods with the same amount of starch (Englyst & Englyst, 2005) and, principally, food processing affects this response. As an example, for many years it has been suggested that starch gelatinization affects the glycemic response (Björck et al., 1994).

In summary, food microstructure affects the nutritive value or quality of food products. However, the health benefit of different nutrients after absorption is not only conditioned by the food matrix, but also by the regular and harmonic functioning of the GIT. As mentioned before, designed food matrices can help in the prevention of diet-related diseases through through the controlled released of macronutrients or bioactive compounds. The next section deals with the latest research aiming to improve the nutritional performance of foods.

5. Structuring food matrices to improve bioavailability

The bioavailability of macro and micronutrients may be either increased or decreased by manipulating the microstructure of the foods that contain them. The food manufacturing processes affect the resulting structures and properties such as appearance, texture, taste, etc. However, food is ultimately going to be eaten. The ingestion of food leads to its breakdown and deconstruction through the different processes involved in the digestive stage (Lundin et al, 2008). Hence, food structures can be designed to control not only material properties before ingestion but also to control digestive breakdown rate and extent in the GIT. This section deals with the main aspects of designing protein, lipid and starch matrices in order to facilitate the rational design and fabrication of functional foods for improved health and wellness.

5.1 Protein systems

The food industry has widely used proteins in food formulations because possess unique functional properties. In addition to their contribution to the nutritional properties of foods through provision of AAs (that are essential for human growth and maintenance), proteins impart the structural basis for various functional properties of foods, such as water binding, gelling, foaming, and emulsifying. In particular, their ability to form gels and emulsions allow them to be an ideal material for controlling the rate of digestion and releasing of bioactive compounds at specific sites in the GIT.

Protein gels are undoubtedly the most convenient and widely used matrix in food applications. Food gels can be defined as three-dimensional continuous polymeric network holding large quantities of aqueous solution that shows mechanical rigidity during the observation time (Aguilera & Stanley, 1999). Gelation of food proteins and particularly of globular proteins (*e.g.*, egg white, soybean, and whey proteins) has received much attention lately among food scientists because they are a useful way to modulate texture and sensory perception of foods. Various processing routes, such as control of ionic strength, change in pH, heating, high pressure and enzymes are used to produce gels with different microstructural properties, which are strongly related to their aggregate molecular structure.

Two types of globular protein gels can be described depending on their microstructure: (i) fine-stranded gels, composed of more or less flexible linear strands making up a regular network characterized by its elastic behavior and high resistance to rupture, and (ii) particulate gels, composed of large and almost spherical aggregates, characterized by their lower elastic behavior and lower rupture resistance. Fine-stranded gels are created by linear aggregation of structural units maintained by hydrophobic interactions, whereas the particulate gels are produced by random aggregation of structural units essentially controlled by van der Waals interactions.

Protein gels can behave as pH-sensitive matrices through the presence of acidic (*e.g.*, carboxylic) or basic (*e.g.*, ammonium) groups in proteins chains, which either accept or release protons in response to changes in pH of the medium. This behavior could strongly influence the rate of molecule release by gels exposed to either gastric (low pH) or intestinal medium (neutral pH) by decreasing polypeptide chain interaction and thereby uptaking water inside the network and allowing the diffusion of molecules outward by osmotic pressure.

Gels of diverse mechanical and microstructural properties can be formed by controlling the assembly of protein molecular chains. Among the factors that modulate the release of protein molecules from gels, factors that promote gel breakdown play a major role.

Remondetto et al. (2004) showed that the release of nitrogen from cold-induced β-lg gels, as an index of matrix degradation, was slightly higher for particulate than for filamentous gels. In addition, the release of iron from these matrices was not correlated with the food matrix degradation at GIT conditions, suggesting that gel microstructure and not proteolysis influences the release of iron. In the case of particulate gels, the release of iron was lower due to its location inside aggregates that associate to form networks; in contrast iron is located at the outer surface of the linear aggregates of filamentous gels which it facilitates its release. Both types of gels may protect iron from gastric environment, but filamentous gels would release a larger amount of iron in the small intestine favoring the increase in iron absorption. In agreement with previous results, Maltais et al., (2009) found that fine-stranded cold-induced soy protein isolate (SPI) gels degraded more slowly than particulate gels, probably because their lower porosity slows down the digestion. Besides, the release of riboflavin was delayed from fine-stranded gels, because of the lower rate of protein digestion. From this study it was demonstrated that fine-stranded and particulate SPI protein gels were able to protect riboflavin from gastric conditions and release it under intestinal conditions. SPI films (*i.e.*, thin dehydrated gels) degraded more slowly as cross-linking density increases. Protein matrix undergoes a so-called first-order degradation during gastric and intestinal enzyme digestion. The behavior of the films was attributed to a more rigid structure and greater entanglement of the polypeptide chains in the SPI films due to increased cross-linking, leading to decreased penetration of digestive enzymes into the network (Chen et al., 2008).

Protein gel micro-beads have been recently used to encapsulate microorganisms with probiotic activity. The main challenge of this technique is to improve the survival of probiotics in the human digestive environment. Hébrard et al., (2006) produced whey protein isolate (WPI) beads containing yeasts from cold-induced gelation. Beads were resistant to acidification and pepsin attack during simulated human gastric digestion. Only about of 2% of initially entrapped yeasts were recovered in the gastric medium. In addition, WPI micro-beads were stable after incubation in simulated gastric juice in the presence of pepsin, allowing a targeted disintegration of protein matrices under physiological intestinal conditions (Doherty et al., 2011). Both results suggest that WPI beads might cross the gastric barrier and can deliver probiotics in the small intestine.

In conclusion, microstructure of protein gels largely affects the digestion of proteins and the release of bioactive compounds. If a digestive enzyme needs to degrade the gel matrix, then a large porosity and a and a high-pore interconnectivity would increase the rate of digestion. In addition, to design efficient systems for specific intestinal absorption of bioactive compounds or probiotics, gels should be gastroresistant. Hence, controlling microstructure and tailoring the porous structure of protein gels, it would give more opportunities for protection or release of a nutrient or a physiologically bioactive component in the GIT.

5.2 Lipid matrices

Many food scientists are currently focusing their efforts on developing novel lipid structures to decrease or increase the bioavailability of food lipids. In this context, much attention has been paid to the formulation of emulsion-based food systems to encapsulate, protect and release lipophilic constituents. A variety of emulsion-based delivery systems are already available (Figure 5), including conventional emulsions, nanoemulsions, solid lipid particles, multiple emulsions, and multilayer emulsions. Nevertheless, the structural design of these systems is still very far from an exact science due to their compositional and structural complexity. Further research is still needed to achieve a detailed understanding of the molecular characteristics, structural organization, physicochemical properties, and functional performance of these delivery systems (McClements et al., 2009).

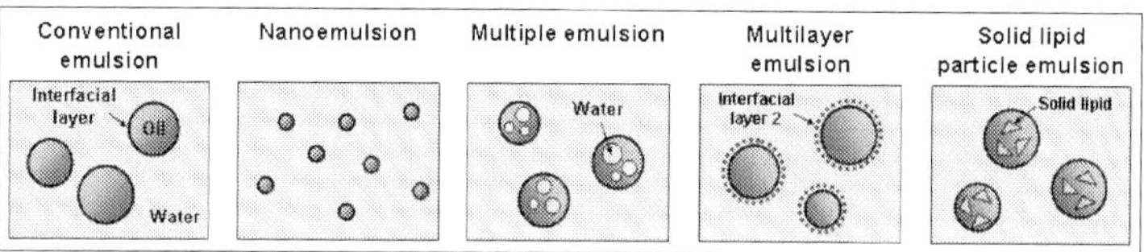

Fig. 5. Examples of different emulsions-based delivery systems

Conventional emulsions consist of oil droplets (~µm) dispersed in an aqueous medium, with the oil droplets being surrounded by emulsifier molecules. Oil-in-water (O/W) emulsions contain a non-polar region (the oil phase), a polar region (the aqueous phase), and an amphiphilic region (the interfacial layer). Then, within O/W emulsions it is possible to incorporate functional agents that are polar, non-polar, and amphiphilic (McClements, 2005). Nevertheless, conventional emulsions have some limitations (*e.g.*, physical instability) that promote the development of more sophisticated structured systems. Like conventional emulsions, nanoemulsions consist of small oil droplets (radius < 100 nm) dispersed within a continuous phase, with each droplet being surrounded by a protective coating of emulsifier molecules. However, nanoemulsions do have a number of advantages over conventional emulsions for certain applications due to their relatively small particle size: (i) they scatter light weakly and so tend to be transparent; (ii) they have high physical stability; (iii) they have unique rheological characteristics; and, (iv) they can greatly increase the bioavailability of encapsulated lipophilic components (Mason, et al., 2006). Additionally, the very small oil droplet size in nanoemulsions increases the digestion rate and the total amount of FFAs released during digestion in comparison with conventional emulsions (Li & McClements, 2010). This last characteristic may be of interest for the realization of lipid food for humans with disorders that prevent efficient digestion or absorption of lipids (*e.g.*, cystic fibrosis or pancreatitis) (Fave et al., 2004).

Multiple emulsions are systems in which dispersed droplets contain smaller droplets inside (Figure 5). Particularly, water-oil-water (W/O/W) emulsions consist of small water droplets contained within larger oil droplets that are dispersed in an aqueous continuous phase. Within these systems, functional components can potentially be located in a number of different ways. Water soluble components can be incorporated into the inner or outer water phase, while oil soluble components can be incorporated into the oil phase. Potential advantages of the multiple emulsions as delivery systems over conventional emulsions are: (i) functional ingredients could be trapped inside the inner water droplets and released at a controlled rate in the GIT; (ii) functional ingredients could be protected from chemical degradation; and (iii) reduction of the overall fat content of food products by loading the oil phase with water droplets (McClements et al., 2009).

The quality and functional performance of conventional O/W emulsions can be improved by the formation of multilayered interfaces using the layer-by-layer electrostatic deposition technique (Guzey & McClements, 2006). Multilayered emulsion delivery systems (Figure 5) may have a number of advantages over conventional single-layered emulsions: (i) improved physical stability to environmental stresses; (ii) greater control over the release rate of functional agents due to the ability to manipulate the thickness and permeability of the laminated interfacial coating; and (iii) ability to trigger release of functional agents in response to specific changes of environmental conditions in the GIT, such as dilution and/or pH (McClements et al., 2009).

Finally, it is possible to control the physical location of a lipophilic component and to slow down molecular diffusion processes using conventional emulsions by crystallizing the lipid phase. These systems are known as solid lipid particle emulsions, which consist of emulsifier coated (partially) solid lipid particles dispersed in an aqueous continuous phase (Figure 5) (Videira et al., 2002). By controlling the morphology of the crystalline lipid matrix it is possible to obtain more precise control over the release kinetics of functional compounds. Additionally, both lipophilic and hydrophilic bioactives can be incorporated within the same system using solid lipid particle emulsions (McClements et al., 2009). Hence, there are a large number of different ways that can be used by food manufacturers to embed lipophilic compounds within food matrices with different degradation rates. However, the selection of a particular matrix is a matter of functional preference.

5.3 Starch

Elevated glucose levels and high postprandial blood glucose cause a metabolic stress concentrations associated with increased risk of diseases, such as type II diabetes, cardiovascular disease and obesity (Kim et al., 2008; Parada & Aguilera, 2011a). The slow digestion of starch, in comparison with simple carbohydrates (*e.g.*, glucose, fructose), involves a gradual release of glucose to the bloodstream, thus producing a low GR.

Different starchy foods (pasta, baked foods, among others) elicit different GRs. The extent of starch digestion within the small intestine is variable, depending on food physical form, and a substantial amount of undigested starch enters the colon, where may be fermented by bacteria or simply appear in faeces (Troncoso & Aguilera, 2009).

The starch in its native state is resistant to enzymatic digestion and the availability of starch chains to the digestive enzymes is increased as gelatinization progresses. Gelatinization breaks the weak hydrogen bonds between polymer chains, so the exposed polar groups can further interact with water molecules. Due to this biophysical phenomenon, the granule structure changes from a crystalline to a disordered structure that is more easily accessible for the digestive enzymes (Parada & Aguilera, 2011a). Digestibility is strongly influenced by shearing and heating of starch samples. Both heat and shear during the preparation of starch suspensions alter the progression towards gelatinization, which increases the availability of polysaccharide starch chains to digestive enzymes, thus affecting the rates of hydrolysis (Dona et al., 2010).

The degree of gelatinization of starch granules and the recrystallization of the released polymers is of vital importance in the breakdown of starch molecules into sugars during digestion, because both phenomena influence the susceptibility to enzymatic degradation. Hence, a nomenclature has emerged describing the susceptibility of starch to digestion by enzymes in the small intestine: rapidly digestible starch (RDS), slowly digestible starch (SDS) that has slow but complete digestion, while resistant starch (RS) is resistant to digestion (Parada & Aguilera, 2011a).

RS obtained through appropriate physical or chemical modification can withstand the environmental changes in the upper GIT and can be rapidly degraded by the enzymes produced by the colonic microbiota. The incomplete digestion of starch is related to the matrix surrounding starch, the nature and physicochemical properties of the starch *per se* at the granule and molecular levels (*e.g.* granule size and amylose/amylopectin ratio), and the presence of other dietary components (*e.g.* sugar, dietary fiber and lipid) (Troncoso & Aguilera, 2009).

Starch granules in a real food are not isolated but exist within a three-dimensional matrix structure formed by proteins or other biopolymeric materials (such as fiber), which affects both the degree of gelatinization during processing and the digestion of starch (Giacco et al., 2001). If a food matrix is relatively impervious to enzymes and the granules are not properly exposed to their action, the digestion could be limited. The "encapsulation" of starch granules by a protein network is an important factor in explaining the slow degradation of starch by α-amylase in cooked pasta, the microstructure of the protein matrix as well as the physical state of starch (degree of gelatinization and retrogradation, amylose-to-amylopectin ratio, etc.) may explain the differences in the *in vivo* and *in vitro* enzymatic susceptibility of starch (Petitot et al., 2009).

The presence of a structured and continuous protein network is an important factor in explaining the slow degradation of starch in pasta. It had been shown that the degree of gelatinization and the interaction starch-gluten are a key factor during digestion of starch. Over-mixing could disrupt the gluten matrix producing an augmented rate of digestion; whereas higher heating temperatures produced a more compact structure (higher denaturation of gluten matrix) delaying digestion (Parada & Aguilera, 2011b). In agreement with these results, Kim et al. (2008) found that the disruption of the starch-coating protein matrix could be responsible for the increase in starch digestibility in pasta.

The degree of gelatinization and the gluten matrix conformation affect the digestibility of starch; the degree of gelatinization probably predominates in the extent of starch digestion, while the state of the gluten matrix is more related to the rate of digestion of starch.

Starch retrogradation and water-limited gelatinization should present an opportunity to redesign starchy products aiming at reducing the GR. Moreover, understanding the mechanisms by which the protein matrix slows down starch digestion could help to a rational design of new products with controlled starch digestion.

6. Conclusion

Recent knowledge supports the hypothesis that, beyond considering nutritional composition, food matrix microstructure may modulate various physiological functions in the human body and play detrimental or beneficial roles in some diet-related diseases. Advanced concepts in physical chemistry and material science provide a convenient and powerful framework for developing an understanding of the interactions and assemblies of food components into microstructure, which influence food macrostructure and functional properties. In turn, medical research coupled with food engineering approaches allow to obtain a deep comprehension of how foods behave during the digestive process. Advances in technologies for producing food matrices with tailored properties make possible the production of foods that have potential impact on the human health. This presents a challenge for the scientific community and the food industry due to the necessity of considering relationships between targets for functional food science and gastrointestinal behavior.

7. Acknowledgment

Author Zúñiga acknowledge to CIEN Austral for the financial support of this chapter.

8. References

Aguilera, J.M. (2005). Why food microstructure? *Journal of Food Engineering*, Vol. 67, No. 1-2, (March 2005), pp. 3-11, ISSN 0260-8774

Aguilera, J.M & Stanley, D.W. (1999). *Microstructural Principles of Food Processing and Engineering 2nd Edition*, Aspen Publishers Inc., ISBN 978-083-4212-56-0, Gaitherburg, MA, USA

Almaas, H.; Cases, A-L.; Devold, T.G.; Holm, H.; Langsrud, T.; Aabakken, L.; Aadnoey, T. & Vegarud, G.E. (2006). In Vitro Digestion of Bovine and Caprine Milk by Human Gastric and Duodenal Enzymes. *International Dairy Journal*, Vol. 16, No. 9, (September 2006), pp. 961-968, ISSN 0958-6946

American Dietetic Association. (2005). Position of the American Dietetic Association: Fat Replacers. *Journal of the American Dietetic Association*, Vol. 105, No. 2, (February 2005), pp. 266-275, ISSN 0002-8223

Armand, M. (2007). Lipases and Lipolysis in the Human Digestive Tract: Where Do We Stand? *Current Opinion in Clinical Nutrition and Metabolic Care*, Vol. 10, No. 2, (March 2007), pp. 156-164, ISSN 1473-6519

Armand, M. ; Pasquier, B. ; André, M. ; Borel, P. ; Senft, M. ; Peyrot, J. ; Salducci, J. ; Portugal, H. ; Jaussan, V. & Lairon, D. (1999). Digestion and Absorption of 2 Fat Emulsions with Different Droplet Sizes in the Human Digestive Tract. *American Journal of Clinical Nutrition*, Vol. 70, No. 6, (December 1999), pp. 1096-1106, ISSN 1938-3207

Bateman, L.; Ye, A. & Singh, H. (2010). In Vitro Digestion of β-Lactoglobulin Fibrils Formed by Heat Treatment at Low pH. *Journal of Agricultural and Food Chemistry*, Vol. 58, No. 17, (August 2010), pp. 9800-9808, ISSN 0021-8561

Bauer, E.; Jakob, S. & Mosenthin, R. (2005). Principles of Physiology of Lipid Digestion. *Asian-Australasian Journal of Animal Sciences*, Vol. 18, No. 2, (January 2005), pp. 282-295, ISSN 1976-5517

Björck, I.; Granfeldt, Y.; Liljeberg, H.; Tovar, J. & Asp, N.-G. (1994). Food Properties Affecting the Digestion and Absorption of Carbohydrates. *American Journal of Clinical Nutrition*, Vol. 59, No. 3, (March 1994), pp. 699-705, ISSN 1938-3207

Centers for Disease Control and Prevention. (July 2011). U. S. Obesity Trends, In: *Overweight and Obesity*, 29.07.2011, Available from
http://www.cdc.gov/obesity/data/trends.html

Chen, L.; Remondetto, G.; Rouabhia, M. & Subirade, M. (2008). Kinetics of the Breakdown of Cross-Linked Soy Protein Films for Drug Delivery. *Biomaterial*, Vol. 29, No. 27, (September 2008), pp. 3750-3756, ISSN 0142-9612

Chicón, R.; Belloque, J.; Alonso, E. & López-Fadiño, R. (2008). Immunoreactivity and Digestibility of High-Pressure-Treated Whey Proteins. *International Dairy Journal*, Vol. 18, No. 4, (April 2008), pp. 367-376, ISSN 0958-6946

Cummings, J. & Englyst, H. (1995). Gastrointestinal Effects of Food Carbohydrate. *American Journal of Clinical Nutrition*, Vol. 61, No. 4, (April 1995), pp. 938-945, ISSN 1938-3207

Doherty, S.B.; Gee, V.L.; Ross, R.P.; Stanton, C.; Fitzgerald, G.F. & Brodkorb, A. (2011). Development of Whey Protein Micro-Beads as Potential Matrices for Probiotic Protection. *Food Hydrocolloids*, Vol. 25, No. 6, (August 2011), pp. 1604-1617, ISNN 0268-005X

Dona, A. C; Pages, G.; Gilbert, R.G & Kuchel, P. W. (2010). Digestion of Starch: In Vivo and In Vitro Kinetic Models Used to Characterize Oligosaccharide or Glucose Release. *Carbohydrate Polymers*, Vol. 80, No. 3, (May 2010), pp. 599-617, ISSN 0144-8617

Dupont, D.; Mandalari, G.; Mollé, D.; Jardin, J.; Rolet-Répécaud, O.; Duboz, G.; Léonil, J.; Mills, C.E.N. & Mackie, A.R. (2010). Food Processing Increases Casein Resistance to Simulated Infant Digestion. *Molecular Nutrition and Food Research*, Vol. 54, No. 11, (December 2010), pp. 1677-1689, ISSN 1613-4133

Englyst, K. & Englyst, H. (2005). Carbohydrate Bioavailability. *British Journal of Nutrition*, Vol. 94, No. 1, (August 2005), pp. 1–11, ISSN 1475-2662

Erickson, R. & Sim, Y. (1990). Digestion and Absorption of Dietary Protein. *Annual Review of Medicine*, Vol. 41, No. 1, (February 1990), pp. 133-139, ISSN 0066-4219

Fave, G.; Coste, T. & Armand, M. (2004). Physicochemical Properties of Lipids: New Strategies to Manage Fatty Acid Bioavailability. *Cellular and Molecular Biology*, Vol. 50, No. 7, (December 2004), pp. 815-831, ISSN 1165-158X

Fernandes, G.; Velangi, A. & Wolever, T. (2005). Glycemic Index of Potatoes Commonly Consumed in North America. *Journal of the American Dietetic Association*, Vol. 105, No. 4, (April 2005), pp. 557–562, ISSN 0002-8223

Friedman, M. (1996). Nutritional Value of Proteins from Different Food Sources. A Review. *Journal of Agricultural and Food Chemistry*, Vol. 44, No. 1, (January 1996), pp. 6-29, ISSN 1520-5118

German, J. & Dillard, C. (2006). Composition, Structure and Absorption of Milk Lipids: a Source of Energy, Fat-Soluble Nutrients and Bioactive Molecules. *Critical Reviews in Food Science and Nutrition*, Vol. 46, No. 1, (January 2006), pp. 57–92, ISSN 1549-7852

German, J.B. & Watzke, H.J. (2004). Personalizing Food for Health and Delight. *Comprehensive Reviews in Food Science and Food Safety*, Vol. 3, No. 4, (October 2004), pp. 145-151, ISSN 1541-4337

Giacco, R.; Brighenti, F.; Parillo, M.; Capuano, M.; Ciardullo, A.; Rivieccio, A.; Rivellese, A. & Riccardi, G. (2001). Characteristics of Some Wheat-Based Foods of the Italian Diet in Relation to their Influence on Postprandial Glucose Metabolism in Patients with Type 2 Diabetes. *British Journal of Nutrition*, Vol. 85, No. 1, (January 2001), pp. 33–40, ISSN 1475-2662

Goetze, O.; Steingoetter, A.; Menne, D.; van der Voort, I.R.; Kwiatek, M.A.; Boesiger, P.; Weishaupt, D.; Thumshirn, M; Fried, M. & Schwizer, W. (2007). The Effect of Macronutrients on Gastric Volume Responses and Gastric Emptying in Humans: a Magnetic Resonance Imaging Study. *American Journal of Physiology, Gastrointestinal and Liver Physiology*, Vol. 292, No. 1, (January 2007), pp. G11-G17, ISSN 0193-1857

Goodman, B.E. (2010). Insights into Digestion and Absorption of Major Nutrients in Humans. *Advances Physiology Education*, Vol. 34, No. 2, (June 2010), pp. 44-53, ISNN 1043-4046

Guzey, D. & McClements, D. (2006). Formation, Stability and Properties of Multilayer Emulsions for Application in the Food Industry. *Advances in Colloid and Interface Science*, Vol. 128-130, No. 1, (December 2006), pp. 227-248, ISSN 0001-8686

Hébrard, G.; Blanquet, S.; Beyssac, E.; Remondetto, G.; Subirade, M. & Alric, M. (2006). Use of Whey protein Beads as a New Carrier System for Recombinant Yeast in the Human Digestive Tract. *Journal of Biotechnology*, Vol. 127, No. 1, (December 2006), pp. 151-160, ISSN 0168-1656

16

Epidemiology of Foodborne Illness

Saulat Jahan
Research and Information Unit
Primary Health Care Administration, Qassim
Ministry of Health
Kingdom of Saudi Arabia

1. Introduction

Foodborne illnesses comprise a broad spectrum of diseases and are responsible for substantial morbidity and mortality worldwide. It is a growing public health problem in developing as well as developed countries. It is difficult to determine the exact mortality associated with foodborne illnesses (Helms et al., 2003). However, worldwide an estimated 2 million deaths occurred due to gastrointestinal illness, during the year 2005 (Fleury et al., 2008). More than 250 different foodborne illnesses are caused by various pathogens or by toxins (Linscott, 2011). Foodborne illnesses result from consumption of food containing pathogens such as bacteria, viruses, parasites or the food contaminated by poisonous chemicals or bio-toxins (World Health Organization [WHO], 2011c). Although majority of the foodborne illness cases are mild and self-limiting, severe cases can occur in high risk groups resulting in high mortality and morbidity in this group. The high risk groups for foodborne diseases include infants, young children, the elderly and the immunocompromised persons (Fleury et al., 2008).

There are changes in the spectrum of foodborne illnesses along with demographic and epidemiologic changes in the population. A century ago, cholera and typhoid fever were prevalent foodborne illnesses, globally. During last few decades, other foodborne infections have emerged, such as diarrheal illness caused by the parasite *Cyclospora*, and the bacterium *Vibrio parahemolyticus*. The newly identified microbes pose a threat to public health as they can easily spread globally and can mutate to form new pathogens. In the United States, 31 different pathogens are known to cause foodborne illness, however, numerous episodes of foodborne illnesses and hospitalizations are caused by unspecified agents (Centers for Disease Control and Prevention [CDC], 2011a).

Although foodborne illnesses cause substantial morbidity in the developed countries, the main burden is borne by developing countries. These illnesses are an obstacle to global development efforts and in the achievement of the Millennium Development Goals (MDGs) (WHO, 2011c). There is an impact of foodborne illnesses on four out of the eight MDGs. These include MDG 1(Eradication of extreme poverty); MDG 3 (Reduction in child mortality); MDG 5 (Improvement of maternal health); MDG 6 (Combating HIV/AIDS and other illnesses). The population in developing countries is more prone to suffer from foodborne illnesses because of multiple reasons, including lack of access to clean water for food preparation; inappropriate transportation and storage of foods; and lack of awareness regarding safe and hygienic food practices (WHO, 2011c). Moreover, majority of the developing countries have limited capacity to implement rules and regulations regarding food safety. Also, there is lack of effective surveillance and monitoring systems for foodborne illness, inspection systems for food safety, and educational programs regarding awareness of food hygiene (WHO, 2011a).

Foodborne illnesses have an impact on the public health as well as economy of a country (Helms et al., 2003). They have a negative impact on the trade and industries of the affected countries. Identification of a contaminated food product can result in recalling of that specific food product leading to economic loss to the industry. Foodborne outbreaks may lead to closure of the food outlets or food industry resulting in job losses for workers, affecting the individuals as well as the communities. Moreover, local foodborne illness outbreaks may become a global threat. The health of people in many countries can be affected by consuming contaminated food products, and may negatively impact a country's tourist industry. The foodborne illness outbreaks are reported frequently at national as well as international level underscoring the importance of food safety(WHO, 2011c).

Increasing commercialization of food production has resulted in the emergence and dissemination of previously unknown pathogens, and resulted in diseases such as bovine spongiform encephalopathy (BSE). BSE is a variant of Creutzfeldt-Jakob disease (vCJD) which affected human population in UK during the 1990s (WHO, 2011c).

During last few decades there are advances in technology, regulation, and awareness regarding food safety but new challenges have emerged because of mass production, distribution, and importation of food and emerging foodborne pathogens (Scallan, 2007; Taormina & Beuchat, 1999). One of the major issues of public health importance is the increasing resistance of foodborne pathogens to antibiotics (WHO, 2011c).

For the sake of public health, it is important to understand the epidemiology of foodborne illnesses because it will help in prevention and control efforts, appropriately allocating resources to control foodborne illness, monitoring and evaluation of food safety measures, development of new food safety standards, and assessment of the cost-effectiveness of interventions. The purpose of this chapter is to describe the epidemiology of foodborne illness in order to determine the magnitude of the problem, risk factors, monitoring and surveillance and measures of control.

2. Impact of foodborne illness

2.1 Public health impact

Foodborne illnesses are prevalent (Hoffman et al., 2005) but the magnitude of illness and associated deaths are not accurately reflected by the data available in both developed and developing countries. To fill the current data gap, the World Health Organization (WHO) has taken initiative for estimation of the global burden of foodborne illnesses (Kuchenmüller et al., 2009). World Health Organization and the US Centers for Disease Control and Prevention (CDC) report every year a large number of people affected by foodborne illnesses (Busani, Scavia, Luzzi, & Caprioli, 2006). Globally, an estimated 2 million people died from diarrheal diseases in 2005; approximately 70% of diarrheal diseases are foodborne. It is estimated that up to 30% of the population suffer from foodborne illnesses each year in some industrialized countries (WHO, 2011c). According to the estimation by CDC in 1999, around 76 million foodborne illnesses occur annually, resulting in 325,000 hospitalizations and 5200 deaths in the United States (Buzby & Roberts, 2009). However, a decrease in the incidence rates of notified foodborne illness was noticed from 1996 to 2005, but these rates have remained static since 2005 (Anderson, Verrill, & Sahyoun, 2011). There is a 20% reduction in illnesses caused by the specific pathogens tracked by FoodNet system, over the past 10 years. There are many explanations for this decrease in foodborne illness. It may be due to improved food safety because of regulatory and industry efforts or because of better detection, prevention, education, and control efforts (CDC, 2011a). According to 2011 estimates of CDC, annually 48 million Americans get sick, there are 128,000 hospitalizations, and 3,000 deaths due to foodborne illnesses in the US (CDC, 2011b). In Canada, an estimated 1.3 episodes per person-year of enteric disease occur (Fleury et al., 2008). In New Zealand, there are an estimated 119,320 episodes of foodborne illnesses each year, accounting for a rate of 3,241 per 100,000 population (Scott et al., 2000).

Foodborne-illness outbreaks are under-reported and it is estimated that 68% of foodborne-illness outbreaks are notified to the Centers for Disease Control and Prevention (Jones et al., 2004). Even during foodborne illness outbreaks, only a small proportion of the total number of cases is reported. In the United States, during 1993–1997, an average of 550 foodborne illness outbreaks was reported annually. Each outbreak had an average of 31 cases (Jones et al., 2004). Foodborne illnesses also play an important role in new and emerging infections. It is estimated that during past 60 years, an estimated 30% of all emerging infections comprised of pathogens transmitted through food leading to foodborne illness (Kuchenmüller et al., 2009).

2.2 Economic impact of foodborne illness

Every illness has an economic cost and same is the case with foodborne illness. However, the economic cost of health losses related to foodborne illnesses has not been extensively studied. There are few studies available which provide either incomplete cost estimates or their estimates are based on limiting assumptions (Buzby & Roberts, 2009). In the United States, data from Foodborne Diseases Active Surveillance Network (FoodNet) and other related studies contributed to estimates of the economic cost of foodborne illness (Angulo & Scallan, 2007).

The annual economic cost of foodborne illness is calculated by multiplying the cost per case with the expected annual number of foodborne illnesses experienced. It was estimated that in 1999, the US government spent $1 billion on food safety efforts at federal level, an additional $300 million were spent by state governments. Moreover, it is estimated that a total of $152 billion a year is spent on foodborne illness in the U.S. (Scharff, 2010). The foodborne illness also bears substantial economic burden at regional level. The annual estimated economic cost of foodborne illness for Ohio is between $1.0 and $7.1 billion i-e., cost of $91 to $624 per Ohio resident (Scharff, McDowell, & Medeiros, 2009).

A retrospective study performed in Uppsala, Sweden during 1998–99, estimated average costs per foodborne illness as $246 to society and $57 to the patient. An estimated $123 million was the annual cost of foodborne illnesses in Sweden (Lindqvist et al., 2001).

In New Zealand, the total cost of foodborne illness cases was estimated to be $55.1 million, accounting for $462 per case. The direct medical costs were calculated as $2.1 million while direct non-medical costs were $0.2 million (Scott et al., 2000). The estimated total costs were $161.9 million including government outlays of $16.4 million, industry costs of $12.3 million and $133.2 million for incident case costs of disease associated with treatment, loss of output and residual lifestyle loss (New Zealand Food Safety Authority, 2010).

2.3 The unknown burden

Data from surveillance systems and sentinel sites show only the tip of the iceberg and do not reflect true disease burden of foodborne illness (Figure 1). The reasons are that the sick persons may not seek medical care, may not be tested at laboratory or may not be notified to the relevant health authorities (WHO, 2011b).

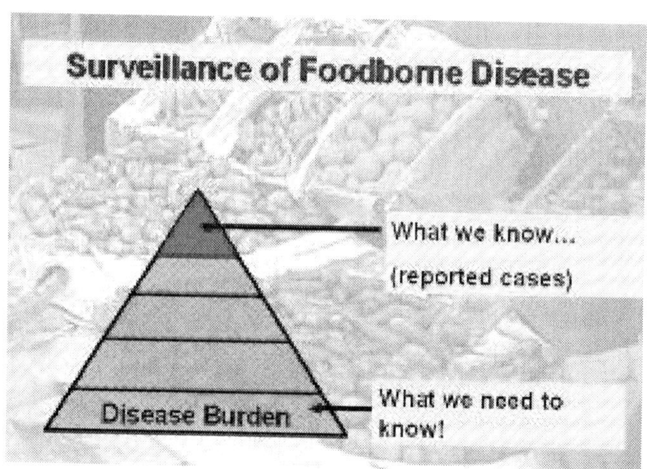

Fig. 1. The Unknown burden of foodborne illness. Adapted from WHO. Available at: http://www.who.int/foodsafety/foodborne_disease/ferg/en/index1.html Accessed August 04, 2011.

3. Risk factors

Factors such as socioeconomic status, malnutrition, micronutrient deficiencies, genetic factors and immunity play an important role in the causation of foodborne illnesses. The risk factors may be related to the host as well as to the environment.

3.1 Host risk factors

The risk factors for the foodborne illness caused by common pathogens have been determined by various studies. A case-control study conducted in the United States determined risk factors for Campylobacter infection. Traveling abroad and consumption of chicken or non-poultry meat prepared at a restaurant were stated as the main risk factors (Friedman et al., 2004). In Australia, annually an estimated 50,500 cases of *Campylobacter* infection in persons more than 5 years of age are associated with consumption of chicken (Stafford et al., 2008).

The risk factors for *Salmonella* Enteritidis have been explored in various studies conducted in different regions of the world. Foodborne Diseases Active Surveillance Network (FoodNet) in US, conducted a population-based case-control study for 12-month period during 2002–2003, to determine the risk factors associated with sporadic *Salmonella* Enteritidis infection. The study enrolled 218 cases and 742 controls. Travelling outside US five days before the onset of illness was found to be strongly associated with *Salmonella* Enteritidis infection [odds ratio (OR) 53, 95% Confidence Interval (CI) 23–125]. In addition, eating chicken cooked outside the home, consuming undercooked eggs and contact with birds and reptiles were associated with *Salmonella* Enteritidis infections (Marcus et al., 2007). In Denmark, a

prospective case-control study of sporadic *Salmonella* Enteritidis infection was conducted in 1997–1999. It was revealed that foreign travel had a positive association with the disease. Foreign travelers were approximately three times more likely to suffer from *Salmonella* Enteritidis infection as compared to the non-travelers. Among non-travelers, the risk factors were eating eggs or dishes containing raw or undercooked eggs. The study concluded that eggs are the main source of *Salmonella* Enteritidis in Denmark (Mølbak & Neimann, 2002). A case-control study to determine risk factors for salmonellosis was conducted in the Netherlands. Consumption of raw eggs (OR 3.1, 95% CI 1.3–7.4) and products containing raw eggs (OR 1.8, 95% CI 1.1–3.0) were found to be the risk factors associated with *Salmonella* Enteritidis infection. The risk factors for *Salmonella* Typhimurium infection included contact with raw meat (OR 3.0, 95% CI 1.1–7.9), consumption of undercooked meat (OR 2.2, 95% CI 1.1–4.1) and recent antibiotic use (OR 1.9, 95% CI 1.0–3.4) (Doorduyn, Van Den Brandhof, Van Duynhoven, Wannet, & Van Pelt, 2006). In Canada, a matched case-control study was conducted to determine risk factors for sporadic *Salmonella* Typhimurium infection. The study was conducted during 1999 and 2000. *Salmonella* Typhimurium phage-type 104 (DT104) and non-DT104 diarrheal illness in Canada, were studied. The risk factors associated with DT104 illness were recent antibiotic use (OR 5.2, 95% CI 1.8–15.3), residence on a livestock farm (OR 4.9, 95% CI 1.9–18.9), and travel outside Canada in the recent past (OR 4.1, 95% CI 1.2–13.8) (Doré et al., 2004). To determine the risk factors for typhoid fever, a community-based case-control study was conducted during 2001- 2003, in Jakarta, Indonesia. The risk factors for typhoid fever were recent typhoid fever in the household (OR 2.38, 95% CI 1.03-5.48); not using soap for hand washing (OR 1.91, 95% CI 1.06-3.46); eating and sharing food in the same dish (OR 1.93, 95% CI 1.10-3.37), and non availability of appropriate sanitation services (OR 2.20, 95% CI 1.06-4.55). Typhoid fever was also found to be associated with young age and female gender (Vollaard et al., 2004).
There is a well-established association between *E. coli* O157:H7 infections and consumption of contaminated ground beef. Sporadic *E. coli* O157:H7 infection has been found to be linked to eating undercooked hamburgers and exposure to cattle on farms (Allos, Moore, Griffin, & Tauxe, 2004). A case-control study of sporadic Shiga toxin-producing *Escherichia coli* O157 (STEC O157) infections was conducted in 1999–2000, in US. In this 12-month study, 283 patients and 534 controls participated. The risk factors for STEC O157 infection included consumption of pink hamburgers, intake of unsafe surface water, and exposure to cattle (Voetsch et al., 2007).

3.2 Environmental risk factors

The environmental risk factors, food industry, food manufacturing, retail and catering sectors play important role in foodborne illness. Various studies of both sporadic and outbreak-associated illness involving different geographic areas, varied study designs, and a variety of pathogens have revealed that restaurants play an important role as source of foodborne illness in the US. Many factors may contribute to an increased risk of foodborne illness due to foods consumed in restaurants such as cross-contamination events within restaurants (Jones & Angulo, 2006). Research has shown that populations with low socioeconomic status have less access to high-quality food products, resulting in reliance on small markets that may sell foods of poorer quality. These populations have less access to supermarkets selling a variety of fresh fruits and vegetables and low-fat foods. Consumption of a poorer- quality diet along with other risk factors may lead to foodborne illness.

4. Causative agents

A variety of agents can cause foodborne illnesses. Although most of the foodborne illnesses are caused by bacterial or viral pathogens, there are certain non-infectious causes such as, chemicals and toxins. Approximately 67% of all foodborne illnesses caused by pathogens have viral etiology. Most commonly implicated viruses in foodborne illnesses are norovirus, hepatitis A virus, hepatitis E virus, rotavirus, and astrovirus (Atreya, 2004).

4.1 Foodborne pathogens

Various bacterial, viral and parasitic agents causing foodborne illness of public health importance are listed below (Nelson & Williams, 2007; The Food Safety and Inspection Service [FSIS], 2008):

Bacterial Agents
Listeria monocytogenes
Staphylococcus aureus
Bacillus cereus
Bacillus anthracis
Clostridium botulinum
Clostridium perfringens
Clostridium difficile
Salmonella spp
Shigella spp
Campylobacter spp
Escherichia coli 0157:H7
Yersinia enterocolitica
Brucella spp
Vibrio Cholerae
Viral Agents
Norovirus
Hepatitis A
Hepatitis E
Adenovirus (Enteric)
Rotaviruses
Parasitic Agents
Cryptosporidium sp.
Cyclospora cayetanensis
Giardia lamblia
Entamoeba histolytica
Balantadium coli

5. Clinical manifestations

The clinical features of foodborne illness include nausea, vomiting, diarrhea, abdominal pain and fever (McCulloch, 2000). Symptoms of foodborne illnesses may persist for two to three days. In majority of patients, the illness is mild in nature; however, severe complications may occur in some cases. The serious complications of foodborne illness include hypovolemic shock, septicemia, hemolytic uremic syndrome, reactive arthritis, and

Guillan- Barré syndrome. Severe complications may occur in immunocompromised cases, patients with co-morbidities, and in very young or elderly patients. Several cases of foodborne illnesses remain undiagnosed as the patient may not seek medical attention. If a person reports to a health care facility, generally, stool specimens are taken and tested for bacteria and parasites (Linscott, 2011).

Table 1 displays the common causative agents of foodborne illnesses along with their incubation period, clinical features, possible contaminants, diagnostic procedures and steps for prevention (Iowa State University, 2010; Linscott, 2011; Nelson & Williams, 2007).

Organism	Incubation Period	Clinical Features	Possible contaminants	Laboratory diagnosis	Preventive measures
Bacillus cereus	30 minutes to 15 hours	Sudden onset of nausea and vomiting, abdominal cramps with or without diarrhea	Cooked but not properly refrigerated foods, such as vegetables, fish, rice, potatoes and pasta.	Stool test Food sources may be tested.	Careful attention to food preparation, cooking and storage standards
Brucella sp.	7 to 21 days	Fever, night sweats, backache, muscle aches, diarrhea	Unpasteurized dairy foods and meat	Serology, blood culture.	Consumption of pasteurized dairy products. Cooking meat thoroughly
Campylobacter jejuni	1 to 7 days	Fever, headache, nausea, abdominal cramps, diarrhea. Guillian-Barre syndrome may occur in some patients.	Raw and undercooked poultry, eggs, raw beef, unpasteurized milk, contaminated water	Stool culture, rapid immune-chromogenic tests, molecular assays.	Pasteurization of milk, cooking foods properly, prevention of cross-contamination
Clostridium botulinum – preformed toxin	12 to 72 hours	Abdominal cramps, nausea, vomiting, diarrhea, diplopia, blurred vision, headache, dryness of mouth, muscle paralysis, respiratory failure, nerve damage.	Inappropriately canned foods, meats, sausage, fish.	Detection of botulinum toxin in serum, stool, or patient's food	Proper canning of foods, proper cooking of foods
Clostridium perfringen- toxin	8 to 22 hours	Nausea, abdominal cramps and diarrhea, dehydration in some cases	Meat, poultry, gravy, inadequately reheated food	Stool test for enterotoxin	Maintenance of proper cooking temperatures

Organism	Incubation Period	Clinical Features	Possible contaminants	Laboratory diagnosis	Preventive measures
Cryptosporidium sp.	2 to 10 days	Nausea, loss of appetite, watery diarrhea accompanied by mild abdominal cramp; severity depends on host immune status.	Undercooked food or food contaminated by ill food handler, contaminated drinking water or milk	Stool test for detecting oocysts by modified acid-fast stain, direct fluorescent antibodies, or with immunoassays	Avoidance of contaminated water or food, washing hands after using the toilet and before handling food.
Cyclospora cayetanensis	1 to 14 days	Watery diarrhea, abdominal cramps, nausea, anorexia, and weight loss	Fresh fruits and vegetables	Stool test for detecting oocysts by modified acid-fast stain	Hygienic practices in the agricultural setting and in the individual's environment.
Entero-hemorrhagic E.coli – 0157;H7 and other Shiga toxins	1 to 8 days	Bloody diarrhea, vomiting, abdominal cramps, fever, hemorrhagic colitis, hemolytic uremic syndrome	Undercooked ground beef, unpasteurized milk, fruit juices, raw fruits, and vegetables	Stool culture for isolating the organism (Sorbitol MacConkey or CHROMagar media), Antisera or latex agglutination, Immunoassays.	Cooking meat thoroughly, prevention of cross-contamination
Hepatitis A virus (HAV)	15 to 50 days	Anorexia, nausea, abdominal discomfort, malaise, diarrhea, fever, dark colored urine, and jaundice. Arthritis, urticarial rash and aplastic anemia in rare cases	Consumption of contaminated water or food. Shellfish, clams, oysters, fruits, vegetables, iced drinks, lettuce, and salads.	Serum test for IgM antibodies to HAV, Serum test for Alanine transaminase (ALT), Aspartate aminotransferase (AST) and bilirubin	Washing hands after using toilet and before preparing food
Listeria monocytogenes	2 days to 3 weeks	Diarrhea, nausea, fever, muscular pain, flu-like symptoms in pregnant females (may result in preterm birth or still birth). Meningitis and septicemia may occur in older age or immunocompromised patients	Unpasteurized milk, cheese prepared with unpasteurized milk, vegetables, meats, hotdogs, and seafood	Blood culture, cerebrospinal fluid cultures, detection of antibodies to listerolysin O.	Pasteurization of dairy products, cooking foods properly, prevention of cross-contamination.

Organism	Incubation Period	Clinical Features	Possible contaminants	Laboratory diagnosis	Preventive measures
Noroviruses	Between 12 and 48 hours (average, 36 hours); duration, 12-60 hours	Nausea, vomiting, diarrhea, abdominal cramps, headache, and fever	Raw food products, contaminated food or drinking water. Shellfish, oysters and salads.	Stool culture, nucleic acid hybridization, Reverse transcription polymerase chain reaction (PCR), Electron microscopy, Enzyme , and Immunoassays (ELISA).	Proper sewage disposal, water chlorination, restricting infected food handlers from handling food
Salmonella sp.	Non-Typhi: 1 to 3 days Typhi: 3 to 60 days	Non- Tyhi: Nausea, diarrhea, abdominal pain, and fever Typhi: fever, headache, shivering, loss of appetite, malaise, constipation, and muscular pain	Contaminated egg; poultry; meat; unpasteurized milk, dairy foods and fruit juice; raw fruits and vegetables.	Non –Typhi: Stool culture Typhi: Stool culture and blood culture	Cooking food thoroughly, prevention of cross-contamination.
Shigella sp.	12 to 50 hours	Vomiting, abdominal pain, diarrhea with blood and mucus, and fever	Contaminated food or drinking water. Raw vegetables, salads, dairy foods, and poultry.	Stool culture	Practicing proper washing and hygienic techniques in food preparation
Staphylococcus aureus (preformed enterotoxin)	1 to 6 hours	Nausea, vomiting, abdominal pain, fever and diarrhea.	Inappropriately refrigerated foods such as meat, salads, salad dressing, cream pastries, cream-filled baked products, poultry, gravy, and sandwich fillings	Stool or vomitus test for detection of toxin. Suspected food may be tested to detect toxin	Refrigerating foods properly, using hygienic practices
Vibrio cholerae (O1, O139) (non- O1 or non O139)	4 hours to 4 days	Severe watery diarrhea, abdominal cramps, nausea vomiting, headache, fever, and chills. Severe dehydration and death may occur.	Contaminated water, shellfish and crustaceans	Stool culture	Cooking food thoroughly

Organism	Incubation Period	Clinical Features	Possible contaminants	Laboratory diagnosis	Preventive measures
Vibrio parahaemolyticus	2 to 48 h	Watery diarrhea, abdominal pain, nausea, and vomiting	Raw or undercooked seafood from contaminated seawater	Stool culture	Cooking food thoroughly, practicing hygienic techniques in food preparation and storing food at the appropriate temperatures
Yersinia enterocolitica	1 to 3 days	Diarrhea, vomiting, fever, and abdominal cramps	Contaminated meat and milk, undercooked pork, unpasteurized milk, and contaminated water.	Stool culture, blood culture.	Pasteurization of milk; Cooking food thoroughly, prevention of cross-contamination, practicing hygienic techniques in food preparation

Table 1. Causative agents, clinical features and preventive measures for common foodborne illnesses

6. Foodborne illness outbreaks

Manifestation of the same symptoms or illness by two or more of the individuals after consumption of the same contaminated food, is labeled as an outbreak of foodborne illness (The Food Safety and Inspection Service (FSIS), 2008). The outbreak can be recognized when several people who ate together at an occasion become sick and have similar clinical manifestations. Most of the outbreaks are localized, such as an outbreak after eating a catered meal or at a restaurant. However, more widespread outbreaks also occur that affect people in various places, and may last for several weeks (FSIS, 2008).

An outbreak is investigated by describing it systematically and trying to find out the causative agent. The description of outbreak includes time, place, and person distribution. Data is collected by interviews, by testing suspected food source, and by gathering other related information (FSIS, 2008). It is important that foodborne illness outbreaks are investigated timely and proper environmental assessments are done so that appropriate prevention strategies are identified (Lynch, Tauxe, & Hedberg, 2009). According to CDC, the etiology of majority (68%) of reported foodborne-illness outbreaks is unknown. The causative agent of many of the outbreaks cannot be determined because of certain issues related to outbreak investigations e.g. lack of timely reporting, lack of resources for investigations and other priorities in health departments. In addition, ill persons who do not seek health care and limited testing of specimens are also the contributory factors in failure to determine the cause of foodborne illness outbreak (Lynch et al., 2009).

A number of foodborne illness outbreaks are reported from various parts of the world. Analysis of foodborne outbreak data helps in the estimation of the proportion of human cases of specific enteric diseases attributable to a specific food item (Greig & Ravel, 2009). Worldwide, a total of 4093 foodborne outbreaks occurred between 1988 and 2007. It was found that *Salmonella* Enteritidis outbreaks were more common in the EU states and eggs were the most frequent vehicle of infection. Poultry products in the EU and dairy products in the United States, were related to *Campylobacter* associated outbreaks. In Canada, *Escherichia coli* outbreaks were associated with beef. In Australia and New Zealand, *Salmonella* Typhiumurium outbreaks were more common (Greig & Ravel, 2009). Daniels and colleague (2002) conducted a study in the United States, to describe the epidemiology of foodborne illness outbreaks in schools, colleges and universities. The data from January 1, 1973, to December 31, 1997 was reviewed. Characteristics of ill persons, the magnitude of foodborne illness outbreaks, causative agents, food vehicles for transmitting infection, place of preparing food and contributory factors for occurrence of outbreaks, were examined. A total of 604 outbreaks of foodborne illness were reported during the study period. In majority (60%) of the outbreaks the etiology was unknown. Among the outbreaks with a known etiology, in 36% of outbreak reports Salmonella was the most commonly identified pathogen. Foods containing poultry, salads, Mexican-style food, beef and dairy foods were the most commonly implicated vehicles for transmission (Daniels et al., 2002). A total of 6,647 outbreaks of foodborne illness were reported during 1998-2002 in U.S (Lynch, Painter, Woodruff, & Braden, 2006). As a result of these outbreaks, 128,370 persons were reported to become ill. The etiology was identified in 33% outbreaks. The largest percentage of outbreaks was caused by bacterial pathogens and *Salmonella* Enteritidis was the most common causative agent. However, the highest mortality was caused by *Listeria monocytogenes*. Viral pathogens were responsible for 33% of the outbreaks. Among the viral pathogens, norovirus was the most common causative agent (Lynch et al., 2006). In 2002, a salmonellosis outbreak occurred in five states of U.S. It occurred after consuming ground beef. During this outbreak, forty seven cases were reported; out of which 17 people were hospitalized and one death was reported (FSIS, 2008).

Systematic surveillance systems are very important to get information about causative organisms, sources of infection and modes of transmission in foodborne illness (O'Brien et al., 2006). An electronic surveillance system (SurvNet) was established for monitoring and investigation of infectious disease outbreaks, in 2001, in Germany (Krause et al., 2007). During 2001–2005, a total of 30,578 outbreaks were reported. These outbreaks ranged in size from 2 to 527 cases. The most common settings for outbreaks in 2004 and 2005 were households, nursing homes, and hospitals (Krause et al., 2007). In England and Wales, systematic national surveillance of outbreaks of infectious intestinal disease (IID) was introduced in 1992. This system provides information on etiologic agent, sources of infection and modes of transmission. Between January 1, 1992 and January 31, 2003, a total of 1763 outbreaks of IID were reported (O'Brien et al., 2006). From 1992 to 2008, 2429 foodborne outbreaks were reported in England and Wales. Approximately half of the outbreaks were caused by *Salmonella* spp. Poultry and red meat was the most commonly implicated foods in the causation of outbreaks. The associated factors in most outbreaks were cross-contamination, lack of adequate heat treatment and improper food storage (Gormley et al., 2011). In central Taiwan, 274 outbreaks of foodborne illness including 12,845 cases and 3 deaths were reported during 1991 to 2000. Majority (62.4%) of the outbreaks were caused by bacterial pathogens. The main etiologic agents were *Bacillus cereus*, *Staphylococcus aureus*, and *Vibrio parahaemolyticus*. The important contributing factor was improper handling of food. The implicated foods included seafood, meat products and cereal products (Chang & Chen, 2003).

A review of *E. coli* O157 outbreaks reported to CDC from 1982 to 2002, was conducted (Sparling, Crowe, Griffin, Swerdlow, & Rangel, 2005). During the study period, 49 states reported 350 outbreaks. During these outbreaks 8,598 cases were reported; out of which 1,493 (17%) were hospitalized, and 40 (0.5%) died. Transmission route for 183 (52%) was foodborne and the food vehicle for 75 (41%) foodborne outbreaks was ground beef (Sparling et al., 2005). In a review of all confirmed foodborne outbreaks reported to the Centers for Disease Control and Prevention (CDC) from 1982 to 1997, a total of 2,246 foodborne outbreaks were reported; 697 (31%) were of known etiology. *Salmonella* was responsible for 65% of outbreaks with a known etiology (Hedberg et al., 2008). An analysis of national foodborne outbreak data from 1973 through 2001for determining the proportion of *Salmonella* Heidelberg outbreaks and its causes was conducted (Chittick, Sulka, Tauxe, & Fry, 2006). Out of 6,633 outbreaks with known etiology, *Salmonella* Heidelberg was responsible for 184 (3%) outbreaks. 53 outbreaks were poultry or egg-related (Chittick et al., 2006). Because of increasing consumption of fresh produce, changes in production and distribution, there is an increase in foodborne outbreaks caused by contaminated fresh produce (Lynch et al., 2009; Sivapalasingam et al., 2004). An analysis of data for 1973 through 1997 from the Foodborne Outbreak Surveillance System was conducted. A total of 190 outbreaks were reported including 16,058 illnesses, 598 hospitalizations, and eight deaths. There was an increase in the proportion of produce-associated outbreaks, rising from 0.7% in the 1970s to 6% in the 1990s. The food items most frequently implicated in the outbreaks associated with fresh produce include salad, lettuce, juice, melon, sprouts, and berries. In this analysis, the bacterial pathogens were the most common causative agents. Among the bacterial pathogens *Salmonella* was the commonest causative agent (Sivapalasingam et al., 2004). In a study carried out from October 2004 to October 2005 in Catalonia, Spain, 181 outbreaks were reported; 72 were caused by *Salmonella* and 30 by norovirus (NoV); 66.7% of NoV outbreaks occurred in restaurants. Hospitalizations were reported more commonly in outbreaks caused by bacterial pathogens as compared to those caused by NoV (Martinez et al., 2008). In Spain, 971 and 1227 outbreaks were reported in 2002 and 2003, respectively. A substantial proportion of outbreaks were associated with consumption of eggs and egg products (Crespo et al., 2005). A study was conducted in northeastern Spain to investigate the trend of foodborne *Salmonella*-caused outbreaks and number of cases, hospitalizations, and deaths. A review of the information on reported outbreaks of foodborne disease from 1990 to 2003 found a total of 1,652 outbreaks. 1,078 outbreaks had a known etiologic agent; out of which 871 (80.8%) were caused by *Salmonella*, with 14,695 cases, 1,534 hospitalizations, and 4 deaths. Forty-eight percent of *Salmonella*-caused outbreaks were associated with consumption of eggs. The study concluded that to prevent and control these outbreaks, health education programs are needed to create awareness regarding risk of consuming raw or undercooked eggs (Domínguez et al., 2007). In 2002, in the Netherlands a national study of foodborne illness outbreaks was performed (van Duynhoven et al., 2005). A total of 281 foodborne illness outbreaks were included. Most of these outbreaks were reported from nursing homes, restaurants, hospitals and day-care centers. In restaurant outbreaks, food was the mode of transmission in almost 90% outbreaks. The causative agents included norovirus (54%), Salmonella spp. (4%), rotavirus (2%), and Campylobacter spp. (1%) (van Duynhoven et al., 2005). The Danish and Dutch investigation reported foodborne outbreaks caused by the same strain of *Salmonella* serotype Typhimurium DT104 during the year 2005. These were two distinct outbreaks caused by the same organism. The outbreaks were traced to two kinds of raw beef provided by the same supplier (Ammon & Tauxe, 2007).

A study conducted in Qassim province, Saudi Arabia, analyzed the foodborne illness surveillance data for the year 2006. During the study period, 31 foodborne illness outbreaks comprising of 251 cases, were reported. The most common etiologic agent was *Salmonella* spp, followed by *Staphylococcus* aureus. Commercially prepared foods were consumed by the majority (68.9%) of the cases. Meat and Middle Eastern meat sandwich were the commonly implicated food vehicles (Al-Goblan & Jahan, 2010).

The globalization of the food industry has resulted in occurrence of international viral foodborne outbreaks (Verhoef et al., 2009). Although mostly the bacterial pathogens are targeted for prevention of foodborne illness, yet it is suspected that noroviruses (NoV) are the most common cause of gastroenteritis. The role of NoV in foodborne illness has become evident because of new molecular assays. An analysis of 8, 271 foodborne outbreaks reported to CDC from 1991 to 2000 showed an increase in the proportion of NoV-confirmed outbreaks. It rose from 1% in 1991 to 12% in 2000. In addition, NoV outbreaks were larger in magnitude as compared to the bacterial outbreaks (Widdowson et al., 2005).

The Belgian data for foodborne norovirus (NoV) outbreaks showed that in 2007, 10 NoV foodborne outbreaks were reported accounting for 392 cases in Belgium. NoV was the most commonly identified agent in foodborne outbreaks. Sandwiches were the most commonly implicated food. A review of forty outbreaks due to NoV, from 2000 to 2007 revealed that the food handler was responsible for the outbreak in 42.5% of the cases. The source of transmission was water (27.5%), shellfish (17.5%) and raspberries (10.0%) in remaining cases (Baert et al., 2009).

The contributory factors observed in most of the foodborne illness outbreaks included poor personal hygiene, inadequate holding times and temperatures, cross-contamination, lack of adequate heat treatment and improper food storage (Chang & Chen, 2003; Gormley et al., 2011; Hedberg et al., 2008).

7. Emerging foodborne infections

The changes in the methods of food production have affected the epidemiology of foodborne infection (Tassios & Kerr, 2010). Various new pathogens have emerged due to changing production processes in food industry. Some of these are new pathogens and were unknown previously, others are emerging pathogens for foodborne infections, and some others are evolving pathogens that have become more potent (Mor-Mur & Yuste, 2009). These pathogens include *Campylobacter jejuni, Salmonella* Typhimurium DT104, enterohemorrhagic *Escherichia coli, Listeria monocytogenes, Arcobacter butzleri, Mycobacterium avium* subsp. *Paratuberculosis,* etc. *Campylobacter jejuni* O:19 and other serotypes may cause neuropathy called Guillain–Barré syndrome. Multi-drug resistant *Salmonella* Typhimurium DT104 and other serotypes may cause salmonellosis leading to chronic reactive arthritis. Enterohemorrhagic *Escherichia coli* infection can cause complications such as hemolytic uremic syndrome and thrombotic thrombocytopenic purpura. *Listeria monocytogenes* causes listeriosis, a public health problem of major concern due to severe non-enteric nature of the disease i-e, meningitis or meningoencephalitis, and septicemia. *Arcobacter butzleri* is isolated from raw poultry, meat, and meat products and is considered a potential foodborne pathogen. *Mycobacterium avium* subsp. *paratuberculosis* may contribute to Crohn's disease (Mor-Mur & Yuste, 2009).

The new emerging foodborne infections are also related to food handling practices. Foodborne infections caused by *Escherichia coli* O157, *Yersinia enterocolitica, Campylobacter* and *Vibrio*, are associated with processing and packaging of food, or the importation of certain food from a new geographical area (Robinson, 2007). The well known foodborne pathogens may develop antimicrobial resistance and generate new public health challenges (Newell et al., 2010). Antimicrobial resistance has developed in many bacterial foodborne pathogens, such as *Salmonella, Shigella, Vibrio* spp., *Campylobacter*, methicillin resistant *Staphylcoccus aureas, E. coli* and *Enterococci*.

Previously un-recognized foodborne pathogens are constantly emerging. The emerging viral foodborne infections include norovirus, hepatitis A, rotaviruses and SARS (Newell et al., 2010). Currently, the norovirus (NoV) and hepatitis A virus (HAV) are considered the most important human foodborne pathogens. NoV and HAV are highly infectious and may cause large outbreaks (Koopmans & Duizer, 2004). Hepatitis E causes substantial morbidity in many developing countries while in developed countries, it was previously thought to be confined to travelers returning from endemic areas (Dalton, Bendall, Ijaz, & Banks, 2008). However, this concept is changing based on the evidence available. In developed countries, Autochthonous hepatitis E is found to be quite common and affects older age groups leading to morbidity and mortality (Dalton et al., 2008).

8. Monitoring and surveillance of foodborne illness

Worldwide, there are various surveillance systems to monitor, investigate, control and prevent illness. To assess and monitor morbidity and mortality in the United States, surveillance activities are conducted by several systems in collaboration with federal agencies and health departments (McCabe-Sellers & Beattie, 2004). Some surveillance systems are specific for foodborne illnesses. In addition to monitoring the foodborne illness, these surveillance activities also help in evaluating the safety of the food supply (Allos et al., 2004). Some of these surveillance systems are discussed below:

8.1 FoodNet (Foodborne disease active surveillance network)

FoodNet is the surveillance system in the United States. For Foodnet, CDC has collaborated with ten Emerging Infections Program (EIP) sites (California, Colorado, Connecticut, Georgia, New York, Maryland, Minnesota, Oregon, Tennessee and New Mexico), the US Department of Agriculture, and the Food and Drug Administration. It performs active surveillance for foodborne illnesses and also conducts epidemiologic studies to determine the changing epidemiology of foodborne illnesses. It responds to new and emerging foodborne illnesses, monitors the burden of foodborne illnesses, and identifies their sources (FSIS, 2008).

Figure 2 shows the burden of illness pyramid. It helps in understanding foodborne disease reporting in FoodNet surveillance system. It shows steps involved in the registration of an episode of foodborne illness in the population. Moving from bottom to top of pyramid, the steps are: exposure of some individuals in the general population to an organism; out of the exposed some persons become ill; out of all ill some seek medical care; a specimen is obtained from some of these persons and sent to a clinical laboratory; some of these specimens are tested for a specific pathogen; the causative organism is identified in some of these tested specimens; a local or state health department receives the report of the laboratory-confirmed case (CDC, 2011c).

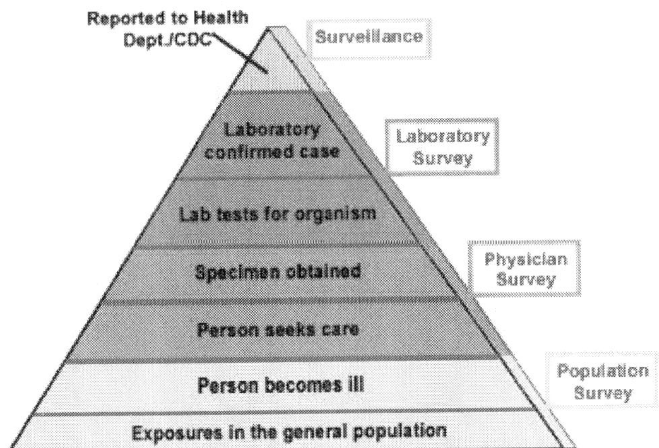

Fig. 2. FoodNet surveillance, burden of illness pyramid. Adapted from CDC. Available at: http://www.cdc.gov/foodnet/surveillance_pages/burden_pyramid.htm. Accessed August 04, 2011.

8.2 PulseNet (The molecular subtyping network for foodborne bacterial disease surveillance)

In the United States, PulseNet has created a national framework for pathogen-specific surveillance (Li et al., 2010). PulseNet is responsible for molecular subtyping of foodborne illness surveillance. It helps in detecting widespread foodborne outbreaks by comparing strains of bacterial pathogens from all over the United States. It performs DNA "fingerprinting" on foodborne bacteria by pulsed-field gel electrophoresis. By identifying and labeling each "fingerprint" pattern, it is possible to rapidly compare these patterns through an electronic database at the CDC, thus identifying related strains (FSIS, 2008).

8.3 Electronic Foodborne Outbreak Reporting System (eFORS)

The Electronic Foodborne Outbreak Reporting System's (eFORS) database is a surveillance system that collects reports on foodborne outbreaks. It requires specialized knowledge and expertise to appropriately analyze and interpret the data (Middaugh, Hammond, Eisenstein, & Lazensky, 2010). Various studies are conducted by analyzing the data collected within the Electronic Foodborne Outbreak Reporting System (eFORS) in various settings, such as schools in order to examine the magnitude of foodborne illness, their etiologies and to provide recommendations for prevention of foodborne illness (Venuto, Halbrook, Hinners, & Mickelson, 2010).

9. Strategies for control of foodborne illness

The contamination of food is influenced by multiple factors and may occur anywhere in the food production process (Newell et al., 2010). However, most of the foodborne illnesses can be traced back to infected food handlers. Therefore, it is important that strict personal hygiene measures should be adopted during food preparation. To prevent foodborne infections in children, educational measures are needed for parents and care-takers. The interventions should focus on avoiding exposure to infectious agents and on preventing cross-contamination (Marcus, 2008).

Good agriculture practice and good manufacturing practice should be adopted to prevent introduction of pathogens into food products (Koopmans & Duizer, 2004). In order to control foodborne viral infections, it is important to increase awareness of food handlers regarding the presence and spread of these viruses. In addition, standardized methods for the detection of foodborne viruses should be utilized and laboratory-based surveillance should be established for early detection of outbreaks (Koopmans & Duizer, 2004).

To prevent food-related zoonotic diseases, collaboration between public health, veterinary and food safety experts should be established. This collaboration will help in monitoring trends in the existing diseases and in detecting emerging pathogens. It will help in developing effective prevention and control strategies (Newell et al., 2010). The control strategies should be based on creating awareness among the consumers, farmers and those raising farm animals. The improvement of farming conditions, the development of more sensitive methods for detection of pathogens in slaughtered animals and in food products, and proper sewage disposal are other intervention strategies (Pozio, 2008). Hygienic measures are required throughout the continuum from "farm to fork". Further research is also required to explore pathways of the foodborne illness and to determine the vehicles of the greatest importance (Unicomb, 2009).

In a study conducted in Turkey, knowledge, attitudes, and practices about food safety among food handlers, were explored. The study revealed that food handlers in Turkish food industry often lacked knowledge regarding basic food hygiene. The authors concluded that the food handlers must be educated regarding safe food handling practices (Bas, Safak Ersun, & KIvanç, 2006). For the prevention of foodborne outbreaks, training of food handlers, regarding appropriate preparation and storage of food is required. In addition, effective environmental cleaning and disinfection, excluding infected staff, implementing hand hygiene principles, and preventing cross-contamination are recommended (Greig & Lee, 2009).

Proper processing of food is necessary to ensure the reduction or elimination of the growth of harmful microorganisms. Pasteurization of milk and dairy products and hygienic manufacturing processes for canned foods will help reduce the cases of food-borne illnesses. Food irradiation is a recent technology for prevention of food-borne illnesses. The food irradiation methods include Gamma irradiation, Electron beam irradiation, and X-irradiation. Irradiation destroys the organism's DNA and prevents DNA replication. Food irradiation could eliminate *E. coli* in ground beef, *Campylobacter* in poultry, *Listeria* in food and dairy products, and *Toxoplasma gondii* in meat. However, all food products cannot be irradiated (Linscott, 2011).

The consumers should also take precautions to prevent foodborne illnesses. These include cooking meat, poultry, and eggs at appropriate temperatures; proper refrigeration and storage of foods at recommended temperatures; prevention of cross-contamination of food; use of clean slicing boards and utensils while cooking; and washing hands often while preparing food (Linscott, 2011).

10. Conclusion

Globally, foodborne illnesses are responsible for substantial morbidity and mortality. Although it is difficult to determine the exact mortality associated with foodborne illnesses, worldwide an estimated 2 million deaths occurred due to gastrointestinal illness during the year 2005. Foodborne illnesses result from consumption of food containing pathogens or the food contaminated by poisonous chemicals or toxins. Foodborne illnesses also play an important role in new and emerging infections. Various new pathogens have emerged due to changing dynamics of food industry. It is estimated that during past 60 years, approximately 30% of all emerging infections comprised of pathogens transmitted through food.

It is important to monitor and investigate the foodborne illnesses in order to control and prevent them. For this purpose various surveillance systems are established in different parts of the world. FoodNet, PulseNet and EFORS are the surveillance systems for monitoring foodborne illness in the United States. These surveillance systems play vital role in prevention, early detection and control of foodborne illness outbreaks.

The contamination of food is influenced by multiple factors and may occur anywhere along the food chain. Good agriculture practice and good manufacturing practice should be adopted to prevent introduction of pathogens into food products. Most foodborne illnesses can be tracked to infected food handlers. Therefore, it is important that strict personal hygiene measures should be adopted during food preparation. The consumers should also take precautions for prevention of foodborne illness. These include cooking food at appropriate temperatures and following standard hygiene practices, proper storage and prevention of cross-contamination of food. Thus integrated intervention strategies are required to prevent foodborne illness at community level. Successful implementation of these interventions requires inter-sectoral collaboration including agriculture industry, food industry and health care sector.

11. References

Al-Goblan, A. S., & Jahan, S. (2010). Surveillance for foodborne illness outbreaks in Qassim, Saudi Arabia, 2006. *Foodborne Pathogens and Disease*, 7(12), 1559-1562. doi:10.1089/fpd.2010.0638

Allos, B. M., Moore, M. R., Griffin, P. M., & Tauxe, R. V. (2004). Surveillance for Sporadic Foodborne Disease in the 21st Century: The FoodNet Perspective. *Clinical Infectious Diseases*, 38(s3), S115-S120. doi:10.1086/381577

Anderson, A. L., Verrill, L. A., & Sahyoun, N. R. (2011). Food Safety Perceptions and Practices of Older Adults. *Public Health Reports*, 126(2), 220-227.

Ammon. A., & Tauxe. R. V. (2007). Investigation of multi-national foodborne outbreaks in Europe: some challenges remain, 135(6), 887-889. doi:10.1017/S0950268807008898

Angulo, F. J., & Scallan, E. (2007). Activities, Achievements, and Lessons Learned during the First 10 Years of the Foodborne Diseases Active Surveillance Network: 1996–2005. *Clinical Infectious Diseases*, 44(5), 718 -725. doi:10.1086/511648

Atreya, C. D. (2004). Major foodborne illness causing viruses and current status of vaccines against the diseases. *Foodborne Pathogens and Disease*, 1(2), 89-96. doi:10.1089/153531404323143602

Baert, L., Uyttendaele, M., Stals, A., VAN Coillie, E., Dierick, K., Debevere, J., & Botteldoorn, N. (2009). Reported foodborne outbreaks due to noroviruses in Belgium: the link between food and patient investigations in an international context. *Epidemiology and Infection*, 137(3), 316-325. doi:10.1017/S0950268808001830

Bas, M., Safak Ersun, A., & KIvanç, G. (2006). The evaluation of food hygiene knowledge, attitudes, and practices of food handlers' in food businesses in Turkey. *Food Control, 17*(4), 317-322. doi:16/j.foodcont.2004.11.006

Busani, L., Scavia, G., Luzzi, I., & Caprioli, A. (2006). Laboratory surveillance for prevention and control of foodborne zoonoses. *Annali dell'Istituto Superiore Di Sanità, 42*(4), 401-404.

Buzby, J. C., & Roberts, T. (2009). The Economics of Enteric Infections: Human Foodborne Disease Costs. *Gastroenterology, 136*(6), 1851-1862. doi:10.1053/j.gastro.2009.01.074

Centers for Disease Control and Prevention [CDC]. (2011a). CDC 2011 Estimates: Findings. Retrieved June 26, 2011, from http://www.cdc.gov/foodborneburden/2011-foodborne-estimates.html

Centers for Disease Control and Prevention [CDC]. (2011b). Estimates of Foodborne Illness in the United States. Retrieved June 26, 2011, from http://www.cdc.gov/foodborneburden/index.html

Centers for Disease Control and Prevention [CDC]. (2011c). Foodnet surveillance: Burden of illness. Retrieved August 18, 2011, from http://www.cdc.gov/foodnet/surveillance_pages/burden_pyramid.htm

Chang, J. M., & Chen, T. H. (2003). Bacterial Foodborne Outbreaks in Central Taiwan, 1991-2000. *Journal of Food and Drug Analysis, 11*(1), 53-59.

Chittick, P., Sulka, A., Tauxe, R. V., & Fry, A. M. (2006). A Summary of National Reports of Foodborne Outbreaks of Salmonella Heidelberg Infections in the United States: Clues for Disease Prevention. *Journal of Food Protection, 69,* 1150-1153.

Crespo, P. S., Hernández, G., Echeíta, A., Torres, A., Ordóñez, P., & Aladueña, A. (2005). Surveillance of foodborne disease outbreaks associated with consumption of eggs and egg products: Spain, 2002-2003. *Euro Surveillance: Bulletin Européen Sur Les Maladies Transmissibles = European Communicable Disease Bulletin, 10*(6), E050616.2.

Dalton, H. R., Bendall, R., Ijaz, S., & Banks, M. (2008). Hepatitis E: an emerging infection in developed countries. *The Lancet Infectious Diseases, 8*(11), 698-709. doi:16/S1473-3099(08)70255-X

Daniels, N. A., MacKinnon, L., Rowe, S. M., Bean, N. H., Griffin, P. M., & Mead, P. S. (2002). Foodborne disease outbreaks in United States schools. *The Pediatric Infectious Disease Journal, 21*(7), 623-628. doi:10.1097/01.inf.0000019885.31694.9e

Domínguez, A., Torner, N., Ruiz, L., Martínez, A., Bartolomé, R., Sulleiro, E., Teixidó, A., et al. (2007). Foodborne Salmonella-Caused Outbreaks in Catalonia (Spain), 1990 to 2003. *Journal of Food Protection, 70,* 209-213.

Doorduyn, Y., Van Den Brandhof, W. E., Van Duynhoven, Y. T. H. P., Wannet, W. J. B., & Van Pelt, W. (2006). Risk Factors for Salmonella Enteritidis and Typhimurium (DT104 and Non-DT104) Infections in The Netherlands: Predominant Roles for Raw Eggs in Enteritidis and Sandboxes in Typhimurium Infections. *Epidemiology and Infection, 134*(03), 617-626. doi:10.1017/S0950268805005406

Doré, K., Buxton, J., Henry, B., Pollari, F., Middleton, D., Fyfe, M., Ahmed, R., et al. (2004). Risk Factors for Salmonella Typhimurium DT104 and Non-DT104 Infection: A Canadian Multi-Provincial Case-Control Study. *Epidemiology and Infection, 132*(03), 485-493. doi:10.1017/S0950268803001924

Fleury, M.D., Stratton, J., Tinga, C., Charron, D.F., & Aramini J. (2008). A descriptive analysis of hospitalization due to acute gastrointestinal illness in Canada, 1995-2004. *Canadian Journal of Public Health*, 99 (6), 489-93.

Friedman, C. R., Hoekstra, R. M., Samuel, M., Marcus, R., Bender, J., Shiferaw, B., Reddy, S., et al. (2004). Risk Factors for Sporadic *Campylobacter* Infection in the United States: A Case-Control Study in FoodNet Sites. *Clinical Infectious Diseases*, *38*(s3), S285-S296. doi:10.1086/381598

Gormley, F. J., Little, C. L., Rawal, N., Gillespie, I. A., Lebaigue, S., & Adak, G. K. (2011). A 17-year review of foodborne outbreaks: describing the continuing decline in England and Wales (1992-2008). *Epidemiology and Infection*, *139*(5), 688-699. doi:10.1017/S0950268810001858

Greig, J. D., & Lee, M. B. (2009). Enteric outbreaks in long-term care facilities and recommendations for prevention: a review. *Epidemiology and Infection*, *137*(2), 145-155. doi:10.1017/S0950268808000757

Greig, J. D., & Ravel, A. (2009). Analysis of foodborne outbreak data reported internationally for source attribution. *International Journal of Food Microbiology*, *130*(2), 77-87. doi:16/j.ijfoodmicro.2008.12.031

Hedberg, C. W., Palazzi-Churas, K. L., Radke, V. J., Selman, C. A., & Tauxe, R. V. (2008). The use of clinical profiles in the investigation of foodborne outbreaks in restaurants: United States, 1982–1997. *Epidemiology and Infection*, *136*(01), 65-72. doi:10.1017/S0950268807008199

Helms, M., Vastrup, P., Gerner-Smidt, P., & Mølbak, K. (2003). Short and long term mortality associated with foodborne bacterial gastrointestinal infections: registry based study. *BMJ (Clinical Research Ed.)*, *326*(7385), 357.

Hoffman, R. E., Greenblatt, J., Matyas, B. T., Sharp, D. J., Esteban, E., Hodge, K., & Liang, A. (2005). Capacity of State and Territorial Health Agencies To Prevent Foodborne Illness. *Emerging Infectious Diseases*, *11*(1), 11-16.

Jones, T. F., & Angulo, F. J. (2006). Eating in restaurants: a risk factor for foodborne disease? *Clinical Infectious Diseases: An Official Publication of the Infectious Diseases Society of America*, *43*(10), 1324-1328. doi:10.1086/508540

Iowa State University. (2010). Common foodborne pathogens. Retrieved July 23, 2011, from http://www.extension.iastate.edu/foodsafety/pathogens/index.cfm?parent=37

Jones, T. F., Imhoff, B., Samuel, M., Mshar, P., McCombs, K. G., Hawkins, M., Deneen, V., et al. (2004). Limitations to Successful Investigation and Reporting of Foodborne Outbreaks: An Analysis of Foodborne Disease Outbreaks in FoodNet Catchment Areas, 1998–1999. *Clinical Infectious Diseases*, *38*(s3), S297-S302. doi:10.1086/381599

Krause, G., Altmann, D., Faensen, D., Porten, K., Benzler, J., Pfoch, T., Ammon, A., et al. (2007). SurvNet Electronic Surveillance System for Infectious Disease Outbreaks, Germany, *13*(10), 1548-1555. doi:10.3201/eid1310.070253

Koopmans, M., & Duizer, E. (2004). Foodborne viruses: an emerging problem. *International Journal of Food Microbiology*, *90*(1), 23-41. doi:16/S0168-1605(03)00169-7

Kuchenmüller, T., Hird, S., Stein, C., Kramarz, P., Nanda, A., & Havelaar, A. H. (2009). Estimating global burden of foodborne diseases – a collaborative effort. *Eurosurveillance, 14* (18), 191-95.

Li, J., Smith, K., Kaehler, D., Everstine, K., Rounds, J., & Hedberg, C. (2010). Evaluation of a Statewide Foodborne Illness Complaint Surveillance System in Minnesota, 2000 through 2006. *Journal of Food Protection, 73*(11), 2059-2064.

Lindqvist, R., Andersson, Y., Lindbäck, J., Wegscheider, M., Eriksson, Y., Tideström, L., Lagerqvist-Widh, A., et al. (2001). A one-year study of foodborne illnesses in the municipality of Uppsala, Sweden. *Emerging Infectious Diseases, 7*(3 Suppl), 588-592.

Linscott, A. J. (2011). Food-Borne Illnesses. *Clinical Microbiology Newsletter, 33*(6), 41-45. doi:10.1016/j.clinmicnews.2011.02.004

Lynch, M. F., Tauxe, R. V., & Hedberg, C. W. (2009). The Growing Burden of Foodborne Outbreaks Due to Contaminated Fresh Produce: Risks and Opportunities. *Epidemiology and Infection, 137*(Special Issue 03), 307-315. doi:10.1017/S0950268808001969

Lynch, M., Painter, J., Woodruff, R., & Braden, C. (2006). Surveillance for foodborne-disease outbreaks--United States, 1998-2002. *MMWR. Surveillance Summaries: Morbidity and Mortality Weekly Report. Surveillance Summaries / CDC, 55*(10), 1-42.

Marcus, R. (2008). New information about pediatric foodborne infections: the view from FoodNet. *Current Opinion in Pediatrics, 20*(1), 79-84. doi:10.1097/MOP.0b013e3282f43067

Marcus, R., Varma, J. K., Medus, C., Boothe, E. J., Anderson, B. J., Crume, T., Fullerton, K. E., et al. (2007). Re-Assessment of Risk Factors for Sporadic Salmonella Serotype Enteritidis Infections: A Case-Control Study in Five FoodNet Sites, 2002–2003. *Epidemiology and Infection, 135*(01), 84-92. doi:10.1017/S0950268806006558

Martinez, A., Dominguez, A., Torner, N., Ruiz, L., Camps, N., Barrabeig, I., Arias, C., et al. (2008). Epidemiology of foodborne Norovirus outbreaks in Catalonia, Spain. *BMC Infectious Diseases, 8*(1), 47. doi:10.1186/1471-2334-8-47

McCulloch, J.E. (ed) (2000). Infection Control: Science, Management and Practice. London: Whurr Publishers.

McCabe-Sellers, B. J., & Beattie, S. E. (2004). Food safety: Emerging trends in foodborne illness surveillance and prevention. *Journal of the American Dietetic Association, 104*(11), 1708-1717. doi:10.1016/j.jada.2004.08.028

Middaugh, J. P., hammond, R. M., Eisenstein, L., & Lazensky, R. (2010). Advancement of the science. Using the Electronic Foodborne Outbreak Reporting System (eFORS) to Improve Foodborne Outbreak Surveillance, Investigations, and Program Evaluation. *Journal of Environmental Health, 73*(2), 8-11.

Mølbak, K., & Neimann, J. (2002). Risk Factors for Sporadic Infection with Salmonella Enteritidis, Denmark, 1997–1999. *American Journal of Epidemiology, 156*(7), 654 -661. doi:10.1093/aje/kwf096

Mor-Mur, M., & Yuste, J. (2009). Emerging Bacterial Pathogens in Meat and Poultry: An Overview. *Food and Bioprocess Technology, 3*(1), 24-35. doi:10.1007/s11947-009-0189-8

Nelson, K., & Williams, C. (Ed.) Second Edition, 2007. *Infectious Disease Epidemiology, Theory and Practice.* New York: Aspen Publishers.

Newell, D. G., Koopmans, M., Verhoef, L., Duizer, E., Aidara-Kane, A., Sprong, H., Opsteegh, M., et al. (2010). Food-borne diseases -- The challenges of 20 years ago still persist while new ones continue to emerge. *International Journal of Food Microbiology, 139*(Supplement 1), S3-S15. doi:16/j.ijfoodmicro.2010.01.021

New Zealand Food Safety Authority. (2010). The economic cost of foodborne disease in New Zealand. Retrieved July 19, 2011, from http://www.foodsafety.govt.nz/elibrary/industry/economic-cost-foodborne-disease/foodborne-disease.pdf

O'Brien, S. J., Gillespie, I. A., Sivanesan, M. A., Elson, R., Hughes, C., & Adak, G. K. (2006). Publication Bias in Foodborne Outbreaks of Infectious Intestinal Disease and Its Implications for Evidence-Based Food Policy. England and Wales 1992–2003. *Epidemiology and Infection, 134*(04), 667-674. doi:10.1017/S0950268805005765

Pozio, E. (2008). Epidemiology and control prospects of foodborne parasitic zoonoses in the European Union. *Parassitologia, 50*(1-2), 17-24.

Robinson, R. K. (2007). Emerging Foodborne Pathogens. *International Journal of Dairy Technology, 60*(4), 305-306. doi:10.1111/j.1471-0307.2007.00331.x

Scallan, E. (2007). Activities, Achievements, and Lessons Learned during the First 10 Years of the Foodborne Diseases Active Surveillance Network: 1996-2005. *Clinical Infectious Diseases, 44*(5), 718-725. doi:10.1086/511648

Scharff, R, L. (2010). Health-related costs from foodborne illness in the United States. Retrieved July 19, 2011, from http://www.marlerblog.com/uploads/image/PSP-Scharff%20v9.pdf

Scharff, R. L., McDowell, J., & Medeiros, L. (2009). Economic Cost of Foodborne Illness in Ohio. *Journal of Food Protection, 72,* 128-136.

Scott, W. G., Scott, H. M., Lake, R. J., & Baker, M. G. (2000). Economic cost to New Zealand of foodborne infectious disease. *The New Zealand Medical Journal, 113*(1113), 281-284.

Sivapalasingam, S., Friedman, C. R., Cohen, L., & Tauxe, R. V. (2004). Fresh produce: a growing cause of outbreaks of foodborne illness in the United States, 1973 through 1997. *Journal of Food Protection, 67*(10), 2342-2353.

Sparling, P. H., Crowe, C., Griffin, P. M., Swerdlow, D. L., & Rangel, J. M. (2005). Epidemiology of *Escherichia coli* O157:H7 Outbreaks, United States, 1982–2002. *Public Health Resources.* Retrieved from http://digitalcommons.unl.edu/publichealthresources/73

Stafford, R. J., Schluter, P. J., Wilson, A. J., Kirk, M. D., Hall, G., & Unicomb, L. (2008). Population-Attributable Risk Estimates for Risk Factors Associated with Campylobacter Infection, Australia. *Emerging Infectious Diseases, 14*(6), 895-901. doi:10.3201/eid1406.071008

Taormina, P. J., & Beuchat, L. R. (1999). Infections Associated with Eating Seed Sprouts: An International Concern. *Emerging Infectious Diseases, 5*(5), 626-634.

Tassios, P. T., & Kerr, K. G. (2010, January). Hard to swallow—emerging and re-emerging issues in foodborne infection. *Clinical Microbiology & Infection,* 1-2

The Food Safety and Inspection Service [FSIS].(2008). Disposition/Food Safety: Overview of Food Microbiology. Retrieved July 23, 2011, from www.fsis.usda.gov/PDF/PHVt-Food_Microbiology.pdf

Unicomb, L. E. (2009). Food Safety: Pathogen Transmission Routes, Hygiene Practices and Prevention, 27(5), 599-601.

van Duynhoven, Y. T. H. P., de Jager, C. M., Kortbeek, L. M., Vennema, H., Koopmans, M. P. G., van Leusden, F., van der Poel, W. H. M., et al. (2005). A one-year intensified study of outbreaks of gastroenteritis in The Netherlands. *Epidemiology and Infection, 133*(1), 9-21.

Venuto, M., Halbrook, B., Hinners, M., & Mickelson, A. L. S. (2010). Analyses of the eFORS (Electronic Foodborne Outbreak Reporting System) Surveillance Data (2000--2004) in School Settings. *Journal of Environmental Health, 72*(7), 8-13.

Verhoef, L. P. B., Kroneman, A., van Duynhoven, Y., Boshuizen, H., van Pelt, W., & Koopmans, M. (2009). Selection Tool for Foodborne Norovirus Outbreaks. *Emerging Infectious Diseases, 15*(1), 31-38. doi:10.3201/eid1501.080673

Widdowson, M.-A., Sulka, A., Bulens, S. N., Beard, R. S., Chaves, S. S., Hammond, R., Salehi, E. D. P., et al. (2005). Norovirus and foodborne disease, United States, 1991-2000. *Emerging Infectious Diseases, 11*(1), 95-102.

Vollaard, A. M., Ali, S., van Asten, H. A. G. H., Widjaja, S., Visser, L. G., Surjadi, C., & van Dissel, J. T. (2004). Risk Factors for Typhoid and Paratyphoid Fever in Jakarta, Indonesia. *JAMA: The Journal of the American Medical Association, 291*(21), 2607 -2615. doi:10.1001/jama.291.21.2607

Voetsch, A. C., Kennedy, M. H., Keene, W. E., Smith, K. E., Rabatsky-Ehr, T., Zansky, S., Thomas, S. M., et al. (2007). Risk Factors for Sporadic Shiga Toxin-Producing Escherichia Coli O157 Infections in FoodNet Sites, 1999–2000. *Epidemiology and Infection, 135*(06), 993-1000. doi:10.1017/S0950268806007564

World Health Organization [WHO]. (2011a). Food Safety. Retrieved June 26, 2011, from http://www.who.int/foodsafety/foodborne_disease/ferg1/en/index.html

World Health Organization [WHO]. (2011b). Initiative to estimate the Global Burden of Foodborne Diseases. Retrieved June 26, 2011, from http://www.who.int/foodsafety/foodborne_disease/ferg/en/index1.html

World Health Organization [WHO]. (2011c). Initiative to estimate the Global Burden of Foodborne Diseases: Information and publications. Retrieved June 26, 2011, from http://www.who.int/foodsafety/foodborne_disease/ferg/en/index7.html

17

Allium Species, Ancient Health Food for the Future?

Najjaa Hanen[1], Sami Fattouch[2], Emna Ammar[2] and Mohamed Neffati[1]
[1]*Laboratoire d'Ecologie Pastorale, Institut des Régions Arides*
[2]*National Institute of Applied Sciences and Technology (INSAT);*
University of Carthage, Tunis
[3]*UR Study & Management of Urban and Coastal Environments,*
National Engineering School in Sfax
Tunisia

1. Introduction

Allium is the largest and the most important representative genus of the *Alliaceae* family that comprises 700 species, widely distributed in the northern hemisphere, North America, North Africa, Europe and Asia (Tsiaganis et al., 2006). Besides the well known garlic and onion, several other species are widely grown for culinary use and for folk medicine including leek (*Allium porrum* L.), scallion (*Allium fistulosum* L.), shallot (*Allium ascalonicum* Hort.), wild garlic (*Allium ursinum* L.), garlic (*Allium sativum*) and onion (*Allium cepa*) (Lanzotti, 2006; Tsiaganis et al., 2006). Its consumption is attributed to several factors, mainly heavy promotion that links flavour and health. The powerful and unusual flavors of many of these plants and their possible nutritional impact and medical applications have attracted the attention of plant physiologists, chemists, nutritionists, and medical researchers (Graham and Graham, 1987).

Allium roseum is a very polymorphous, widespread species that is represented by 12 different taxa: 4 varieties, 4 subvarieties and 4 forms in North Africa (Cuénod, 1954; Le Floc'h, 1983). In Tunisia, the same authors mentioned the presence of only three varieties: var. *grandiflorum*, var. *perrotii* and var. *odoratissimum*. The *odoratissimum* variety is an endemic taxon in North Africa and a perennial spontaneous weed (Cuenod, 1954). Its flowering stem is about 30-60 cm, leaves are fleshy and very small, flowers are wide, rosy or white coloured and its odour is eyelet (Jendoubi et al., 2001).

In Southern Tunisia, local people on the extension area where *A. roseum* or rosy garlic occurs have extensively developed uses for this species both as a cooking ingredient and a sauce (Najjaa et al., 2011a). *A. roseum* leaves are the main edible part, with a distinctive pungent odour and strong flavour. Besides its culinary use, rosy garlic is also used in folk medicine. Le Floc'h (1983) reported its use for the treatment of headaches and rheumatism. It is also used for the treatment of bronchitis, colds as an inhalation, fever diminution and as an appetizer.

While several studies have provided information about *A. roseum*, detailed studies documenting compositional, nutritional and functional properties are very limited, if not lacking. The objective of the present study was to characterize chemical composition, nutritional properties, bioactive components, and antioxidant and antimicrobial activities of *A. roseum* grown in Tunisia and to infer their role in human nutrition.

All samples were collected on the same day. To preserve freshness, *A. roseum* samples were transported to the laboratory on ice. Upon arrival, leaves of *A. roseum* samples were cleaned to remove all foreign matter and washed with distilled water. For the analysis of moisture content, pH, carotenoids, vitamin C, and anthocyanidins, samples were frozen and stored at –80 °C until analysis. A second portion of samples was kept fresh, and the leaves were extracted with methanol, then evaporated at 40 °C using a rotary evaporator under high vacuum. The resulting crude extract was used for total phenolic content and flavonoids composition analysis. Fresh leaves were also used to prepare an aqueous extract for allicin determination. The third portion of *A. roseum* samples was air dried in the shade for 24 h at 45 °C. From these air-dried samples, plants were chopped into pieces $\leq$ to 5 cm and ground in a Sorvall Omnimixer into a fine powdery consistency and used for other chemical analyses. The powder sample was packed in a hermetic glass vessel and stored at +4 °C for subsequent analyses. In this study, all methodologies used and data presented are in accordance with FAO standards (Greenfield and Southgate, 2003).

The *A. roseum* methanolic extracts were tested against a panel of spoilage and pathogenic bacteria strains, including the Gram positive *Staphylococcus aureus* ATCC 25923, and *Enterococcus fecalis* ATCC 29212, and the Gram negative *Escherichia coli* ATCC 25922, as well as the yeast *Candida albicans* ATCC 90028, *Candida glabrata* ATCC 90030, *Candida kreusei* ATCC 6258, *Candida parapsilosis* ATCC 22019. The cultures were incubated at 37°C (27°C for the yeast) for 18 h and then diluted in Nutrient Broth to obtain 10^6 CFU/mL.

The recommended methods of the Association of Official Analytical Chemists (AOAC, 1995) were adopted to determine the level of crude protein, water content, ash, carbohydrates, fibres and lipids. Nitrogen content was determined using the Kjeldahl method (AOAC, 1995) and multiplied by a factor of 6.25 to determine the total protein content. Water content was estimated by drying the sample to a constant weight at 70±2 °C (AOAC, 2002). Ash was determined by the incineration of 1.0 g sample in a muffle furnace, at 550 °C for 6 h (Alfawz, 2006). Fibres content was quantified using 2 g sample previously boiled with diluted H_2SO_4 (0.3) using Wende method (AOAC, 1995). Soluble carbohydrates were determined by the phenol-sulphuric acid colorimetric method (AOAC, 1995). Total carbohydrates were calculated by difference as follows: Carbohydrates (%) = 100 - [Proteins (%) + Lipids (%) + Ash (%) + Fibres (%)], according to Alfawz (2006). The ash obtained underwent an acidic hydrolysis and the minerals (Ca, Na, K, Mg, Fe, Zn, Cu, Ni, Mn, Pb, Cd and Cr) were determined separatively, using an atomic absorption spectrophotometer (Hitachi Z6100). Phosphorus content was determined using a spectrophotometer method, based on phosphoric molybdovanadate absorption at 730 nm according to Falade, Otemuyiwa, Oladipo, Oyedapo, Akinpelu and Adewusi (2005).

Fatty acids composition was analysed by gas chromatography. The fatty acids were previously methylated to esters using a born trifluoride methanol complex (14% w/v). The mixture was held one hour at 100 °C.

For the bioactive compound, the total phenolic content of *A. roseum* methanolic extract was determined using Folin-Ciocalteu reagent (Fattouch et al., 2007). Total flavonoids were measured by a colorimetric assay according to Galvez, Martin-Cordero, Houghton and Ayuso (2005). The vitamin C content was analyzed with 2, 6-dichloroindophenol titrimetric method. Total anthocyanidins were determined using the Reay, Fletcher and Thomas (1998) method. Colorimetric quantification of total carotenoids was determined, as described by Mackinney (1941). Allicin determination was based on Miron et al. (2002) method.

Various concentrations (1 µg/ml up to 10 mg/ml) of *A. roseum* extracts were used to determine the antimicrobial and antifungal activity. Minimum inhibitory concentration (MIC) values were determined by a micro-titre plate dilution method.

The assessment of radical scavenging activity was determined using ABTS (2, 2'-Azino-(bis-3-ethylbenzthiazoline-6-sulfonic acid) di-ammonium salt) radical scavenging activity of the methanolic extracts was determined according to Re, Pellegrini, Proreggente, Pannala, Yang and Rice-Evans (1999). The results were expressed in terms of Trolox equivalent antioxidant capacity (TEAC). The antioxidant activity of the extracts was also evaluated using the DPPH (2, 2-diphenyl-2-picrylhydrazyl) free radical specrophotometrically according to Fattouch et al. (2007).

All analysis mentioned were effected with quality control, than a proper sampling plan was followed with representative samples from the geographic area studied and sufficient replications of the sample were used to ensure statistically reliable and valid data. The analyses of the nutrient contents samples were made in our laboratory where ISO/CEI 17025 (2005) was respected to assure the quality of results.

2. Results and discussion

2.1 Nutritional composition

The proximate chemical and nutritional composition of *A. roseum* edible part collected from Tunisia is listed in Table 1.

Components	Mean Value*
Soluble carbohydrates (g/100 g DW[a])	32.80 ± 0.21
Protein (g/100 g DW[a])	22.70 ± 1.51
Fibre (g/100 g DW[a])	12.30 ± 0.05
Ash (g/100 g DW[a])	7.20 ± 1.31
Fat (g/100 g DW[a])	3.60 ± 0.29

*Values are means ± SD, n = 3.
[a]: Dry weight

Table 1. Content of soluble carbohydrates, protein, fibre, ash and fat in *Allium roseum* L. expressed as g/100 g of dry weight basis. (Moisture content 81.2 ± 2.6 g/100 g fresh weight).

2.1.1 Water content and pH

Water content is important because it affects the plant's properties. Compared to related vegetables, the *A. roseum* water content is lower than that of *A. porrum* varing from 83 to 89% (Tirilly and Bourgeois, 1999) and the *A. cepa* (89%) (Dini et al., 2008). The *A. roseum* is rather neutral (pH = 6.80 ± 0.05) compared to that of garlic (pH = 6.05) (Haciseferoğullari et al., 2005).

2.1.2 Sugars, proteins, fibres and lipids

Soluble carbohydrates represent the most abundant *A. roseum* leaves nutrients class (> 30%); as has also been observed in onion bulbs (Moreau et al., 1996) and garlic (Haciseferoğullari et al., 2005). The total carbohydrates content in this species calculated by difference is 54.2 g/100 g DW. Compared to the carbohydrates content of *A. porrum* (5 to 11%) (Tirilly and Bourgeois, 1999) and to aerial parts of other *Alliums* (5 to 12%) (Brewster, 1994), the leaves of *A. roseum* are rich sources of soluble carbohydrates. Dietary fibres are considered as unavailable carbohydrates, but nonetheless they still play a very important role in maintaining good health. Interestedly, A. *roseum* aerial part fibres content was higher than that reported for *A. cepa* bulb (1.7%), the edible part of the vegetable. Rosy garlic leaves proteins rate is relatively high compared to *A. sativum* bulbs (9.3%) and *A. cepa* (1.7%) (Haciseferoğullari et al., 2005; Dini et al., 2008). Fats accounted for 0.68% of the fresh weight of *A. roseum*, making them the least abundant class of nutrients. Yet this was higher than typical values of <0.5% fresh weight basis for most plant tissues, and also compared to most *Allium* plants. Where onion, leek and garlic contain 0.15%, 0.25% and 0.42%, respectively, as reported by Haciseferoğullari et al., (2005) and Tsiaganis et al., (2006).

2.1.3 Minerals

A. roseum is characterized by high ash content (Table 1) including macro and micro elements (Table 2). *Allium* ash content ranges from 0.6 and 1.0% and higher values are associated to high dry matter content (Brewster, 1994). The mineral element composition of *A. roseum* exhibited a higher concentration of potassium than calcium and magnesium (Table 2). Minerals are important as constituents of bones, teeth, soft tissues, haemoglobin, muscle, blood and nerve cells and are vital to overall mental and physical well being (Jouanny, 1988). The high content of potassium in *A. roseum* is nutritionally significant in since it contributes to the control of hypertension which results in excessive excretion of potassium (Dini et al., 2008). Calcium is found at relatively high concentration in *A. roseum* (Table 2). Onion leaf calcium concentration (2540 mg/100 g fresh weight) (Boukari et al., 2001) is much higher than that of *A. roseum* leaves but bulb calcium concentration (45 mg/100 g fresh weight) (Adrian et al., 1995) is much lower. Therefore, calcium concentration in *A. roseum* is between that of onion leaves and bulbs. *A. roseum* can be considered as a source of calcium for human nutrition. This is important since calcium mineral deficiency is a world-wide problem; particularly in developing countries where the daily average intake is very low, ranging between 300 and 500 mg for adults (Boukari et al., 2001). The low sodium content of *A. roseum* and consequently low Na/K ratio (0.03) is another indication that *A. roseum* consumption would reduce the incidence of hypertension (Iqbal et al., 2006). *A. roseum* leaves also contains several oligo-elements including iron, zinc, copper and manganese. These values are similar to, but higher than, those of Haciseferoğullari et al., (2005) in *A. sativum*. The iron content of 'rosy garlic' was somewhat higher than that of *A. cepa* (8.1 mg/ 100 g) (Moreau et al., 1996). The magnesium, iron and phosphorous levels are adequate. Cadmium, lead and chromium were below the detection limit, as observed by Moreau et al., (1996) for onion.

Component	Concentration*
Major elements	
Potassium	1530.500 ± 0.036
Calcium	712.500 ± 0.048
Magnesium	101.900 ± 0.007
Sodium	46.500 ± 0.003
Na:K ratio	0.030
Anions	
Chlorides	724.00 ± 0.01
Sulfates	437.00 ± 0.03
Phosphates	219.00 ± 6.40
Nitrates	< 0.03
Heavy metals	
Iron	10.110 ± 0.002
Manganese	2.000 ± 0.001
Zinc	1.800 ± 0.001
Copper	1.100 ± 0.006
Nickel	< 0.013
Lead	< 0.035
Cadmium	< 0.007
Chromium	< 0.006

*Values are means ± SD, n = 3.

Table 2. The mineral content of *Allium roseum* expressed in mg/100 g of dry weight basis

2.1.4 Fatty acid composition

The fatty acid composition of *A. roseum* leaves is given in Table 3. Chromatographic analysis revealed twelve compounds. Unsaturated fatty acids accounted for most of the fatty acids (85 %) and were represented mainly by linolenic, linoleic, oleic and gadoleic. Five saturated acids (palmitic, stearic, myristic, arachidic and margaric), accounted for ~15% of the total fatty acids. Myristoleic, palmitoleic and heptadecanoic acids were found as minor compounds. The overall fatty acid profile of *A. roseum* reveals a good source of the nutritionally essential linolenic and oleic acids (Zia-Ul-Haq et al., 2007). Linoleic and linolenic acids are the most important essential fatty acids required for growth, physiological functions and maintenance (Pugalenthi et al., 2004) While the major fatty acid in *A. roseum* was linolenic acid, linoleic acid was most abundant in onion, garlic and leek where it represents about 50% of the total (Tsiaganis et al., 2006). The same authors demonstrated that garlic oils contain relatively high levels of linoleic acid, and that myristoleic acid ($C_{14:1}$) was absent in onion. As a consequence, the *A. roseum* fatty acid composition quality is comparable to that of *A. sativum* (Tsiaganis et al., 2006). It could be concluded that fatty acid composition varies within the species. We may note that the most abundant fatty acids are similar to those found in the oil of other *Allium* species. The less abundant fatty acids are present in *A. roseum* but at a lower concentration than reported by Tirilly and Bourgeois (1999) in *A. porrum*, Moreau et al. (1996) in *A. cepa* and Tsiaganis et al. (2006) in *A. sativum*. Overall, the *A. roseum* fatty acid composition was not qualitatively different from that of the other species.

Fatty acid	Percentage*
Myristic ($C_{14:0}$)	0.78 ± 0.11
Myristoleic ($C_{14:1}$)	0.12 ± 0.08
Palmitic ($C_{16:0}$)	12.82 ± 0.33
Palmitoleic ($C_{16:1}$)	0.50 ± 0.18
Margaric ($C_{17:0}$)	0.16 ± 0.13
Heptadecanoic ($C_{17:1}$)	0.15 ± 0.12
Stearic ($C_{18:0}$)	0.98 ± 0.23
Oleic ($C_{18:1}$)	2.87 ± 0.70
Linoleic ($C_{18:2}$)	25.68 ± 0.44
Linolenic ($C_{18:3}$)	52.68 ± 0.41
Arachidic ($C_{20:0}$)	0.19 ± 0.17
Gadoleic ($C_{20:1}$)	2.55 ± 0.32
Total	
Saturated	14.93 ± 0.11
Monounsaturated	6.19 ± 0.11
Polyunsaturated	78.37 ± 0.11

*Values are means ± SD, $n = 3$.

Table 3. Fatty acid composition of *Allium roseum*

2.2 Bioactive compounds

The content of potential health-promoting substances, flavonoid, total phenolic content, vitamin C, tannin, anthocyanidin, carotenoids and allicin in the wild *A. roseum* growing in the arid region of Tunisia is listed in Table 4.

2.2.1 Phenolic compounds, flavonoids, anthocyanidins, carotenoids and vitamin C contents

Total phenolic content of *A. roseum* expressed in equivalent catechol was higher than that reported for garlic (61.8 mg/100 g FW) and onion (31.0 mg/100 g FW) (Kaur and Kapoor, 2002). Although, shallots had the highest total phenolic content (114.7 mg/100 g) among the bulb onion varieties tested by Lanzotti (2006). However, this content is lower than that of rosy garlic leaves. Significant correlations were observed between the total phenolic content of *A. roseum*, and antioxidant activity, suggesting that phenolic compounds would be the major contributors to the antioxidant capacity of *A. roseum* (Najjaa et al., 2011a). Moreover, tannins and flavonoids were detected by several other authors and are usually less abundant in several other species of *Allium* (*A. cepa, A. ascalonicum, A. sativum*) (Bozin et al., 2008; Zielinskaa et al., 2008). Flavonoids are important secondary plant metabolites. The flavonoids content of rosy garlic is seven time that of garlic (0.5 mg/100 g) (Miron et al., 2002). *Allium* species are among the richest sources of dietary flavonoids and contribute significantly to the overall intake of flavonoids (Slimestad et al., 2007). *In vitro* and *in vivo* pharmacological tests have shown that flavonoids exhibit the following variety of actions: (i) antioxidative (Boyle et al., 2000); (ii) reduction of cardiovascular disease (Janssen et al., 1998) and (iii) reduction of carcinogenic activity (Steiner, 1997).

The high *A. roseum* vitamin C content (1523.35 mg/100 g DW) may be an important reason that it has been reputedly used as a traditional Tunisian medicine for treating rheumatism and cold. Furthermore, its high vitamin C content confers considerable nutritional value. *A. roseum* leaves had high anthocyanidin content (1239.62 µg/100 g DW). Much is known about the anthocyanins of *A. cepa* bulbs, and leaves of *A. victorialis* and *A. schoenoprasum* (Terahara et al., 1994; Fossena et al., 2000; Slimestad et al., 2007). Moreover, *A. roseum* had a typical carotenoids content (Table 4) of leafy vegetables, which is higher than those of legumes and fruits (Combris et al., 2007).

Substances	Mean value*
Phenolic compounds (mg CA/100g DW [ab])	736.65 ± 1.51
Flavonoids (mg CE/g DW [ac])	3.37 ± 0.32
Anthocyanidin (µg CE/100 g DW [ac])	1239.62 ± 6.79
Vitamin C (mg/100 g DW [a])	1523.35 ± 74.72
Total Carotenoids (µg/100 g DW[a])	242.25 ± 48.84
Allicin (mg / 100g DW [a])	657.00 ± 0.49

*Values are means ± standard deviations of triplicate determination (Mean ± SD (n = 3)).
[a] DW = dry weight
[b] Total phenolic contents expressed as as mg catechol (CA) equivalents per gram of dry weight
[c] Total flavonoid and anthocyanidin content were expressed as mg catechin (CE)/100 g dry weight

Table 4. *Allium roseum* L. var. *odoratissimum* bioactive substances content.

2.2.2 Allicin content

Garlic antibacterial bioactive principal was identified as diallylthiosulphinate and was given allicin as trivial name since 1944. This bioactive substance is also detected in *A. roseum* with a concentration equivalent to 0.0328 µg/mL. This result is similar to that mentioned by Miron et al. (2002) in garlic (0.0308 µg/mL). Allicin (diallylthiosulfinate) is the most abundant organosulfurous compound, representing about 70% of the overall thiosulfinates formed upon garlic cloves crushing (Miron et al., 2002).

2.3 Antioxidant activity

The antioxidant activities of leaf extracts were assessed and confirmed using two functional analytical methods based on the radicals (ABTS and DPPH) scavenging potential, as recommended by Sànchez-Alonso et al., (2007). A good correlation was found between DPPH and ABTS methods (R^2–0.827), indicating that these two methods gave consistent results. The extracts obtained were all able to inhibit the DPPH, as well as ABTS radicals (Table 5). The antioxidant potential was 378.89 mg Trolox/100g DW with the DPPH method, and 399.99 mg Trolox/100g DW with the ABTS. In comparison to previous data based on the ABTS scavenging capacity, *A. roseum* leaf extracts were comparable or higher than other investigated species known to be rich in antioxidants including strawberry (25.9), raspberry (18.5), red cabbage (13.8), broccoli (6.5), and spinach (7.6) (Proteggente et al., 2002). Significant correlations were observed between the TPC of *A. roseum*, and antioxidant activity (R^2=0.828 for TPC vs. DPPH and R^2=0.925 for TPC vs. ABTS), suggesting that polyphenolic compounds are the major contributors to the antioxidant capacity of *A. roseum*.

Regarding the favourable redox potentials and relative stability of their phenoxyl radical, these biomolecules are considered to be human health promoting antioxidants (Acuna et al., 2002).

Extracts	DPPH (mg Trolox /100g DW)	ABTS (mg/100g DW)
Methanol (75%)	378.80±5.55	399.90± 4.59

Table 5. Free radical scavenging activity of *A. roseum*

2.4 Antibacterial activity

The *in vitro* antibacterial effects of the *A. roseum* extracts obtained with the methanolic extract values are presented in Table 6. The results showed that *A. roseum* extracts have great potential as antimicrobial agent against the tested bacteria. *C. albicans* and *C. glabrata*, were the most sensitive tested organisms to the extract with the MIC values were 0.63 and 2.5 mg/ml, respectively.

The strong antifungal activity was observed against *C. albicans* and *C. glabrata* may be related to the high level of polyphenols content. Cai et al. (2000) showed that several classes of polyphenols such as phenolic acids, flavonoids and tannins serve as plant defence mechanism against pathogenic microorganisms. In fact, the site and the number of hydroxyl groups on the phenol components increased the toxicity against the microorganisms.

Strains	MIC (mg/ml)
Escherichia Coli ATCC 25922	10±1.20
Enterococcus Faecalis ATCC29212	10±0.57
Staphylococcus aureus ATCC 25923	10±0.60
Candida albicans ATCC 90028	0,63±1.85
Candida glabrata ATCC 90030	2,5±1.20
Candida kreusei ATCC 6258	10±2.13
Candida parapsilosis ATCC 22019	10±1.41

MIC, Minimum Inhibitory Concentrations as (mg ml^{-1}).

Table 6. Minimal inhibitory concentrations of extracts of A. roseum on bacterial growth

3. Conclusion

This study revealed that *A. roseum* var. *odoratissimum* growing in Tunisia had a high soluble carbohydrates, crude protein and dietary fibre contents, compared to other *Alliums*. Its mineral content was high in potassium, and calcium. The mineral composition of 'rosy garlic' is sufficient in Ca, P, K, Cu, Fe, Zn and Mg so that it can meet many macronutrient and micronutrient requirements of the human diets. As a consequence, a diet based on *A. roseum* would help in preventing deficiencies in potassium, calcium, iron and magnesium. Furthermore, edible part oil included 15% saturated and 85% unsaturated fatty acids. Linolenic acid and palmitic acid were the most abundant unsaturated and saturated fatty acids, respectively. This fatty composition confers to the *A. roseum* oil considerable nutritional value, acting on physiological functions and reducing cardiovascular, cancer and arthroscleroses diseases occurrence risk. The most abundant phytonutrients found in *A.*

roseum (polyphenolic compounds, flavonoids, anthyacinidins, vitamin C and allicin) exhibit a positive effect on human health as antioxidants and antibacterial compounds. Since the chemical composition of *A. roseum* has not been reported before, this report provides a starting point for comparison to the other *Allium* genus vegetables and it confirms the potentially important positive nutritional value that *A. roseum* can have on human health. Since *A. roseum* is a rich source of many important nutrients and bioactive compounds responsible for many promising health beneficial physiological effects, it may be considered a nutraceutical that serves as a natural source of necessary components to help fulfil our daily nutritional needs and as a functional food as well as in ethnomedecine .

4. References

Acuna, U.M., Atha, D.E., Nee, M.H., & Kennelly, E. J. (2002). Antioxidant capacities often edible North American plants. *Phytotherapy Research, 16*, 63-65.

Adrian, J., Potus, J., & Frangne, R. (1995). La science alimentaire de a à z. Paris: Lavoisier, 2ème Edition.

Alfawz, M.A. (2006). Chemical composition of hummayd (*Rumex visicarius*) grown in Saudi Arabia. *Journal of Food Composition and Analysis, 19*, 552-555.

AOAC. (1995). Official Methods of Analysis, 16th Edition. AOAC International. (Chapter 12, Tecn. 960.52).Washington: Cuniff, P. (Ed.), p. 7

AOAC. (2002). Official methods of analysis of AOAC International. 17th edition. Gaithersburg , MD, USA, Association of Analytical Communities.

Besbes, S., Blecker, C., Deroanne, C., Drira, N.E., & Attia, H. (2004). Date seeds: chemical composition and characteristic profiles of the lipid fraction. *Food Chemistry, 84*, 577-584.

Boukari, I., Shier, N.W., Xinia, E.F.R., Frisch, J., Bruce, A.W., Pawloski, L., & Fly, A.D. (2001). Calcium analysis of selected Western African foods. *Journal of Food Composition and Analysis, 14*, 37-42.

Boyle, S.P., Dobson, V.L., Duthie, S.J., Kyle, J.A.M., & Collins, A.R. (2000). Absorption and DNA protective effects of flavonoid glycosides from onion meal. *European Journal of Nutrition, 39*, 213-223.

Bozin, B., Mimica-Dukic, N., Samojlik, I., Goran, A., & Igic, R. (2008). Phenolics as antioxidants in garlic (*Allium sativum* L. Alliaceae). *Food Chemistry, 111*, 925-929.

Brewster, J.L. (1994). Onions and other vegetable *Alliums*. Wallingford: CAB International, UK.

Cai, Y., Luo, Q., Sun, M.,& Carke, H. (2002). Antioxidant activity and phenolic compounds of 112 food and plants. *Journal of chromatography A, 975*, 71-93.

Combris, P., Amiot-Carlin, M., & Cavaillet, F. (2007). Les fruits et légumes dans l'alimentation. Enjeux et déterminants de la consommation. Rennes, INRR : Expertise scientifique collective Inra.

Cuénod, A. (1954). Flore analytique et synoptique de la Tunisie. Cryptogames vasculaires, Gymnospermes et Monocotylédones, Tunis: S.E.F.A.N.

Dini, I. Carlo Tenore, G., & Dini, A. (2008). Chemical composition, nutritional value and antioxidant properties of *Allium caepa* L. Var. tropeana (red onion) seeds. *Food Chemistry, 107*, 613-621.

Doner, G., & Ege, A. (2004). Evaluation of digestion procedures for the determination of iron and zinc in biscuits by flame atomic absorption spectrometry. *Analytica Chimica Acta, 520*, 217-222.

Falade, O.S., Otemuyiwa, I.O., Oladipo, A., Oyedapo, OO.., Akinpelu, B.A., & Adewusi, S.R.A. (2005). The chemical composition and membrane stability activity of some herbs used in local therapy for anemia. *Journal of Ethnopharmacology, 102*, 15-22.

Fattouch, S., Caboni, P., Coroneo, V., Tuberoso, C.I.G., Angioni, A., Dessi, S, Marzouki, N., & Cabras, P. (2007). Antimicrobial activity of Tunisian Quince. (Cydonia oblonga miller) Pulp and peel polyphenolic extracts. *Journal of Agriculture and Food Chemistry, 55*, 963-969.

Fossen, T., Slimestad, R., Overstedal D.O., & Andersena, M. (2000). Covalent anthocyanin-favonol complexes from flowers of chive, *Allium schoenoprasum. Phytochemistry, 54*, 317-323.

Gálvez, C.J., Martin-Cordero, C., Houghton, A.J., &Ayuso, M.J. (2005). Antioxidant Activity of methanol extracts obtained from *Plantago* species. *Journal of Agricultural and Food Chemistry, 53*, 1927–1933.

Graham, H.D., &Graham, E.J.F. (1987). Inhibiton of *Aspergillus parasiticus* growth and toxin production by garlic. *Journal of Food Safety, 8*, 101-108.

Greenfield, H., & Southgate, D.A.T, (2003). Food composition data: production, management and use. Rome: Part 1. 2nd Edition. FAO.

Haciseferoğullari, H., Özcan, M., Demir, F., & Çalişir, S. (2005). Some nutritional and technological properties of garlic (*Allium sativum* L.). *Journal of Food Engineering, 68*, 463-469.

International Standardization Organization (ISO). (2005). Exigences générales concernant la compétence des laboratoires d'étalonnages et d'essais (ISO/CEI 17025), ISO, Genève, 29 p.

Iqbal, A., Khalil, I.A., Ateeq, N., & Khan, M.S. (2006). Nutritional quality of important food legumes. *Food Chemistry, 97*, 331-335.

Janssen, K., Mensink, R.P., Cox, F.J., Harryvan, J.L., Hovenier, R., Hollman, P.C., & Katan, M.B. (1998). Effects of the flavonoids quercetin and apigenin on hemostasis in healthy volunteers: results from an *in vitro* and a dietary supplemented study. *American Journal of Clinical Nutrition, 67*, 255-262.

Jendoubi, R., Neffati, M., Henchi, B., &Yobi, A. (2001). Système de reproduction et variabilité morphologique chez *Allium roseum* L. *Plant Genetic Research Newsletter, 127*, 29-34.

Jouanny, J.(1988). Notions essentielles de matières médicales homéopathiques. France : Bouen.

Kaur, C., & Kapoor, H.C. (2002). Antioxidant activity and total phenolic content of some Asian vegetables. *International Journal of Food Science and Technology, 37*, 153-161.

Lanzotti, V. (2006). The analysis of onion and garlic. *Journal of Chromatography A, 1112*, 3-22.

Le Floc'h, E. (1983). Contribution à une étude ethnobotanique de la flore tunisienne. Programme flore et végétation tunisienne. Tunis : Ministère de l'Enseignement Supérieur et de la Recherche Scientifique.

Mackinney, G. (1941). Absorption of light by chlorophyll solution. *Journal of Biological Chemistry, 140*, 315–322.

Messiaen, C.M., Cohat, J., Pichon, M., Leroux, J.P., &Beyries, A., (1993). Les *Allium* alimentaires reproduits par voie végétative, INRA, Paris, pp. 150-225.

Miron, T., Shin, I., Feigenblat, G., Weiner, L., Mirelman, D., Wilchek, M., & Rbinkov, A. (2002). A spectrophotometric assay far allicin, alliin, and alliinase (alliin lyase) with a chromogenic thiol: reaction of 4–mercaptopyridine with thiosulfinate. *Analytical Biochemistry, 307,* 76-83.

Moreau, B., Le Bohec, J., & Guerber–Cahuzac, B. (1996). L'oignon de garde. Paris: Lavoisier,.

Najjaa, H., Neffati, M., Zouari, S., & Ammar, E. (2007). Essential oil composition and antibacterial activity of different extracts of a North African endemic species. *Comptes Rendus de Chimie, 10,* 820-826.

Najjaa, H., Zerria, K., Fattouch, S., Ammar, E., & Neffati, M. (2009). Antioxidant and antimicrobial activities of *Allium roseum* "Lazoul", a wild edible endemic species in North Africa. *International Journal of Food Properties* (LJFP-2009-0049.R2) (In press).

Najjaa, H., Zouari, S., Ammar, E., & Neffati, M. (2009b). Phytochemical screening and antibacterial properties of *Allium roseum* L. a wild edible species in North Africa. *Journal of Food Biochemistry,* 35: 699–714.

Proteggente, A.R., Pannala, A.S, Pagana, G., Van Buren, L., Wagner, E., & Wiseman, S. (2002). The antioxidant activity of regularly consumed fruit and vegetables reflects their phenolic and vitamin C composition. *Free Radical research, 36,* 217-233.

Pugalenthi, M., Vadivel, V., Gurumoorthi, P., & Janardhanan, K. (2004). Comparative nutritional evaluation of little known legumes, *Tamarindus indica, Erythrina indica* and *Sesbania bispinosa. Tropical and Subtropical Agroecosystems, 4,* 107–123.

Re, R. Pellegrini, N., Proreggente, A. , Pannala, A. , Yang, A. , & Rice-Evans, C. (1999). Antioxidant activity applying an improved ABTS radical cation decolorizing assay. *Free radicals in biology and medicine, 26,* 1231-1237.

Reay, P.F., Fletcher, R.H., &Thomas, V.J.G. (1998). Chlorophylls, carotenoids and anthocyanin concentrations in the skin of "Gala" apples during maturation and the influence of foliar applications of nitrogen and magnesium. *Journal of the Science of Food and Agricultural, 76,* 63–71.

Sànchez-Alonso, I., Jimenez-Escrig, A., Saura-Calixto, F., & Borderias, A. J. (2007). Effect of grape antioxidant dietary fibre on the prevention of lipid oxidation in minced fish: evaluation by different methodologies. *Food Chemistry, 101,* 372-378.

Slimestad, R., Fossen, T., &Vagen, I.M. (2007). Onions: a source of unique dietary flavonoids. *Journal of Agricultural and Food Chemistry, 55,* 10067–10080.

Steiner, M. (1997). The role of flavonoids and garlic in cancer prevention. In H. Ohigashi, *Food Factors for Cancer Prevention,* (pp. 222-225). New York: Springer.

Terahara, N., Yamaguchi, M., & Honda, T. (1994). Malonylated anthocyanins from bulbs of red onions, *Allium cepa* L. *Bioscience Biotechnology Biochemistry, 58,* 1324–1325.

Tirilly, Y., Bourgeois, C.M. (1999). Technologie des légumes. Paris : Lavoisier.

Trémolières, A. (1998). Les lipides végétaux : voies de biosynthèse des glycérolipides. Paris : Université De Boeck.

Tsiaganis, M.C., Laskari, K., & Melissari, E. (2006). Fatty acid composition of *Allium* species lipids. *Journal of Food Composition and Analysis, 19,* 620-627.

Zia-Ul-Haq, M., Iqbal, S., Ahmad, S., Imran, M., Niaz, A., & Bhanger, M.I. (2007). Nutritional and compositional study of Desi chickpea (*Cicer arietinum* L.) cultivars grown in Punjab. *Pakistan. Food Chemistry, 105*, 1357–1363.

Zielinskaa, D., Nagels, L., & Piskuła, M.K. (2008). Determination of quercetin and its glucosides in onion by electrochemical methods. *Analytica Chimica Acta, 617*, 22–31.

Zou, ZM., Yu, D.Q., & Cong, P.Z. (1999). Research progress in the chemical constituents and pharmacological actions of *Allium* species. *Acta Pharmacologica Sinica, 34*, 395-400.

18

Starch: From Food to Medicine

Emeje Martins Ochubiojo[1] and Asha Rodrigues[2]
[1]National Institute for Pharmaceutical Research and Development,
[2]Physical and Materials Chemistry Division,
National Chemical Laboratory,
[1]Nigeria
[2]India

1. Introduction

Starch is a natural, cheap, available, renewable, and biodegradable polymer produced by many plants as a source of stored energy. It is the second most abundant biomass material in nature. It is found in plant leaves, stems, roots, bulbs, nuts, stalks, crop seeds, and staple crops such as rice, corn, wheat, cassava, and potato. It has found wide use in the food, textiles, cosmetics, plastics, adhesives, paper, and pharmaceutical industries. In the food industry, starch has a wide range of applications ranging from being a thickener, gelling agent, to being a stabilizer for making snacks, meat products, fruit juices (Manek, et al., 2005). It is either used as extracted from the plant and is called "native starch", or it undergoes one or more modifications to reach specific properties and is called "modified starch". Worldwide, the main sources of starch are maize (82%), wheat (8%), potatoes (5%), and cassava (5%). In 2000, the world starch market was estimated to be 48.5 million tons, including native and modified starches. The value of the output is worth €15 billion per year (Le Corre, et al., 2010). As noted by Mason (2009), as far back as the first century, Celsus, a Greek physician, had described starch as a wholesome food. Starch was added to rye and wheat breads during the 1890s in Germany and to beer in 1918 in England. Also, Moffett, writing in 1928, had described the use of corn starch in baking powders, pie fillings, sauces, jellies and puddings. The 1930s saw the use of starch as components of salad dressings in mayonnaise. Subsequently, combinations of corn and tapioca starches were used by salad dressing manufacturers. (Mason, 2009). Starch has also find use as sweetners; sweeteners produced by acid-catalyzed hydrolysis of starch were used in the improvement of wines in Germany in the 1830s. Between 1940 and 1995, the use of starch by the US food industry was reported to have increased from roughly 30 000 to 950 000 metric tons. The leading users of starch were believed to be the brewing, baking powder and confectionery industries. Similar survey in Europe in 1992, showed that, 2.8 million metric tons of starch was used in food. Several uses of starch abound in literature and the reader is advised to refer to more comprehensive reviews on the application of starch in the food industry. In fact, the versatility of starch applications is unparalleled as compared to other biomaterials.

It is obvious that, the need for starch will continue to increase especially as this biopolymer finds application in other industries including medicine and Pharmacy. From serving as food for man, starch has been found to be effective in drying up skin lesions (dermatitis), especially where there are watery exudates. Consequently, starch is a major component of dusting powders, pastes and ointments meant to provide protective and healing effect on skins. Starch mucilage has also performed well as emollient and major base in enemas. Because of its ability to form complex with iodine, starch has been used in treating iodine poisoning. Acute diarrhea has also been effectively prevented or treated with starch based solutions due to the excellent ability of starch to take up water. In Pharmacy, starch appears indispensable; It is used as excipients in several medicines. Its traditional role as a disintegrant or diluent is giving way to the more modern role as drug carrier; the therapeutic effect of the starch-adsorbed or starch-encapsulated or starch-conjugated drug largely depends on the type of starch.

2. The role of excipients in drug delivery

The International Pharmaceutical Excipient Council (IPEC) defines excipients as substances, other than the active pharmaceutical ingredient (API) in finished dosage form, which have been appropriately evaluated for safety and are included in a drug delivery system to either aid the processing or to aid manufacture, protect, support, enhance stability, bioavailability or patient acceptability, assist in product identification, or enhance other attributes of the overall safety and effectiveness of the drug delivery system during storage or use (Robertson, 1999). They can also be defined as additives used to convert active pharmaceutical ingredients into pharmaceutical dosage forms suitable for administration to patients. Excipients no longer maintain the initial concept of —Inactive support; because of the influence they have over both biopharmaceutical aspects and technological factors (Jansook and Loftsson, 2009; Killen and Corrigan, 2006; Langoth, et al., 2003; Lemieux, et al., 2009; Li, et al., 2003; Massicotte, et al., 2008; Munday and Cox, 2000; Nykänen, et al., 2001; Williams, et al.) The desired activity, the excipient's equivalent of the active ingredients efficacy, is called its functionality. The inherent property of an excipient is its functionality in the dosage form. In order to deliver a stable, uniform and effective drug product, it is essential to know the properties of the active pharmaceutical ingredient alone and in combination with all other ingredients based on the requirements of the dosage form and process applied. This underscores the importance of excipients in dosage form development.

The ultimate application goal of any drug delivery system including nano drug delivery, is to develop clinically useful formulations for treating diseases in patients (Park, 2007). Clinical applications require approval from FDA. The pharmaceutical industry has been slow to utilize the new drug delivery systems if they include excipients that are not generally regarded as safe. This is because, going through clinical studies for FDA approval of a new chemical entity is a long and costly process; there is therefore, a very strong resistance in the industry to adding any untested materials that may require seeking approval. To overcome this reluctant attitude by the industry, scientists need to develop not only new delivery systems that are substantially better than the existing delivery systems (Park, 2007), but also seek for new ways of using old biomaterials. The use of starch (native or modified) is an important strategy towards the attainment of this objective. This is because starch unlike synthetic products is biocompatible, non toxic, biodegradable, eco-friendly and of low prices. It is generally a non-polluting renewable source for sustainable supply of cheaper pharmaceutical products.

3. What is starch?

Starch, which is the major dietary source of carbohydrates, is the most abundant storage polysaccharide in plants, and occurs as granules in the chloroplast of green leaves and the amyloplast of seeds, pulses, and tubers (Sajilata, et al., 2006). Chemically, starches are polysaccharides, composed of a number of monosaccharides or sugar (glucose) molecules linked together with α-D-(1-4) and/or α-D-(1-6) linkages. The starch consists of 2 main structural components, the amylose, which is essentially a linear polymer in which glucose residues are α-D-(1-4) linked typically constituting 15% to 20% of starch, and amylopectin, which is a larger branched molecule with α-D-(1-4) and α-D-(1-6) linkages and is a major component of starch. Amylose is linear or slightly branched, has a degree of polymerization up to 6000, and has a molecular mass of 105 to 106 g/mol. The chains can easily form single or double helices. Amylopectin on the other hand has a molecular mass of 107 to 109 g/mol. It is highly branched and has an average degree of polymerization of 2 million, making it one of the largest molecules in nature. Chain lengths of 20 to 25 glucose units between branch points are typical. About 70% of the mass of starch granule is regarded as amorphous and about 30% as crystalline. The amorphous regions contain the main amount of amylose but also a considerable part of the amylopectin. The crystalline region consists primarily of the amylopectin (Sajilata, et al., 2006).

Starch in the pharmaceutical industry

During recent years, starch has been taken as a new potential biomaterial for pharmaceutical applications because of the unique physicochemical and functional characteristics (Cristina Freire, et al., 2009; Freire, et al., 2009; Serrero, et al.).

3.1 Starch as pharmaceutical excipient

Native starches were well explored as binder and disintegrant in solid dosage form, but due to poor flowability their utilization is restricted. Most common form of modified starch i.e. Pre-gelatinized starch marketed under the name of starch 1500 is now a day's most preferred directly compressible excipients in pharmaceutical industry. Recently modified rice starch, starch acetate and acid hydrolyzed dioscorea starch were established as multifunctional excipient in the pharmaceutical industry. The International Joint Conference on Excipients rated starch among the top ten pharmaceutical ingredients (Shangraw, 1992).

3.2 Starch as tablet disintegrant

They are generally employed for immediate release tablet formulations, where drug should be available within short span of time to the absorptive area. Sodium carboxymethyl starch, which is well established and marketed as sodium starch glycolate is generally used for immediate release formulation. Some newer sources of starch have been modified and evaluated for the same.

3.3 Starch as controlled/sustained release polymer for drugs and hormones

Modified starches in different forms such as Grafted, acetylated and phosphate ester derivative have been extensively evaluated for sustaining the release of drug for better patient compliances. Starch-based biodegradable polymers, in the form of microsphere or hydrogel, are suitable for drug delivery (Balmayor, et al., 2008), (Reis, et al., 2008). For example, high amylose corn starch has been reported to have good sustained release properties and this has been attributed to its excellent gel-forming capacity (Rahmouni, et al., 2003; Te Wierik, et al., 1997). Some authors (Efentakis, et al., 2007; Herman and Remon, 1989; Michailova, et al., 2001) have explained the mechanism of drug release from such gel-forming matrices to be a result of the controlled passage of drug molecules through the obstructive gel layer, gel structure and matrix.

3.4 Starch as plasma volume expander

Acetylated and hydroxyethyl starch are now mainly used as plasma volume expanders. They are mainly used for the treatment of patients suffering from trauma, heavy blood loss and cancer.

3.5 Starch in bone tissue engineering

Starch-based biodegradable bone cements can provide immediate structural support and degrade from the site of application. Moreover, they can be combined with bioactive particles, which allow new bone growth to be induced in both the interface of cement-bone and the volume left by polymer degradation (Boesel, et al., 2004). In addition, starch-based biodegradeable polymer can also be used as bone tissue engineering scaffold (Gomes, et al., 2003).

3.6 Starch in artificial red cells

Starch has also been used to produce a novel and satisfactory artificial RBCs with good oxygen carrying capacity. It was prepared by encapsulating hemoglobin (Hb) with long-chain fatty-acids-grafted potato starch in a self-assembly way (Xu, et al., 2011).

3.7 Starch in nanotechnology

Starch nanoparticles, nanospheres, and nanogels have also been applied in the construction of nanoscale sensors, tissues, mechanical devices, and drug delivery system. (Le Corre, et al., 2010).

3.7.1 Starch microparticles

The use of biodegradable microparticles as a dosage form for the administration of active substances is attracting increasing interest, especially as a means of delivering proteins. Starch is one of the polymers that is suitable for the production of microparticles. It is biodegradable and has a long tradition as an excipient in drug formulations. Starch microparticles have been used for the nasal delivery of drugs and for the delivery of vaccines administered orally and intramuscularly. Bioadhesive systems based on polysaccharide microparticles have been reported to significantly enhance the systemic absorption of conventional drugs and polypeptides across the nasal mucosa, even when devoid of absorption enhancing agents. A major area of application of microparticles is as dry powder inhalations formulations for asthma and for deep-lung delivery of various agents. It has also been reported that, particles reaching the lungs are phagocytosed rapidly by alveolar Macrophages. Although phagocytosis and sequestration of inhaled powders may be a problem for drug delivery to other cells comprising lung tissue, it is an advantage for chemotherapy of tuberculosis. Phagocytosed microparticles potentially can deliver larger amounts of drug to the cytosol than oral doses. It is also opined strongly that, microparticles have the potential for lowering dose frequency and magnitude, which is especially advantageous for maintaining drug concentrations and improving patient compliance. This is the main reason this dosage form is an attractive pulmonary drug delivery system. (Le Corre, et al., 2010).

3.7.2 Starch microcapsules

Microencapsulation is the process of enclosing a substance inside a membrane to form a microcapsule. it provides a simple and cost-effective way to enclose bioactive materials within a semi-permeable polymeric membrane. Both synthetic/semi-synthetic polymers and natural polymers have been extensively utilized and investigated as the preparation materials of microcapsules. Although the synthetic polymers display chemical stability, their unsatisfactory biocompatibility still limits their potential clinical applications. Because the natural polymers always show low/non toxicity, low immunogenicity and thereafter good biocompatibility, they have been the preferred polymers used in microencapsulation systems. Among the natural polymers, alginate is one of the most common materials used to form microcapsules, however, starch derivatives are now gaining attention. For instance starch nasal bioadhesive microspheres with significantly extended half-life have been reported for several therapeutic agents including insulin. Improved bioavailability of Gentamycin-encapsulated starch microspheres as well as magnetic starch microspheres for parenteral administration of magnetic iron oxides to enhance contrast in magnetic resonance imaging has been reported. (Le Corre, et al., 2010).

3.7.3 Starch nanoparticles

Nanoparticles are solid or colloidal particles consisting of macromolecular substances that vary in size from 10-1000 nm. The drug may be dissolved, entrapped, adsorbed, attached or encapsulated into the nanoparticle matrix. The matrix may be biodegradable materials such as polymers or proteins or biodegradable/biocompatible/bioasborbable materials such as starch. Depending on the method of preparation, nanoparticles can be obtained with different physicochemical, technical or mechanical properties as well as modulated release characteristics for the immobilized bioactive or therapeutic agents. (Le Corre, et al., 2010).

4. Application of modified starches in drug delivery

Native starch irrespective of their source are undesirable for many applications, because of their inability to withstand processing conditions such as extreme temperature, diverse pH, high shear rate, and freeze thaw variation. To overcome this, modifications are usually done to enhance or repress the inherent property of these native starches or to impact new properties to meet the requirements for specific applications. The process of starch modification involves the destructurisation of the semi-crystalline starch granules and the effective dispersion of the component polymers. In this way, the reactive sites (hydroxyl groups) of the amylopectin polymers become accessible to electrophilic reactants (Rajan, et al., 2008). Common modes of modifications useful in pharmaceuticals are chemical, physical and enzymatic with, a much development already seen in chemical modification. Starch modification through chemical derivation such as etherification, esterification, cross-linking, and grafting when used as carrier for controlled release of drugs and other bioactive agents. It has been shown that, chemically modified starches have more reactive sites to carry biologically active compounds, they become more effective biocompatible carriers and can easily be metabolized in the human body (Prochaska, et al., 2009; Simi and Emilia Abraham, 2007).

4.1 Chemical modification of starch

There are a number of chemical modifications made to starch to produce many different functional characteristics. The chemical reactivity of starch is controlled by the reactivity of its glucose residues. Modification is generally achieved through etherification, esterification, crosslinking, oxidation, cationization and grafting of starch. However, because of the dearth of new methods in chemical modifications, there has been a trend to combine different kinds of chemical treatments to create new kinds of modifications. The chemical and functional properties achieved following chemical modification of starch, depends largely on the botanical or biological source of the starch,, reaction conditions (reactant concentration, reaction time, pH and the presence of catalyst), type of substituent, extent of substitution (degree of substitution, or molar substitution), and the distribution of the substituent in the starch molecule (Singh, et al., 2007). Chemical modification involves the introduction of functional groups into the starch molecule, resulting in markedly altered physico-chemical properties. Such modification of native granular starches profoundly alters their gelatinization, pasting and retrogradation behavior (Choi and Kerr, 2003; Kim, et al., 1993) (Perera, et al.) and (Liu, et al., 1999) (Seow and Thevamalar, 1993). The rate and efficiency of the chemical modification process depends on the reagent type, botanical origin of the starch and on the size and structure of its granules (Huber and BeMiller, 2001).This also includes the surface structure of the starch granules, which encompasses the outer and inner surface, depending on the pores and channels (Juszczak, 2003).

4.1.1 Carboxymethylated starch

Starches can have a hydrogen replaced by something else, such as a carboxymethyl group, making carboxymethyl starch (CMS). Adding bulky functional groups like carboxymethyl and carboxyethyl groups reduces the tendency of the starch to recrystallize and makes the starch less prone to damage by heat and bacteria. Carboxymethyl starch is synthesized by reacting starch with monochloroacetic acid or its sodium salt after activation of the polymer with aqueous NaOH in a slurry of an aqueous organic solvent, mostly an alcohol. The total degree of substitution (DS), that is the average number of functional groups introduced in the polymer, mainly determines the properties of the carboxymethylated products (Heinze, 2005). The functionalization influences the properties of the starch. For example, CMS have been shown to absorb an amount of water 23 times its initial weight. This high swelling capacity combined with a high rate of water permeation is said to be responsible for a high rate of tablet disintegration and drug release from CMS based tablets. CMS has also been reported to be capable of preventing the detrimental influence of hydrophobic lubricants (such as magnesium stearate) on the disintegration time of tablets or capsules.. Some of the recent use of carboxymethylated starch in pharmaceuticals are summarised in Table 1.

Study Title	Methodology	Drug used	Summary	References
Synthesis and in vitro evaluation of carboxymethyl starch-chitosan nanoparticles as drug delivery system to the colon.	Complex coacervation process.	5-aminosalicylic acid	Chitosan-carboxymethyl starch nanoparticles developed based on the modulation of ratio show promise as a system for controlled delivery of drugs to the colon.	(Saboktakin, et al., 2010)
Carboxymethyl-Starch Excipients for Gastrointestinal Stable Oral Protein Formulations Containing Protease Inhibitors.	Monochloroacetic acid Semisynthesis	Protease inhibitors	The feasibility to deliver stable bioactive proteins into intestinal environment using CMS was demonsttrated. In addition to its protective role, the CMS allows the release of the bioactive agents in less than 6 h, fitting well with the upper intestinal transit. Such stability is needed for formulation of oral vaccines with specific antigens.	(Patrick De Koninck, 2010)
Novel Polymeric Biomaterial Based on Carboxymethyl starch and its Application in Controlled Drug Release.	Blender, mixing and sieving	Acetylsalicylic acid	Matrix (CMS) releases the enclosed drug at a much faster rate in neutral and alkaline pH than in acidic pH, thus holding the promise of being developed into a vehicle for targeted drug delivery (oral route) to the lower gastrointestinal tract.	(Sen and Pal, 2009b)
Carboxymethyl high amylose starch as excipient for controlled drug release: Mechanistic study and the influence of degree of substitution.	Non aqueous medium synthesis, protanated using acid treatment.	Acetaminophen	The results showed that, CMHAS(Na) with DS between 0.1 and 0.2 can be used as sustained release excipients, while those with high DS, between 0.9 and 1.2, are better as delayed release excipient.	(Lemieux, et al., 2009)
High-amylose sodium carboxymethyl starch matrices for oral, sustained drug-release: Formulation aspects and in vitro drug-release evaluation.	Spray Drying	Acetaminophen	The results proved that the new Spray Drying process developed for High-amylose sodium carboxymethyl starch (HASCA) manufacture is suitable for obtaining similar-quality (HASCA) in terms of drug release and compression performances.	(Brouillet, et al., 2008)
Carboxylated high amylose starch as pharmaceutical excipients and formulation Structural insights of pancreatic enzymes.	Carboxylation and protonation	Pancreatic enzymes	High loading capacity (up to 70–80% enzymes) was obtained. An advantage of these formulations is that gastroprotection is afforded by the carboxylated matrices (carboxylic groups), without enteric coating.	(Massicotte, et al., 2008)

High-amylose carboxymethyl starch matrices for oral sustained drug-release: In vitro and in vivo evaluation.	Blending of HASCA with NaCl	Acetaminophen	In vitro drug-release from an optimized HASCA formulation was not affected by either acidic pH value or acidic medium residence time. Compressed blend of HASCA with an optimized quantity of sodium chloride provided a pharmaceutical sustained-release tablet with improved integrity for oral administration. In vivo studies demonstrated extended drug absorption.	(Nabais, et al., 2007)
Physicochemical and Pharmaceutical Properties of Carboxymethyl Rice Starches Modified from Native Starches with Different Amylose Content.	Monochloroacetic acid Reaction	Material science	Carboxymethyl rice starches (CMRS) can function as tablet binder in the wet granulation of both water-soluble and water-insoluble diluents. The tablets compressed from these granules showed good hardness with fewer capping problems compared with those prepared using the pregelatinized native rice starch as a binder.	(Kittipongpa tana, et al., 2007)
An Aqueous Film-coating Formulation based on Sodium Carboxymethyl Mungbean Starch	Petri dish method	Material scienc	Carboxymethyl mungbean starch (SCMMSs) exhibited the ability to form a clear, thin film with greater flexibility and strength than that of the native starch. This study reports the potential of SCMMS as tablet film coating agent	(Kittipongpa tana, et al., 2006)

Table 1. Use of carboxymethylated starch in drug delivery

4.1.2 Acetylated starch

Acetylated starch has also been known for more than a century. Starches can be esterified by modifications with an acid. When starch reacts with an acid, it loses a hydroxyl group, and the acid loses hydrogen. An ester is the result of this reaction. Acetylation of cassava starch has been reported to impart two very important pharmaceutical characters to it; increased swelling power (Rutenberg, 1984) and enhanced water solubility of the starch granules (Aziz, 2004). Starch acetates and other esters can be made very efficiently on a micro scale without addition of catalyst or water simply by heating dry starch with acetic acid and anhydride at 180°C for 2-10 min (Shogren, 2003). At this temperature, starch will melt in acetic acid (Shogren, 2000)and thus, a homogeneous acetylation would be expected to occur. Using acetic acid, starch acetates are formed, which are used as film-forming polymers for pharmaceutical products. A much recent Scandium triflate catalyzed acetylation of starch at low to moderate temperatures is reported by (Shogren, 2008). Generally, starch acetates have a lower tendency to create gels than unmodified starch. Acetylated starches are distinguishable through high levels of shear strength. They are particularly stable against heat and acids and are equally reported to form flexible, water-soluble films. Some of the recent uses of acetylated starch in pharmaceuticals are summarized in Table 2.

Study Title	Methodology	Drug used	Summary	References
An Oral Colon-Targeting Controlled Release System Based on Resistant Starch Acetate: Synthetization, Characterization, and Preparation of Film-Coating Pellets.	Acetic anhydride synthesis	5-ASA, BSA, HGF, and insulin	The study suggests that an oral colon-targeting controlled release system based on resistant starch acetate (RSA) as a film-coating material has an excellent colon-targeting release performance and the universality for a wide range of bioactive components.	(Pu, 2011)
Preparation and Properties of Starch Acetate Fibers for Potential Tissue Engineering Applications	Extruded as fibers onto a rotating drum with variable speed using a syringe and needle	Tissue engineering scaffolds	The starch acetate fibers support the adhesion of fibroblasts demonstrating that the fibers would be suitable for tissue engineering and other medical applications. They possess better mechanical properties and water stability than most polysaccharide-based biomaterials and protein fibers used in tissue engineering.	(Narendra Reddy, 2009)
Acetylated starch-based biodegradable materials with potential biomedical applications as drug delivery systems.	Acetyl esterification	Bovine serum albumin	Drug release studies show that the starch acetate coated tablets could deliver the drug to the colon suggesting that it can be a potential drug delivery carrier for colon-targeting.	(Chen, et al., 2007)
Optimization and characterization of controlled release multi-particulate beads coated with starch acetate	Organic synthesis	Dyphylline	Starch acetate-coated beads provided controlled release of dyphylline.	(Nutan, et al., 2005)
Drug release from starch-acetate microparticles and films with and without incorporated alpha-amylose.	Acetyl Esterification /aqueous synthesis	Timolol calcein and bovine serum albumin	This study demonstrates the achievement of slow release of different molecular weight model drugs from the starch acetate microparticles and films as compared to fast release from the native starch preparations.	(Tuovinen, et al., 2004a)
Starch acetate microparticles for drug delivery into retinal pigment epithelium-in vitro study.	Modified water-in-oil-in-water double-emulsion technique.	Calcein	The study indicates that the natural enzyme-sensitive starch acetate is suitable for drug delivery into retinal pigment epithelium (RPE). The starch acetate microparticles were easily taken up by cultured human RPE cells without significant toxicity.	(Tuovinen, et al., 2004b)

Starch acetate as a tablet matrix for sustained drug release	Monolithic matrix system	Propranolol hydrochloride	The drug release was considerably slower from sodium acetate tablet matrices. Also, a decrease in starch acetate concentration increased the drug release rate. Crack formation increased area available for Fickian diffusion, which caused slow attenuation of drug release rate.	(Pohja, et al., 2004)
Drug release from starch-acetate films.	Monolithic matrix system	BSA, FITC-dextran timolol and sotalol–HCl	The results showed that acetylation of potato starch can substantially retard drug release thus allowing sustained drug delivery. The drug release profiles may be controlled by the degree of substitution.	(Tuovinen, et al., 2003)
Acetylation enhances the tabletting properties of starch	Acetic Anhydride synthesis	materials science	The acetate moiety, perhaps in combination with existing hydroxyl groups, was a very effective bond-forming substituent. The formation of strong molecular bonds increased, leading to a very firm and intact tablet structure.	(Raatikainen, et al., 2002)

Table 2. Some acetylated starches and their application in drug delivery.

4.1.3 Hydroxypropylated starch

Hydroxypropyl groups introduced into starch chains are said to be capable of disrupting inter- and intra-molecular hydrogen bonds, thereby weakening the granular structure of starch leading to an increase in motional freedom of starch chains in amorphous regions (Choi and Kerr, 2003; Seow and Thevamalar, 1993; Wootton and Manatsathit, 1983). Such chemical modification involving the introduction of hydrophilic groups into starch molecules improves the solubility of starch and the functional properties of starch pastes, such as its shelf life, freeze/thaw stability, cold storage stability, cold water swelling, and yields reduced gelatinization temperature, as well as retarded retrogradation. Owing to these properties, hydroxypropylated starches is gaining interest in medicine.

Study Title	Methodology	Drug used	Summary	References
Hydroxypropylated starches of varying amylose contents as sustained release matrices in tablets	Monolithic matrix tablet formulation	Propranolol hydrochloride	Hydroxypropylation improved the sustained release ability of amylose-containing starch matrices, and conferred additional resistance to the hydrolytic action of pancreatin under simulated gastrointestinal conditions.	(Onofre and Wang, 2010)

Table 3. Hydroxyl-propylated starch in drug delivery.

4.1.4 Succinylated starch

Modification of starch by Succinylation has also been found to modify its physicochemical properties, thereby widening its applications in food and non-food industries like pharmaceuticals, paper and textile industries. Modification of native starch to its succinate derivatives reduces its gelatinisation temperature and the retrogradation, improves the freeze-thaw stability as well as the stability in acidic and salt containing medium (Trubiano Paolo, 1997; Trubiano, 1987; Tukomane and Varavinit, 2008). Generally, succinylated starch can be prepared by treating starches with different alkenyl succinic anhydride, for example, dodecenyl succinic anhydride, octadecenyl succinic anhydride or octenyl succinic anhydride. The incorporation of bulky octadecenyl succinic anhydride grouping to hydrophilic starch molecules has been found to confer surface active properties to the modified starch (Trubiano Paolo, 1997). Unlike typical surfactants, octadecenyl succinic anhydride starch, forms strong films at the oil–water interface giving emulsions that are resistant to reagglomeration. A recent application of succinylated starch in pharmaceuticals are summarized in Table 4.

Study Title	Methodology	Drug used	Summary	References
Preparation and characterisation of octenyl succinate starch as a delivery carrier for bioactive food components	Pyridine-catalyzed esterification	Bovine serum albumin/ASA	Ocetyl succcinate starch was found to be a potential carrier for colon-targeted drug delivery	(Wang, et al., 2011)

Table 4. A recent application of succinylated starch in drug delivery

4.1.5 Phosphorylated starch

Phosphorylation was the earliest method of starch modification. The reaction gives rise to either monostarch phosphate or distarch phosphate (cross-linked derivative), depending upon the reactants and subsequent reaction conditions. Phosphate cross-linked starches show resistance to high temperature, low pH, high shear, and leads to increased stability of the swollen starch granule. The presence of a phosphate group in starch increases the hydration capacity of starch pastes after gelatinization and results in the correlation of the starch phosphate content to starch paste peak viscosity , prevents crystallization and gel-forming capacity (Nutan, et al., 2005). These new properties conferred on starch by phosphorylation, makes them useful as disintegrants in solid dosage formulations and as matrixing agents. Interestingly, it has been documented that, the only naturally occurring covalent modification of starch is phosphorylation. Traditionally, starch phosphorylation is carried out by the reaction of starch dispersion in water with reagents like mono- or di sodium orthophosphates, sodium hexametaphosphate, sodium tripolyphosphate (STPP), sodium trimetaphosphate (STMP) or phosphorus oxychloride. Alternative synthetic methods such as extrusion cooking, microwave irradiation and vacuum heating have been reported (A. N. Jyothi, 2008; Sitohy and Ramadan, 2001). Some of the recent uses of phosphorylated starch in pharmaceuticals are summarized in Table 5.

Study Title	Methodology	Drug used	Summary	References
Starch Phosphate: A Novel Pharmaceutical Excipient For Tablet Formulation.	Phosphorylation using mono sodium phosphate dehydrate	Ziprasidone	At low concentration, starch phosphate proved to be a better disintegrant than native starch in tablet formulation.	(N.L Prasanthi, 2010)
Starch phosphates prepared by reactive extrusion as a sustained release agent.	Reactive extrusion	Metoprolol tartrate	Starch phosphate prepared by reactive extrusion produced stronger hydrogels with sustained release properties as compared with native starch.	(O'Brien, et al., 2009)

Table 5. Use of phosphorylated starch in Pharmaceuticals.

4.1.6 Co-polymerized starch

Chemical modification of natural polymers by grafting has received considerable attention in recent years because of the wide variety of monomers available. Graft copolymerization is considered to be one of the routes used to gain combinatorial and new properties of natural and synthetic polymers. In graft copolymerization the guest monomer benefits the host polymer with some novel and desired properties in which the resultant copolymer gains characteristic properties and applications (Fares, et al., 2003). As a rule, graft copolymerization produces derivatives of significantly increased molecular weight. Starch grafting usually entails etherification, acetylation, or esterification of the starch with vinyl monomers to introduce a reaction site for further formation of a copolymeric chain. Such a chain would typically consist of either identical or different vinyl monomers (block polymers), or it may be grafted onto another polymer altogether. Graft copolymers find application in the design of various stimuli-responsive controlled release systems such as transdermal films, buccal tablets, matrix tablets, microsphers/hydrogel bead system and nanoparticulate system (Sabyasachi Maiti, 2010). Some of the recent uses of graft co-polymerized starch in pharmaceuticals is summarized in Table 6.

Study Title	Methodology	Drug used	Summary	References
Grafted Starch-Encapsulated Hemoglobin (GSEHb) Artificial Red Blood Cells Substitutes.	Self –assembly	Artificial RBCs	Satisfactory artificial RBCs with good oxygen carrying capacity obtained.	(Xu, et al., 2011)
Preparation and using of acrylamide grafted starch as polymer drug carrier.	Chemical grafting of acrylamide on the starch polymer	Ceftriaxone Sodium	Acrylamide grafted starch showed higher uptake of water compared with native starch. The starch exhibited good controlled release properties.	(Al-Karawi and Al-Daraji, 2010)
Graft copolymers of ethyl methacrylate on waxy maize starch derivatives as novel excipients for matrix tablets: Physicochemical and technological characterisation	Free radical polymerization and alternatively dried in a vacuum oven (OD) or freeze-dried (FD)	Material science	The copolymers could be used as excipients in matrix tablets and have potentials for use as controlled release materials.	(Marinich, et al., 2009)
Physical blends of starch graft copolymers as matrices for colon targeting drug delivery systems.	Synthesis by Cerium iv ion method	Theophylline, procaine and bovineserum	The physical blend offered good controlled release of drugs, as well as of proteins and presented suitable properties for use as hydrophilic matrices for colon-specific drug delivery.	(Silva, et al., 2009)
Microwave initiated synthesis of polyacrylamide grafted carboxymethylstarch (CMS-g-PAM): application as a novel matrix for sustained drug release.	Microwave initiated synthesis	5-amino salicylic acid	'In vitro' release of a model drug (5-amino salicylic acid) from CMS-g-PAM matrix showed a sustained drug release. In this matrix, the rate of release of the enclosed drug could be precisely programmed simply by adjustment of percentage grafting during synthesis.	(Sen and Pal, 2009a)
Starch–Acrylics Graft Copolymers and Blends: Synthesis, Characterization, and Applications as Matrix for Drug Delivery	Polymerization reaction	Paracetamol	The graft copolymers, provided a pH sensitive matrix system for site-specific drug delivery. The authors concluded that graft copolymers may be a useful tool to overcome the harsh environment of the stomach and can possibly be used in future as excipient for colon-targeted drug delivery.	(Shaikh and Lonikar, 2009)

Characterization and in vitro evaluation of starch based hydrogels as carriers for colon specific drug delivery systems	γ-rays induced polymerization and crosslinking	Ketoprofen	Hydrogels prepared showed pH responsive property; preventing drug release pH 1, but released it at pH 7.	(El-Hag Ali and AlArifi, 2009)
Hydrophobic grafted and crosslinked starch nano particles for drug delivery.	Grafted using potassium persulphate as catalyst	Indomethacin	Fatty acid grafted starch nano particle with high swelling power was obtained and was found to be a good vehicle for oral controlled drug delivery	(Simi and Emilia Abraham, 2007)
Bioadhesive grafted starch copolymers as platforms for peroral drug delivery: a study of theophylline release.	Radiation of starch and acrylic acid mixtures with ^{60}Co	Theophylline	Results show that,the release of theophylline from the graft copolymer tablets was practically independent of the pH of the dissolution medium and the type of starch used for grafting. Incorporation of divalent cations into the graft copolymers led to a significant decrease in swelling and retardation of drug release.	(Geresh, et al., 2004)
Ethyl Methacrylate grafted on two starches as polymeric matrices for drug delivery	Ceric ion redox initiation method	Theophylline and procaine hydrochloride	The HS-EMA and S-EMA were found to be efficient matrices for insoluble drugs	(Echeverria, et al., 2005)

Table 6. Use of graft co-polymerized starch in Pharmaceuticals.

4.2 Physical modification of starch

Physical modification of starch is mainly applied to change the granular structure and convert native starch into cold water-soluble starch or small-crystallite starch. The major methods used in the preparation of cold water-soluble starches involve instantaneous cooking–drying of starch suspensions on heated rolls (drum-drying), puffing, continuous cooking–puffing–extruding, and spray-drying (Jarowrenko, 1986). A method for preparing granular cold water-soluble starches by injection and nozzle-spray drying was described by (Pitchon & Joseph 1981). Among the physical processes applied to starch modification, high pressure treatment of starch is considered an example of 'minimal processing'(Stute, et al., 1996). A process of iterated synersis applied to the modification of potato, tapioca, corn and wheat starches resulted in a new type of physically modified starches (Lewandowicz and Soral-Smietana, 2004). Some of the recent uses of physically modified starch in pharmaceuticals are summarized in Table 7.

Study Title	Methodology	Drug used	Summary	References
Microwave Assisted Modification of Arrowroot Starch for Pharmaceutical Matrix Tablets	Microwave Assisted Modification	Theophylline	The modified arrowroot starch, demonstrated promising properties as hydrophilic matrix excipients for sustained release tablets	(Pornsak Sriamornsak 2010)
Effect of heat moisture treatment on the functional and tabletting properties of corn starch Gelatinized/freeze-dried starch as excipient in sustained release tablets.	Heat moisture treatment	Material science	Heat moisture treatment (HMT) of corn starch could be useful when fast disintegration, lower swelling power, and lower crushing strength are desired in tablet.	(Iromidayo Olu-Owolabi.B, 2010)
Pregelatinized glutinous rice starch as a sustained release agent for tablet preparations	Heat treatment and spray drying	Propranolol HCl	In this study glutinous rice starch slurry was physically modified by heat and then dried by spray drying. The tablet containing pregelatinized glutinous starch and propranolol HCl were prepared by wet granulation method. The mechanisms of drug release from the matrices were anomalous (non-Fickian) diffusion in both hydrochloric buffer (pH 1.2) and phosphate buffer media (pH 6.8).	(Peerapattan a, et al.)
A Novel Pregelatinized Starch as a Sustained-Release Matrix Excipient.	Controlled thermal pregelatinization And spray-drying	Ethenzamide acetaminophen and sodium salicylic acid	The study shows that highly functional pregelatinized starch is a new matrix excipient that can control drug-release profiles from first- to zero-order sustained release and enables drug release independent of drug solubility and external conditions.	(Masaaki Endo, 2009)
Effects of Drying Process for Amorphous Waxy Maize Starch on Theophylline Release from Starch-Based Tablets.	Oven drying and freeze drying procedure	Theophylline	The drying method was found to affect the morphology and drug release profiles of the compressed tablets.	(Yoon, et al., 2007)

Gelatinized/freeze-dried starch as excipient in sustained release tablets.	Gelatinization and freeze-drying	Material science	A new technique for the production of cold water-swellable starch using gelatinization and freeze-drying processes was obtained and the matrices containing different modified starch -hydroxypropyl methylcellulose mixtures possess good sustained release properties.	(Sánchez, et al., 1995)
Modified starches as hydrophilic matrices for controlled oral delivery III. Evaluation of sustained-release theophylline formulations based on thermal modified starch matrices in dogs	Thermal Modification	Theophylline	Several thermally modified starch matrices evaluated in dogs demonstrated good sustained-release performance	(Herman and Remon, 1990)
Modified starches as hydrophilic matrices for controlled oral delivery. II. In vitro drug release evaluation of thermally modified starches.	Thermal Modification	Theophylline	Thermally modified starches containing a low amount of amylose (25% and lower) revealed promising properties as directly compressible tabletting excipients for sustained release purposes.	(Herman and Remon, 1989)

Table 7. Some of the recent uses of physically modified starch in medicine.

Study Title	Methodology	Drug used	Summary	References
Enzymatic modification of cassava starch by fungal lipase.	Esterification using fungal lipase	Material science	Esterification of starch using fungal lipase with long chain fatty acids like palmitic acid gives thermoplastic starch which has got wide use in plastic industry, pharmaceutical industries, and in biomedical applications such as materials for bone fixation and replacements, carriers for controlled release of drugs and other bioactive agents. Unlike chemical esterification, enzymatic esterification is ecofriendly and avoids the use of nasty solvents.	(Rajan, et al., 2008)
Enzyme-Catalyzed Regioselective Modification of Starch Nanoparticles	Etherification of starch nanoparticles catalyzed by Candida antartica Lipase B	Material science	Selective etherification of starch nanoparticles catalyzed by Candida antartica Lipase B (CAL-B) in its immobilized (Novozyme 435) and frees (SP-525) forms. The starch nanoparticles were made accessible for acylation reactions by formation of Aerosol-OT (AOT, bis[2-ethylhexyl]sodium sulfosuccinate) stabilized microemulsions. The close proximity of the lipase and substrates promotes the modification reactions.	(Chakraborty, et al., 2005)

Table 8. Some enzymatically modified starches and potential medical/pharmaceutical applications

4.3 Enzymatic modification of starch

An alternative to obtaining modified starch is by using various enzymes. These include enzymes occurring in plants, e.g pullulanase and isoamylase groups. Pullulanase is a 1,6-α-glucosidase, which statistically impacts the linear α-glucan, a pullulan which releases maltotriose oligomers. This enzyme also hydrolyses α-1,6-glycoside bonds in amylopectin and dextrines when their side-chains include at least two α-1,4-glycoside bonds. Isoamylase is an enzyme which totally hydrolises α-1,6-glycoside bonds in amylopectin, glycogen, and some branched maltodextrins and oligosaccharides, but is characterised by low activity in relation to pullulan (Norman 1981). In a study (Kim and Robyt, 1999) starch granules was modified in situ by using a reaction system in which glucoamylase reacts inside starch granules to give conversions of 10–50% D-glucose inside the granule. Enzymatic modification of starch still needs to be explored and studied. Some of the recent uses of enzymatically modified starch in pharmaceuticals is summarized in Table 8.

Study Title	Methodology	Drug used	Summary	References
Evaluation of glutinous rice starch based matrix microbeads using scanning electron microscopy	Micro orifice ionotropic-getation method	Material science	Microbeads were prepared by ionotropic-getation method using glutinous rice starch from Assam Bora rice, and sodium alginate backbone with different crosslinking agents. Photomicrographs provides information on the surface texture, size, mechanistic properties, suitability of drying condition and mechanism of drug release from the prepared micro devices.	(Nikhil K Sachan, 2010)
Development of porous HAp and β-TCP scaffolds by starch consolidation with foaming method and drug-chitosan bilayered scaffold based drug delivery system.	Starch consolidation with foaming method	Ceftriaxone	This study confirmed the ability of starch consolidation scaffolds to release drugs suitable for treating osteomyelitis.	(Kundu, et al.)
Preparation of starch-based scaffolds for tissue engineering by supercritical immersion precipitation.	Supercritical immersion precipitation	Scaffolds	Highly porous and interconnected scaffolds were obtained in this report which made use of polymeric blend of starch and poly (L-lactic acid) for tissue engineering purposes. Good porosity is critical in scaffolds technology	(Duarte, et al., 2009)
Porous scaffold of gelatin-starch with nanohydroxyapatite composite processed via novel microwave vaccum drying.	Microwave vaccum drying	Hydroxyapatite	Gelatin blended with starch results in scaffold composites with enhanced mechanical properties. A gelatin–starch blend reinforced with HA nanocrystals (nHA) gave biocompatible composites with enhanced mechanical properties.	(Sundaram, et al., 2008)

Novel Starch-Based Scaffolds for Bone Tissue Engineering: Cytotoxicity, Cell Culture, and Protein Expression	Melt-based technology	Scaffolds	A study on 50/50 (wt%) blend of corn starch/ethylene-vinyl alcohol (SEVA-C) led these authors to conclude that, starch-based scaffolds should be considered as an alternative for bone tissue-engineering applications in the near future.	(Salgado, et al., 2004)
Microwave Processing of Starch-Based Porous Structures for Tissue Engineering Scaffolds	Microwave Processing	Scaffolds	This study reports the suitability of different types of starch-based polymers; Potato, sweet potato, corn starch, and nonisolated amaranth and quinoa starch in preparing porous structures.	(Torres, et al., 2007)
Scaffold development using 3D printing with a starch- based polymer.	Rapid prototyping	Scaffolds	In this study, a unique blend of starch-based polymer powders (corn starch, dextran and gelatin) was developed for the 3DP process. Cylindrical scaffolds of five different designs were fabricated and found to possess enhanced mechanical and chemical properties.	(Lam, et al., 2002)
New partially degradable and bioactive acrylic bone cements based on starch blends and ceramic fillers	Free radical polymerization of methyl methacrylate and acrylic acid	Acrylic bone cements	This work reports the development of new partially biodegradable acrylic bone cements based on corn starch/cellulose acetate blends (SCA).	(Espigares, et al., 2002)
Porous starch-based drug delivery systems processed by a microwave route	Microwave baking method	Non-steroid anti-inflammatory agent	A new simple processing route to produce starch-based porous materials was developed for the delivery of non-steroid anti-inflammatory agents.	(Malafaya, 2001)

Table 9. Other starch derivatives and starch scaffolds with potential medical/pharmaceutical applications

5. Conclusions

It is obvious that starch has moved from its traditional role as food to being an indispensable medicine. The wide use of starch in the medicine is based on its adhesive, thickening, gelling, swelling and film-forming properties as well as its ready availability, low cost and controlled quality. From the foregoing, to think that starch is still ordinary inert excipients is to be oblivious of the influence this important biopolymer plays in therapeutic outcome of bioactive moieties. Starch has proven to be the formulator's "friend" in that, it can be utilized in the preparation of various drug delivery systems with the potential to achieve the formulator's desire for target or protected delivery of bioactive agents. It is important to note that apart from the low cost of starch, it is also relatively pure and does not need intensive purification procedures like other naturally occurring biopolymers, such as celluloses and gums. A major limitation to starch use appears to be its higher sensitivity to the acid attack; however, modification has been proved to impart acid-resistance to the product. It is important to optimize the process of transition of starch granules from its native micro- to the artificial submicron levels in greater detail and also pay greater attention to its toxicological profiles especially when it is desired to be used at nanoscale. Although starch is generally regarded as safe, its derivatives and in fact at submicron levels it may pose some safety challenges especially as carriers in drug delivery systems. It is possible to conclude that, although starch is food, it is also medicine.

6. References

A. N. Jyothi, M.S.S., S. N. Moorthy, J. Sreekumar and K. N. Rajasekharan, (2008). 'Microwave-assisted Synthesis of Cassava Starch Phosphates and their Characterization'. *Journal of Root Crops*, 34 (1):34-42.

Al-Karawi, A.J.M. and Al-Daraji, A.H.R., (2010). 'Preparation and using of acrylamide grafted starch as polymer drug carrier'. *Carbohydrate Polymers*, 79 (3):769-774.

Aziz, A., R. Daik, M.A. Ghani, N.I.N. Daud and B.M. Yamin, , (2004). 'Hydroxypropylation and acetylation of sago starch'. *Malaysian J. Chem.*, 6 (48-54).

Balmayor, E., Tuzlakoglu, K., Marques, A., Azevedo, H. and Reis, R., (2008). 'A novel enzymatically-mediated drug delivery carrier for bone tissue engineering applications: combining biodegradable starch-based microparticles and differentiation agents'. *Journal of Materials Science: Materials in Medicine*, 19 (4):1617-1623.

Boesel, L.F., Mano, J.F. and Reis, R.L., (2004). 'Optimization of the formulation and mechanical properties of starch based partially degradable bone cements'. *Journal of Materials Science: Materials in Medicine*, 15 (1):73-83.

Brouillet, F., Bataille, B. and Cartilier, L., (2008). 'High-amylose sodium carboxymethyl starch matrices for oral, sustained drug-release: Formulation aspects and in vitro drug-release evaluation'. *International Journal of Pharmaceutics*, 356 (1-2):52-60.

Chakraborty, S., Sahoo, B., Teraoka, I., Miller Lisa, M. and Gross Richard, A., (2005). 'Enzyme-Catalyzed Regioselective Modification of Starch Nanoparticles'. *Polymer Biocatalysis and Biomaterials*: American Chemical Society, 246-265.

Chen, L., Li, X., Li, L. and Guo, S., (2007). 'Acetylated starch-based biodegradable materials with potential biomedical applications as drug delivery systems'. *Current Applied Physics*, 7 (Supplement 1):e90-e93.

Choi, S.G. and Kerr, W.L., (2003). 'Water mobility and textural properties of native and hydroxypropylated wheat starch gels'. *Carbohydrate Polymers*, 51 (1):1-8.

Cristina Freire, A., Fertig, C.C., Podczeck, F., Veiga, F. and Sousa, J., (2009). 'Starch-based coatings for colon-specific drug delivery. Part I: The influence of heat treatment on the physico-chemical properties of high amylose maize starches'. *European Journal of Pharmaceutics and Biopharmaceutics*, 72 (3):574-586.

Duarte, A.R.C., Mano, J.F. and Reis, R.L., (2009). 'Preparation of starch-based scaffolds for tissue engineering by supercritical immersion precipitation'. *The Journal of Supercritical Fluids*, 49 (2):279-285.

Echeverria, I., Silva, I., Goñi, I. and Gurruchaga, M., (2005). 'Ethyl methacrylate grafted on two starches as polymeric matrices for drug delivery'. *Journal of Applied Polymer Science*, 96 (2):523-536.

Efentakis, M., Pagoni, I., Vlachou, M. and Avgoustakis, K., (2007). 'Dimensional changes, gel layer evolution and drug release studies in hydrophilic matrices loaded with drugs of different solubility'. *International Journal of Pharmaceutics*, 339 (1-2):66-75.

El-Hag Ali, A. and AlArifi, A., (2009). 'Characterization and in vitro evaluation of starch based hydrogels as carriers for colon specific drug delivery systems'. *Carbohydrate Polymers*, 78 (4):725-730.

Espigares, I., Elvira, C., Mano, J.F., Vázquez, B., San Román, J. and Reis, R.L., (2002). 'New partially degradable and bioactive acrylic bone cements based on starch blends and ceramic fillers'. *Biomaterials*, 23 (8):1883-1895.

Fares, M.M., El-faqeeh, A.S. and Osman, M.E., (2003). 'Graft Copolymerization onto Starch– I. Synthesis and Optimization of Starch Grafted with N-tert-Butylacrylamide Copolymer and its Hydrogels'. *Journal of Polymer Research*, 10 (2):119-125.

Freire, C., Podczeck, F., Veiga, F. and Sousa, J., (2009). 'Starch-based coatings for colon-specific delivery. Part II: Physicochemical properties and in vitro drug release from high amylose maize starch films'. *European Journal of Pharmaceutics and Biopharmaceutics*, 72 (3):587-594.

Geresh, S., Gdalevsky, G.Y., Gilboa, I., Voorspoels, J., Remon, J.P. and Kost, J., (2004). 'Bioadhesive grafted starch copolymers as platforms for peroral drug delivery: a study of theophylline release'. *Journal of Controlled Release*, 94 (2-3):391-399.

Gomes, M.E., Sikavitsas, V.I., Behravesh, E., Reis, R.L. and Mikos, A.G., (2003). 'Effect of flow perfusion on the osteogenic differentiation of bone marrow stromal cells cultured on starch-based three-dimensional scaffolds'. *Journal of Biomedical Materials Research Part A*, 67A (1):87-95.

Heinze, T., (2005). 'CARBOXYMETHYL ETHERS OF CELLULOSE AND STARCH – A REVIEW'. *Chemistry of plant raw materials.*, 3:13-29.

Herman, J. and Remon, J.P., (1989). 'Modified starches as hydrophilic matrices for controlled oral delivery. II. In vitro drug release evaluation of thermally modified starches'. *International Journal of Pharmaceutics*, 56 (1):65-70.

Herman, J. and Remon, J.P., (1990). 'Modified starches as hydrophilic matrices for controlled oral delivery III. Evaluation of sustained-release theophylline formulations based on thermal modified starch matrices in dogs'. *International Journal of Pharmaceutics*, 63 (3):201-205.

Huber, K.C. and BeMiller, J.N., (2001). 'Location of Sites of Reaction Within Starch Granules1'. *Cereal Chemistry*, 78 (2):173-180.

Iromidayo Olu-Owolabi.B, A.A.T.K.A.O., (2010). 'Effect of heat moisture treatment on the functional and tabletting properties of corn starch'. *African Journal of Pharmacy and Pharmacology*, 4 (7):498-510.

Jansook, P. and Loftsson, T., (2009). 'CDs as solubilizers: Effects of excipients and competing drugs'. *International Journal of Pharmaceutics*, 379 (1):32-40.

Jarowrenko, W., (1986). 'Pregelatinised starches'. *In O. B. Wurzburg (Ed.), Modified starches: Properties and uses.Boca Raton, FL: CRC Press:*71.

Juszczak, (2003). 'Surface of triticale starch granules – NC-AFM observations'. *Electronic Journal of Polish Agricultural Universities, Food Science and Technology*, 6.

Killen, B.U. and Corrigan, O.I., (2006). 'Effect of soluble filler on drug release from stearic acid based compacts'. *International Journal of Pharmaceutics*, 316 (1-2):47-51.

Kim, H.R., Muhrbeck, P. and Eliasson, A.-C., (1993). 'Changes in rheological properties of hydroxypropyl potato starch pastes during freeze—thaw treatments. III. Effect of cooking conditions and concentration of the starch paste'. *Journal of the Science of Food and Agriculture*, 61 (1):109-116.

Kim, Y.-K. and Robyt, J.F., (1999). 'Enzyme modification of starch granules: in situ reaction of glucoamylase to give complete retention of -glucose inside the granule'. *Carbohydrate Research*, 318 (1-4):129-134.

Kittipongpatana, O.S., Chaichanasak, N., Kanchongkittipoan, S., Panturat, A., Taekanmark, T. and Kittipongpatana, N., (2006). 'An Aqueous Film-coating Formulation based on Sodium Carboxymethyl Mungbean Starch'. *Starch - Stärke*, 58 (11):587-589.

Kittipongpatana, O.S., Chaitep, W., Kittipongpatana, N., Laenger, R. and Sriroth, K., (2007). 'Physicochemical and Pharmaceutical Properties of Carboxymethyl Rice Starches Modified from Native Starches with Different Amylose Content'. *Cereal Chemistry*, 84 (4):331-336.

Kundu, B., Lemos, A., Soundrapandian, C., Sen, P., Datta, S., Ferreira, J. and Basu, D., 'Development of porous HAp and β-TCP scaffolds by starch consolidation with foaming method and drug-chitosan bilayered scaffold based drug delivery system'. *Journal of Materials Science: Materials in Medicine*, 21 (11):2955-2969.

Lam, C.X.F., Mo, X.M., Teoh, S.H. and Hutmacher, D.W., (2002). 'Scaffold development using 3D printing with a starch-based polymer'. *Materials Science and Engineering: C*, 20 (1-2):49-56.

Langoth, N., Kalbe, J. and Bernkop-Schnürch, A., (2003). 'Development of buccal drug delivery systems based on a thiolated polymer'. *International Journal of Pharmaceutics*, 252 (1-2):141-148.

Le Corre, D., Bras, J., Dufresne, A., (2010). 'Starch Nanoparticles: A Review'. *Biomacromolecules 11, 1139–1153.*

Le Corre, D.b., Bras, J. and Dufresne, A., (2010). 'Starch Nanoparticles: A Review'. *Biomacromolecules*, 11 (5):1139-1153.

Lemieux, M., Gosselin, P. and Mateescu, M.A., (2009). 'Carboxymethyl high amylose starch as excipient for controlled drug release: Mechanistic study and the influence of degree of substitution'. *International Journal of Pharmaceutics*, 382 (1-2):172-182.

Lewandowicz, G. and Soral-Smietana, M., (2004). 'Starch modification by iterated syneresis'. *Carbohydrate Polymers*, 56 (4):403-413.

Li, S., Lin, S., Daggy, B.P., Mirchandani, H.L. and Chien, Y.W., (2003). 'Effect of HPMC and Carbopol on the release and floating properties of Gastric Floating Drug Delivery System using factorial design'. *International Journal of Pharmaceutics*, 253 (1-2):13-22.

Liu, H., Ramsden, L. and Corke, H., (1999). 'Physical properties and enzymatic digestibility of hydroxypropylated ae, wx, and normal maize starch'. *Carbohydrate Polymers*, 40 (3):175-182.

Malafaya, P.B.E., C.; Gallardo, A.; San Román, J.; Reis, R.L., (2001). 'Porous starch-based drug delivery systems processed by a microwave route'. *Journal of Biomaterials Science, Polymer Edition*, 12:1227-1241.

Manek, R.V., Kunle, O.O., Emeje, M.O., Builders, P., Rao, G.V.R., Lopez, G.P. and Kolling, W.M., (2005). 'Physical, Thermal and Sorption Profile of Starch Obtained from Tacca leontopetaloides'. *Starch - Stärke*, 57 (2):55-61.

Marinich, J.A., Ferrero, C. and Jiménez-Castellanos, M.R., (2009). 'Graft copolymers of ethyl methacrylate on waxy maize starch derivatives as novel excipients for matrix tablets: Physicochemical and technological characterisation'. *European Journal of Pharmaceutics and Biopharmaceutics*, 72 (1):138-147.

Masaaki Endo, K.O., Yoshihito Yaginuma, (2009). 'A Novel Pregelatinized Starch as a Sustained-Release Matrix Excipient'. *Pharmaceutical Technology*.

Mason, W.R., (2009). ' Starch Use in Foods. In Starch: Chemistry and Technology, Third Edition; chapter 20'.746 - 772

Massicotte, L.P., Baille, W.E. and Mateescu, M.A., (2008). 'Carboxylated high amylose starch as pharmaceutical excipients: Structural insights and formulation of pancreatic enzymes'. *International Journal of Pharmaceutics*, 356 (1-2):212-223.

Michailova, V., Titeva, S., Kotsilkova, R., Krusteva, E. and Minkov, E., (2001). 'Influence of hydrogel structure on the processes of water penetration and drug release from mixed hydroxypropylmethyl cellulose/thermally pregelatinized waxy maize starch hydrophilic matrices'. *International Journal of Pharmaceutics*, 222 (1):7-17.

Munday, D.L. and Cox, P.J., (2000). 'Compressed xanthan and karaya gum matrices: hydration, erosion and drug release mechanisms'. *International Journal of Pharmaceutics*, 203 (1-2):179-192.

N.L Prasanthi, N.R.a.R., (2010). *Starch Phosphate: A Novel Pharmaceutical Excipient For Tablet Formulation*.

Nabais, T., Brouillet, F., Kyriacos, S., Mroueh, M., Amores da Silva, P., Bataille, B., Chebli, C. and Cartilier, L., (2007). 'High-amylose carboxymethyl starch matrices for oral sustained drug-release: In vitro and in vivo evaluation'. *European Journal of Pharmaceutics and Biopharmaceutics*, 65 (3):371-378.

Narendra Reddy, Y.Y., (2009). 'Preparation and Properties of Starch Acetate Fibers for Potential Tissue Engineering Applications'. *Biotechnology and Biengineering*, 103 (5).

Nikhil K Sachan, S.K.G.a.A.B., (2010). 'Evaluation of glutinous rice starch based matrix microbeads using scanning electron microscopy'. *Journal of Chemical and Pharmaceutical Research*, 2 (3):433-452.

Norman 1981, B.E., 'New developments in starch syrup technology'. *w: Enzymes and Food Processing, G.G. Birch, N. Blakebrough, K.J. Parker eds., Applied Science Publishers, London,*:15-50.

Nutan, M.T.H., Soliman, M.S., Taha, E.I. and Khan, M.A., (2005). 'Optimization and characterization of controlled release multi-particulate beads coated with starch acetate'. *International Journal of Pharmaceutics*, 294 (1-2):89-101.

Nykänen, P., Lempää, S., Aaltonen, M.L., Jürjenson, H., Veski, P. and Marvola, M., (2001). 'Citric acid as excipient in multiple-unit enteric-coated tablets for targeting drugs on the colon'. *International Journal of Pharmaceutics*, 229 (1-2):155-162.

O'Brien, S., Wang, Y.-J., Vervaet, C. and Remon, J.P., (2009). 'Starch phosphates prepared by reactive extrusion as a sustained release agent'. *Carbohydrate Polymers*, 76 (4):557-566.

Onofre, F.O. and Wang, Y.J., (2010). 'Hydroxypropylated starches of varying amylose contents as sustained release matrices in tablets'. *International Journal of Pharmaceutics*, 385 (1-2):104-112.

Park, K., (2007). 'Nanotechnology: What it can do for drug delivery'. *Journal of Controlled Release*, 120 (1-2):1-3.

Patrick De Koninck, D.A., Francine Hamel, Fathey Sarhan and Mircea Alexandru Mateescu, (2010). 'Carboxymethyl-Starch Excipients for Gastrointestinal Stable Oral Protein Formulations Containing Protease Inhibitors'. *Journal of Pharmacy and Pharmaceutical Sciences*, 13 (1):78-92.

Peerapattana, J., Phuvarit, P., Srijesdaruk, V., Preechagoon, D. and Tattawasart, A., 'Pregelatinized glutinous rice starch as a sustained release agent for tablet preparations'. *Carbohydrate Polymers*, 80 (2):453-459.

Perera, C., Hoover, R. and Martin, A.M., 'The effect of hydroxypropylation on the structure and physicochemical properties of native, defatted and heat-moisture treated potato starches'. *Food Research International*, 30 (3-4):235-247.

Pitchon. E, O.R.J.D., & Joseph T. H (1981). 'Process for cooking or gelatinizing materials'. *US Patent 4280 851*.

Pohja, S., Suihko, E., Vidgren, M., Paronen, P. and Ketolainen, J., (2004). 'Starch acetate as a tablet matrix for sustained drug release'. *Journal of Controlled Release*, 94 (2-3):293-302.

Pornsak Sriamornsak , M.J., Suchada Piriyaprasarth, (2010). 'Microwave-Assisted Modification of Arrowroot Starch for Pharmaceutical Matrix Tablets'. *Advanced Materials Research*, 93-94:358-361.

Prochaska, K., Konowal, E., Sulej-Chojnacka, J. and Lewandowicz, G., (2009). 'Physicochemical properties of cross-linked and acetylated starches and products of their hydrolysis in continuous recycle membrane reactor'. *Colloids and Surfaces B: Biointerfaces*, 74 (1):238-243.

Pu, H.C., Ling Li, Xiaoxi Xie, Fengwei Yu, Long Li, Lin, (2011). 'An Oral Colon-Targeting Controlled Release System Based on Resistant Starch Acetate: Synthetization, Characterization, and Preparation of Film-Coating Pellets'. *Journal of Agricultural and Food Chemistry*, 59 (10):5738-5745.

Raatikainen, P., Korhonen, O., Peltonen, S. and Paronen, P., (2002). 'Acetylation Enhances the Tabletting Properties of Starch'. *Drug Development and Industrial Pharmacy*, 28 (2):165-175.

Rahmouni, M, Lenaerts, V, Leroux and J, C , (2003) *Drug permeation through a swollen cross-linked amylose starch membrane*. Paris, FRANCE: Editions de santé.

Rajan, A., Sudha, J.D. and Abraham, T.E., (2008). 'Enzymatic modification of cassava starch by fungal lipase'. *Industrial Crops and Products*, 27 (1):50-59.

Reis, A.V., Guilherme, M.R., Moia, T.A., Mattoso, L.H.C., Muniz, E.C. and Tambourgi, E.B., (2008). 'Synthesis and characterization of a starch-modified hydrogel as potential carrier for drug delivery system'. *Journal of Polymer Science Part A: Polymer Chemistry*, 46 (7):2567-2574.

Robertson, M.I., (1999). 'Regulatory issues with excipients'. *International Journal of Pharmaceutics*, 187 (2):273-276.

Rutenberg, M.W.a.D.S., Whistler, R.L., J.N. BeMiller and E.F. Paschall (Eds.), (1984). 'Starch Derivatives: Production and Uses. In: Starch: Chemistry and Technology, '. *Academic Press, New York,*:312-388.

Saboktakin, M.R., Tabatabaie, R.M., Maharramov, A. and Ramazanov, M.A., (2010). 'Synthesis and in vitro evaluation of carboxymethyl starch-chitosan nanoparticles as drug delivery system to the colon'. *International Journal of Biological Macromolecules,* 48 (3):381-385.

Sabyasachi Maiti, S.R., Biswanath Sa, (2010). 'Polysaccharide-Based Graft Copolymers in Controlled Drug Delivery'. *International Journal of PharmTech Research,* 2 (2):1350-1358.

Sajilata, M.G., Singhal, R.S. and Kulkarni, P.R., (2006). 'Resistant Starch–A Review'. *Comprehensive Reviews in Food Science and Food Safety,* 5 (1):1-17.

Salgado, A.J., Coutinho, O.P. and Reis, R.L., (2004). 'Novel Starch-Based Scaffolds for Bone Tissue Engineering: Cytotoxicity, Cell Culture, and Protein Expression'. *Tissue Engineering,* 10 (3-4):465-474.

Sánchez, L., Torrado, S. and Lastres, J., (1995). 'Gelatinized/freeze-dried starch as excipient in sustained release tablets'. *International Journal of Pharmaceutics,* 115 (2):201-208.

Sen, G. and Pal, S., (2009a). 'Microwave initiated synthesis of polyacrylamide grafted carboxymethylstarch (CMS-g-PAM): Application as a novel matrix for sustained drug release'. *International Journal of Biological Macromolecules,* 45 (1):48-55.

Sen, G. and Pal, S., (2009b). 'A novel polymeric biomaterial based on carboxymethylstarch and its application in controlled drug release'. *Journal of Applied Polymer Science,* 114 (5):2798-2805.

Seow, C.C. and Thevamalar, K., (1993). 'Internal Plasticization of Granular Rice Starch by Hydroxypropylation: Effects on Phase Transitions Associated with Gelatinization'. *Starch - Stärke,* 45 (3):85-88.

Serrero, A.l., Trombotto, S.p., Cassagnau, P., Bayon, Y., Gravagna, P., Montanari, S. and David, L., 'Polysaccharide Gels Based on Chitosan and Modified Starch: Structural Characterization and Linear Viscoelastic Behavior'. *Biomacromolecules,* 11 (6):1534-1543.

Shaikh, M.M. and Lonikar, S.V., (2009). 'Starch–acrylics graft copolymers and blends: Synthesis, characterization, and applications as matrix for drug delivery'. *Journal of Applied Polymer Science,* 114 (5):2893-2900.

Shangraw, R.F., (1992). 'International harmonization of compendia standards for pharmaceutical excipients'. *D.J.A. Crommelin, K.Midha (Eds.), Topics in Pharmaceutical Sciences, MSP, Stuttgart, Germany*:205-223.

Shogren, R., (2008). 'Scandium triflate catalyzed acetylation of starch at low to moderate temperatures'. *Carbohydrate Polymers,* 72 (3):439-443.

Shogren, R.L., (2000). 'Modification of maize starch by thermal processing in glacial acetic acid'. *Carbohydrate Polymers,* 43 (4):309-315.

Shogren, R.L., (2003). 'Rapid preparation of starch esters by high temperature/pressure reaction'. *Carbohydrate Polymers,* 52 (3):319-326.

Silva, I., Gurruchaga, M. and Goñi, I., (2009). 'Physical blends of starch graft copolymers as matrices for colon targeting drug delivery systems'. *Carbohydrate Polymers,* 76 (4):593-601.

Simi, C. and Emilia Abraham, T., (2007). 'Hydrophobic grafted and cross-linked starch nanoparticles for drug delivery'. *Bioprocess and Biosystems Engineering*, 30 (3):173-180.

Singh, J., Kaur, L. and McCarthy, O.J., (2007). 'Factors influencing the physico-chemical, morphological, thermal and rheological properties of some chemically modified starches for food applications--A review'. *Food Hydrocolloids*, 21 (1):1-22.

Sitohy, M.Z. and Ramadan, M.F., (2001). 'Degradability of Different Phosphorylated Starches and Thermoplastic Films Prepared from Corn Starch Phosphomonoesters'. *Starch - Stärke*, 53 (7):317-322.

Stute, R., Heilbronn, Klingler, R.W., Boguslawski, S., Eshtiaghi, M.N. and Knorr, D., (1996). 'Effects of High Pressures Treatment on Starches'. *Starch - Stärke*, 48 (11-12):399-408.

Sundaram, J., Durance, T.D. and Wang, R., (2008). 'Porous scaffold of gelatin-starch with nanohydroxyapatite composite processed via novel microwave vacuum drying'. *Acta Biomaterialia*, 4 (4):932-942.

Te Wierik, G.H.P., Eissens, A.C., Bergsma, J., Arends-Scholte, A.W. and Bolhuis, G.K., (1997). 'A new generation starch product as excipient in pharmaceutical tablets: III. Parameters affecting controlled drug release from tablets based on high surface area retrograded pregelatinized potato starch'. *International Journal of Pharmaceutics*, 157 (2):181-187.

Torres, F.G., Boccaccini, A.R. and Troncoso, O.P., (2007). 'Microwave processing of starch-based porous structures for tissue engineering scaffolds'. *Journal of Applied Polymer Science*, 103 (2):1332-1339.

Trubiano Paolo, C., (1997). 'The Role of Specialty Food Starches in Flavor Encapsulation'. *Flavor Technology*: American Chemical Society, 244-253.

Trubiano, P.C., (1987). ' Succinate and substituted succinate derivatives of starch. ' *Modified starches: Properties and uses, CRC Press, Boca Raton, Florida*, In: Wurzburg, O.B., Editor, 1987:131-148.

Tukomane, T. and Varavinit, S., (2008). 'Influence of Octenyl Succinate Rice Starch on Rheological Properties of Gelatinized Rice Starch before and after Retrogradation'. *Starch - Stärke*, 60 (6):298-304.

Tuovinen, L., Peltonen, S. and Järvinen, K., (2003). 'Drug release from starch-acetate films'. *Journal of Controlled Release*, 91 (3):345-354.

Tuovinen, L., Peltonen, S., Liikola, M., Hotakainen, M., Lahtela-Kakkonen, M., Poso, A. and Järvinen, K., (2004a). 'Drug release from starch-acetate microparticles and films with and without incorporated [alpha]-amylase'. *Biomaterials*, 25 (18):4355-4362.

Tuovinen, L., Ruhanen, E., Kinnarinen, T., Rönkkö, S., Pelkonen, J., Urtti, A., Peltonen, S. and Järvinen, K., (2004b). 'Starch acetate microparticles for drug delivery into retinal pigment epithelium--in vitro study'. *Journal of Controlled Release*, 98 (3):407-413.

Wang, X., Li, X., Chen, L., Xie, F., Yu, L. and Li, B., (2011). 'Preparation and characterisation of octenyl succinate starch as a delivery carrier for bioactive food components'. *Food Chemistry*, 126 (3):1218-1225.

Williams, H.D., Ward, R., Culy, A., Hardy, I.J. and Melia, C.D.2011, 'Designing HPMC matrices with improved resistance to dissolved sugar'. *International Journal of Pharmaceutics*, 401 (1-2):51-59.

Wootton, M. and Manatsathit, A., (1983). 'The Influence of Molar Substitution on the Water Binding Capacity of Hydroxypropyl Maize Starches'. *Starch - Stärke*, 35 (3):92-94.

Xu, R., Feng, X., Xie, X., Xu, H., Wu, D. and Xu, L., (2011). 'Grafted Starch-Encapsulated Hemoglobin (GSEHb) Artificial Red Blood Cells Substitutes'. *Biomacromolecules*: null-null.

Yoon, H.-S., Kweon, D.-K. and Lim, S.-T., (2007). 'Effects of drying process for amorphous waxy maize starch on theophylline release from starch-based tablets'. *Journal of Applied Polymer Science*, 105 (4):1908-1913.

19

Antihypertensive and Antioxidant Effects of Functional Foods Containing Chia (*Salvia hispanica*) Protein Hydrolysates

Ine M. Salazar-Vega, Maira R. Segura-Campos,
Luis A. Chel-Guerrero and David A. Betancur-Ancona
*Facultad de Ingeniería Química, Campus de Ciencias Exactas e Ingenierías,
Universidad Autónoma de Yucatán, Yucatán,
México*

1. Introduction

High blood pressure increases the risk of developing cardiovascular diseases such as arteriosclerosis, stroke and myocardial infarction. Angiotensin I-converting enzyme (ACE, dipeptidylcarboxypeptidase, EC 3.4.15.1) is a multifunctional, zinc-containing enzyme found in different tissues (Bougatef et al., 2010). Via the rennin-angiotensin system, ACE plays an important physiological role in regulating blood pressure by converting angiotensin I into the powerful vasoconstrictor angiotensin II and inactivating the vasodilator bradykinin. ACE inhibition mainly produces a hypotensive effect, but can also influence regulatory systems involved in immune defense and nervous system activity (Haque et al., 2009). Commercial ACE-inhibitors are widely used to control high blood pressure, but can have serious side-effects. Natural ACE-inhibitory peptides are a promising treatment alternative because they do not produce side-effects, although they are less potent (Cao et al., 2010).

Oxidation is a vital process in organisms and food stuffs. Oxidative metabolism is essential for cell survival but produces free radicals and other reactive oxygen species (ROS) which can cause oxidative changes. An excess of free radicals can overwhelm protective enzymes such as superoxide dismutase, catalase and peroxidase, causing destruction and lethal cellular effects (e.g., apoptosis) through oxidization of membrane lipids, cellular proteins, DNA, and enzymes which shut down cellular processes (Haque et al., 2009). Synthetic antioxidants such as butylatedhydroxyanisole (BHA) and butylatedhydroxytoluene (BHT) are used as food additives and preservatives. Antioxidant activity in these synthetic antioxidants is stronger than that found in natural compounds such as α-tocopherol and ascorbic acid, but they are strictly regulated due to their potential health hazards. Interest in the development and use of natural antioxidants as an alternative to synthetics has grown steadily; for instance, hydrolyzed proteins from many animal and plant sources have recently been found to exhibit antioxidant activity (Lee et al., 2010).

Native to southern Mexico, chia (*Salvia hispanica*) was a principal crop for ancient Mesoamerican cultures and has been under cultivation in the region for thousands of years. A recent evaluation of chia's properties and possible uses showed that defatted chia seeds have fiber (22 g/100 g) and protein (17 g/100 g) contents similar to those of other oilseeds currently used in the food industry (Vázquez-Ovando et al., 2009). Consumption of chia seeds provides numerous health benefits, but they are also a potential source of biologically-active (bioactive) peptides. Enzymatic hydrolysis is natural and safe, and effectively produces bioactive peptides from a variety of protein sources, including chia seeds. Chia protein hydrolysates with enhanced biological activity could prove an effective functional ingredient in a wide range of foods. The objective of present study was to evaluate ACE inhibitory and antioxidant activity in food products containing chia (*Salvia hispanica* L.) protein hydrolysates.

2. Material and methods

2.1 Materials

Chia (*S. hispanica*, L.) seeds were obtained in Yucatan state, Mexico. Reagents were analytical grade and purchased from J.T. Baker (Phillipsburg, NJ, USA), Sigma (Sigma Chemical Co., St. Louis, MO, USA), Merck (Darmstadt, Germany) and Bio-Rad (Bio-Rad Laboratories, Inc. Hercules, CA, USA). The Alcalase® 2.4L FG and Flavourzyme® 500MG enzymes were purchased from Novo Laboratories (Copenhagen, Denmark). Alcalase 2.4L is an endopeptidase from *Bacillus licheniformis*, with subtilisin Carlsberg as the major enzyme component and a specific activity of 2.4 Anson units (AU) per gram. One AU is the amount of enzyme which, under standard conditions, digests hemoglobin at an initial rate that produces an amount of thrichloroacetic acid-soluble product which produces the same color with Folin reagent as 1 meq of tyrosine released per minute. Optimal endopeptidase activity was obtained by application trials at pH 7.0. Flavourzyme 500 MG is an exopeptidase/endoprotease complex with an activity of 1.0 leucine aminopeptidase unit (LAPU) per gram. One LAPU is the amount of enzyme that hydrolyzes 1 mmol of leucine p-nitroanilide per minute. Optimal exopeptidase activity was obtained by application trials at pH 7.0.

2.2 Protein-rich fraction

Flour was produced from 6 Kg chia seed by first removing all impurities and damaged seeds, crushing the remaining sound seeds (Moulinex DPA 139) and then milling them (Krups 203 mill). Standard AOAC procedures were used to determine nitrogen (method 954.01), fat (method 920.39), ash (method 925.09), crude fiber (method 962.09), and moisture (method 925.09) contents in the milled seeds (AOAC, 1997). Nitrogen (N_2) content was quantified with a Kjeltec Digestion System (Tecator, Sweden) using cupric sulfate and potassium sulfate as catalysts. Protein content was calculated as nitrogen x 6.25. Fat content was obtained from a 1 h hexane extraction. Ash content was calculated from sample weight after burning at 550 °C for 2 h. Moisture content was measured based on sample weight loss after oven-drying at 110 °C for 2 h. Carbohydrate content was estimated as nitrogen-free extract (NFE). Oil extraction from the milled seeds was done with hexane in a Soxhlet system for 2 h. The remaining fraction was milled with 0.5 mm screen (Thomas-Wiley®, Model 4, Thomas Scientific, USA) and AOAC (1997) procedures used to determine proximate composition of the remaining flour. The defatted chia flour was dried in a Labline stove at 60 °C for 24 h. Defatted flour mill yield was calculated with the equation:

$$\text{Mill yield} = \frac{\text{Weight of 0.5 mm particle size flour}}{\text{Total weight of defatted flour}} \times 100 \tag{1}$$

Extraction of the protein-rich fraction was done by dry fractionation of the defatted flour according to Vázquez-Ovando et al. (2010). Briefly, 500 g flour was sifted for 20 min using a Tyler 100 mesh (140 μm screen) and a Ro-Tap® agitation system. Proximate composition was determined following AOAC (1997) procedures and yield calculated with the equation:

$$\text{Protein rich fraction yield} = \frac{\text{Protein rich fraction weight}}{0.5 \text{ particle size flour weight}} \times 100 \qquad (2)$$

2.3 Enzymatic hydrolysis of protein-rich fraction

The chia protein-rich fraction (44.62% crude protein) was sequentially hydrolyzed with Alcalase® for 60 min followed by Flavourzyme® for a total of up to 150 min. Degree of hydrolysis was recorded at 90, 120 and 150 min. Three hydrolysates were generated with these parameters: substrate concentration, 2%; enzyme/substrate ratio, 0.3 AU g^{-1} for Alcalase® and 50 LAPU g^{-1} for Flavourzyme®; pH, 7 for Alcalase® and 8 for Flavourzyme®; temperature, 50 °C. Hydrolysis was done in a reaction vessel equipped with a stirrer, thermometer and pH electrode. In all three treatments, the reaction was stopped by heating to 85 °C for 15 min, followed by centrifuging at 9880 xg for 20 min to remove the insoluble portion (Pedroche et al., 2002).

2.4 Degree of hydrolysis

Degree of hydrolysis (DH) was calculated by determining free amino groups with o-phthaldialdehyde following Nielsen et al. (2001): DH = h/h$_{tot}$× 100; where h$_{tot}$ is the total number of peptide bonds per protein equivalent, and h is the number of hydrolyzed bonds. The h$_{tot}$ factor is dependent on raw material amino acid composition.

2.5 In Vitro biological activities

ACE inhibitory and antioxidant activities were evaluated in the chia (*S. hispanica*) protein hydrolysates. Hydrolysate protein content was previously determined using the bicinchoninic acid method (Sigma, 2006).

2.5.1 ACE inhibitory activity

Hydrolysate ACE inhibitory activity was analyzed with the method of Hayakari et al. (1978), which is based on the fact that ACE hydrolyzes hippuryl-L-histidyl-L-leucine (HHL) to yield hippuric acid and histidyl-leucine. This method relies on the colorimetric reaction of hippuric acid with 2,4,6-trichloro-s-triazine (TT) in a 0.5 mL incubation mixture containing 40 μmol potassium phosphate buffer (pH 8.3), 300 μmol sodium chloride, 40 μmol 3% HHL in potassium phosphate buffer (pH 8.3), and 100 mU/mL ACE. This mixture was incubated at 37 °C/45 min and the reaction terminated by addition of TT (3% v/v) in dioxane and 3 mL 0.2 M potassium phosphate buffer (pH 8.3). After centrifuging the reaction mixture at 10,000 $x\,g$ for 10 min, enzymatic activity was determined in the supernatant by measuring absorbance at 382 nm. All runs were done in triplicate. ACE inhibitory activity was quantified by a regression analysis of ACE inhibitory activity (%) versus peptide concentration, and IC$_{50}$ values (i.e. the peptide concentration in μg protein/mL required to produce 50% ACE inhibition under the described conditions) defined and calculated as follows:

$$\text{ACE inhibitory activity (\%)} = \frac{A-B}{A-C} \times 100 \qquad (3)$$

Where: A represents absorbance in the presence of ACE and sample; B absorbance of the control and C absorbance of the reaction blank.

$$IC_{50} = \frac{50-b}{m} \tag{4}$$

Where b is the intersection and m is the slope.

2.5.2 ABTS$^{•+}$ (2,2'-azino-bis(3-ethylbenzothiazoline-6-sulfonic acid) decolorization assay

Antioxidant activity in the hydrolysates was analyzed following Pukalskas et al. (2002). ABTS$^{•+}$ radical cation was produced by reacting ABTS with potassium persulfate. To prepare the stock solution, ABTS was dissolved at a 2 mM concentration in 50 mL phosphate-buffered saline (PBS) prepared from 4.0908 g NaCl, 0.1347 g KH_2PO_4, 0.7098 g Na_2HPO_4, and 0.0749 g KCl dissolved in 500 mL ultrapure water. If pH was lower than 7.4, it was adjusted with NaOH. A 70 mM $K_2S_4O_8$ solution in ultrapure water was prepared. ABTS radical cation was produced by reacting 10 mL of ABTS stock solution with 40 µL $K_2S_4O_8$ solution and allowing the mixture to stand in darkness at room temperature for 16-17 h before use. The radical was stable in this form for more than 2 days when stored in darkness at room temperature.

Antioxidant compound content in the hydrolysates was analyzed by diluting the ABTS$^{•+}$ solution with PBS to an absorbance of 0.800 ± 0.030 AU at 734 nm. After adding 990 µL of diluted ABTS$^{•+}$ solution (A 734 nm= 0.800 ± 0.030) to 10 µL antioxidant compound or Trolox standard (final concentration 0.5 -3.5 mM) in PBS, absorbance was read at ambient temperature exactly 6 min after initial mixing. All analyses were run in triplicate. The percentage decrease in absorbance at 734 nm was calculated and plotted as a function of the Trolox concentration for the standard reference data. The radical scavenging activity of the tested samples, expressed as inhibition percentage, was calculated with the equation:

$$\% \, Inhibition = \frac{A_B - A_A}{A_B} X \, 100 \tag{5}$$

Where A_B is absorbance of the blank sample (t=0), and A_A is absorbance of the sample with antioxidant after 6 min.

The Trolox equivalent antioxidant coefficient (TEAC) was quantified by a regression analysis of % inhibition versus Trolox concentration using the following formula:

$$TEAC = \frac{\%I_M - b}{m} \tag{6}$$

Where b is the intersection and m is the slope.

2.6 White bread and carrot cream containing chia protein hydrolysates

To test if the chia protein hydrolysates increased biological potential when added to food formulations, those were used as ingredients in preparing white bread and carrot cream, and the ACE inhibitory and antioxidant activity of these foods evaluated.

2.6.1 Biological potential and sensory evaluation of white bread containing chia protein hydrolysates

White bread was prepared following a standard formulation (Table 1) (Tosi et al., 2002), with inclusion levels of 0 mg (control), 1 mg and 3 mg chia protein hydrolysate/g flour.

Hydrolysates produced at 90, 120 and 150 min were used. Treatments (two replicates each) were formed based on inclusion level and hydrolysate preparation time (e.g. 1 mg/90 min, etc.), and distributed following a completely random design. Each treatment was prepared by first mixing the ingredients (Farinograph Brabender 811201) at 60 rpm for 10 min, simultaneously producing the corresponding farinograph. The "work input" value, or applied energy required (Bloksma, 1984), was calculated from the area under the curve (in which 1 cm^2 was equivalent to 454 J/kg). The resulting doughs were placed in a fermentation chamber at 25 °C and 75% relative humidity for 45 min. Before the second fermentation, the dough for each treatment was divided into two pieces (approximately 250 g) and each placed in a rectangular mold; each piece was treated as a replicate. The second fermentation was done for 75 min under the same temperature and humidity conditions. Finally, the fermented doughs were baked at 210 °C for 25 min.

Sensory evaluation of the baked white bread loaves was done by judges trained in evaluating baked goods. Evaluation factors and the corresponding maximum scores were: specific volume (15 points); cortex (15 points); texture (15 points); color (10 points); structure (10 points); scent (15 points); and flavor (20 points). Overall score intervals were: 40-50 "very bad"; 50-60 "bad"; 60-70 "regular"; 70-80 "good"; 80-90 "very good"; and 90-100 "excellent". For total nitrogen content, ACE inhibitory activity and antioxidant activity analyses, the bread was sliced, dried at 40 °C for 48 h and milled. Total nitrogen content was determined following the applicable AOAC (1997) method (954.01). To analyze ACE inhibitory activity, 10, 20, 30, 40 and 50 mg of milled bread were dissolved in 1 mL buffer mixture and centrifuged at 13,698 x g for 10 min. The supernatant (40 µl) was taken from each lot and processed according to Hayakari et al. (1978). After adding 990 µl diluted ABTS$^{\bullet+}$ solution to 50 mg of milled bread in PBS, antioxidant activity was determined according to Pukalskas et al. (2002).

Ingredients	Control (%)	Hydrolysate (%)	
		(1 mg/g)	(3 mg/g)
Flour	56.51	56.47	56.40
Water	33.33	33.31	33.28
Sugar	3.39	3.39	3.38
Yeast	2.82	2.82	2.82
Fat	1.69	1.69	1.69
Powdered milk	1.13	1.13	1.13
Salt	1.13	1.13	1.13
Hydrolysate	0.00	0.06	0.17

Table 1. Formulation of white bread made according to a standard formula (control) and with chia protein hydrolysate added at two levels (1 and 3 mg/g).

2.6.2 Biological potential and sensory evaluation of carrot cream containing chia protein hydrolysates

Carrot cream was prepared following a standard formulation (Table 2), with inclusion levels of 0 mg/g (control), 2.5 mg/g and 5 mg/g carrot. Hydrolysates produced at 90, 120 and 150 min were used. Treatments were formed based on inclusion level and hydrolysate preparation time (e.g. 2.5 mg/90 min, etc.), and distributed following a completely random design. Two replicates consisting of 330 g carrot cream were done per treatment. The carrots were washed, peeled and cooked in water at a 1:4 (p/v) ratio for 40 min. Broth and butter were dissolved in low fat milk and liquefied with the cooked carrots and the remaining ingredients. Finally, the mixture was boiled at 65 °C for 3 min.

Viscosity was determined for a commercial product (Campbell's[®]) and the hydrolysate-containing carrot creams using a Brookfield (DV-II) device with a No. 2 spindle, 0.5 to 20 rpm deformation velocity (γ) and a 24 °C temperature. A viscosity curve was generated from the γ log versus viscosity coefficient log (η), while the consistency index (k) and fluid behavior (n) were quantified by applying the potency law model: log η = log k + (n-1) log γ. Brightness L* and chromaticity a*b* were determined with a Minolta colorimeter (CR200B). Differences in color (ΔE^*) between the control and hydrolysate-supplemented carrot creams was calculated with the equation (Alvarado & Aguilera, 2001): $\Delta E^* = [(\Delta L^*)^2 + (\Delta a^*)^2 + (\Delta b)^2]^{0.5}$. Biological potential was analyzed by first centrifuging the samples at 13,698 $x\ g$ for 30 min and then determining total nitrogen content (AOAC, 1997)(954.01 method), ACE inhibitory and antioxidant activity in the supernatant.

Using a completely random design, sensory evaluation was done of the control product and the hydrolysate-containing carrot creams with the highest biological activity. Acceptance level was evaluated by 80 untrained judges who indicated pleasure or displeasure levels along a 7-point hedonic scale including a medium point to indicate indifference (Torricella et al., 1989).

Ingredients	Control (%)	Hydrolysate (%)	
		2.5 mg/g	5 mg/g
Carrot	40.12	40.08	40.04
Low fat milk	38.58	38.54	38.50
Purified water	19.29	19.27	19.25
Butter	1.16	1.16	1.16
Broth	0.85	0.85	0.85
Hydrolysate	0	0.10	0.20

Table 2. Formulation of carrot cream made according to a standard formula (control) and with chia protein hydrolysate added at two levels (2.5 and 5 mg/g).

2.7 Statistical analysis

All results were analyzed using descriptive statistics with a central tendency and dispersion measures. One-way ANOVAs were run to evaluate protein extract hydrolysis data, *in vitro* ACE inhibitory, antioxidant and antimicrobial activities, and the sensory scores. A Duncan multiple range test was applied to identify differences between treatments. All analyses were done according to Montgomery (2004) and processed with the Statgraphics Plus ver. 5.1 software.

3. Results and discussion

3.1 Proximate composition

Proximate composition analysis showed that fiber was the principal component in the raw chia flour (Table 3), which coincides with the 40% fiber content reported elsewhere (Tosco, 2004). Its fat content was similar to the 33% reported by Ixtaina et al. (2010), and its protein and ash contents were near the 23% protein and 4.6% ash contents reported by Ayerza & Coates (2001). Nitrogen-free extract (NFE) in the raw chia flour was lower than the 7.42% reported by Salazar-Vega et al. (2009), probably due to the 25.2% fat content observed in that study. In the defatted chia flour, fiber decreased to 21.43% and fat to 13.44%, while protein content increased to 34.01%: as fat content decreased, crude protein content increased. Mill yield (0.5 mm particle size) from the defatted chia flour was 84.33%, which is lower than the 97.8% reported by Vázquez-Ovando et al. (2010). Dry fractionation yield of the defatted chia flour was 70.31% particles >140 µm and 29.68% particles <140 µm. Protein-rich fraction yield was higher than reported elsewhere (Vázquez-Ovando et al., 2009), probably due to lower initial moisture content in the processed flour, which increases the tendency to form particle masses and thus retain fine particles. The 44.62% protein content of the protein-rich fraction was higher than observed in the raw chia flour (23.99%) and defatted chia flour (34.01%).

Components	Chia flour	Defatted chia flour	Protein-rich fraction
Moisture	6.32[a]	6.17[a]	7.67[b]
Ash	4.32[a]	5.85[b]	8.84[c]
Crude fiber	35.85[b]	21.43[a]	11.48[c]
Fat	34.88[c]	13.44[b]	0.54[a]
Protein	23.99[a]	34.01[b]	44.62[c]
NFE	0.96[a]	25.27[b]	34.52[c]

Table 3. Proximate composition of chia (*Salvia hispanica* L.) flour, defatted flour and protein-rich fraction.[a-b] Different superscript letters in the same row indicate statistical difference (P < 0.05). Data are the mean of three replicates (% dry base).

3.2 Enzymatic hydrolysis of protein-rich fraction

The protein-rich fraction used to produce the protein hydrolysates was isolated by alkaline extraction and acid precipitation of proteins as described above. This fraction proved to be good starter material for hydrolysis. Production of extensive (i.e. >50% DH) hydrolysates requires use of more than one protease because a single enzyme cannot achieve such high DHs within a reasonable time period. For this reason, an Alcalase®-Flavourzyme® sequential system was used in the present study to produce an extensive hydrolysate. Protease and peptidase choice influences DH, peptide type and abundance, and consequently the amino acid profile of the resulting hydrolysate. The bacterial endoprotease Alcalase® is limited by its specificity, resulting in DHs no higher than 20 to 25%, depending on the substrate, but it can attain these DHs in a relatively short time under moderate conditions. In the present study, Alcalase® exhibited broad specificity and produced hydrolysates with 23% DH during 60 min reaction time. The fungal protease Flavourzyme® has broader specificity, which, when combined with its exopeptidase activity, can generate DH values as high as 50%. The highest DH in the present study (43.8%) was attained with Flavourzyme® at 150 min (Table 4), made possible in part by predigestion with Alcalase®, which increases the number of N-terminal sites, thus facilitating hydrolysis by Flavourzyme®. The 43.8% DH obtained here with the defatted chia hydrolysate was lower than the 65% reported by Pedroche et al. (2002) in chickpea hydrolysates produced sequentially with Alcalase® and Flavourzyme® at 150 min. Likewise, Clemente et al. (1999) reported that the combination of these enzymes in a two-step hydrolyzation process (3 h Alcalase® as endoprotease; 5 h Flavourzyme® as exoprotease) of chickpea produced DH >50%. In this study, the globular structure of globulins in the isolated protein limited the action of a single proteolytic enzyme, which is why sequential hydrolysis with an endoprotease and exoprotease apparently solves this problem. Cleavage of peptide bonds by the endopeptidase increases the number of peptide terminal sites open to exoprotease action. Imm & Lee (1999) reported that when using Flavourzyme® more efficient hydrolysis and higher DH can be achieved by allowing pH to drift. They suggested that a more effective approach would be initial hydrolysis with Alcalase® under optimum conditions followed by Flavourzyme® with pH being allowed to drift down to its pH 7.0 optimum. Using this technique for hydrolysis of rapeseed protein, Vioque et al. (1999) attained a 60% DH.

Hydrolysate (min)	DH (%)	IC$_{50}$ mg/mL	TEAC (Mm/mg)
90	37.5[a]	44.01[a]	7.31[a]
120	40.5[b]	20.76[b]	4.66[b]
150	43.8[c]	8.86[c]	4.49[c]

Table 4. Degree of hydrolysis (DH), ACE inhibitory and antioxidant activities of chia (*Salvia hispanica*) protein hydrolysates produced at three hydrolysis times. [a-b]Different superscript letters in the same column indicate statistical difference (P < 0.05).

Controlled release of bioactive peptides from proteins via enzymatic hydrolysis is one of the most promising techniques for producing hydrolysates with potential applications in the pharmaceutical and food industries: hydrolysates with >10% DH have medical applications while those with <10% DH can be used to improve functional properties in flours or protein isolates (Pedroche et al. (2003). Several biological properties have been attributed to low-molecular-weight peptides, although producing them normally requires a combination of commercial enzyme preparations (Gilmartin & Jervis, 2002). When hydrolyzed sequentially with Alcalase® and Flavourzyme®, chia *S. hispanica* is an appropriate substrate for producing bioactive peptides with high DH (43.8%).

3.3 ACE inhibitory activity

ACE inhibitory activity of the chia protein hydrolysates produced with an Alcalase®-Flavourzyme® sequential system at 90, 120 and 150 min was measured and calculated as IC$_{50}$ (Table 4). The fact that the alkaline proteases Alcalase® and Flavourzyme® have broad specificity and hydrolyze most peptide bonds, with a preference for those containing aromatic amino acid residues, has led to their use in producing protein hydrolysates with better functional and nutritional characteristics than the original proteins, and in generating bioactive peptides with ACE inhibitory activity (Segura-Campos et al., 2010). The chia protein hydrolysates produced with this sequential system exhibited ACE inhibitory activity, suggesting that the peptides released from the proteins are the agents behind inhibition. ACE inhibitory activity in the analyzed hydrolysates depended significantly on hydrolysis time, and therefore on DH. Bioactivity was highest in the hydrolysate produced at 150 min (IC$_{50}$ = 8.86 µg protein/mL), followed by those produced at 120 min (IC$_{50}$ = 20.76 µg/mL) and at 90 min (IC$_{50}$ = 44.01 µg/mL). Kitts & Weiler (2003) found that peptides with antihypertensive activity consist of only two to nine amino acids and that most are di- or tripeptides, making them resistant to endopeptidase action in the digestive tract. The ACE inhibitory activity in the hydrolysates studied here was higher than reported by Segura et al. (2010) for *V. unguiculata* hydrolysates produced using a 60 min reaction time with Alcalase® (2564.7 µg/mL), Flavourzyme® (2634.3 µg/mL) or a pepsin-pancreatin sequential system

(1397.9 µg/mL). It was also higher than the 191 µg/mL reported by Pedroche et al. (2002) for chickpea protein isolates hydrolyzed sequentially with Alcalase® and Flavourzyme®. The chia protein hydrolysates' ACE inhibitory activity was many times higher than reported for *Phaseolus lunatus* and *Phaseolus vulgaris* hydrolysates produced with Alcalase® at 15 (437 and 591µg/mL), 30 (569 and 454 µg/mL), 45 (112 and 74µg/mL), 60 (254 and 61µg/mL), 75 (254 and 98 µg/mL) and 90 min (56 and 394 µg/mL), and with Flavourzyme® at 15 (287 and 401 µg/mL), 30 (239 and 151 µg/mL), 45 (265 and 127µg/mL), 60 (181 and 852 µg/mL) and 75 min (274 and 820 µg/mL). However, the *P. lunatus* hydrolysate produced with Flavourzyme® at 90 min had a lower IC_{50} value (6.9 µg/mL) and consequently higher ACE inhibitory activity than observed in the present study (Torruco-Uco et al., 2009).

The *in vitro* biological potential observed here in the enzymatically hydrolyzed chia proteins was higher than the 140 µg/mL reported by Li et al. (2007) for a rice protein hydrolysate produced with Alcalase®. After a single oral administration in spontaneously hypertensive rats (SHR), this rice hydrolysate exhibited an antihypertensive effect, suggesting its possible use as a physiologically functional food with potential benefits in the prevention and/or treatment of hypertension. Enzymatic hydrolysates from different protein sources, and IC_{50} values ranging from 200 to 246700 µg/mL, have also been shown to have *in vitro* ACE inhibitory activity as well as antihypertensive activity in SHR (Hong et al., 2005). Matsufuji et al. (1994) reported that peptides produced by enzymes such as Alcalase®, and which exhibit ACE inhibitory activity, may resist digestion by gastrointestinal proteases and therefore be absorbed in the small intestine, a quality also reported in a number of SHR studies. Based on the above, it is probable that the chia protein hydrolysates produced here with Alcalase®-Flavourzyme®, which exhibit ACE inhibitory activity, are capable of resisting gastrointestinal proteases and are therefore appropriate for application in food systems (e.g. functional foods) focused on people suffering arterial hypertension disorders. Further research will be needed, however, to determine if the peptide mixture exerts an *in vivo* antihypertensive effect because peptide ACE inhibitory potencies do not always correlate with their antihypertensive activities in SHR.

3.4 ABTS$^{\bullet+}$ (2,2'-azino-bis(3-ethylbenzothiazoline-6-sulfonic acid) decolorization assay

Antioxidant activity of the chia protein hydrolysates, quantified and calculated as TEAC values (mM/mg), decreased as DH increased (Table 4). The highest TEAC value was for the hydrolysate produced at 90 min (7.31 mM/mg protein), followed by those produced at 120 min (4.66 mM/mg protein) and 150 min (4.49 mM/mg protein); the latter two did not differ (P<0.05). Increased antioxidant activity in hydrolyzed proteins has also been reported for dairy, soy, zein, potato, gelatin and egg yolk among other proteins. This increase has been linked to greater solvent exposure of amino acids (Elias et al., 2008), in other words, enzymatic hydrolysis probably increased exposure of antioxidant amino acids in the chia proteins, consequently providing them greater antioxidant activity. Extensive proteolysis of the chia protein hydrolysates at 120 and 150 min resulted in lower antioxidant activity because it may have generated free amino acids, which are not effective antioxidants. The increased antioxidant activity of peptides is related to unique properties provided by their chemical composition and physical properties. Peptides are potentially better food antioxidants than amino acids due to their higher free radical scavenging, metal chelation and aldehyde adduction activities. An increase in the ability of a protein hydrolysate to lower a free radical's reactivity is related to an increase in amino acid exposure. This leads to increased peptide-free radical reactions and an energy decrease in the scavenged free radical, both of which compromise a free radical's ability to oxidize lipids (Elias et al., 2008). The present results are lower than reported for *P. lunatus* hydrolysates produced with Alcalase® at 90 (9.89 mM/mg) or Flavourzyme® at 90 (11.55 mM/mg), and *P. vulgaris* hydrolysates produced with Alcalase® at 60 min (10.09 mM/mg) or Flavourzyme® at 45 min (8.42 mM/mg)(Torruco-Uco et al., 2009). They are also lower than *V. unguiculata* protein hydrolysates produced with Alcalase® (14.7 mM/mg), Flavourzyme® (14.5 mM/mg) or pepsin-pancreatin (14.3 mM/mg) for 90 min. However, the present results were higher than the 0.016 mM/mg reported by Raghavan et al. (2008) for tilapia protein hydrolysates. The above results show that chia protein hydrolysates undergo single-electron transfer reactions in the ABTS•+ reduction assay, which effectively measures total antioxidant activity of dietary antioxidants and foods. Under the analyzed conditions, the chia protein hydrolysates may have acted as electron donors and free radical sinks thus providing antioxidant protection. However, this purported antioxidant action needs to be confirmed for each peptide in different oxidant systems and under *in vitro* and *in vivo* conditions.

No relationship was observed between antioxidant activity and the hydrolysates with the highest ACE inhibitory activity. This suggests that peptide antioxidant activity may depend on the specific proteases used to produce them; the DH attained; the nature of the peptides released (e.g. molecular weight, composition and amino acid sequence); as well as the combined effects of their properties (e.g. capacity for free radical location, metallic ion chelation and/or electron donation) (Tang et al., 2009). Peptide size may also play a role since antihypertensive peptides are short, with only two to nine amino acids (are di- or tri-peptides), whereas antioxidant peptides contain from three to sixteen amino acid residues (Kitts & Weiler, 2003).

3.5 White bread and carrot cream containing chia protein hydrolysates
3.5.1 Biological potential and sensory evaluation of white bread containing chia protein hydrolysates

Addition of the chia protein hydrolysates (90, 120 and 150 min) to white bread resulted in products with higher ACE inhibitory activity than the control treatment. Bioactivity was higher (i.e. lower IC$_{50}$ values) in the bread containing the hydrolysates produced at either 90 or 120 min, than in that containing the hydrolysate produced at 150 min. Hydrolysate inclusion level (i.e. 1 or 3 mg/g) had no effect (P>0.05) on product biological potential. Hydrolysate bioactivity (8.86-44.01µg protein/mL) declined notably after incorporation into the white bread (141.29-297.68 µg protein/mL), suggesting that fermentation and high temperatures during baking hydrolyzed the ACE inhibitory peptides and generated peptide fractions with lower antihypertensive potential. In contrast, antioxidant activity was unaffected by addition of the chia protein hydrolysates. As occurred with the IC$_{50}$ values, hydrolysate TEAC values (7.31 mM/mg at 90 min; 4.66 mM/mg at 120 min; 4.49 mM/mg at 150 min) decreased after incorporation into the bread, with levels no higher than approximately 0.53 mM/mg (Table 5). Again, high temperature during baking probably lowered product biological potential by oxidating tryptophan and hystidine, or through methionine desulfuration.

Hydrolysis Time (min)	Inclusion level (mg/g)	IC$_{50}$ (µg protein/ml)	TEAC (mM/mg protein)
Control	0	400.76[a]	0.53[a]
90	1	141.29[b]	0.53[a]
	3	155.88[b]	0.53[a]
120	1	163.14[b]	0.54[a]
	3	159.04[b]	0.53[a]
150	1	237.60[c]	0.53[a]
	3	297.68[c]	0.55[a]

Table 5. ACE inhibitory (IC$_{50}$ values) and antioxidant (TEAC values) activity of white bread containing two levels (1 and 3 mg/g) of chia protein hydrolysates produced at three hydrolysis times (90, 120 and 150 min).[a-c] Different superscript letters in the same column indicate statistical difference (P<0.05). Data are the mean of three replicates.

Kneading of the bread dough containing chia protein hydrolysates required more (P<0.05) applied energy (26.1 to 28.7 kJ/kg) than for the control product (22.9 kJ/kg). Higher applied energy requirements were probably a result of the greater viscoelasticity in the hydrolysate-containing doughs due to the gum residuals, in which the protein-rich chia hydrolysate would have competed for water with the wheat flour protein and starch (Figure 1).

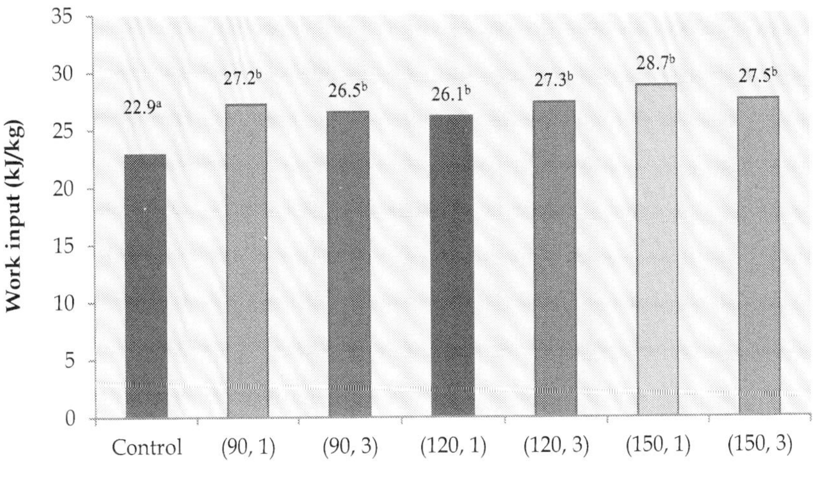

Fig. 1. Applied energy required during kneading of a control white bread and treatments containing different concentrations (1 and 3 mg/g) of chia protein hydrolysates produced at three hydrolysis times. [a-b]Different superscript letters indicate statistical difference (P<0.05)

Sensory evaluation of the hydrolysate-containing bread treatments resulted in scores of 80-90 ("very good") whereas the control was scored as 90-100 ("excellent") (Figure 2).

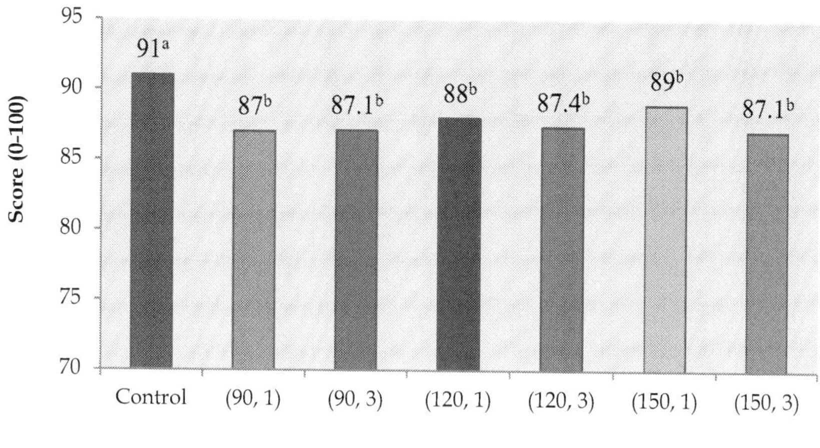

Fig. 2. Scores generated by trained judges for sensory evaluation of a control white bread and treatments containing two concentrations (1 and 3 mg/g) of chia protein hydrolysates produced at three hydrolysis times. [a-b]Different superscript letters indicate statistical difference ($P<0.05$).

Differences in scores were attributed mainly to texture, color and structure factors (Figure 3). Crumbs were stickier in the hydrolysate-containing bread treatments than in the control, a difference which can be attributed to gum content. Crumb color was darker in the hydrolysate-containing bread treatments than in the control, probably due to hydrolysate inclusion level and Maillard reactions. Gum content in the chia protein hydrolysates also affected bread structure by producing a greater number of and larger-sized holes in the crumbs.

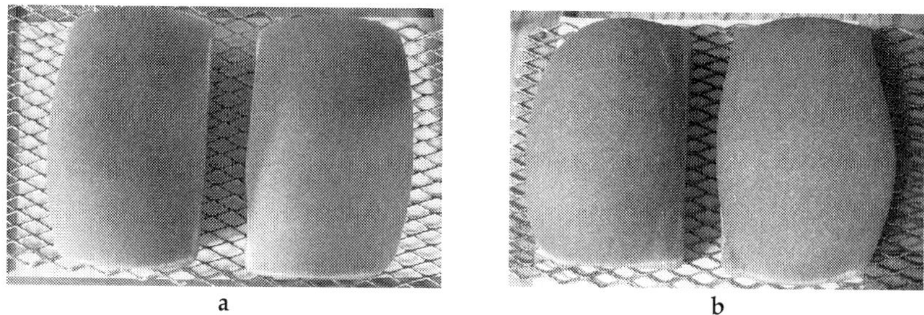

Fig. 3. White bread: a) Control b) White bread containing 3mg/g of chia hydrolysate

3.5.2 Biological potential and sensory evaluation of carrot cream containing chia protein hydrolysates

ACE inhibitory activity in the carrot cream improved markedly with addition of the chia protein hydrolysates (Table 6). An analogous improvement in ACE inhibitory activity was reported by Nakamura et al. (1995) in milk fermented with Calpis sour milk starter containing *Lactobacillus helveticus* and *Saccharomyces cerevisiae*, which they attributed to VPP and IPP peptides. Although biological potential did improve in the carrot creams, neither protein hydrolysate inclusion level (2.5 or 5 mg/g) nor hydrolysis time (90, 120 and 150 min) had a significant (P>0.05) effect. Addition of the chia protein hydrolysates (90 min, 120 min and 150 min) to carrot cream at both inclusion levels (2.5 and 5 mg/g) resulted in IC_{50} values as low as 0.24 µg/mL. These substantially lower values suggest that the peptides released from chia during hydrolysis with the Alcalase®-Flavourzyme® sequential system complemented the peptides (β-casokinins) released from the milk during carrot cream preparation, producing a higher ACE inhibitory activity than in the original hydrolysates or the carrot cream control treatment.

Antioxidant activity increased from 10.21 mM/mg in the carrot cream control treatment to 17.52-18.88 mM/mg in the treatments containing the chia protein hydrolysates. As occurred with ACE inhibitory activity, neither hydrolysate inclusion level (2.5 or 5 mg/g) nor hydrolysis time (90, 120 and 150 min) had a significant effect (P>0.05) on antioxidant activity. Again, this suggests that the higher antioxidant activity in the hydrolysate-containing carrot creams was due to the combined effect of the peptides released during hydrolysis of chia and the antioxidant potential of the carotenoids in the carrots included in the carrot cream.

Hydrolysis time (min)	Inclusion level (mg/g)	IC_{50} (µg protein/ml)	TEAC (mM/mg protein)
Control	0	27.67[a]	10.21[a]
90	2.5	1.23[b]	18.82[b]
	5	1.05[b]	18.54[b]
120	2.5	0.61[b]	18.88[b]
	5	0.24[b]	17.52[b]
150	2.5	1.29[b]	17.58[b]
	5	1.71[b]	18.60[b]

Table 6. ACE inhibitory (IC_{50}) and antioxidant (TEAC values) activities of carrot cream containing two levels (2.5 and 5 mg/g) of chia protein hydrolysates produced at three hydrolysis times (90, 120 and 150 min). [a-b]Different superscript letters in the same column indicate statistical difference (P<0.05). Data are the mean of three replicates.

Fluid behavior (n values) in the carrot creams indicated pseudoplastic properties, suggesting that apparent viscosity depended on deformation velocity rather than tension time (Table 7).

Their higher deformation velocity made these fluids thinner. The pseudoplastic behavior observed here was similar to that reported in other foods such as ice creams, yogurts, mustards, purees or sauces (Alvarado & Aguilera, 2001). No difference (P>0.05) in n and k values was observed between the carrot creams containing 2.5 mg/g hydrolysate (90, 120 or 150 min) and the control product. In contrast, the carrot creams containing 5 mg/g hydrolysate (90, 120 or 150 min) exhibited higher (P<0.05) k values and lower (P<0.05) n values than the control product, indicating that the hydrolysate-containing carrot creams had lower viscosity. This behavior was probably due to the amino acid composition of the chia protein hydrolysates, consisting mainly of hydrophobic residues, which may have limited their interaction with water.

Hydrolysis time (min)	Inclusion level (mg/g)	n	k (Pa sⁿ)
Control	0	0.46[a]	0.54[a]
90	2.5	0.46[a]	0.53[a]
	5	0.54b	0.43b
120	2.5	0.45[a]	0.54[a]
	5	0.56[b]	0.43[b]
150	2.5	0.44[a]	0.53[a]
	5	0.57[b]	0.44[b]
Commercial product	---	0.55[b]	0.41[b]

Table 7. Flow (n) and consistency (k) index values for carrot cream containing two levels (2.5 and 5 mg/g) of chia protein hydrolysates produced at three hydrolysis times (90, 120 and 150 min).[a-b] Different superscript letters indicate statistical difference (P<0.05). Data are the mean of three replicates.

Hydrolysis time had no effect (P>0.05) in the color (ΔE) values, but the carrot creams containing 2.5 mg/g hydrolysate exhibited lower (P<0.05) ΔE values than those containing 5 mg/g hydrolysate (Figure 4).

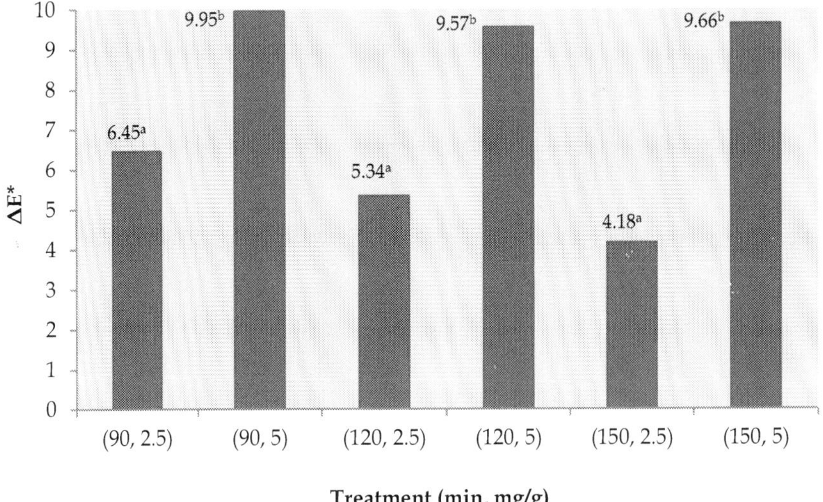

Treatment (min, mg/g)

Fig. 4. Color (ΔE*) values for carrot creams containing two concentrations (2.5 and 5 mg/g) of chia protein hydrolysates produced at three hydrolysis times. [a-b] Different superscript letters indicate statistical difference (P<0.05).

Because no statistical difference (P<0.05) was observed in the biological potential of the hydrolysate-containing carrot cream treatments (at both concentrations and all three hydrolysis times), sensory evaluation was done comparing the control product to the carrot creams containing chia protein hydrolysate produced at 90 min and incorporated at 2.5 and

5 mg/g (Figure 5). Control product scores were higher (P<0.05) than those for the carrot cream containing 2.5 mg/g hydrolysate, but not different (P>0.05) from those for the carrot cream containing 5 mg/g hydrolysate (Figure 6).

Fig. 5. Carrot creams: a) Control, b) Carrot cream containing 2.5 mg/g of chia protein hydrolysate at 90 min, c) Carrot cream containing 5 mg/g of chia protein hydrolysate at 90 min.

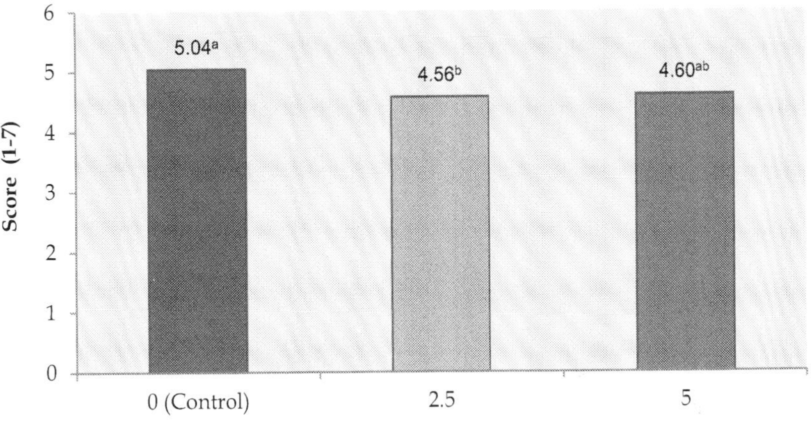

Hydrolysate concentration (mg/g)

Fig. 6. Scores generated by untrained judges for sensory evaluation of carrot cream containing three concentrations (0, 2.5 and 5 mg/g) of chia protein hydrolysates. [a-b] Different superscript letters indicate statistical difference (P<0.05).

4. Conclusions

Inclusion of the studied chia protein hydrolysates in white bread and carrot cream increased product biological potential without notably affecting product quality. Hydrolysis of a protein-rich fraction from *S. hispanica* with the Alcalase®-Flavourzyme® sequential system generated extensive hydrolysates with potential biological activity. This hydrolysis system produces low-molecular-weight hydrolysates, probably peptides, with ACE inhibitory and antioxidant activities and commercial potential as "health-enhancing ingredients" in the production of functional foods such as white bread and carrot cream.

5. Acknowledgments

The authors take this opportunity to thank the following persons for their special contributions: Ing. Hugo Sanchez, MC. Carlos Osella, Instituto de Tecnología de alimentos, Universidad Autónoma del litoral, Argentina.

6. References

Alvarado, J. & Aguilera, J. (2001). *Métodos para medir propiedades físicas en industrias de alimentos (2th Ed.)*. Acribia, S.A., ISBN 84-200-0956-3, Zaragoza, España.

AOAC., (1997). *Official methods of analysis of AOAC international,* (17th ed.). AOAC International, ISSN 1080-0344, Gaithersburg, M.D.

Ayerza, R., & Coates, W. (2001). *Chia seeds: new source of omega-3 fatty acids, natural antioxidants, and dietetic fiber*. Southwest Center for Natural Products Research & Commercialization, Office of Arid Lands Studies, Tucson, Arizona, USA.

Bloksma, A. (1984). *The Farinograph Handbook (3rd ed.)*. American Association of Cereal Chemists Inc., St. paul, Minnesota, USA.

Bougatef, A., Balti, R., Nedjar-Arroume, N., Ravallec, R., Adje, E.Y., Souissi, N., Lassoued, I., Guillochon, D., Nasri, M. (2010). Evaluation of angiotensin I-converting enzyme (ACE) inhibitory activities of smooth hound (*Mustelus mustelus*) muscle protein hydrolysates generated by gastrointestinal proteases: identification of the most potent active peptide. *European Food Research and Technology*, Vol. 231, No.1, pp. 127-135, ISSN 1438-2377.

Cao, W., Zhang, C., Hong, P., Ji, H., Hao, J. (2010). Purification and identification of an ACE inhibitory peptide from the peptic hydrolysates of *Acetes chinensis* and its antihypertensive effects in spontaneously hypertensive rats. *International Journal of Food Science and Technology*, Vol.45, No.1, pp. 959-965, ISSN 0950-5423.

Clemente, A., Vioque, J., Sánchez-Vioque, R., Pedroche, J., Millán, F. (1999). Production of extensive chickpea (*Cicer arietinum* L.) protein hydrolysates with reduced antigenic activity. *Journal of Agricultural and Food Chemistry*, Vol.47, No.9, pp. 3776-3781, ISSN 0021-8561.

Elias, R.J., Kellerby, S.S, Decker, E.A. (2008). Antioxidant activity of proteins and peptides. *Critical Reviews in Food Science and Nutrition*, Vol. 48, No.5, pp.430-411, ISSN 1040-8398.

Gilmartin, L. & Jervis, L. (2002). Production of cod (*Gadus morhua*) muscle hydrolysates: influence of combinations of commercial enzyme preparation on hydrolysate peptide size range. *Journal of Agricultural and Food Chemistry*, Vol.50, No.19, pp. 5417-5423, ISSN 0021-8561.

Haque, E., Chand, R., Kapila, S. (2009). Biofunctional properties of bioactive peptides of milk origin. *Food Reviews International*, Vol.25, No. 1, pp. 28-43, ISSN 8755-9129.

Hayakari, M., Kondo, Y., Izumi, H. (1978). A rapid and simple espectrophotometric assay of angiotensin-converting enzyme. *Analytical Biochemistry*, Vol.84, No.1, pp.361-369, ISSN 0003-2697.

Hong, L.G., Wei, L., Liu, H., Hui, S.Y. (2005). Mung-bean protein hydrolysates obtained with Alcalase® exhibit angiotensin I-converting enzyme inhibitory activity. *Food Science and Technology International*, Vol.11, No.4, pp. 281-287, ISSN 1082-0132.

Imm, J.Y., & Lee, C. M. (1999).Production of seafood flavor from Red Hake (*Urophycis chuss*) by enzymatic hydrolysis. *Journal of Agricultural and Food Chemistry*, Vol. 47, No.6, pp. 2360-2366, ISSN 0021-8561.

Ixtaina, V.Y., Vega, A., Nolasco, S.M., Tomás, M.C., Gimeno, M., Bárzana, E., Tecante, A. (2010). Supercritical carbón dioxide extraction of oil from Mexican chia seed (Salvia hispanica L.); chracterization and process optimization. *Journal of Supercritical Fluids*, Vol.55, No.1, pp.192-199, ISSN 0896-8446.

Kitts, D., & Weiler, K. (2003). Bioactive proteins and peptides form food sources. Applications of bioprocesses used in isolation and recovery. *Current Pharmaceutical Design*, Vol.9, No.16, pp. 1309-1325, ISSN 1381-6128.

Lee, J.K., Yun, J.H., Jeon, J.K., Kim, S.K., Byun, H.G. (2010). Effect of antioxidant peptide isolated from *Brachionus calciflorus*. *Journal of the Korean Society for Applied Biological Chemistry*, Vol.53, No.2, pp.192-197, ISSN 1738-2203.

Li, G.H., Qu, M.R., Wan, J.Z., You, J.M. (2007). Antihypertensive effect of rice protein hydrolysate with in vitro angiotensin I-converting enzyme inhibitory activity in spontaneously hypertensive rats. *Asia Pacific Journal of Clinical Nutrition*, Vol.16, No.1, pp.275-280, ISSN 0964-7058 .

Matsufuji, H., Matsui, T., Seki, E., Osajima, K., Nakashima, M., Osajima, Y. (1994). Angiotensin I-converting enzyme inhibitory peptides in an alkaline protease hydrolysate derived from sardine muscle. *Bioscience, Biotechnology and Biochemistry*, Vol.58, No.1, pp.2244–2245, ISSN 0916-8451.

Montgomery, D. (2004). *Diseño y análisis de experimentos (2th Ed.)*. Limusa-Wiley, ISBN 968-18-6156-6, México, D.F.

Nakamura, Y., Yamamoto, N., Sakai, K., Okubo, A., Yamazaki, S., Takano, T. (1995). Purification and characterization of angiotensin I-converting enzyme inhibitors from sour milk. *Journal of Dairy Science*,Vol.78, No.1, pp. 777-783, ISSN 0022-0302.

Nielsen, P., Petersen, D., Dammann, C. (2001). Improved method for determining food protein degree of hydrolysis. *Journal of Food Science*, Vol.66, No.5, pp. 642-646, ISSN 0022-1147.

Pedroche, J., Yust, M.M, Girón-Calle, J., Alaiz, M., Millán, F., Vioque, J. Utilisation of chickpea protein isolates for production of peptides with angiotension I-converting enzyme (ACE) inhibitory activity. *Journal of the Science of Food and Agriculture*, Vol.82, No.1, pp. 960-965, ISSN 0022-5142.

Pedroche, J., Yust, M., Girón-Calle, J., Vioque, J., Alaiz, M., Millán, F. (2003). Plant protein hydrolysates and tailor-made foods. *Electronic Journal of Environmental, Agricultural and Food Chemistry*, Vol.2, No.1, pp. 233-235, ISSN 1579-4377.

Pukalskas, A., Van Beek, T., Venskutonis, R., Linssen, J., Van Veldhuizen, A., Groot, A. (2002). Identification of radical scavengers in sweet grass (*Hierochloe odorata*). *Journal of Agricultural and Food Chemistry*, Vol.50, No.10, pp. 2914-2919, ISSN 0021-8561.

Raghavan, S., Kristinsson, G.H., Leewengurgh, C. (2008). Radical scavenging and reducing ability of tilapia (*Oreichromis niloticus*) protein hydrolysates. *Journal of Agricultural and Food Chemistry*, Vol.56, No. 21, pp.10359-10367, ISSN 0021-8561.

Salazar-Vega, I.M., Rosado-Rubio, G., Chel-Guerrero, L.A., Betancur-Ancona, D.A., & Castellanos-Ruelas, A.F. (2009). Composición en ácido graso alfa linolénico (ϖ3) en el huevo y carne de aves empleando chia (*Salvia hispanica*) en el alimento. *Interciencia*, Vol.34, No.3, pp. 209-213, ISSN 0378-1844.

Segura-Campos, M.R., Chel-Guerrero, L.A. Betancur-Ancona, D.A. (2010). Angiotensin-I converting enzyme inhibitory and antioxidant activities of peptide fractions extracted by ultrafiltration of cowpea *Vigna unguiculata* hydrolysates. *Journal of the Science of Food and Agriculture*,Vol.90, No.1, pp. 2512-2518, ISSN 0022-5142.

Sigma (2006).*Technical bulletin OF Sigma BCA assay*. Product Code BCA, B9643, Available from http:// www. Sigmaaldrich.com/sigma/bulletin/b9643bul.pdf.

Tang, C.H., Peng, J., Zhen, D. W., Chen, Z. (2009). Physicochemical and antioxidant properties of buckwheat (*Fagopyrum esculentum Moench*) protein hydrolysates. *Food Chemistry*, Vol.115, No.2, pp. 672-687, ISSN 0308-8146.

Torricella, M., Zamora U., Pulido A. (1989). *Evaluación Sensorial aplicada al desarrollo de la calidad en la industria sanitaria (1th ed.)*. Editorial universitaria, Instituto de Investigaciones para la Industria Alimentaria, ISBN 978-959-16-0577-1, Habana, Cuba.

Torruco-Uco, J., Chel-Guerrero, L., Martínez-Ayala, A., Dávila-Ortíz, G., Betancur-Ancona, D. (2009). Angiotensin-I converting enzyme inhibitory and antioxidant activities of protein hydrolysates from *Phaseolus lunatus* and *Phaseolus vulgaris*. LWT-Food Science and Technology, Vol.42, No. 1, pp. 1597-1604, ISSN 0023-6438.

Tosco, G. (2004). Los beneficios de la chía en humanos y animales. Nutrimentos de la semilla de chía y su relación con los requerimientos humanos diarios. *Actualidades Ornitológicas*, No.119, pp.1-70.

Tosi, E., Ré, E., Masciarelli, R., Sánchez, H., Osella, C., De la Torre, M. (2002). Whole and defatted hyperproteic amaranth flours tested as wheat flour supplementation in mold breads. *LWT-Food Science and Technology*, Vol.35, No.5, pp. 472-475, ISSN 0023-6438.

Vioque, J., Sánchez-Vioque, R., Clemente, A., Pedroche, J., Bautista, J., Millan, F. (1999). Production and characterization of an extensive rapeseed protein hydrolysate. *Journal of the American Oil Chemists Society*, Vol.76, No.7, pp. 819-823, ISSN 0003-021X.

Vázquez-Ovando, A., Rosado-Rubio, G., Chel-Guerrero, L., Betancur-Ancona, D.A. (2009). Physicochemical properties of a fibrous fraction from chia (*Salvia hispánica* L.). *LWT-Food Science and Technology*, Vol.42, No.1, pp. 168-173, ISSN 0023-6438.

Vázquez-Ovando, J.A., Rosado-Rubio, J.G., Chel-Guerrero, L.A., Betancur-Ancona, D.A. (2010). Dry processing of chia (*Salvia hispanica* L.) flour: chemical and characterization of fiber and protein. *CYTA-Journal of Food*, Vol.8, No.2, pp. 117-127, ISSN 1947-6345.

20

Wine as Food and Medicine

Heidi Riedel, Nay Min Min Thaw Saw, Divine N. Akumo,
Onur Kütük and Iryna Smetanska
Technical University Berlin, Department of Food Technology and Food Chemistry,
Methods of Food Biotechnology
Germany

"Wine is the most civilized thing in the world." - Ernest Hemingway

1. Introduction

The name "Wine" is derived from the Latin word vinum, "wine" or "(grape) vine". Wine is the earliest domesticated fruit crops and is defined as an alcoholic beverage which is produced by the fermentation of grape juice. Grapes are small berries with a semi-translucent flesh, whitish bloom and a smooth skin, whereby some berries contain edible seeds and others are seedless. Grape berries have a natural chemical balance which allows a completely fermentation without the addition of sugar, acid, enzymes or other nutrients. It is a rich source of vitamins, many essential amino acids, minerals, fatty acid and others. Grapevine, botanically called *Vitis vinifera*, has a wide range of different species whereby wine is a mixture of one or more varieties (Bouquet et al. 2006). Pinot Noir, Chardonnay, or Merlot for example are predominated by grapes with a minimum of 75 or 85% grape by law and the result is a varietal as opposed to a blended wine. Nevertheless, blended wines are not of minor value compared to varietal wines. Wines from the Bordeaux, Rioja or Tuscany regions are one of the most valuable and expensive wines which are a mixture of many different grape varieties of the same vintage. Wines of high quality are not permitted to be labeled as varietal names because of the 'cépage' (grape mix) which is restricted by law. Red Bordeaux wines are a composition of four different grapes including, but not exclusively, Cabernet Sauvignon and Merlot. Red and white Burgundy are made from a single grape variety and they use their regional label because of marketing strategies and historical reasons. To name some few native North American grapes like *Vitis labrusca, Vitis aestivalis, Vitis rupestris, Vitis rotundifolia* and *Vitis riparia* which are usually used for eating as fruit or made into grape juice, jam, or jelly sometimes into wine for example Concord wine (*Vitis labrusca species*). The most common vineyards worldwide are planted with the European vinifera vines that have been grafted with native species of North America. This is because grape species from North America are resistant against phylloxera (Granett et al. 2001). The theory of "terroir" is defined by the variety of the grape, orientation and topography of the vineyard, elevation, type and chemistry of soil, the climate and seasonal conditions under which grapes are grown. Among wine products, there is a high variety which is due to the fermentation and aging processes. Many winemakers with small production volume prefer growing and using production methods that preserve the unique sensory properties like aroma and the taste of their terroir.

The main challenge of the producers is to reduce differences in sources of grapes by using wine making technology such as micro-oxygenation, tannin filtration, cross-flow filtration, thin film evaporation, and spinning cone. Grapevine is one of the major fruits in the world which is grown in temperate regions on the northern and southern hemisphere mostly between the 30th and 50th parallel. Grapevine is cultivated in large fields because of their economic value (Bouquet et al. 2006). The most popular wine regions of the world are France, Italy, Northern California, Germany, Australia, South Africa, Chile and Portugal respectively. The genus grape (*Vitis vinifera*) contains around 60 species which are common in the temperate zones with some species growing in the tropical region. In 2009 the production of grape was about 69 million tons (especially for wine, juice and raisins) compared to data from 1995 with only 55 million tons (Data from the Organisation Internationale de la Vigne et du Vin (OIV)). Grapes need a minimum of 1500 hours of sunshine to ripen fully, red wine needs more radiation than white.

1.1 Some historical facts of wine

Wine and grape are one of the oldest fruits and the traditional winemaking processes are an ancient art, which began as early as 1,000 B.C. Archaeological investigations and discoveries attest that the wine production by fermenting processes, took place from early as 6000 BC. Other studies from China show that grapes were used together with rice to produce fermented juices as early as 7000 BC. Some research studies document the origin home countries of wine to the Balkan Range along the coast of the Black Sea. Wine is mentioned in historical literature documents as Iliad and Odyssey by Homer. In Greco-Roman mythology, Dionysius is adored as the god of wine. He is also known as Bacchus whereby Dionysius is regarded as the patron of vine events. One of the most important grape wine producers in Europe is Turkey as well as other neighboring countries around the Mediterranean Sea. Especially Anatolia was described as the origin place of viticulture and wine making. One of the first traces of the cultivation of grape wine was in the Early Bronze Age around the Mediterranean basin (Gorny 1996). Many archaeological investigations prove the early domestication of wine in the East (This et al. 2006). During the Bronze Age in the Mediterranean were olive, fig and grape the most common fruits. Scientists discovered many evidences like grape pips in a shrine, wine shop with jars and cups from the Bronze Age (Refai 2002). In Europe around the Mediterranean area between Black Sea and Caspic Sea grape was cultivated and used for winemaking 4000 BC (Monti 1999). The wild grapevine specie Vitis vinifera ssp. Sylvestris Gmelin was grown from Portugal to Turkmenistan and the north of Tunesia. About 8400 years seeds of the oldest wild grape in Turkey were discovered in a valley near Urfa (This et al. 2006). Specific investigations of the chlorotype showed a higher diversity of the wild grape population in the central and eastern parts than in the western areas of the Mediterranean (Arroyo-Garcia et al. 2006). All domesticated grapevine species originated from the wild type Vitis vinifera ssp. sylvestris whereby Vitis vinifera L. is the only native species of Eurasia and appeared 65 million years ago. To enhance the yield of grapevine hermaphrodite genotypes were selected for domestication procedures with intensive pigmentation and techniques for propagation (Terral et al.).Nowadays the skills of the winemaking process are considered for intellectual persons. Special famous events only about wine are exhibits, expos, and auctions worldwide. The Boston Wine Expo is one major annual convention where top wine producer exhibit, sell their goods and show new technologies. Such expos serve as venue for the world's top producers to exhibit and sell their good. Persons with high interests as well as wine collectors attend such exhibitions to exchange ideas and share their passions for wine. Wine is associated with education, lifestyle and class and that is the reason why wine is always included in special occasions. What did early wine producers start out with, and how did they change grapevines in the course of domesticating them? How does the evolutionary history of grapevines affect grape growers today?

2. Classification of wine

The naming of wines has a long tradition and is based on their grape variety or by their place of production. European wines are labeled after their place of production like Bordeaux, Rioja and Chianti as well as the type of grapes used such as Pinot, Chardonnay and Merlot. All other wines from all over the world are generally named for the grape variety. Non- European wine labels become more and more famous and the market recognition get more stability. Some examples include Napa Valley, Barossa Valley, Willamette Valley, Cafayate, Marlborough, and Walla Walla just to name a few. Sensory properties like the taste of a wine depends on the grape species and the blend, and furthermore on the ground and climatic conditions (terroir).

2.1 Red wine

The color of wine is caused by the presence or absence of the grape skin during the fermentation process. Grapes with colored juice like Alicante Bouchet became popular as colorants so called "teinturier". The basic natural products of red wine are red or black grapes, but the intensive red color originates from maceration, which is a process whereby the skin is left in contact with the juice during fermentation. In the following table 1 are listed some red wine varieties, their country origin and characteristics.

Wine variety	Country of origin	Characteristics
Aleatico	Italy	Dark skinned grape, fragrant, very rare
Alicante Bouschet	France	Red skinned grape
Cabernet Sauvignon	France (Bordeaux)	principal component of Bordeaux reds
Concord	America	Most important variety in US, belongs to Vitis labrusca
Dolcetto	Italy	Wine is soft and fruity
Pinotage	South Africa	Wild grown hybrid variety

Table 1. Varieties of Red Wine

Dependent on the grape specie, climatic conditions during the ripening process and many other external factors can influence the sugar and alcohol content of the wine. In the following table 2 is shown the nutritional value of red table wine.

Energy 80 Kcal, 360 KJ	
Carbohydrates	26 g
Sugar	0.6 g
Fat	0.0 g
Protein	0.1 g
Alcohol	10.6 g

Table 2. Nutritional value of red table wine per 100 g

2.2 White wine

White wine can be produced from any color of grape as the skin is separated from the juice during fermentation. The following table 3 shows some examples white wine, their country origin and characteristics.

Wine variety	Country of origin	Characteristics
Chardonnay	France	Widely grown throughout the world.
Frontignac	Greece	highly fragrant
Gewürztraminer	France (Alsace)	highly fragrant and spicy
Muscadelle	France (Bordeaux)	An aromatic variety.
Picolit	Italy (Friuli region)	Used to make sweet white wine.
Riesling	Germany	noble variety producing some of the world's greatest wines.
Sauvignon Blanc	France (Bordeaux)	A highly aromatic variety.

Table 3. Varieties of White Wine

2.3 Rosé wine

Rosé wine is produced from different very dark red grape-varieties whereby it is not a blending of red and white wine. In recent times many wine dressers mix a special amount of white wine with red wine.

2.4 Sparkling wines

Sparkling wines contains carbon dioxide which is naturally made due to the fermentation process; champagne for example. To achieve this sparkling effect, the wine has to ferment two times. The first time in an uncovered container that carbon dioxide can escape into the environment. In a second step the wine is in a sealed fermentation container so that the gas remains in the wine. In the following table 4 are listed some famous sparkling wines from different countries.

Wine variety	Country of origin	Characteristics
Chardonnay	France	Basic component of champagne
Macabeo	Middle East	Basic component of the Spanish sparkling wine Cava.
Muscat Blanc À Petits Grains	Greece	Used for Italian sparkling wines known as Asti.
Prosecco	Italy (Veneto)	Used for Prosecco, an Italian sparkling wine.

Table 4. Varieties of Sparkling Wines

2.5 Table wine

The characteristic of table wines is that the alcohol content is not higher than 14% in the U.S. whereas in Europe, the alcohol range of light wine must be between 8.5% and 14% by volume. Depending on the color of the wine, table wines are classified as "white", "red" or "rosé".

2.6 Dessert wine

The sugar range in dessert wines can be from slightly sweet (less than 50 g/L sugar) to very sweet wines (more than 400 g/L sugar). For example wines such as Spätlese are produced from grapes harvested after they reached the maximum ripeness. Dried grape wines like Recioto and Vin Santo are made from partially raisined grapes after harvesting. Botrytized wines are produced from grapes infected with Botrytis cinerea; some examples include Sauternes from Bordeaux, Bonnezeaux and Quarts de Chaume, Tokaji Aszú from Hungary, and Beerenauslese from Germany and Austria.

2.7 Fortified wine

Fortified wines are sweet with high alcoholic content because their fermentation process stopped by the addition of spirit like brandy. To popular fortified wines belong Port, Madeira, Tokay and Banyuls.

3. Social and cultural aspects of wine

Wine has a long tradition as cultural beverage and is a popular social gathering since ancient times. Wine was a favorite drink among Roman emperors, Greek scholars, monks living in monasteries and other civilizations. Monks and royalty preferred to drink wine, while beer was only used from the workers. Egyptians investigated the wine regardingly in that quality and developed the first arbors and pruning methods. One path of wine history could follow the developments and science of grape growing and wine production; another might trace the spread of wine commerce through civilization, but there would be many crossovers and detours between them. However the timeline is followed, clearly wine and history have greatly influenced one another. Fossil vines, 60-million-years-old, are the earliest scientific evidence of grapes. The earliest written account of viniculture is in the Old Testament of the Bible which tells us that Noah planted a vineyard and made wine. As cultivated fermentable crops, honey and grain are older than grapes, although neither mead nor beer has had anywhere near the social impact of wine over recorded time. This unique alcoholic drink is enjoyed by people from all walks of life up until contemporary times. The social background of wine includes gatherings, parties, religious rites, special occasions, and even casual events. Wine experts believe that wine is more than a product, it is a culture. It is not just a commodity; it is a collector's item. The main reason why wine is strict regarded to social tools is because of historical distingue purpose. Wine has special characteristics and qualities that make it a favorite among works of art, poetries, and other literary pieces. Winemaking and oenophilists investigate technological novelties and processes are constantly being invented to reach the perfection in wine production.

4. Grapevine in food industry

There are many different ways in which grape fruits can be used and they include; fresh, preserved, dried into raisins or crushed for juice or wine (*Wellness Encyclopedia of Food and Nutrition, 1992*). Grape berries are sensitive fruits and should be carefully handled during the winemaking process because once in a bottle, it will develop with time. The long period of wine process is affected by different external factors which are listed below. The optimal temperature during the ripening process of the wine should be between 12 and 15 °C as well as the humidity should be between 70 and 80%. The circulation of fresh air in the wine cellar should avert any odour from moisture, chemicals; wood fruits etc. to avoid negative side effects in the wine. Also light, vibration and noise will ruin potentially good wine because it disturbs the development process. Wine matures with time and every wine needs different periods for their developing process. The maximum storage time of wine because after reaching the peak the wine will degrade (table 5).

Wine type	Maximum storage time (years)
Dry White	1-8
Sweet White	2-8
Rosé	1-3
Young Red	0-4
Mature Red	1-20
Champagne	3-10
Sparkling Wine	1-6
Sherry & Port	1-20

Table 5. Different types of wine according to their storage time

5. Factors influencing on the phenolic synthesis and wine quality

"The sun, with all those planets revolving around it and dependent on it, can still ripen a bunch of grapes as if it had nothing else in the universe to do."
Galilée (Galileo Galilei), 1564-1642

The cultivation and winemaking process needs a long term experience and optimal conditions for the production of a high quality wine. The quality of wine depends on the maturity of berries or the so called "sugar ripeness", that is the content of soluble substances in the fruit as well as sugar/ acid balance (Kliewer 1964, Coombe 1960). Thanks to new investigations with the use of analytical chemistry like gas chromatography, volatile compounds most especially aromas which are typical for Sauvignon blanc wine can be analyzed. As earlier mentioned, grape is a rich source for phenolic substances and these compounds can be measured since the 80s by HPLC. The initial fruit chemistry measurements can be differentiated from the sensory perception of wine because of the biochemical and chemical transformation of compounds during fermentation. Because many factors influence the entire process of winemaking, thus for a good quality a long term experience is needed to go through all the steps of winery and vineyard respectively. Since thousands of years, humans have tried to manipulate the natural growth habits of grapevines to make the plant productive and economically effective for agriculture. The main focus was in breeding new varieties of grapes to maximize production, fruit quality and economic efficient (Coombe 1960, Bogs et al. 2006).

Different kind of factors such as genotype, environment and cultural practices have an influence on the phenolic biosynthesis and accumulation through the ripening process of grape berries. The quantity of phenolic compounds and also the composition has an influence on the wine quality. One characteristic of grape are the high concentrations of anthocyanin which can be used as chemical marker for the classification of red-grape varieties and wines. Furthermore the intravarietal heterogeneity can be used as characteristic which induces a very different behavior among the different clones. The research interest focuses on high productive clones with a loss of color and phenolic compounds. Rootstock genotypes are associated with water, gas exchange status and canopy growth as well as yield (Koundouras et al. 2006). That is why rootstocks have an influence on the composition of phenolic compounds and also on the time of harvesting (Koundouras et al. 2006).

Grape and the quality of wine is also influences by many different environmental factors such as topography, agro-pedology, climate), which are described as "*terroir*". The amount and composition of phenolics depends also on the sunlight exposure and the temperature which acts on the grape berries during the ripening process (Cohen et al. 2008). Low temperatures at night have a positive effect on the anthocyanin accumulation whereas high temperatures cause a decrease of their concentration. Otherwise the accumulation of colored pigments seems to be increased linearly with increasing sunlight exposure. Berries which are exposed ultraviolet light and temperature are related with metabolic reactions and alterations in anthocyanin composition (Joscelyne et al. 2007). Water availability is one of the most important factors which are responsible for the wine quality. Vine water status cause accumulation of phenolic compounds in grape berries with positive reactions of water deficit on berry phenolic composition (Qian et al. 2009). Environmental factors like rainfall, soil water storage capacity, as well as soil characteristics such as soil depth, structure, texture and fertility affect phenolic composition (Mateus et al. 2003). Different agricultural practices during the berry ripening process are also responsible for the synthesis of phenolic substances. Another interesting aspect are cultural aspects such as training system, row vine spacing, pruning, bunch thinning, bud and leaf removal and also the management of fertilization and water irrigation (Poni et al. 2009) with their special influence on phenolic biosynthesis and accumulation. Another interesting effect on the phenolic compounds during the ripening of grape berries are different agronomic techniques and growth conditions such as conventional, organic or biodynamic systems (Vian et al. 2006). Finally, other the vine age and pathogenesis (Amati et al. 1996).

6. Management of fruit quality

Considering the fact that grapes grow in a wide range of temperate climates on the northern and southern hemisphere, special modifications in breeding resistant and tolerant to environmental factors have been achieved to extend the margins of production. However the standard approach to grapes production is intensive agricultural management to balance between climate and site (soil conditions, topography) on one hand and vine biology and preferred fruit quality on the other hand. For high quality end products, the choice of cultivar (clone, rootstock) as well as the genetic potential of the grape plants are from high importance and can influence the quality of fruits. The yield and growth of wine depends on many environmental factors like climate, soil, water, nutrients, just to name a few. In some areas, vineyards have to be watered by irrigation which is by inter- annual changes in soil water. Nevertheless, the yield and fruit quality is inconsistent, most especially red varieties are growing under the principle "regulated deficit irrigation" which suppress the vegetative growth and influence directly the fruit quality. That means water deficit cause smaller berries, early sugar maturity and modifications in phenolic contents. Temperature, solar radiation, intensity of sunlight reaching the fruit and air movement have an influence on the metabolism in grape berries and furthermore the fruit quality. The orientation of the vineyards should be north-south because of the daily solar cycle. Recent investigations showed that the amount for vines in a loose vertical canopy and under water stress can be about 42 to 45 % of the daily total solar radiation. On the other hand fruits of east-west oriented vineyards are either shaded north-facing fruits or south-facing fruits with high sunlight intensities during the day.

7. Polyphenolics in wine

Grapevine and their wide range of phenolic compounds represent a large group of biomolecules with an important role in enology. Sensory properties of wine such as taste, astringency, bitterness as well as color are caused by many different phenolic substances accumulated in grape (Kammerer et al. 2004). High antioxidant potential of grapevine is also related with other health promoting compounds and that is why the consumption amounts of red wine is increasing (Pitsavos et al. 2005). Wine "experts" explain this phenomenon as the "French paradox" which describes the high life expectancy of French people in regard to their diets, exceptionally high in fats. Many nutritionists believe that this phenomenon is caused by the high consumption of red wine which contains high amounts of antioxidants and flavonoids (Renaud and de Lorgeril 1992). In a recent study the blood drawn of 20 probands before and after drinking wine were analyzed. After drinking it, higher levels of nitric oxide was found (nitric oxide reduce clots), as well as a reduction in platelet aggregation. Furthermore an increase of alpha-tocopherol, which is connected with the antioxidant vitamin E, and the total amount of antioxidants in blood were found to be 50% higher. The consumption of wine causes other effects like the protection of LDL cholesterol from oxidation. The negative properties of oxidized LDL are responsible for arteries damaging their walls and an increased risk of atherosclerosis (Iriti and Faoro 2006). Actual investigations found that some phenolic compounds which are accumulated in grape skins inhibit protein tyrosine kinases. These enzymes regulate cells, inhibiting production of endothelin-1 which seems to be a key component in several heart conditions (da Luz et al. 1999). Furthermore scientists got a special interest to investigate the composition of phenolic compounds in grape because of their anticarcenogenic properties (Block 1992) as well as neuroprotective effect (Ma et al.). (Monagas et al. 2005)) explains the acidy character of the phenolic function and due to the nucleophilic characteristics of the aromatic group is responsible for their reactivity.

Phenolic compounds are a large group of secondary metabolites which can be classified in various ways whereby the most common separation is in flavonoids and non-flavonoids (Table 6). There are more specific and detailed families in each group whereby the chemicals structure of the compound is responsible for properties such as color, aroma and taste (Fournand et al. 2006).

Phenolics are formed from the essential amino acid, phenylalanine within the phenylpropanoid biosynthetic pathway. More than 4,000 phenolic compounds have been identified whereby their role in plants is linked to several functions: protection from UV light, pigmentation, defense against pathogens (antifungal properties), nodule production, attraction of pollinators as well as dispersion of seed (Gould et al. 2000). Grapes accumulate a high variety of polyphenolics as describes below. Phenolic acids are classified into hydroxybenzoic and hydroxycinnamic acids whereby hydroxybenzoic acids derive from benzoic acid. Grapes pulp accumulates high amounts of gallic acid (Lu and Serrero 1999) as well as flavan-3-ols in grape (Singleton and Esau 1969). In the following table 7 are listed more phenolic acids as and their derivates (Monagas et al. 2005, Rentzsch et al. 2007).

Phenolic compounds	
Flavonoids	Non- Flavonoids
	Phenolic acids
Flavonols	• Benzoic acids
	• Cinnamic acids
Flavononols and flavones	Stilbenes
Flavanols	
• Catechins	
• Condensed Tannins	
• Procyanidins	
• Prodelphinidins	
Anthocyanins	

Table 6. Classification and composition of phenolic compounds in grape (adapted from (Kontoudakis et al.)

Benzoic acid	Hydroxycinnamic acid	Hydroxycinnamic ester
p-Hydroxybenzoic	Sinapic	Trans-feruloyltartaric acid (fertaric acid)
Protocatechuic	p-Cumaric	Trans-p-coumaroyltartaric acid (coutaric acid)
Vanillic	Caffeic	Trans-caffeoyltartaric acid (caftaric acid)
Gentisic	Ferulic	
Syringic		
Gallic		
Salicilic		

Table 7. Common phenolic acids in grape and derivates

Hydroxicinnamic acids are located and accumulated in the vacuoles of the skin and pulp as tartaric esters (Ribereau Gayon 1965). The highest amounts of principal hydroxicinnamic acids found in grapes are the caftaric, cutaric and fertaric acids in *trans* form as well as lower contents of the *cis* form (Singleton and Esau 1969). Further bioactive important compound in grape are stilbenes whereby *trans*-resveratrol is the most abundant. Stilbenes in grape can occur in oligomeric and polymeric form (Rentzsch et al. 2007). Plants synthesize stilbenes as defense reaction against infections by fungis or UV irradiation especially in leaves, roots and skin. *Vitis rotundifolia* seeds accumulated high contents of stilbene (Langcake and Pryce 1977, Adrian et al. 2000). Stilbenes play no important role in the organoleptic characteristics of wine but their importance in human health due to their antioxidative, anticarcinogenic potential and neuroprotective effects is from high interest (Nassiri-Asl and Hosseinzadeh 2009). Flavonoids are another large group of compounds synthesizes in grape and are divided in four subclasses. The class name depends on the base of the oxidation state of the pyran ring: the flavonols, the flavanonols and flavones, the flavanols and the anthocyanins (Souquet et al. 2000). Flavanols and anthocyanins have the highest concentrations in wine and play an important role of the quality of red wine. Flavonols are yellow colored pigments which are responsible for the protection against UV light and are mainly accumulated in grape skin (Mane et al. 2007) but also detected in grape pulp (Pereira et al. 2006). In table 8 are listed the most abundant flavonols in grape.

The most abundant phenolics of the group flavanonols and flavones are astilbin and engeletin. They were found in high amounts in the skin and wine of white grapes, grape pomace and in stems (Souquet et al. 2000) but also in red wine (Vitrac et al. 2000). The chemical structure of flavones is similar with flavonols. Some examples of flavones in grape are apigenin, baicalein and luteolin (Wang and Huang 2004). Grape contains also high amounts of flavanols or flavan-3-ols in seed, skin and stem (Gomez-Miguez et al. 2006, Souquet et al. 2000). Flavanols were found in the monomeric, oligomeric and polymeric form and are responsible for organoleptic characteristics in grape wine. The common name of flavan-3-ols monomers is "catechin"(Escribano-Bailon et al. 2001). Anthocyanins (greek: purple flower) are a very well investigated large group of phenolics, are mainly accumulated in skin of red grapes and are responsible for red wine color (Castillo-Munoz et al. 2009). Cyanidin, Delphinidin, Peonidin, Petunidin and Malvidin are the most important anthocyanins in grape (Monagas et al. 2005).

Flavonol	
Kaempferol	Myricetin
Kaempferol-3-O -glucoside	Myricetin-3-O -glucoside
Kaempferol-3-O -galactoside	Myricetin-3-O -glucuronide
Kaempferol-3-O -glucuronide	Isorhamnetin-3-O -glucoside
Quercetin	Isorhamnetin
Quercetin-3-O -glucoside	Quercetin-3-O -glucoronide

Table 8. Classification of flavonols

8. Accumulation of metabolic compounds during ripening

Basically, the maturity of grape berries depends on solar radiation which is indirectly by driving temperature of berries. That means that the sugar accumulation of berries is slowly under low temperatures and the ripening process will be slow. Organic acids will be metabolized more slowly and that cause high concentrations of acids. High temperatures cause various biochemical thresholds which limit metabolic reactions and is commonly called "the vine shuts down". Grape berries are a rich source of phenolic compounds which are important plant secondary metabolites and are produced through photosynthesis (primary metabolites). The composition of phenolic substances is responsible for sensory properties in plant derived food including grapes and wine (Sandhu and Gu). Every wine contains a typical aroma, color, taste and mouth feel. The three major phenolics in grape are anthocyanins and flavonols (Ferrer et al. 2008). Flavonols are useful for UV radiation absorption and anti- microbial properties as response to wounds. The synthesis of these substances in berries is promoted in the vineyard due to solar radiation that is why they are accumulated in the exocarp (surface layer) (Versari et al. 2001). Anthocyanins cause the color modifications of fruits during ripening with variations from red, purple to black what is typical for the color of red wine. Anthocyanins form complexes with tannins and flavonols that contribute in the stability of the pigments. They are also synthesized in darkness but berries exposed for a long time to solar radiation during the ripening period will synthesize and accumulate higher contents of plant pigments. Astringent sensory properties of wine are the consequence of condensed tannins which are typical compounds of seed and skin of grapes. The decision on the optimal harvesting time depends on the climate, fruit developmental stage because high contents of condensed tannins in young berries have a higher biological benefit because of their bitterness and astringency.

9. Health promoting compounds in grape

The interests of scientists in the potential for specific phytochemicals increased in the last years because of their medicinal importance. Actual investigations show that the consumption of fresh fruits and vegetables are important to prevent chronic diseases and provides essential nourishment to humans. Plant geneticists around the world are able to breed special cultivars with higher nutraceuticals value to increase the concentration of health-promoting compounds in grape (God et al. 2007). Grape has many different varieties in health promoting compounds and is rich in bioactive metabolites like phenolic compounds that act as antioxidants or resveratrol which serve as chemo-preventative. Plant phenolics are attractive for researchers and the industry due to their techno-functional and putative beneficial bio-functional properties (Kammerer et al. 2004). The health benefit in wine comes from their high content in flavonoids and phytonutrients which are responsible for the color in grapes. Quercetin and Resveratrol are two important stilbenes in grapes which are responsible for heart- protection effects (Frankel et al. 1993). These bioactive phenolic compounds can reduce blood clotting due to their antiaggregant effect and protect low-density lipoproteins (LDL) cholesterol from oxidative reactions which can cause arterial damage (Albini et al.). The resveratrol (3-4'-5-trihydroxystilbene), is a bioactive secondary metabolite with high importance in medicine and pharmaceutics. Many studies showed that *trans*-resveratrol has a high antioxidative potential and reduces the risks of coronary heart disease and prevent any formation of cancer cells (Cui et al. 2002). The results of different investigations demonstrated that the concentrations of anthocyanins, phenolics and resveratrol differ significantly among cultivars and breeding lines. Grape (*Vitis vinifera* L.) has large amounts of phenolic compounds whereby the highest accumulation is in the skins and seeds (Rodriguez *et al.*, 2006; Poudel *et al.*, 2008). The most abundant phenolic compounds in grape skins are flavonols. The seeds contain high amounts of monomeric phenolic compounds like (+)-catechins, (-)-epicatechin and (-)-epicatechin-3-Ogallate as well as dimeric, trimeric and tetrameric procyanidins which do have antimutagenic and antiviral effects (Kammerer et al. 2004).

In *in vitro* tests, it was verified that phenolic compounds inhibit the oxidation of low-density lipoproteins (LDL) (Fauconneau et al. 1997). Phenolic substances are from high importance in the quality of grapes and wines. They can be classified in two groups: non-flavonoid (hydroxybenzoic, hydroxycinnamic acids, stilbenes) and flavonoid compounds (anthocyanins, flavan-3-ols, flavonols). Anthocyanins are a large group within the family of phenolics which are responsible for the pigmentation- coloration in grapes. Anthocyanins produce in a reaction with flavanols are more stable pigments (Butkhup and Samappito 2008). Flavan- 3-ols (monomeric catechins and proanthocyanidins) create a further large class of phenolic compounds which are responsible for astringent and bitter properties. They are responsible for the browning process in grape (Macheix et al. 1991). Special phenolic compounds are part in the phenomenon of co-pigmentation. Another large group of flavonoids are flavonols like quercetin, myricetin, kaempferol, isorhamnetin as well as their glycosides which have a high potential such as antioxidants. Grape became very attractant because of their high contents of phenolics especially antioxidant properties and their beneficial effects on human health (Vitseva et al. 2005). Scientific investigations verified the health benefits of catechins and procyanidins and the use of grape extract as an antioxidant supplement in the dietary food (Guendez et al. 2005). Special antioxidants can be used as food preservation due to their protective effects against microorganisms (Vattem et al. 2005). Phenolic compounds which have microbiological effects are found in seeds, skins and stem extracts of *Vitis vinifera* (Chidambara Murthy et al. 2002). Further phytonutrient in grape with health benefits are saponins which reduce blood cholesterol by binding its molecules and preventing their absorption, as well as reducing inflammation and cancer-protecting effect. Scientists from the University of California found that alcohol improve the bioavailability of saponins that means making them more easily absorbed in wine. The content of saponins in red wines is 3 to 10 times higher than the amount of saponins contained in white wines. Besides there is a positive association between the alcohol concentration and saponin concentration; that means stronger wines having more saponins. The highest content of saponins was found in red Zinfandel (16% alcohol) followed by Syrah, Pinot noir and Cabernet Sauvignon. White varieties like Sauvignon blanc and Chardonnay contain much less saponins. Furthermore Grape berries contain Pterostilbene with antioxidant activity which has also positive effects against cancer and cholesterol.

10. Resveratrol

Resveratrol is in the human diet and daily life of high importance because of their protection against benzopyrene which is the main substance in cigarette smoke and provokes lung cancer. Resveratrol inhibits the cell receptor aryl hydrocarbon receptor (AhR) on which benzopyrene and carcinogenic polycyclic aromatic hydrocarbons are not able to bind to cell membranes; because by binding they cause an expression of several cancer-promoting genes (Halls and Yu 2008). Researchers found out that Resveratrol inhibits the production of endothelin-1 and influence hearth cells by inhibiting antiotensin II which is a very powerful vasocontricting hormone. They are equally known to prevention the differentiation of fibroblasts into myofibroblasts because of the collagen production. Resveratrol is one of the main components in red wine but they contain several other phytochemicals like catechins and epicatechins in high concentrations (Anekonda 2006). These two phenolic compounds have a high potency to reduce the activity of COX-1 and COX-2. The anti- microbial effect of Resveratrol was carried out in Turkey where an extract from grape seeds, skin and stems showed anti-microbial effects against 14 bacteria like Escherichia coli and Staphylococcus aureus with the only resistant bacterium being Yersinia enterocolica at a concentration of 2.5%.

11. Byproducts of winery

Many byproducts and wastes contain polyphenols with a high potential for the application as food antioxidants and preventive agents against several diseases as well as due to a cost-efficient recovery. Valorization from grape by-products contains different bioactive substances like pigments, flavonoids, stilbenes and phenolic acids which could be used as natural antioxidants or colorants. Polyphenolic substances are extracted into wine, but the highest concentration remains in the vinification waste (pomace, stems, and seeds), which form over 13% of the processed grape weight (Torres et al. 2002). The majority of grape byproducts remains in the waste during the vinification process, it is also an animal food as source of some products such as ethanol, alcoholic beverages, tartaric and citric acid, grape seed oil and dietary fiber (Torres et al. 2002). Residues of the winemaking process so called pomace a very attractive residual sources of valuable bioactive compounds, even though it is still underutilized (Kammerer et al. 2004). For example fractions of Parellada grape (Vitis vinifera) containing oligomers with galloylation ca. 30% which are the most potential free radical scavengers and antioxidants (Torres et al. 2002). Several substances from grape pomace like ethanol, tartaric, malic and citric acids as well as of seed oil can be recovered. Winery byproducts are from high importance in the food industry and also for agricultural purposes because pomace is used as soil conditioner or for compost production. Red and white grape varieties were investigated in their phenolic potential of the press residues. The results showed over 35 different compounds especially anthocyanins, phenolic acids, non-anthocyanin, flavonoids and stilbenes. Anthocyanins were found in high concentrations in red grape pomace up to 132 g/kg dry matter. Some compounds found in pomace showed significant differences in their content in dependence of the cultivar-specific differences, grape ripening stage, microclimatic and phytosanitary conditions (Kammerer et al. 2004). Seed extracts of *Vitis vinifera* contain high contents of flavan-3-ols and their derivatives. The main compounds of pomace and stem extracts are significant amounts of flavonoids, stilbenes, and phenolic acids. Stems contain high concentrations of *trans*-resveratrol and ε-viniferin (Anastasiadi et al. 2009). Pomace was also used for the recovery of phenolics on laboratory scale. Until now for the extraction of anthocyanin was acted using acidified and sulfited solvents. Sulfite cannot be quantitatively removed and allergenic reaction after the consumption of sulfited food was visible. A new pectinolytic and cellulolytic method of pomace enhance the release of phenolic compounds (Maier et al. 2006). By the optimization of this extraction process the yields of extracted phenolic acids, non-anthocyanin flavonoids and anthocyanins reaching 91.9, 92.4 and 63.6 %. Pomace is also a rich source for the edible oil from the seed and rich in polyphenols with high antioxidant activity. To obtain the phenolic compounds from the press residues in high contents is very easy, whereby they can be applied as supplements of functional or enriched foods (Maier et al. 2006). A new technology with resin adsorption allowes a high level of purification and concentration of anthocyanin extracts from a Cabernet Mitos´ grape pomace. These processes are common in industrial processes, for example to debitter citrus juices or to stabilize and standardize juice concentrates (Kammerer et al. 2004). Furthermore, resin adsorption can be also an effective method to concentrate plant phenolics, to fractionate extracts and to enrich some compounds. This novel technique can also contribute to the production of valuable plant extracts with health-beneficial properties. High-speed countercurrent chromatography (HSCCC) is also a new technique which is useful for the analytical and preparative part in winemaking process as well as for the isolation of bioactive compounds from crude plant extracts. By using the HSCCC method phenolic extracts from a ´Lemberger´ grape pomace were analyzed and some compounds like caftaric, coutaric and fertaric acids were isolated. Purified extracts of caftaric, coutaric, and fertaric acids were up to 97.0 %, 97.2 % and 90.4 % and the end product from 10 g of grape pomace were 62, 48 and 23 %. These byproducts especially from grape waste can be used for the recovery of bioactive beneficial compounds to increase the profit of conventional processing techniques and to maximize sustainable agricultural production (Pinelo et al. 2005). Grape has a wide variety of polyphenols whereby flavonoids are the best investigated compounds and known to have antibacterial activities. These metabolites are produced and accumulated due to their interaction with extracellular soluble proteins and/or bacterial cell walls (Cowan 1999). Furthermore catechins inhibit in vitro the action of many bacteria like *Vibrio cholerae*, *Streptococcus mutans*, and *Shigella*, whereas (-)-epigallocatechin gallate is a potent Gram-positive bactericidal that acts by damaging the respective bacterial membranes (Ikigai et al. 1993).

12. Grape phenolics and their nutritional pharmacological effects

Recent investigations show that the addition of biomolecules in food supports human health. Especially glycosylated, esterified, thiolyated, or hydroxylated forms of bioactive compounds displays their health benefit in metabolic activities combined with several diseases. All these bioactive plant food components are mainly found in whole grains, fruits and vegetables (Klotzbach-Shimomura 2001). There are thousands of bioactive food compounds derived from plants like the polyphenols, phytosterols, carotenoids, tocopherols, tocotrienols, isothiocyanates and diallyl- (di, tri)sulfide compounds, fiber, and fruto-ogliosaccharide. Grape is a well known and investigated plant because of their social importance and useful metabolites especially flavonoids in humans health. Polyphenols belong to the major substances in grape which are a widely distributed group of biomolecules. Bioflavonoids belongs to health promoting compounds in grape and their recent scientific interest confirm the importance in our daily healthy diet. They play an important role in longlivety, cancer prevention and heart disease. Many studies investigated the content of phenolics and their effect on human cancer cells. (Yi et al. 2005) reported that muscadine grapes are rich in phenolic compounds and show a high potential on human liver cancer cells HepG2. The phenolic content of four cultivars of muscadine grapes ('Carlos', 'Ison', 'Noble', and 'Supreme') were investigated and separated into phenolic acids, tannins, flavonols, and anthocyanins. Extracted phenolic acids of muscadine grapes inhibited the HepG2 cell population growth in about 50% at concentration of 1–2 mg/mL. Anthocyanins showed the greatest positive effects regardingly apoptosis as well as cell viability at concentrations of 70–150 and 100–300 µg/mL .

13. Bioavailability of phenolic compounds

Polyphenols are very important micronutrients in our diet, and play a key role in the prevention of cancer and cardiovascular diseases. The health benefit of polyphenol depends on the consumed amount as well as their bioavailability (Manach et al. 2003). The main interests are the antioxidant properties of polyphenols as well as their appearence in the human diet and their role in the prevention of cancer, cardiovascular and neurodegenerative diseases (Scalbert et al. 2005). Polyphenols are responsible for the activity of a wide range of enzymes and cell receptors (Middleton et al. 2000). Polyphenols are not absorbed with equal efficacy because they are extensively metabolized by intestinal and hepatic enzymes as well as the intestinal micro flora.

14. Wine in biotechnological approach

Plant cell cultures are a potential alternative to traditional agriculture for the industrial production of valuable bioactive secondary metabolites. Especially pharmaceutical compounds and food additives like flavors, fragrances and colorants), perfumes and dyes are produced and accumulated in plant cell cultures (DiCosmo and Misawa 1995). Thereby the anthocyanin production can explain the basic mechanisms of biosynthesis of secondary metabolites, their transport as well as their accumulation in specific plant tissue. Plant pigments like anthocyanins are the large group of water-soluble pigments which are responsible for many colors. These pigments are also used in acidic solutions for the stabilization of the red color in soft drinks, sugar confectionary, jams and bakery products. As mentioned before grape pomaces is the major source of anthocyanins for commercial purposes and wastes from juice and wine industries (Curtin et al. 2003). Crude preparations of anthocyanins are relatively inexpensive and are used extensively in the food industry. The costs of pure anthocyanins are around US$ 1,250–2,000/kg. Cell suspension cultures of *Vitis vinifera* produce high contents of anthocyanins after cessation of cell division (Kakegawa et al. 2005). Furthermore the external influence and stimulation of the anthocyanin biosynthesis is regulated by the amino acid phenylalanine which is accumulated in the plant cells (Shimada et al. 2005). End products of the flavonoid biosynthesis pathway include anthocyanin pigments. Plant pigment extracts contain mixtures of various anthocyanin molecules, which differ in their levels of hydroxylation, methylation, and acylation. The major anthocyanins which are produced and accumulated in *V. vinifera* cell culture are cyanidin 3-glucoside (Cy3G), peonidin 3-glucoside (Pn3G), malvidin 3-glucoside (Mv3G) as well as the acylated versions of these compounds, cyaniding 3-p-coumaroylglucoside (Cy3CG), peonidin 3-p-coumaroylglucoside (Pn3CG) and malvidin 3-coumaroylglucoside (Mv3CG) (Conn et al. 2003). The production of anthocyanins by plant in- vitro cultures has been estimated in different plant species. Many investigation studies are using a cell line as model system for the production of secondary metabolites especially anthocyanin, because of the color that enables production. (Sano et al. 2005), of Nippon Paint Co. in Japan, investigated the production of anthocyanins. High osmotic potential in *Vitis vinifera* L. (grape) cell suspension cultures caused an increase in the anthocyanin production. By addition of sucrose or mannitol in the medium the osmotic pressure and the anthocyanin concentration accumulated was increased (Zhao et al.).

15. References

Adrian, M., Jeandet, P., Douillet-Breuil, A. C., Tesson, L. & Bessis, R. (2000). Stilbene content of mature Vitis vinifera berries in response to UV-C elicitation. J Agric Food Chem, 48, 6103-5.

Albini, A., Pennesi, G., Donatelli, F., Cammarota, R., De Flora, S. & Noonan, D. M. Cardiotoxicity of anticancer drugs: the need for cardio-oncology and cardio-oncological prevention. J Natl Cancer Inst, 102, 14-25.

Amati, A., Piva, A., Castellari, M. & Arfelli, G. (1996). Preliminary studies on the effect of Oidium tuckeri on the phenolic composition of grapes and wine. . Vitis, 35, 149-150.

Anastasiadi, M., Chorianopoulos, N. G., Nychas, G. J. & Haroutounian, S. A. (2009). Antilisterial activities of polyphenol-rich extracts of grapes and vinification byproducts. J Agric Food Chem, 57, 457-63.

Anekonda, T. S. (2006). Resveratrol--a boon for treating Alzheimer's disease? Brain Res Rev, 52, 316-26.

Arroyo-Garcia, R., Ruiz-Garcia, L., Bolling, L., Ocete, R., Lopez, M. A., Arnold, C., Ergul, A., Soylemezoglu, G., Uzun, H. I., Cabello, F., Ibanez, J., Aradhya, M. K., Atanassov, A., Atanassov, I., Balint, S., Cenis, J. L., Costantini, L., Goris-Lavets, S., Grando, M. S., Klein, B. Y., Mcgovern, P. E., Merdinoglu, D., Pejic, I., Pelsy, F., Primikirios, N., Risovannaya, V., Roubelakis-Angelakis, K. A., Snoussi, H., Sotiri, P., Tamhankar, S., This, P., Troshin, L., Malpica, J. M., Lefort, F. & Martinez-Zapater, J. M. (2006).

Multiple origins of cultivated grapevine (Vitis vinifera L. ssp. sativa) based on chloroplast DNA polymorphisms. Mol Ecol, 15, 3707-14.

Block, G. (1992). The data support a role for antioxidants in reducing cancer risk. Nutr Rev, 50, 207-13.

Bogs, J., Ebadi, A., Mcdavid, D. & Robinson, S. P. (2006). Identification of the flavonoid hydroxylases from grapevine and their regulation during fruit development. Plant Physiol, 140, 279-91.

Bouquet, A., Torregrosa, L., Iocco, P. & Thomas, M. R. (2006). Grapevine (Vitis vinifera L.). Methods Mol Biol, 344, 273-85.

Butkhup, L. & Samappito, S. (2008). An analysis on flavonoids contents in Mao Luang fruits of fifteen cultivars (Antidesma bunius), grown in northeast Thailand. Pak J Biol Sci, 11, 996-1002.

Castillo-Munoz, N., Fernandez-Gonzalez, M., Gomez-Alonso, S., Garcia-Romero, E. & Hermosin-Gutierrez, I. (2009). Red-color related phenolic composition of Garnacha Tintorera (Vitis vinifera L.) grapes and red wines. J Agric Food Chem, 57, 7883-91.

Chidambara Murthy, K. N., Singh, R. P. & Jayaprakasha, G. K. (2002). Antioxidant activities of grape (Vitis vinifera) pomace extracts. J Agric Food Chem, 50, 5909-14.

Cohen, S. D., Tarara, J. M. & Kennedy, J. A. (2008). Assessing the impact of temperature on grape phenolic metabolism. Anal Chim Acta, 621, 57-67.

Conn, S., Zhang, W. & Franco, C. (2003). Anthocyanic vacuolar inclusions (AVIs) selectively bind acylated anthocyanins in Vitis vinifera L. (grapevine) suspension culture. Biotechnol Lett, 25, 835-9.

Coombe, B. G. (1960). Relationship of Growth and Development to Changes in Sugars, Auxins, and Gibberellins in Fruit of Seeded and Seedless Varieties of Vitis Vinifera. Plant Physiol, 35, 241-50.

Cowan, M. M. (1999). Plant products as antimicrobial agents. Clin Microbiol Rev, 12, 564-82.

Cui, J., Tosaki, A., Bertelli, A. A., Bertelli, A., Maulik, N. & Das, D. K. (2002). Cardioprotection with white wine. Drugs Exp Clin Res, 28, 1-10.

Curtin, C., Zhang, W. & Franco, C. (2003). Manipulating anthocyanin composition in Vitis vinifera suspension cultures by elicitation with jasmonic acid and light irradiation. Biotechnol Lett, 25, 1131-5.

Da Luz, P. L., Serrano Junior, C. V., Chacra, A. P., Monteiro, H. P., Yoshida, V. M., Furtado, M., Ferreira, S., Gutierrez, P. & Pileggi, F. (1999). The effect of red wine on experimental atherosclerosis: lipid-independent protection. Exp Mol Pathol, 65, 150-9.

Dicosmo, F. & Misawa, M. (1995). Plant cell and tissue culture: alternatives for metabolite production. Biotechnol Adv, 13, 425-53.

Escribano-Bailon, T., Alvarez-Garcia, M., Rivas-Gonzalo, J. C., Heredia, F. J. & Santos-Buelga, C. (2001). Color and stability of pigments derived from the acetaldehyde-mediated condensation between malvidin 3-O-glucoside and (+)-catechin. J Agric Food Chem, 49, 1213-7.

Fauconneau, B., Waffo-Teguo, P., Huguet, F., Barrier, L., Decendit, A. & Merillon, J. M. (1997). Comparative study of radical scavenger and antioxidant properties of phenolic compounds from Vitis vinifera cell cultures using in vitro tests. Life Sci, 61, 2103-10.

Ferrer, J. L., Austin, M. B., Stewart, C., Jr. & Noel, J. P. (2008). Structure and function of enzymes involved in the biosynthesis of phenylpropanoids. Plant Physiol Biochem, 46, 356-70.

Fournand, D., Vicens, A., Sidhoum, L., Souquet, J. M., Moutounet, M. & Cheynier, V. (2006). Accumulation and extractability of grape skin tannins and anthocyanins at different advanced physiological stages. J Agric Food Chem, 54, 7331-8.

Frankel, E. N., Waterhouse, A. L. & Kinsella, J. E. (1993). Inhibition of human LDL oxidation by resveratrol. Lancet, 341, 1103-4.

God, J. M., Tate, P. & Larcom, L. L. (2007). Anticancer effects of four varieties of muscadine grape. J Med Food, 10, 54-9.

Gomez-Miguez, M., Gonzalez-Manzano, S., Escribano-Bailon, M. T., Heredia, F. J. & Santos-Buelga, C. (2006). Influence of different phenolic copigments on the color of malvidin 3-glucoside. J Agric Food Chem, 54, 5422-9.

Gorny, R. T. (1996). Viniculture and Ancient Anatolia. In: The Origins and Ancient History of Wine, Mc Govern, P.E., S.J. Fleming and S.H. Katz (Eds.). Gordon and Breach Publisher, Amsterdam, 133.

Gould, K. S., Markham, K. R., Smith, R. H. & Goris, J. J. (2000). Functional role of anthocyanins in the leaves of Quintinia serrata A. Cunn. J Exp Bot, 51, 1107-15.

Granett, J., Walker, M. A., Kocsis, L. & Omer, A. D. (2001). Biology and management of grape phylloxera. Annu Rev Entomol, 46, 387-412.

Guendez, R., Kallithraka, S., Makris, D. P. & Kefalas, P. (2005). An analytical survey of the polyphenols of seeds of varieties of grape (Vitis vinifera) cultivated in Greece: implications for exploitation as a source of value-added phytochemicals. Phytochem Anal, 16, 17-23.

Halls, C. & Yu, O. (2008). Potential for metabolic engineering of resveratrol biosynthesis. Trends Biotechnol, 26, 77-81.

Ikigai, H., Nakae, T., Hara, Y. & Shimamura, T. (1993). Bactericidal catechins damage the lipid bilayer. Biochim Biophys Acta, 1147, 132-6.

Iriti, M. & Faoro, F. (2006). Grape phytochemicals: a bouquet of old and new nutraceuticals for human health. Med Hypotheses, 67, 833-8.

Joscelyne, V. L., Downey, M. O., Mazza, M. & Bastian, S. E. (2007). Partial shading of Cabernet Sauvignon and Shiraz vines altered wine color and mouthfeel attributes, but increased exposure had little impact. J Agric Food Chem, 55, 10888-96.

Kakegawa, K., Ishii, T. & Matsunaga, T. (2005). Effects of boron deficiency in cell suspension cultures of Populus alba L. Plant Cell Rep, 23, 573-8.

Kammerer, D., Claus, A., Carle, R. & Schieber, A. (2004). Polyphenol screening of pomace from red and white grape varieties (Vitis vinifera L.) by HPLC-DAD-MS/MS. J Agric Food Chem, 52, 4360-7.

Kliewer, W. M. (1964). Influence of Environment on Metabolism of Organic Acids and Carbohydrates in Vitis Vinifera. I. Temperature. Plant Physiol, 39, 869-80.

Klotzbach-Shimomura, K. (2001). Herbs and health: safety and effectiveness. J Nutr Educ, 33, 354-5.

Kontoudakis, N., Esteruelas, M., Fort, F., Canals, J. M. & Zamora, F. Comparison of methods for estimating phenolic maturity in grapes: correlation between predicted and obtained parameters. Anal Chim Acta, 660, 127-33.

Koundouras, S., Marinos, V., Gkoulioti, A., Kotseridis, Y. & Van Leeuwen, C. (2006). Influence of vineyard location and vine water status on fruit maturation of nonirrigated cv. Agiorgitiko (Vitis vinifera L.). Effects on wine phenolic and aroma components. J Agric Food Chem, 54, 5077-86.

Langcake, P. & Pryce, R. J. (1977). A new class of phytoalexins from grapevines. Experientia, 33, 151-2.

Lu, R. & Serrero, G. (1999). Resveratrol, a natural product derived from grape, exhibits antiestrogenic activity and inhibits the growth of human breast cancer cells. J Cell Physiol, 179, 297-304.

Ma, L., Cao, T. T., Kandpal, G., Warren, L., Fred Hess, J., Seabrook, G. R. & Ray, W. J. Genome-wide microarray analysis of the differential neuroprotective effects of antioxidants in neuroblastoma cells overexpressing the familial Parkinson's disease alpha-synuclein A53T mutation. Neurochem Res, 35, 130-42.

Macheix, J. J., Sapis, J. C. & Fleuriet, A. (1991). Phenolic compounds and polyphenoloxidase in relation to browning in grapes and wines. Crit Rev Food Sci Nutr, 30, 441-86.

Maier, T., Sanzenbacher, S., Kammerer, D. R., Berardini, N., Conrad, J., Beifuss, U., Carle, R. & Schieber, A. (2006). Isolation of hydroxycinnamoyltartaric acids from grape pomace by high-speed counter-current chromatography. J Chromatogr A, 1128, 61-7.

Manach, C., Morand, C., Gil-Izquierdo, A., Bouteloup-Demange, C. & Remesy, C. (2003). Bioavailability in humans of the flavanones hesperidin and narirutin after the ingestion of two doses of orange juice. Eur J Clin Nutr, 57, 235-42.

Mane, C., Souquet, J. M., Olle, D., Verries, C., Veran, F., Mazerolles, G., Cheynier, V. & Fulcrand, H. (2007). Optimization of simultaneous flavanol, phenolic acid, and anthocyanin extraction from grapes using an experimental design: application to the characterization of champagne grape varieties. J Agric Food Chem, 55, 7224-33.

Mateus, N., Silva, A. M., Rivas-Gonzalo, J. C., Santos-Buelga, C. & De Freitas, V. (2003). A new class of blue anthocyanin-derived pigments isolated from red wines. J Agric Food Chem, 51, 1919-23.

Middleton, E., Jr., Kandaswami, C. & Theoharides, T. C. (2000). The effects of plant flavonoids on mammalian cells: implications for inflammation, heart disease, and cancer. Pharmacol Rev, 52, 673-751.

Monagas, M., Bartolome, B. & Gomez-Cordoves, C. (2005). Updated knowledge about the presence of phenolic compounds in wine. Crit Rev Food Sci Nutr, 45, 85-118.

Monti, A. (1999). [Vine and wine in history and in law from the 11th to the 19th centuries]. Arch Stor Ital, 157, 357-65.

Nassiri-Asl, M. & Hosseinzadeh, H. (2009). Review of the pharmacological effects of Vitis vinifera (Grape) and its bioactive compounds. Phytother Res, 23, 1197-204.

Pereira, G. E., Gaudillere, J. P., Pieri, P., Hilbert, G., Maucourt, M., Deborde, C., Moing, A. & Rolin, D. (2006). Microclimate influence on mineral and metabolic profiles of grape berries. J Agric Food Chem, 54, 6765-75.

Pinelo, M., Rubilar, M., Jerez, M., Sineiro, J. & Nunez, M. J. (2005). Effect of solvent, temperature, and solvent-to-solid ratio on the total phenolic content and antiradical activity of extracts from different components of grape pomace. J Agric Food Chem, 53, 2111-7.

Pitsavos, C., Makrilakis, K., Panagiotakos, D. B., Chrysohoou, C., Ioannidis, I., Dimosthenopoulos, C., Stefanadis, C. & Katsilambros, N. (2005). The J-shape effect of alcohol intake on the risk of developing acute coronary syndromes in diabetic subjects: the CARDIO2000 II Study. Diabet Med, 22, 243-8.

Poni, S., Bernizzoni, F., Civardi, S. & Libelli, N. (2009). Effects of pre-bloom leaf removal on growth of berry tissues and must composition in two red vitis vinifera L. cultivars. Australian Journal of Grape and Wine Research 15 (2), 185-193.

Qian, M. C., Fang, Y. & Shellie, K. (2009). Volatile composition of Merlot wine from different vine water status. J Agric Food Chem, 57, 7459-63.

Refai, M. (2002). Incidence and control of brucellosis in the Near East region. Vet Microbiol, 90, 81-110.

Renaud, S. & De Lorgeril, M. (1992). Wine, alcohol, platelets, and the French paradox for coronary heart disease. Lancet, 339, 1523-6.

Rentzsch, M., Schwarz, M., Winterhalter, P. & Hermosin-Gutierrez, I. (2007). Formation of hydroxyphenyl-pyranoanthocyanins in Grenache wines: precursor levels and evolution during aging. J Agric Food Chem, 55, 4883-8.

Ribereau Gayon, P. (1965). [Identification of Esters of Cinnamic Acid and Tartaric Acid in the Limbs and Berries of V. Vinifera]. C R Hebd Seances Acad Sci, 260, 341-3.

Sandhu, A. K. & Gu, L. Antioxidant capacity, phenolic content, and profiling of phenolic compounds in the seeds, skin, and pulp of Vitis rotundifolia (Muscadine Grapes) As determined by HPLC-DAD-ESI-MS(n). J Agric Food Chem, 58, 4681-92.

Sano, T., Oda, E., Yamashita, T., Naemura, A., Ijiri, Y., Yamakoshi, J. & Yamamoto, J. (2005). Anti-thrombotic effect of proanthocyanidin, a purified ingredient of grape seed. Thromb Res, 115, 115-21.

Scalbert, A., Manach, C., Morand, C., Remesy, C. & Jimenez, L. (2005). Dietary polyphenols and the prevention of diseases. Crit Rev Food Sci Nutr, 45, 287-306.

Shimada, S., Inoue, Y. T. & Sakuta, M. (2005). Anthocyanidin synthase in non-anthocyanin-producing Caryophyllales species. Plant J, 44, 950-9.

Singleton, V. L. & Esau, P. (1969). Phenolic substances in grapes and wine, and their significance. Adv Food Res Suppl, 1, 1-261.

Souquet, J. M., Labarbe, B., Le Guerneve, C., Cheynier, V. & Moutounet, M. (2000). Phenolic composition of grape stems. J Agric Food Chem, 48, 1076-80.

Terral, J. F., Tabard, E., Bouby, L., Ivorra, S., Pastor, T., Figueiral, I., Picq, S., Chevance, J. B., Jung, C., Fabre, L., Tardy, C., Compan, M., Bacilieri, R., Lacombe, T. & This, P. Evolution and history of grapevine (Vitis vinifera) under domestication: new morphometric perspectives to understand seed domestication syndrome and reveal origins of ancient European cultivars. Ann Bot, 105, 443-55.

This, P., Lacombe, T. & Thomas, M. R. (2006). Historical origins and genetic diversity of wine grapes. Trends Genet, 22, 511-9.

Torres, J. L., Varela, B., Garcia, M. T., Carilla, J., Matito, C., Centelles, J. J., Cascante, M., Sort, X. & Bobet, R. (2002). Valorization of grape (Vitis vinifera) byproducts. Antioxidant and biological properties of polyphenolic fractions differing in procyanidin composition and flavonol content. J Agric Food Chem, 50, 7548-55.

Vattem, D. A., Ghaedian, R. & Shetty, K. (2005). Enhancing health benefits of berries through phenolic antioxidant enrichment: focus on cranberry. Asia Pac J Clin Nutr, 14, 120-30.

Versari, A., Parpinello, G. P., Tornielli, G. B., Ferrarini, R. & Giulivo, C. (2001). Stilbene compounds and stilbene synthase expression during ripening, wilting, and UV treatment in grape cv. Corvina. J Agric Food Chem, 49, 5531-6.

Vian, M. A., Tomao, V., Coulomb, P. O., Lacombe, J. M. & Dangles, O. (2006). Comparison of the anthocyanin composition during ripening of Syrah grapes grown using organic or conventional agricultural practices. J Agric Food Chem, 54, 5230-5.

Vitrac, X., Larronde, F., Krisa, S., Decendit, A., Deffieux, G. & Merillon, J. M. (2000). Sugar sensing and Ca2+-calmodulin requirement in Vitis vinifera cells producing anthocyanins. Phytochemistry, 53, 659-65.

Vitseva, O., Varghese, S., Chakrabarti, S., Folts, J. D. & Freedman, J. E. (2005). Grape seed and skin extracts inhibit platelet function and release of reactive oxygen intermediates. J Cardiovasc Pharmacol, 46, 445-51.

Wang, S. P. & Huang, K. J. (2004). Determination of flavonoids by high-performance liquid chromatography and capillary electrophoresis. J Chromatogr A, 1032, 273-9.

Yi, W., Fischer, J. & Akoh, C. C. (2005). Study of anticancer activities of muscadine grape phenolics in vitro. J Agric Food Chem, 53, 8804-12.

Zhao, Q., Duan, C. Q. & Wang, J. Anthocyanins Profile of Grape Berries of Vitis amurensis, Its Hybrids and Their Wines. Int J Mol Sci, 11, 2212-28.

Part 4

Quality Control

21

Food Quality Control: History, Present and Future

Ihegwuagu Nnemeka Edith[1] and Emeje Martins Ochubiojo[2]
[1]Agricultural Research Council of Nigeria
[2]National Institute for Pharmaceutical Research and Development
Nigeria

1. Introduction

Food is any substance which when consumed provides nutritional support for the body. It may be of plant or animal origin, containing the known five essential nutrients namely, carbohydrates, fats, proteins, vitamins and minerals. Usually after consumption, food undergoes different metabolic processes that eventually lead to the production of energy, maintenance of life, and/or stimulation of growth (Aguilera 1999). The history of early man shows that, people obtained food substances through hunting, gathering, and agriculture. The assurance and protection of food quality has always been important to man. This is evident from the fact that, one of the earliest laws known to man was that of Food. Right from the Garden of Eden, there was a law guiding the consumption of food. In our time too, governments over many centuries have endeavoured to provide for the safety and wholesomeness of man's food by legal provisions, (Alsberg 1970; Jango-Cohen 2005). In spite of these provisions, adulteration of foods has increased and the detection of these adulterants has proved more difficult, essentially because of the sophisticated methods being used in the adulteration. The birth of modern chemistry in the early nineteenth century made possible the production of materials possessing properties similar to normal foods which, when fraudulently used, did not readily attract the attention of the unsuspecting consumer. However, modern analytical techniques are now available to detect adulterants in foods. In this modern age, most of the food consumed by man and animals alike is supplied by the food industry. The food industry is operated at different levels by either local, national or multinational corporations, that use either subsistence or intensive farming and industrial agriculture to maximize output. Whichever applies the importance of ensuring the quality of food and food products from the food industry cannot be over emphasized. In this chapter, we look at past and present attempts by scientists, individuals and/or governments to control food quality and prospect what the future might be.

2. Food quality control

Quality control is the maintenance of quality at levels and tolerance limits acceptable to the buyer while minimizing the cost for the vendor. Scientifically, quality control of food refers to the utilization of technological, physical, chemical, microbiological, nutritional and sensory parameters to achieve the wholesome food. These quality factors depend on specific attributes such as sensory properties, based on flavor, color, aroma, taste, texture and quantitative properties namely; percentage of sugar, protein, fibre etc. as well as hidden attributes likes peroxides, free fatty acids, enzyme (Adu-Amankwa 1999; Radomir 2004,Raju and Onishi 2002;).

Although, quality attributes are many, not all need to be considered at every point in time for every particular product. It is important to always determine how far relatively a factor is in relation to the total quality of the product. The quality attribute of a particular product is based on the composition of the product, expected deteriorative reactions, packaging used, shelf life required and the type of consumers. The most important element and ultimate goal in food quality control is protecting the consumer. To ensure standardization of these procedures, food laws and regulations cover the related acts affecting the marketing, production, labeling, food additive used, dietary supplements, enforcement of General Manufacturing Practice (GMP), Hazard Analysis and Critical Control Point (HACCP), federal laws and regulations, factory inspections and import/export inspections (Adamson 2004; Gravani 1986, Lasztity et al, 2004, Roe 1956).

3. Importance of food quality control (QC)

The most important quality factor of processed food is safety and reliability followed by "deliciousness" and "appropriate price" (Raju and Onishi 2002). The colossal loss a food industry will record if defective products were rejected or recalled, as well as the damaging effect on the company's image and public trust justifies the need for food quality control. For this reason, quality assurance should be a corporate goal, and should stem from the uppermost management level to the least staff of the industry. The Plan, Do, Check, Action (PDCA) cycle should be used when quality control is implemented.

4. History of food quality control

Many years ago, about 2500 years BC Mosaic and Egyptian laws had provisions to prevent the contamination of meat. Also, more than 2000 years ago, India already had regulations prohibiting the adulteration of grains and edible fats. In the actual sense, the laws of Moses contained decrees on food that are quite similar to certain aspects of modern food laws. Books of the Old Testament prohibited the consumption of meat from animals that died other than those intentionally slaughtered, perhaps consciously or otherwise, this was to ensure that contaminated meats were not consumed. There were also regulated weights and measures in foods and other commodities. Other ancient food regulations are referred to in Chinese, Hindu, Greek, and Roman literatures. Early records according to Lasztity et al (2004), classical writers also referred to the control of beer and inspection of wines in Athens, "to ensure purity and soundness of these products." The Roman government also provided state control over food supplies to protect consumers against bad quality and fraud. Even when most traders prefer to deal honestly and fairly, history has shown the need to ensure quality as evident by laws to protect purchasers and honest traders against those who refuse to adhere to accepted codes of good practice. It has been found (Adamson 2004) that, when food was scarce, resulting in expected increase in demand, fraudulent practices were prevalent.

4.1 The middle ages

This was the era of formation of trade guilds, especially the European communities with their powerful influence on the regulation of commerce (Petró-Turza and Földesi 2004).

These were groups of tradesmen of particular specialties whose purpose was to provide control and general supervision over the honesty and integrity of their members and the quality of their products. For example, in 1419, a proclamation was issued prohibiting the adulteration or mixing of wine from different geographical areas (Adamson 2004). Many countries had their own peculiar way of controlling food quality. For example, in 1649 a Commonwealth statute was enacted to regulate the quality of butter (Petró-Turza and Földesi 2004). In France the most interesting and complete economic document of the Middle Ages available on the subject is the *Livre des Métiers*, which, in the thirteenth century, outlined a code of the comparative practices of the trade guilds of Paris.

By the seventeenth and eighteenth centuries, chemistry was being used as an analytical tool in the fight against food adulteration. Robert Boyle, using the principles of specific gravity, established the foundation for the scientific detection of the adulteration of food (Adamson 2004). Not much has changed, just the degree and level of sophistication of fraud, and the analytical techniques used to detect it.

4.2 Industrial revolution in the nineteenth century

Records (Roe1956; Alsberg 1970) show that, there were some sporadic standardization activities in ancient cultures and during the middle ages, a more organized standardization however, actually developed in the second half of the nineteenth century. This was when the industry reached a state of development and it was obvious that, unification was indispensable for large projects such as the railway. Efforts were made to find uniform measures for length and weight, but the meter and kilogram were not accepted as those standards until 1870s (Roe1956; Alsberg 1970). The period of industrial revolution was a time of tremendous expansion in many fields, which had a particular bearing on food production, regulations, and control too. Rapid changes from a rural to an urbanized society and from a domestic to a factory production system placed a strain on food production and distribution. Expectedly, the period created many public health problems, particularly in the industrialized centres, which were ill prepared to accommodate the masses that flocked to them. There was much poverty, and the uncontrolled development of industrial towns led to appalling conditions similar to those which can still be seen in urban areas in some parts of the world today. There were attendant calls for reform and improvements in matters relating to public health including essential food supplies within the crowded and unhygienic industrial centres (Petró-Turza and Földesi 2004). In England in 1820, Friederich Accum's *A Treatise on Adulteration of Foods and Culinary Poisons* highlighted the fraudulent practices that endangered public health. Unfortunately, at that time, the knowledge and understanding of hygiene and the dangers of food adulteration was so low and limited that this work was not considered. Many earlier papers that tried to draw attention to the adulteration of some foodstuffs were also been disregarded (Industrial Revolution 6th edn. 2004).

Worthy of note however, was the setting up of a municipal service for the control of foodstuffs and beverages in Amsterdam in 1858. This was closely followed in England 1860, by the enactment of the first comprehensive modern food law in the world; "An Act to Prevent the Adulteration of Food and Drink." In addition to this being the first such act, it provided for a scientific approach to food problems by the appointment of an analyst whose sole duty, was "to examine the purity of articles of food and drink." (FAO 1999).

Seven years later in Budapest, Hungary, a municipal service was set up for control of drinking water, this culminated in the establishment of an institute for control of food. During this period, similar types of laws appeared in Belgium, Italy, Austria, Hungary, and the Scandinavian countries. This was also a period of the foundation of institutions serving food inspection and food quality control. To give an idea of the volume of activity of such institutions, it may be mentioned that, according to a report from the Food and Drug Inspection of the State Board of Health of Massachusetts, since the passage of the law in 1882 until 1907, more than 176 000 food samples were controlled, and more than 11 000 were found to be adulterated. Although the main food control activity at this time was in the industrialized nations of Western Europe, many other countries such as Australia, Canada, and the US also enacted food laws. Although Australia did not enact a national food law, each state had the power to enact food laws and has remained ever since (Alsberg 1970; Hutt& Merril 1991).

4.3 The twentieth century

The most prominent and substantive development at this period took place in India. The country amended its food adulteration control to ensure the purity of articles of food sold throughout the country between 1919 and 1941. In 1954, India enacted the Prevention of Food Adulteration Act and this Act, with its later amendments, is still in force (Roe 1956; Lasztity 2004). In the Far East, food control was slow to appear, and it was not until the 1940s or as late as the 1960s that food control measures were introduced. During this period, many Latin American countries also enacted food laws, even though the legal systems of most Latin American nations are based upon those of Spain and Portugal, significant differences have developed in their food laws. Efforts are now being made to harmonize these differences.

In Africa, food laws were of little significance until the second half of the twentieth century Lasztity et al (2004). Independent states that started to emerge in the late 1950s were influenced in many matters, including food control services, by the European countries with which they had been closely associated (Petró-Turza and Földesi 2004). For example, French territories had developed French food enactments, while British territories followed British procedures. Food legislation was often inherited in total by the newly independent states. Although it is obvious that, there is the need for major adjustments today as situations now are quite different from those for which the enactments had originally been designed, several factors ranging from corruption to lack of visionary leadership has contributed largely to the inability of many African countries to make reasonable headway. This is beyond the scope of this book. Worthy of mention however, is the great lack of skilled or trained personnel in developing countries to draw up food laws suited to the nation's particular circumstances, or the scientific or technical staff necessary for food analysis, sampling, and efficient inspection (Lasztity et al, 2004). There is also a shortage of materials and equipment, and many other problems connected with the inauguration and operation of an effective food control system.

5. The food supply chain

The food supply chain is a net chain that food moves along the peasant household, the processing industry, the distribution centre, the wholesaler, the retailer as well as consumer

(Thirupathi 2006). In general, the food supply chain is composed of five links: product material supplying link, the production processing link, the packing storage and transporting link, the sale link and the consumer expending link with each link involving related sub-links and the different organization carriers.

The supply chain is a networked structure, composed of flow of the physical, the information, the finance, the technical, the standardized, the security and the value-added connections (Grunert 2005; Thirupathi 2006).

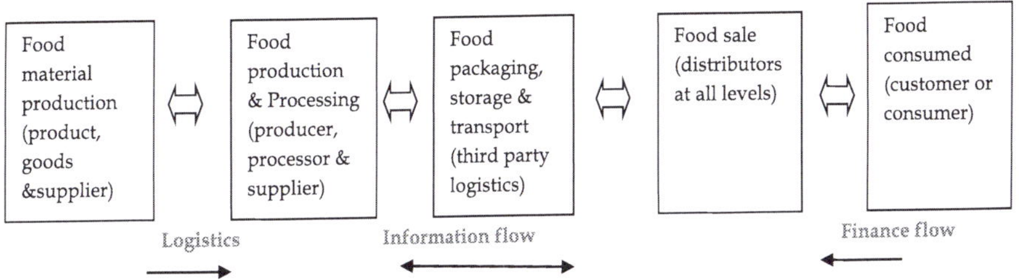

Fig. 1. Food supply chain basic structure (Thirupathi 2006)

5.1 Quality procedures

Along the supply chain, food is unavoidably exposed to numerous hazards. Therefore, knowing the risk factors at each phase of the supply chain assists in ensuring that an effective and comprehensive quality system is put in place. To guarantee food quality, it is important to ensure that, all the steps of the supply chain are carried out strictly, with care, and according to the standard operating procedures. At each phase of the supply chain, the potential risks, responsibilities, and how they can best be addressed must be fully explored.

6. Food contamination

Contamination of food can occur at any of the phases of the food supply chain and these will be expounded under the following broad categories:

1. Physical
2. Chemical
3. Microbiological
4. Other contaminants

6.1 Physical contamination

One of the major physical contaminations is adulteration. It is the mixing of inferior quality material with the superior product, thereby reducing the nature, quality and originality in taste, color, odor and nutritional value and ultimately causing ill effects on the health of the consumer (Thirupathi 2006). The main motive of adulteration and of course the "adultratee" is to gain undue advantage and most often profits. Almost all the food stuffs being sold in the market are prone to adulteration, but main food products that are often heavily adulterated are spices, milk products, edible oil, beverages drinks, sweets, pulses, sugar, processed foods, rice and cereal products like flour. The table below shows some common foodstuff and their adulterants (Lucey 2006).

Types of Foodstuff	Main adulterants
Milk and Milk products	Water, refined oil, separated (cream less) skim milk, solution, flours, animal fats, starch, dextrin.
Butter and Ice cream	Hydrogenated fats and animal fats, Artificial sweeteners,jelling agents, other fats, water nut flour, non permitted colours
Vegetable Oils and Fats	Coffee Powder Exhausted Coffee powder, starch, roasted dates, tamarind seeds, tea residues, other leaves with added color
Dry Beverages	Coffee Powder Exhausted Coffee powder, starch, roasted dates, tamarind seeds, tea residues, other leaves with added color
Wet Beverages	Spurious narcotics and other concentrated liquids, Artificial sweeteners (Saccharin), mineral acids, Colored invert sugar, high concentrated sugar solution, Cheap liquids, wine and beer water and juices.
Spices and Condiments	Whole turmeric Coating with lead chromate or coal tar dye starch or talc colored yellow with coal tar dye, Starch colored brown with coal tar dye, Coriander seed Other green colored seeds Powdered bran or sawdust colored green with dye, Starch colored red with coal tar dye, Argemone seeds, Artificial cumin seed like product, Dried papaya seeds and other plant gums.
Cereals and their products.	Wheat Stones, straws, low variety grains,Talc, chalk powder, tapioca flour, Maize flour, low variety rice and semolina.
Pulses and their products.	Yellow maize flour and Tapioca flour.
Miscellaneous Items	Seeds or nuts broken and colored, Pickle, jam, chutney and squash of cheap products. Also Sweets Artificial sweeteners and tapioca flour.

Table 1. List of common foodstuffs and their adulterants

6.2 Chemical contamination

Chemicals, which elicit harmful effects when consumed by animals or humans, are said to be toxic. The use of chemicals in the production and processing of food and food products not only affects the quality, but also disguises the deterioration and constitutes deliberate adulteration which is potentially very harmful to the health (Wilm 2003). It is advised that food additives like colouring matter, preservatives, artificial sweetening agents, antioxidants, emulsifiers/stabilizer, flavors/flavouring enhancers etc., if used should be of approved quality and processed under good manufacturing practices.

Types of Foodstuff	Main adulterants (Chemical formula and type)
Fish ,Meat & Dairy Products, Eggs and Poultry.	PCBs (polychlorinated biphenyls)
Citrus and other fruits	Juice Production and flavourings (multiple organic acids)
Oil and fat solvents	Dichloropropane ($CH_3CHClCH_3Cl$).
Alcohol Beverages	Ethanol (CH_2CH_2OH).
In Asian-style sauces such as soy, oyster, mushroom sauces, etc.	3-MCPD (3-monochloropropane-1,2-diol)
Potato tubers.	Glycoalkaloids (GA)
Unclean soft wheat for use. in non-staple foods and baby foods	Deoxynivalenol (Vomitoxin)
Fruit brandies and liqueurs, wines, distilled spirits.	Ethyl carbamate

Table 2. List of chemical adulterants in food

6.3 Microbiological contamination

Microbiological contamination of food is perhaps the most prevalent health problem in the contemporary world (Wilm 2003; Thirupathi 2006). To ensure good quality and safe food therefore, microbiological criteria should be established and freedom from pathogenic microorganisms must be ensured, including the raw materials, ingredients and finished products at any stage of production/processing. Accordingly the microbiological examination of the foods products has to be adopted widely. The microbiological criteria must be applied to define the distinction between acceptable and unacceptable foods (Thirupathi 2006). Food poisoning often results from the consumption of old, used, residual, fermented or spoiled food, as these may be contaminated with bacteria or other microorganisms, hence toxic. Infants and children are more susceptible to food poisoning and care should always be taken when giving them food. Gastroenteritis is caused by food contaminated with the *enterococcus, streptococcus faecalis*, which is frequently found in the human intestinal tract (Thirupathi 2006). Food poisoning may also be caused by inadequately refrigerated food contaminated with microorganisms such as *Clostridium perfringens* which grows well in the alimentary canal producing the poisoning within 8-12 hours after the ingestion of contaminated food. (Wilm 2003). *Bacillus cereus*, a gram positive, aerobic, spore-forming organism has been reported to be the etiologic agent in numerous food poisoning outbreaks. Some of the fatal effects of microbial contamination of food include; liver cancer is which is high in some countries due to aflatoxin. Flavism (hemolytic anemia) is caused by eating broad beans or by inhaling the pollen of its flower. In severe cases death may occur within 24-48 hours of the onset of the attack. Lathyrism is a disease, which paralyses the lower limbs (Thirupathi 2006).

6.4 Other contaminants

Metals are one of the many unintentional contaminants of food. When present beyond trace amounts, they are toxic. They find their way into food through air, water, soil, industrial pollution and other routes including food utensils. Common examples include, enamelware of poor quality which contributes antimony and galvanized utensils leaching zinc (Thirupathi 2006). A major source of tin contamination is tin plate, which is used for making containers for all types of processed foods. It has been shown that, a small quantity of metal is added when food is cooked in aluminum utensils (Wilm 2003). Although, copper is an essential trace element required by the human body, but copper-contaminated food is toxic. Fumigants used to sterilize food under conditions in which steam heating is impractical contaminate food. This is because, they may react with food constituents to produce or destroy essential nutrients. For example, ethylene oxide, a commonly used fumigant, reacts with inorganic chloride to form ethylene chloro hydride, which is toxic (Wilm 2003; Thirupathi 2006). Some solvents like trichloro ethylene used for the extraction of oil from oil seeds react with the foodstuff being processed resulting in the formation of toxic products. During processing of food, lipids may undergo numerous changes on prolonged heating, oxidative and polymerization reactions could take place, thereby decreasing the value of the processed products.

Smoking of meat and fish for preservation and flavouring is an old practice. But this processing contaminates the food with polycyclic aromatic hydrocarbons such as benzopyrene, many of which are carcinogenic.

Lubricants, packing materials etc. also contaminate foods.

A number of chemicals are intentionally added to foods to improve their nutritional value, maintain freshness, impact desirable properties or aid in processing. They also contaminate food if excessive in quantity (Thirupathi et al, 2006).

7. Methods of food quality control

In addition to ensuring safe and health food for the consumer, product manufacturers and service industries have realized that competition in a global market require a continual and committed effort towards the improvement of product and service quality. Therefore, they follow the process improvement cycle comprising **PLAN** (plan improvement), **DO** (implement plan for improvement), **CHECK** (analyze collected data) and **ACT** (take action). Quality control process consists of raw materials, in-process, product and service. The major factors in process that cause variability in quality of finished product are people, equipment and methods or technologies employed in the process. Use of proper statistical process control methods is also vital for assurance of the product quality. Usually, the value of quality characteristics is used to provide feedback on how processes may be improved. Statistical quality control comprises the following procedure:

a. Finished product is measured
b. Sampling occurs for days or weeks
c. Lot is either accepted or rejected based on information from sample

Contrary to statistical quality control, statistical process control methods focus on identifying factors in process that cause variability in finished product, eliminating the effect of these factors before worse product is manufactured, and control charts give on-line feedback of information about process. (Raju 2002). Food quality control measures have continuously improved since the 20th century, owing largely to the implementation of good practices, quality systems and increased traceability in food production. Ever since microorganisms were discovered in our environment and linked to typhoid fever and other diseases that have plagued humanity, public health authorities have been concerned with the accumulation of filth and foul odours in urban areas (Raju 2002). The first early inspection systems based on sensory evaluations were legally enforced at the beginning of the 20th century. Initial bacteriological techniques to detect pathogenic bacteria in foods, such as shellfish, appeared soon after (Raju 2002). From that point on, the food and beverage industry has applied stricter product inspection procedures and more and more effective production methods to conserve the freshness of natural raw materials. Today, the establishment of good manufacturing practices (GMP) and good hygienic practices (GHP) in many countries has significantly reduced the risk of spoilage and pathogenic microorganisms in modern food products. In addition to complying with national and international food regulations, food manufacturers are required to follow international quality standards, such as ISO as well as the Hazard Analysis Critical Control Point (HACCP) system.

In recent years, there has been an increasing focus on traceability in food production (Raju 2002). This has followed public concerns arising from cases of food contaminations and the development of foods containing ingredients derived from genetically modified (GM) crops. In the light of the increasing need for food more rapid food testing, it became clear that, the traditional microbiological detection and identification methods for food borne pathogens was no longer effective, because, it was time consuming and laborious to perform, and are increasingly unable to meet the demands for rapid quality control. A rapid method is generally characterized as a test giving quicker results than the standard accepted method of isolation and biochemical and/or serological identification (Raju 2002). Some of the newer and more rapid methods of food quality control are:

7.1 Ion mobility spectrometry (IMS) or differential mobility spectrometry (DMS)

These methods are used in the identification and quantification of analytes with high sensitivity. The selectivity can even be increased - as necessary for the analyses of complex mixtures - using pre-separation techniques such as gas chromatography or Multi-Capillary Columns (MCC). The method is suitable for application in the field of food quality and safety -including storage, process and quality control as well as the characterization of food stuffs (Vautz 2006).

7.2 Electronic-nose

This is an instrument which comprises an array of electronic chemical sensors with partial specificity or broad-band chemical selectivity and an appropriate pattern recognition system, capable of recognizing simple or complex odours (Gardner and Bartlett's 1994). This method is used to detect the bacterial growth on foods such as meat and fresh vegetables. It can also be used to test the freshness of fish. It is used in the process control of cheese, sausage, beer, and bread manufacture as well as for detection of off-flavors in milk and dairy products. This technique has other applications too. The advantages of the electronic nose can be attributed to its rapidity, objectivity, versatility, non requirement for the sample to be pretreated and ease of use Natale et al (1997).

7.3 Immunochemical methods

This method is based on antigen – antibody interaction. The antibodies are highly specific for the antigen (analyte), and secondly, the antigen, the antibody, or an antiglobulin may be conjugated to an enzyme that produces an intensely colored or fluorescent product in the presence of the enzyme substrate to enhance the detectability of the analyte in an amplification step. Toxins produced by *E. coli, Clostridium, Salmonella and Shigella* have also been similarly detected (Mason et al, 1961).

7.4 Enzyme Immunoassay (EIA)

Microorganisms are often characterized and identified by the presence of unique protein carbohydrate markers also called antigens, located within the body or the flagella of the cell (Greiner and Konietzny, 2008). Detection of these unique antigens has been a cornerstone of diagnostic microbiology for many years. In recent years, EIA using monoclonal antibodies have made available rapid and consistent microbiological detection systems. The most widely used systems employ a sandwich technique using antibody attached to a polystyrene matrix to which the sample is added. Post incubation, a second antibody, which is specific for the organism and has been tagged with an enzyme, is added. The addition of enzyme substrate to the mixture completes the EIA. The presence of the specific organism results in a colorimetric change in the enzyme substrate, which may be observed visually or with a spectrophotometer. Most EIA are very specific but lack sensitivity. Normal sensitivity has been reported to be in the range of 106 org/ml (Greiner and Konietzny, 2008).

7.5 Biosensors

Biosensor is usually a device or instrument comprising a biological sensing element coupled to a transducer for signal processing (Songa et al, 2009). Biological sensing elements include enzymes, organelles, antibodies, whole cells, DNA, and tissues. There are different types, conductance bioluminescence enzyme sensors utilizing potentiometric, amperometric, electrochemical, optoelectric, calorimetric, or piezoelectric principles. Basically, all enzyme sensors work by immobilization of the enzyme system onto a transducer (Songa et al, 2009). This technique provides sensitive, miniaturized systems that can be used to detect unwanted microbial activity or the presence of a biologically active compound, such as glucose or a pesticide in food. Immunodiagnostics and enzyme biosensors are two of the leading technologies that have had the greatest impact on the food industry (Greiner and Konietzny, 2008).

7.6 Flow cytometry (FCM)

Specific detection of pathogenic strains can be achieved by Flow cytometry using immunofluorescence techniques, which allow microorganism detection at the single-cell level. Although this technology can be used for food samples, it requires prior isolation of the target organism to generate antibodies (Comas-Riu, 2009). FCM finds wide application in milk and brewing quality control. The advantage of FCM is that it can also differentiate Viable Non-Culturable (VBNC) form of bacteria from healthy cultivable cells (Comas-Riu, 2009). This technology has the ability to detect microorganisms at relatively low concentrations in a short time, while multiple labeling allows the detection of different organisms or different stages in the same sample (Comas-Riu, 2009).

7.7 Polymerase chain reaction (PCR)

This technique involves the following steps: Isolation of DNA from the food, amplification of the target sequences, separation of the amplification products by agarose gel electrophoresis, estimation of their fragment size by comparison with a DNA molecular mass marker after staining with ethidium bromide and finally, a verification of the PCR results by specific cleavage of the amplification products and by restriction endonuclease or southern blot. Alternatively amplification products may be verified by direct sequencing or a second PCR (Greiner and Konietzny, 2008).

7.8 Pulsed-field gel electrophoresis (PFGE)

PFGE is a restriction-based typing method that is considered by many to be the "gold standard" molecular typing method for bacteria (Greiner and Konietzny, 2008). In this electrophoretic approach, DNA fragments are separated under conditions where there is incremental switch of the polarity of the electric field in the running apparatus. This technique allows for the resolution of DNA fragments up to 800 kb in size. When DNA is restricted with a restriction enzyme, PFGE provides a DNA "fingerprint" that reflects the DNA sequence of the entire bacterial genome. PFGE is a widely accepted method for comparing the genetic identity of bacteria (Greiner and Konietzny, 2008). PFGE typing has demonstrated a high level of reproducibility for food borne pathogens. A major advantage of this method is its universal nature making it useful in bacteria sub typing, however, its limitation is that it is time consuming.

7.9 Magnetic separation

One of the typical applications of this technique was reported by Mattingly (1984) & Safarik (1995). The author separated salmonella from food and faecal matter using myeloma protein and hybrid antibody (for O antigen), conjugated to a polycarbonate- coated metal bead. It has also been reported (Haik et al, 2008) that, Food sample like milk, yogurt, meat and vegetables can be tested. The challenge is in detecting E. coli is in the isolation of pathogenic strain from nonpathogenic strains. Immunomagnetic detection of listeria monosytogens has also been reported by Skjerve et al. (1990). While Johne et al (1989) using magnetic beads coated with polyclonal antibodies was able to detect and isolate the specific protein of s. aureus.

8. Near-infrared (NIR) spectroscopy

NIR has proven to be an effective analytical tool in the area of food quality control. The key advantages of NIR spectroscopy are (1) its relatively high speed of analysis, (2) the lack of a need to carry out complex sample preparation or processing, (3) low cost, and (4) suitability for on-line process monitoring and quality Control. The disadvantage of this method includes the requirement for large sample sets for subsequent multivariate analysis. (Michelini 2008; Cozzolino 2008). Recently, researchers at Zhejiang University in Hangzhou, China used Vis/NIR spectroscopy together with multivariate analyses to classify non-transgenic and transgenic tomato leaves (Xie et al, 2007).

8.1 X-ray

This is a relatively newer technology in food quality control. X -rays started making in-roads into the food industry in the early 1990s. The driving force behind this was the increasing number of foreign bodies which could not be identified by metal detectors. Other than contaminants like glass, bone, rubber, stone or plastic, some specific applications are also more challenging for metal detectors, such as fresh meat and poultry, or foil-wrapped products (Ansell 2008). X-ray inspection has considerable advantages in many food and beverage processing environments in that, it is easy to install, safe and simple to use, even without previous experience. It quickly and consistently identifies substandard products, reducing product recall, customer returns and complaints, therefore protecting manufacturers' brands and most importantly, preventing ill health.

8.2 Computer vision

Aguilera et al (2006) reported a computer system that consisted of four basic components: the illumination source, an image acquisition device, the processing hardware, and suitable software modules. Their study was focused on analyzing the relevance of computer vision techniques for the food industry, mainly in Latin America. The authors described how the use of these techniques in the food industry eliminates the subjectivity of human visual inspection, adding accuracy and consistency to the investigation. They also reported that, the technique can provide fast identification and measurement of selected objects, classification into categories, and color analysis of food surfaces with high flexibility. They opined that, since the method was non-contact and non-destructive, temporal changes in properties such as color and image texture can also be monitored and quantified.

9. Some recent reports on food quality control

9.1 Livestock

Nardone (2002) found that the meat and milk characteristics are more related with human health and with some factors affecting their quality. They used various techniques to control meat characteristics. The Molecular biology techniques was of great interest to the author as it gave insight to new product certification viz, species, breed, animal category (age, sex, etc). He also reported that milk quality may be influenced by some innovative technology among which is the Automatic Milking Systems (AMS). AMS increases milk yield and milking frequency from twice to three times or more per day requiring a minimum extra amount of labour, however, contradictory results are reported about the effects of AMS on milk quality. Several authors found that after the introduction of AMS milk quality decreased, particularly fat, proteins percentage while total bacterial plate count, SCC, freezing point and the amount of free fatty acids increased significantly. Nardone (2002). Aguilera et al, (2006) reported the usefulness of vision Q-Lab to assess the quality of food Samples. The system consists of a highly general hardware setting, able to support different applications, and highly modular software, easily adapted to the measurement needs of diverse food products. The main results of this application, was to classify rice grains and lentils (Aguilera et al, 2006). Grain quality attributes are very important for all users and especially the milling and baking industries. An earlier report by Zayas et al (1996) showed the usefulness of machine vision to identify different varieties of wheat and to discriminate wheat from non-wheat components. In his own report, Katsumata (2007) showed that, visible light Photoluminescence (PL) peaking at around λ = 460 nm is characteristic of cereals, such as rice, wheat, barley, millet, flour, corn starch, peanut, under illumination of ultra-violet light at λ = 365 nm. They further reported that Peak intensity of PL and Distribution of PL intensity varies with variety and source of the specimens which was found to be fitted with a Gaussian curve. Visible light PL is suggested to be potentially useful technique for the non-destructive and quick evaluation of the cereals and other starchy products. Songa (2007) reported the use of amperometric nanobiosensor for determination of glyphosate and glufosinate residues in corn and soya bean samples. The author found that biosensor has the features of high sensitivity, fast response time (10 to 20 s) and long term stability at 4^0C (>1month). Detection limits was in the order of 10^{-10} to 10^{-11} M for standard solutions of herbicides and the spiked samples. The author found that herbicide analyses can be spiked on real samples of corn and soy beans, corroborating that the biosensor is sensitive enough to detect herbicides in these matrices.

9.2 Fruits, vegetables and nuts

Narendra and Hareesh (2010) observed that Computer vision has been widely used for the quality inspection and grading of fruits and vegetables. It offers the potential to automate manual grading practices and thus to standardize techniques and eliminate tedious inspection tasks. The capabilities of digital image analysis technology to generate precise descriptive data on pictorial information have contributed to its more widespread and increased use. Aranceta-Garza et al (2011) reported a PCR-SSCP method for the genetic differentiation of canned abalone and commercial gastropods in the Mexican retail market. Their study was aimed at creating molecular tools that can differentiate abalone (Haliotis spp), from other commercial fresh, frozen and canned gastropods based on18S rDNA and also identify specific abalone product at the species level using the lysine gene. The authors found that the methods were reliable and useful for rapid identification of Mexican abalone products and could distinguish abalone at the species level. The methods could genetically identify raw, frozen and canned products and the approach could be used to certify authenticity of Mexican commercial products or identify commercial fraud. Lehotay (2011) reported the qualitative analysis of pesticide residues in fruits and vegetables using fast, low-pressure gas chromatography – time of flight mass spectrometry (LP-GC/MS). The author demonstrated that, to increase the speed of analysis for GC-amenable residues in various foods and provide more advantages over the 40 traditional GC-MS approach, LP-GC/MS on a time-of-flight (ToF) instrument, which provides high sample throughput with <10 min analysis time should be applied. The method had already been validated to be acceptable quantitatively for nearly 150 pesticides, and in this study of qualitative performance, 90 samples in total of strawberry, tomato, potato, orange, and lettuce extracts were analyzed. The extracts were randomly spiked with different pesticides at different levels, both unknown to the analyst, in the different matrices. They compared automated software evaluation with human assessments in terms of false positive and negative results only to found that the result was not significantly different. Mustorp (2011) reported a robust ten-plex quantitative and sensitive ligation-dependent probe amplification method, the allergen- Multiplex, quantitative ligation-dependent probe amplification (MLPA) method, for specific detection of eight allergens: sesame, soy, hazelnut, peanut, lupine, gluten, mustard and celery. Ligated probes were amplified by PCR and amplicons were detected using capillary electrophoresis. Quantitative results were obtained by comparing signals with an internal positive control. The limit of detection varied from approx. 5 to 400 gene copies depending on the allergen. The method was tested using different foods spiked with mustard, celery, soy or lupine flour in the 1-0.001 % range. Depending on the allergen, sensitivities were similar or better than those obtained with PCR.

10. Conclusions

Without doubts, there is a need for continuous improvements in rapid diagnostic methods, analytical techniques as well as visionary and computational equipments required for food quality control of the future. For example, as it stands today, only about 41.5% of microbiology tests utilized rapid methods. The growth of diagnostic industry should result in increased rapid tests in the nearest future and this should result in improved performance. It is expected that, there will be significant economic benefits and the ability to practice proactive and risk prevention food safety programs. In fact various schools of thoughts have it that, by 2015, the companies should be able to utilize automation technology to screen incoming raw materials and in-process parameters with near real-time information; physical, chemical or biological, while utilizing these newer methods. With automation, time is saved and productivity increases. In this regard, the two main traditional and rapid methods: flow cytometry to provide the total microbial count rapidly and polymerase chain reaction (PCR) to detect microorganisms both quickly and specifically will continue to grow in the near future. Although some of these automated newer technologies are extremely rapid, there have been questions about their sensitivity. Some investigators agree that, the instruments can produce results in seconds, but they opine strongly that they are not sensitive enough. As instrument companies today are working to provide the best combination of speed and sensitivity, the challenge of the future still lies with their ability to produce one instrument combining accuracy, rapidity and sensitivity while ensuring that, the entire control process still ensures a minimal risk of contamination. In the nearest future, the mass spectrometry may be needed in food quality control. It is already proofing useful identifying unknown compounds, antibiotics and pesticides in raw materials and in detecting trace metals in foods. DNA-based assays, instruments, and software, all designed to work together including the emerging area of bioinformatics; sequencing, assay design, and chemical analysis are all capable of developing new ways for food quality control in the future. This chapter looked at the history, current and future of food quality control and also revealed a large number of approaches to enhancing food quality. It is therefore safe to conclude that, food quality control is an indispensable tool in the food industry. As enumerated above, the development of adequate, effective, rapid, and sensitive food quality control systems however, faces serious challenges driven by its capital intensive nature and sophisticated adulteration. While it may seem easier for the developed nations to match quality control with adulteration techniques, to make any meaningful progress in resource limited nations of the world, there is the need for collaborations between laboratories around the globe, just as it is necessary for regulatory agencies around the world to also collaborate both in sharing information and in technologies as well as capacity development. The existing legislations seem adequate, but implementation may be weak, there is need for international coordinated efforts to enforce the laws.

11. References

Adamson Melitta Weiss (2004): *Food in medieval times*, pp 64-67; Greenwood Publishing Group, 88 Post Road West, Westport, CT 06881.

Adu-Amankwa Pearl, (1999) Quality and Process Control in the Food Industry Food Research Institute, Published in The Ghana Engineer, http://practicalaction.org/practicalanswers/

Aguilera, Jose Miguel and David W. Stanley,(1999): *Microstructural Principles of Food Processing and Engineering*, Second Edition Springer, ISBN=0834212560.

Alexandrakis D, Downey G, Scannell AG. (2008),Detection and identification of bacteria in an isolated system with near-infrared spectroscopy and multivariate analysis. *J Agric Food Chem*; 56: 3431-7.

Alsberg. CL. (1970),*Progress in Federal Food control* In: Ravenel. MP.ed. A. Half century of Health. New york Times Pp. 211-220.

Ansell Tim: (2008) , X-ray a new force in food quality control, Al Hilal Publishing & Marketing Group.

Augustin Scalbert, Cristina Andres-Lacueva, Masanori Arita, Paul Kroon, Claudine Manach, Mireia Urpi-Sarda, and David Wishart: (2011): Databases on food phytochemicals and their health promoting effects *J. Agric. Food Chem., Just Accepted Manuscript Publication.*

Chaplain CV, (1970).History of state & municipal control of diseases In: Ravenel. MP, ed. A. Half century of Health, New york Arno Press and the New york Times: Pp 133-160

Codex Alimentarius Commission (1997), Report of the Twelfth Session of the Codex Committee on General Principles, ALINORM 97/33.

Comas-Riu Jaume, Núria Rius: (2009), Flow cytometry applications in the food industry; *J Ind Microbiol Biotechnol* 36: 999–1011.

Cozzolino D, Fassio A, Restaino E, Fernandez E, La Manna A. (2008). Verification of silage type using near infrared spectroscopy combined with multivariate analysis. *J Agric Food Chem.*; 56: 79-83

FAO. (1998)*FAO Technical Assistance Programme: Food Quality and Safety. (ESN internal publication)*,Rome.

FAO. (1999). *FAO* trade-related technical assistance and information. Rome. Food and Agriculture Organization of the United Nations. (2005), The State of Food Insecurity in the World.

Gravani, Robert B: (1986), *How to Prepare A Quality Assurance Plan, Food Warehousing.* Department of Health and Human Services, Public Health Service, U.S. Food and Drug Administration, Food Science Facts for the Sanitarian, Dairy and Food Sanitation.

Greiner Ralf and Ursula Konietzny http://www.worldfoodscience.org/cms/?pid=1003869 Modern Molecular Methods (PCR) in Food Control: GMO, Pathogens, Species Identification, Allergens.

Haik Yousef et al (2008); *MagneticTechniques for Rapid Detectionof Pathogens:* Reyad M. Zourob *et al.* (eds.), Principles of Bacterial Detection: Biosensors, Recognition Receptors and Microsystems,. İ *Springer Science+Business Media, LLC 2008.*

Halász,Anna Radomir Lásztity, Tibor Abonyi, and Arpad Bata, (2009): *Decontamination of Mycotoxin-Containing Food* and Feed by Biodegradation Food Reviews International, volume 25, pp:284–298.

Handbook of organic food safety and quality, Edited by J Cooper, C Leifert, Newcastle University, UK and U Niggli, Research Institute of Organic Agriculture (FiBL), Switzerland, Woodhead Publishing Series in Food Science, Technology and Nutrition No. 148 n.d.

http://quality.com/details/print/807365/December/January2009.html

http://www.foodquality.com/mag/06012006_07012006/fq_06012006_SS1.htm

Hussain, M. A. (2010): Future insights: *Proteomics a power technique to ensure food quality and safety;* Proceedings China International Food Safety and Quality Conference, 10-11 Shanghai, PR China. p. 51.

Hutt, Peter Barton and Merrill Richard(1991): *public health and the healthy public:*a communitarian perspective on privatization and the FDA; *Food and drug law,* pp 6-14.

"Industrial Revolution,"(2004), Columbia Encyclopedia, Sixth Edition, Copyright (c).

Jane Byrne: (2008): *RFID temperature logger could enhance cold chain quality control,* News on Food and Beverage Processing and Packaging.

Jango-Cohen, Judith; (2005):*The History of Food;* Twenty-First Century Books, ISBN 0-8225-2484-8, Minneapolis, Minn.

Janni J, Weinstock BA, Hagen L, Wright S. (2008): Novel near-infrared sampling apparatus for single kernel analysis of oil content in maize. *Appl Spectrosc* 62: 423-6.

Johne et al.: (1989);Imunomagnetic separation of s. aureus , *J. Clin Microbiol,* 27(7): pp; 1631-1635.

Katsumata T. , T. Suzuki, H. Aizawa and E. Matashige (2007): Photoluminescence evaluation of cereals for a quality control application , *Journal of Food Engineering, Volume 78, Issue 2,* Pages 588-590.

Klaus G. Grunert (2005): Food quality and safety: consumer perception and demand, European Review of Agricultural Economics Vol. 32 (3) pp. 369–391.

Lehotay Steven J., Urairat Koesukwiwat, Henk van 5 der Kamp, Hans G.J. Mol, and Natchanun Leepipatpiboo (2011): Qualitative Aspects in the Analysis of Pesticide Residues in Fruits and Vegetables using Fast, Low-Pressure Gas Chromatography – Time-of-Flight Mass Spectrometry Low-pressure gas chromatography –mass spectrometry (LP-GC/MS *Journal of Agricultural and Food Chemistry,* just accepted manuscript.

Lucey John: (2006) "Management Should Serve as Role Models for Good Work Habits and Acceptable Hygienic Practices "Food Quality

Mattingly, J.A1. (984).: An enzyme immunoassay for the detection of all Salmonella using a ombination of a myeloma protein and a hybridoma antibody. *Journal of Immunological Methods* 73, 147-156.

Michelini E, Simoni P, Cevenini L, Mezzanotte L, Roda A. (2008): New trends in bioanalytical tools for the detection of genetically modified organisms: an update. *Anal Bioanal Chem* 392: 355-67.

Moros J, Llorca I, Cervera ML, Pastor A, Garrigues S, de la Guardia M. (2008): .Chemometric determination of arsenic and lead in untreated powdered red paprika by diffuse reflectance near-infrared spectroscopy. *Anal Chim Acta;* 613: 196-206

Nardone Alessandro (2002): Evolution of livestock production and quality of animal products, Proc. "39th Annual Meeting of the Brazilian Society of Animal Science" Brazil, 486-513.

Narendra V G and Hareesh K S (2010): Quality Inspection and Grading of Agricultural and Food Products by Computer Vision- A Review: *International Journal of Computer Applications* Volume 2 – No.1.

Natale Corrado Di et al (1997) ; Electronic-nose modelling and data analysis using a self-organizing map; *Meas. Sci. Technol.* 8 1236–1243.

Perrot. N.,. Ioannou, I. Allais, C. CurtJ. Hossenloppand G. Trystram (2006): Fuzzy Concepts Applied to Food Control Quality Control Fuzzy Sets and Systems Volume 157, Issue 9, pp, 1145-1154.

Potter, Norman N. and Joseph H. Hotchkiss, (1995): *Food Science.* 5th Edition. New York: Chapman & Hall. pp. 90-112.

Quality Assurance / Control in Food Processing. Contained in: Food Fortification - Technology and Quality Control. (FAO Food and Nutrition Paper – 60)

R.V. Sudershan, Pratima Rao and Kalpagam Polasa.(2009): Food safety research in India: a review *As. J. Food Ag-Ind.* 2(03), 412-433.

Radomir Lasztity, Marta Petro-Turza, Tamas Foldesi (2004): *History of food quality standards, in Food Quality and Standards,* [Ed. Radomir Lasztity], in *Encyclopedia of Life Support Systems (EOLSS),* Developed under the Auspices of the UNESCO, Eolss Publishers, Oxford ,UK.

Raju K. V. R.and Onishi Yoshihisa. (2002): *Report of the APO Seminar on Quality Control for Processed Food held in the Republic of China,* (02-AG-GE-SEM-02). This report was edited by. Raju K. V. R.
http//foodquality.wfp.org/qualityprocedure/tabid/119/.aspxdefault

Roe R.S. (1956): *The food & Drugs Act- past,present & future* in Welch H marti-Ibunez. F..eds. The impact of the food & Drug Admin. on our society, New york : MD Publications Pp. 15-17.

Rust and Olson (1987): Coping With Recalls, Meat and Poultry Vol. No. 3, March.

Safarik.M, Safarikova,&M.J.Forsethe. (1995): Application of Magnetic Separation in applied Microbiology; *Journal of Applied Microbiology,* 78, 575-585.

"Short Narratives: the Early Modern Agricultural Revolution," World History at KMLA, (2002) ;http://www.zum.de/whkmla/apeur/narratives/NarrativesAgrRev.html

Skjerve E, L M Rørvik and O Olsvik. (1990): Detection of Listeria monocytogenes in foods by immunomagnetic separation. *Appl Environ Microbiol,* 56(11): 3478-3481.

Songa Everlyne A., Vernon S. Somerset, Tesfaye Waryo, Priscilla G. L. Baker, and Emmanuel I. Iwuoha. (2009): Amperometric nanobiosensor for determination of glyphosate and glufosinate residues in corn and soya bean samples; *Pure Appl. Chem.,* Vol. 81, No. 1, pp. 123–139.

Stina Lund Mustorp, Signe Marie Drømtorp, and Askild Lorentz Holck: Multiplex. (2011): Quantitative ligation-dependent probe amplification for determination of allergens in food, *J. Agric. Food Chem.*,Just Accepted Manuscript Publication .

Swetman Tony and Barrie Axtell, (2008) (updated): *Quality Control in Food Processing*, Technical Information Online-practical answers.mht

Thirupathi. V., Viswanathan .R. & Devadas. CT; (2006): *Science Tech Entrepreneur.*

Thomas Weschler, J. Stan Bailey, Jerry Zweigenbaum and Dipankar Ghosh. (2009): food quality magazine, December/January issue.

Vautz W.; D. Zimmermann; M. Hartmann; J. I. Baumbach and J. Nolte; J. Jung. (2006): *Ion mobility spectrometry for food quality and safety: Food Additives & Contaminants*: Part A: Chemistry, Analysis, Control, Exposure & Risk Assessment, Volume 23, Issue 11, Pages 1064 – 1073.

Wilm Karl Heinz. (2003): Chemical Contaminants, Our Food; Food Safety and Control System, www.ourfood.com www.ourfood.com

XIAO Jing and MA Zhongsu: Research on Food Security Risk Early Warning under Supply Chain Environment, research sponsored by Heilongjiang Province science and technology department (Project Grant No.GA06C101-02) n.d.

Xie L, Ying Y, Ying T.(2007): Quantification of chlorophyll content and classification of nontransgenic and transgenic tomato leaves using visible/near-infrared diffuse reflectance spectroscopy. *J Agric Food Chem*; 55: 4645-50.

Zayas, I.Y., Martin, C.R., Steele, J.L., Katsevich, A.(1996): "Wheat classification using image analysis and crush force parameters",Transactions of the *ASAE* 39 (6), pp. 2199-2204.

22

Employment of the Quality Function Deployment (QFD) Method in the Development of Food Products

Caroline Liboreiro Paiva and Ana Luisa Daibert Pinto
University Federal of Minas Gerais
Brazil

1. Introduction

Currently, in a more intensive way, companies have been forced to adapt to the new competitive market. The technological industry changes' occurring since the 80's brought implications for the international competition, specially the demarcation of new areas of global competition. That happens due to the acceleration of technological changes added by the shortening of the life cycle of products and processes, besides the increasing of the products differentiation. In fact, what is observed is that these factors have not only led companies to restructure their production systems and their types of management, but above all, to guarantee the capacity to deliver to a market, products even more sophisticated. The ability to realize alternatives to compete in the market, to develop strategies and to invest in appropriate training that is what has ensured the survival and profitability of organizations in this new structure.

In terms of food industry competition, what is observed is that the integrity of the product has become the main focus. This means that product excellence, in the food industry, goes beyond simply offering goods with basic attributes. These attributes have only become a precondition for the company to keep playing the competitive game. In fact, nowadays, the products must not only satisfy, but above all, surprise their customers. What is realized is that these consumers have accumulated experience with several products and become sensitive to small differences in many ways. This means that innovations in products and processes increase the excellence standard of product, making the process of development an essential factor for the competition of enterprises.

Certainly, the best projects require staff competence, efficiency in the work, on the exchange of information between functions and on the understanding of the market needs into technical language. They also require efficiency in problem solving and in the use of its resources. In this sense the Quality Function Deployment (QFD) method has proven to be efficient in order to translate in a more effectively way the needs and expectations of consumers, to promote greater interaction between the teams involved in the project, to accelerate the solution of problems and to reduce the development time.

The QFD was invented in the late 1960's in Japan. Within the context of TQC (Total Quality Control), the Japanese model of quality management system was responsible to cause a revolution in the production system of that country. All that was possible, due to the emphasis on the product quality considering the point of view of the customer. QFD is the unfolding, step by step, of functions or operations that make up the product quality. The methodology seeks to solve the problems inherent to the product's development process in their early stages, in a way that the critical points that determine the quality of the product and the manufacturing process are established in the phase of their design and controlled during the development stages. The methodology also ensures the achievement of quality because it works with a focus on consumer needs. More specifically, it translates the consumer's requirements into technical language and then ensures their satisfaction along the process of product's development.

The quality matrix is the tool used to organize the consumer's needs into technical information. The matrix goal is to define the pattern, quantitative or qualitative, of each attribute of quality of the final product. The other matrices are due to the quality matrix and aim to detail the project so that all the factors that contribute to the achievement of the final product are designed, as characteristics of the intermediates products, parameters of the manufacturing process, raw materials and inputs.

In addition, the QFD method assists the management of product development process because it coordinates the flow of information and organizes activities in terms of functions. It promotes the functional integration and rapid resolution of problems.

With all that, the purpose of this chapter is to describe the potential use of the QFD method into product development in food companies. The study initially intends to contextualize the management of product development in the food industry and show the QFD method as a tool capable of directing, in a practice way, how to plan and conduct the activities of the process of product development. So the steps for the application of QFD in the development of a food product will be detailed. In addition, support tools within the marketing research and sensory analysis will be suggested.

2. Product differentiation: A strategy adopted by the food industry

The food industry never has launched so many new products as it has in recent years. Due to factors such as technological development, increasing of competitiveness in the sector due to the growth of the competition such in and out of the countries, and greater consumer demand which incorporated new values to its preferences, the shelves of supermarkets receive daily new products (Athayde, 1999).

The focus on markets niche is one of the strongest trends today in the food sector. There is a search for products that provide pleasure to be consumed, such as the sophisticated products with high added value, or looking for fun products aimed at children. Likewise, products that refer to a particular region of the world, or of exotic flavors are searched by another portion of the market of processed foods.

Allied to all that, a strong feature of the new releases is the convenience in food consumption. This requirement is related to changes on consumers' lifestyle. The growing participation of women in the labor market, added by the increasing mobility of consumers, reduced the demand for ingredients to prepare meals at home and increased the offer of practical foods that can be consumed at any time, and of ready to go foods or pre-prepared.

Industry has also been required to apply new technologies in the development of food and beverages, specially the search for new ingredients. The changes in consumption habits it is driven by the concern for the health, aesthetics and environment. It demands food products of low-calorie, healthier and natural and environmentally friendly. A strong trend is the launch of products, which besides the presence of the sensory and nutritional quality, also present health benefits, so-called functional foods.

It is also important to emphasize the growing importance of equipment suppliers. Companies specializing in process engineering, who believe the research as a basis for technological innovation, have an important role in the development of food products (Earle, 1997). New technologies are able to provide new concepts of product, new alternatives for use, and being difficult to be imitated by competitors.

Finally, in addition to the significant number of new products available on the market in recent years, it is worth noting the great contribution of the sector of packaging for the food market, making it possible that these strategies of differentiation, segmentation and consumer convenience can be realized.

3. Stages of product development process

The process of product development, outlined in a model, consists of a sequence of activities ordered in time or a set of tasks that aim to facilitate the management of the process as a whole.

There isn't a standard development model that fits all circumstances and conditions experienced in a company. However, if the company adapts your way of management to a model more suited to its environment, probably the company will get better performance in their innovation processes. Anyway, the consensus is that development must be conducted so that the product reaches the market as quickly as possible, providing to the product the quality expected by customers and having costs optimized.

Students of product development management have different ways of representing the necessary steps to this process. Picture 1 seeks to represent the basic steps; steps that will assist the planning, the development of the product itself and the release of the same. Of course the product will have a greater chance of market success with this process if there is efficient management. For Clark & Wheelwright (1993), this means that the company should have skills to quickly identify opportunities, which often leads them to introduce new products ahead of their competitors. The best projects require also the team's competence, work efficiency, the exchange of information between functions and translation of the markets needs in technical language. They also required efficiency in problem solving and in the use of resources.

In Picture 1, the process of product development is represented by *stage-gates*. The *stages* are the various stages of development and the *gates*, decision points that precede each stage, opening or closing the door to continue the project (Cooper, 2001). These *gates* serve as critical steps for assessing the projects. The results of these evaluations are reflected in the decision to continue the project, drop it, stop it or resume it on another occasion.

Before joining the project into the development phase, the organization must seek the means that will ensure that the product will reach customers needs. Several market research – research of needs and desires of the consumers, competitive analysis and concept testing - will help to define more precisely the concept of the product. The first step in this direction is to translate the information inside and outside of the company in technical language, until you define the product's concept. For this, extract the data through market research, group discussions, customer complaints and tacit knowledge of employees. The various functions involved in the process are then in charge of mapping information and developing the work.

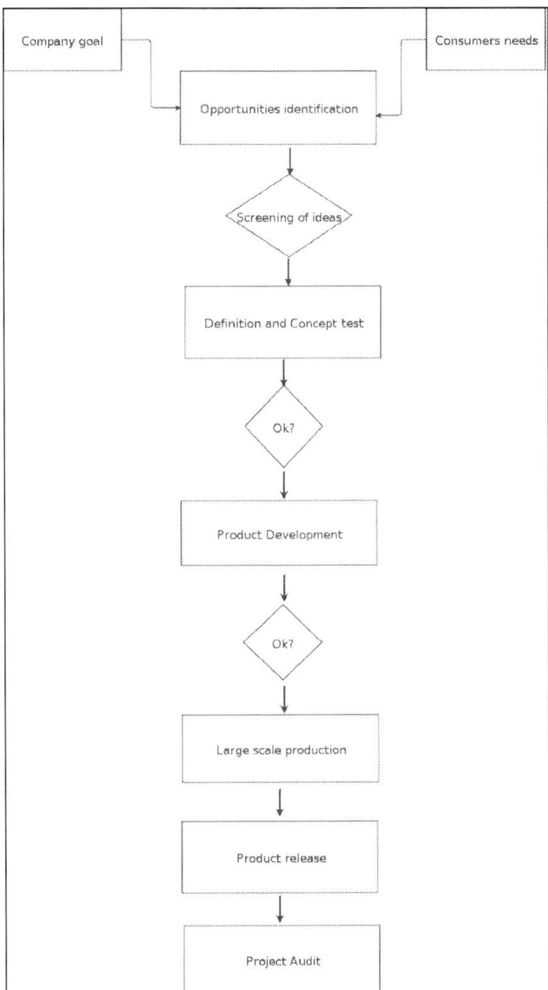

Fig. 1. Stages of the process of product development

In the stage of identifying opportunities, the company must seek ways to generate ideas for the new products. This can be achieved by internal efforts, through research in the departments of R&D, through contests to stimulate ideas for new products, or through the Customer Services, in meetings, using *brainstorming* techniques, or by stimulating a business culture that valorize the opinions and ideas of employees. On the other hand, the ideas for new products can come from external sources such as quantitative or qualitative research with target consumers. Other sources of ideas can come from university research publications or specialized organs, experience and knowledge of sales staff, contact with suppliers and also reverse lookup on products of competitors.

In this step it should be also made a prior assessment of the market for each idea, considering its size, segments and potential. It should also be evaluated the feasibility of manufacturing the product, the ability to be accepted by the market as well as their vulnerability towards competitors products or substitutes.

The ideas should then go through a team that will evaluate and select the ideas by checking out promising, profitable or those that must be rejected. Every idea that is nominated as possible to be developed will then go to the stage of definition and testing of its concept. The concept of a product can be defined as the expected benefits to meet the needs and expectations of consumers. The concept definition phase must determine the target audience, what are the main benefits that the product will present and a more appropriate occasion to consume it.

After the definition of the product's conception is convenient to test it. The test of the concept is a marketing research technique used to assess the market potential of the concept. Provides estimates of intent to purchase and sales volume. Define "who" would use the product, in which "circumstances" and how "often".

Finally, we must make the financial analysis of the project. The size of the market, the expected market share, the price analysis, along with technical cost estimates of equipment and for product launch are the inputs needed to make such an analysis. Once the project is defined, the only thing needed is that top management approves it, so the development of the product can be started.

Only then the product will go into the product development stage itself. However, it is necessary to first make the process and product planning. Regarding the product is necessary to define the product requirements, such as: the ingredients that will be needed, the most suitable additives, the quantity/volume that will be marketed. It is still important to define the requirements of legislation, such as: what will be the standard of identity and quality of the product, if the planned additives are allowed by the competent organizations and what the limit of application, and also the labeling requirements.

In relation to the manufacture of the product, it is necessary to first specify the parameters of the process, which involves the study of manufactured technology and the parameters of quality and of process that need to be controlled in the manufacturing line.

The development process then proceeds to the phase of preparation of the prototypes, usually in an industrial kitchen for the definition of the formulation and of the sensory products parameters. Soon after it should be performed sensory tests in one or more prototypes, if possible with a sample of the target market, in order to verify the acceptance of the product.

Thus, the development of the product passes to the manufacturing phase of the pilot which consists in the manufacture of the product on a small industry scale, in order to define the quality parameters of intermediate products and process parameters that will be monitorized. Likewise, tests should be made of pilot products: sensory tests, again if possible with a sample of the target market, in order to verify if the product remains viable. Only then the company will plan the production on an industrial scale.

Soon afterwards the company can produce on an industrial scale, to launch the new product. In the launch phase is necessary to determine: a release date, geographical location and potential consumers in the target market. It is necessary to establish an advertising plan, that would include promotional and dissemination strategies.

Finally, the last step is to evaluate the project developed. The company has an opportunity to implement its system of product development through learning gained during the implementation of each individual project. The project audit aims to verify the strong and weak points and to define strategies for improving the performance of future projects. It is believed that only a deep understanding of the causes of problems and circumstances in which they occur, will allow the company to improve the performance of development activities, by improving the procedures, processes, management skills, methods, making the company able to develop a faster process, more efficient in the use of resources and in the development of products of higher quality.

4. The QFD Method

The QFD Method, *Quality Function Deployment*, originated in Japan in the late 60's, as a result of the study of the professors Akao and Mizuno (Mizuno, 1969). On this occasion, the movement for the Total Quality in that country had already achieved very significant results. The ideas of quality emerged after World War II starting with the Statistical Process Control (SPC) and evolved in the late 60's, to a much broader approach, in which it was already understood at the system level, and not only in technical terms or isolated functions, but also in management terms, and thus practiced throughout the whole organization. To get an idea, in 1968, the Quality Control (QC) in Japan had already reached the point where virtually all firms made usage of the QC in some way (Mizuno, 1969).

However, there was a gap in establishing the quality into a level of development of products. There were questions about what points should be considered in the design phase of projects that could operationalize the quality planning of both products and processes. There were also difficulties in ensuring that the planned quality was actually executed in the phase of serial production (Mizuno, 1969).

Then arises from these needs, the initial concepts of *Quality Deployment*, and in 1972, after conducting some researches, the ideas become practically implemented. In 1978, it was published the book *"Quality Function Deployment"* which gave a new impetus to the dissemination of QFD, causing it to be quickly implemented in several companies in the country.

Currently, QFD inspires a strong interest in the world, generating ever-new applications, practitioners and researchers each year. This method is in use in several countries in the world such as South Africa, Germany, Australia, Brazil, China, Spain, United States, Italy, India, Japan, Mexico, United Kingdom, Sweden and others, not only in product development, but also in developing manufacturing processes, software, services, etc. (Akao & Mazur, 2003; Chan & Wu, 2002).

In the U.S., QFD has become known in 1983 after conducting a seminar on the subject in Chicago. It was initially introduced in the *3M Corporation*. Currently, the use of QFD in the U.S. is in almost all industry sectors, particularly in the automotive, electronics, software and services industry. It is also used by the space industry.

In 1996 a survey was conducted through a collaboration of Tamagawa University and the University of Michigan on the applications of QFD in the U.S. and Japan. It was selected 400 companies from each country. 146 Japanese companies (37%) and 147 American (37.6%) responded to the survey. According to the results, 31.5% of Japanese companies and 68.5% of Americans use the QFD (Akao & Mazur, 2003).

In Europe, QFD is also well known and many application cases have been reported. In other parts of the world, one can mention the innovative applications of QFD in Australia in the area of strategic planning and development of new business or improving existing business (Melo Filho & Cheng, 2007).

In emerging countries such as Brazil, QFD was introduced in 1989 and the concern now is how to make the method more effective, better understood and applied (Akao & Ohfuji, 1989). In China, the Quality Bureau from the State Bureau of Technical Supervision, a national agency of The People's Republic of China, has invited Professor Akao to give QFD seminars in Peking and Shanghai since 1994. India has shown a strong interest in the application of QFD, specially in software industry and in manufacturing industries such as trucks, automobiles, and farm tractors (Akao & Mazur, 2003).

The true in general is that the QFD method has ensured the achievement of project quality because it relays in one point that is the most cited by scholars of the subject as essential to the success of the product: a focus on customer needs. In addition, assists in managing the development process because it coordinates the flow of information and organizes activities in a function level. Thus promoting cross-functional integration and quick problem solving.

4.1 Method's approach

The QFD method, as it was originally designed by the professors Akao and Mizuno, includes the deployment of information, called the Quality Deployment (QD) and displays of work, addressed as Quality Function Deployment narrowly defined or restricted (QFDr).

In the first approach, QFD works detailing the necessary information to the innovation process. For that, are used tables, matrices, and the conceptual model, called the basic units of the QD. On the tables the data are organized, which in turn will be linked into the matrices. The interaction between the matrices is shown in the conceptual model (Akao, 1996).

The beginning of the process of extracting information in the QFD always starts from a table, so it is considered as the elementary unit of the method. It has the main purpose of deploying the information, always starting from a more general level to a more concrete. Using data from market research or internal information of the company, the work team uses the tables to detail the information, which are then arranged so that they are grouped according to their level of abstraction. Thus, the characteristics, requirements or functions that aren't so explicit, they become more clear for the working group.

The use of matrices in QFD aims to translate succinctly the relationship between two tables. It is a way of storing information and at the same time, to visualize the degree of interaction between each element of a table in relation to all the other elements of the other.

The conceptual model is the structure within the QFD that allows the visualization of the path taken to deploy the information until they get the technical standard processes. According to the sequence of matrices, it is able to verify a relation of cause and effect between the characteristics of the final product, its components, their functions, costs, raw materials and intermediate processes for their manufacture. Thus, it has been stored in a visible and detailed way, all product design and process.

The second approach of the method refers to the deployment of the work (QFDr). The technical and management procedures are established to ensure that all functions involved in the activities have their tasks previously established. The QFDr aims to specify who will do the job and how it will be done. Thus, from this deployment of the work it can be generated a set of documents, such flowchart of product development and a plan to manage the product development activities. The first determines the functional areas involved in each stage of development and the procedures for carrying out the work. The second specifies the schedule for each activity within the project.

4.2 Elaboration of the quality matrix of the final product

The Quality Matrix of the finished product is the first matrix that should be developed within the QFD method. In it are contained all information relating to the finished product. This topic displays an example of developing step-by-step from the quality matrix of a functional ready to bake dough for pies (Pinto & Paiva, 2010). While the development of other matrices of raw materials, intermediate products and processes are not treated in this chapter, the understanding of this first matrix will benefit the reader to understand how the matrices are made in the context of QFD.

4.2.1 Listing of primitive data

The primitive data are informations written in colloquial language, which can be collected through interviews or questionnaires with consumers, through discussion with focus groups or can be extracted from consumer complaints. They may also be got from opinions of company employees and in the news world. When the consumer does not directly express their needs, the imagination of scenes, or occasions of consumption, facilitates the description of the item required.

To meet the needs of the target market related to the dough for pies, there was a market research through semi-structured interviews with a sample of thirty possible consumers of the product. In the interview it was assessed the characteristics that the interviewers hoped to find in the ready dough for pies through the deployment of the scene in the manner, place and circumstances under which they would like to consume the product.

With the primitive information obtained, it was listed the greatest possible number of consumer desires. An example of this conversion is when an interviewee said that "the dough for pie should be used for both pies – sweet and salty," and the translation of primitive data for a required item was that "the dough for pie has to have a neutral flavor."

4.2.2 Establishment of the required qualities

At this stage you just have to format the primitive language, obtained from the market or from the deployment of scenes, observing certain rules. It is important that the terms of the customer requirements are simple, summarized in a single sentence, without explanation and did not have double meaning, making sure that the desired quality is clear. Whenever possible, you should be careful to avoid expressions in the form of denial, employing for this, affirmatives expressions.

Then the customer requirement qualities must be arranged in a table, the table of required qualities. This table is assembled from right to left. From the more concreted level to the more abstract. Generally, for food products, the markets requirements are grouped in terms of looks or appearance, flavor, texture, ease of preparation.

For the elaboration of the table it should be observe the following script: sentences with the same content should be eliminated to avoid repetition. The sentences should be arranged so that they can be viewed in only one frame (tertiary level, Table 1). It must then be joined in groups of four or five sentences with similar content and add expressions of customer requirements that represents the groups formed (secondary level, Table 1). With the phrases of similar content from the previous procedure must be formed other groups and add expressions of customer requirements to represent the groups formed (primary level, Table 1).

Primary Level	Secondary Level	Tertiary Level
Looks nice	Nice texture	Being soft
		Being crunchy
		Being a dough that dissolves easily in the mouth
	Nice color	Have a color next to cream/beige
	Appealing aspect	Have an uniform size
		Have an uniform thickness
Being tasty	Pleasant aroma	Have an appetizing aroma
	Pleasant flavor	Have a neutral taste
Satisfaction of the preparation	Being fully	Being fully
Being safe	Being safe	Being safe
Being healthy	Being healthy	Being functional
		Have a padronized caloric value

Table 1. Customer requirements to the functional dough for pies.

In the example of the functional dough for pies, it was constructed a table of deployment of the required qualities mainly from the joining of different sensory aspects of the product (Table 1).

4.2.3 Establishment of the quality characteristics

From the customer requirement of the tertiary level, must be extracted the technical characteristics of the finished product. At this point you have to convert the world of market into the technological world, drawing as much as possible, technical characteristics that will be easy to be measured. To do this, you should use the following reasoning: "How the required quality could be assessed in the final product?"

Then the table of quality characteristics should be built the same way as the table of the required qualities was. It should be built in groups thinking in the objectives of measurement QFD or types of analysis to be carried out in the final product. For example, in the case of food products, in physico-chemical, microbiological and sensory analysis (Table 2).

Primary Level	Secondary Level	Tertiary Level
Physico-chemical characteristics	Physico-chemical characteristics (cold dough)	Thickness of the dough for pie
		Diameter of the dough for pie
		Ash content
		Moisture of the dough for pie
		Baking time
		Fiber content
		Carbohydrate content
		Protein content
		Fat content
Sensory characteristics	Visual (baked dough)	Color
		Integrity
	Taste (baked dough)	Aroma
		Neutral flavor
		Soft texture
		Crispness
		"Hollow" texture
Microbiological characteristics	Microbiological characteristics (cold dough)	Coliforms at 45°C
		Salmonella sp/25g
		B. cereus
		Estafilococcus coagulase positive

Table 2. Table of quality characteristics of the dough for pies

4.2.4 Establishing the correlations in the quality matrix

In the central part of the matrix it's necessary to make the correlation of each required quality with each characteristic quality. To this must be observed the following rules:

1. Judge each relationship independently.
2. Assign symbols for each correlation which correspond to numeric values. The meanings can be:
   - ◎ or 9: Strong correlation;
   - ○ or 6: There is a correlation;
   - △ or 3: Possible correlation
3. For each required quality should be at least one strong correlation.
4. The symbols can not be concentrated in one place only.

5.   There should not be an item excessively marked with symbols.

With this information it was possible to build the quality matrix of a functional ready to bake dough for pies (Picture 2). There was done a correlation between these required qualities by the market and the quality characteristics of the finished product, assigning values 3, 6 or 9.

4.2.5 Establishment of the planned quality

The first column of the planned quality is the degree of importance. This must be established by the survey with consumers of the target market. When the survey is done, it is necessary to launch the averages values obtained in the matrix.

In the example of the functional dough for pies, the degree of importance of the required qualities has been established through research with thirty-two prospective buyers, where the interviewed indicates how important each characteristic was on a scale from 1 (unimportant) to 5 (very important). The medians for each attribute were also launched into the matrix.

After the development of the prototypes in industrial kitchens or in a pilot plant, it should be performed a search for sensory analysis with a sample of the target market. For this, should perform an affective sensory test with samples of one or more developed prototypes and a competitor's product, if any.

In the survey of the sensory analysis can be used items of the second level of the table of deployment of required qualities. Based on the type of scale used in the sensory test, it should be launched into the matrix the averages or medians of the results of sensory analysis.

In the case of the functional dough for pies, for the sensory analysis of the products developed, it was used the test of acceptance by the hedonic scale, varying gradually from 1 to 9, based on attributes like or dislike. Fifty tasters commented on all the attributes initially listed as important to the market. The medians of the attributes evaluated in the sensory analysis of the product developed were included in the matrix (Picutre 2). To compare the performance of the products developed for each required quality, we used nonparametric statistical test of Mann & Whitney (Siegel & Castellan, 2006) in order to distinguish the preferred.

Then you must define, through consensus among the development team, the column called planned quality, taking into account the degree of importance and the performance of the company and competitors. See the example in Picture 2.

The rate of improvement is established through the ratio between each value of planned quality by the performance of the product in the sensory analysis. Then it should be to establish which required qualities are considered strong, medium or weak selling points, i.e., which attributes will be attractive to the market, which items are attractive to the consumers in an average way and which attributes are obvious or mandatory to the product. In the column of the selling point, items considered to be attractive to the market receiving the note 1.5, intermediate items, the value 1.2 and those considered obvious, is assigned the value 1.0.

To calculate the absolute weight, multiply the degree of importance by the rate of improvement and also by the selling point. The relative weight of each required quality is the corresponding percentage of the absolute weight.

Required Qualities \ Quality Characteristics	Thickness of the dough for pie	Diameter of the dough for pie	Ash content	Moisture of the dough for pie	Fiber content	Carbohydrate content	Protein content	Fat content	Baking time	Color	Integrity	Aroma	Neutral flavor	Soft texture	Crispness	"Hollow" texture	Coliforms at 45°C	Salmonella sp/25g	B cereus	Eataflrococcus coagulase positive	Degree of Importance	Sensory Analysis of prototype A	Sensory Analysis of prototype B	Planned Quality	Rate of Improvement	Selling Point	Absolute weight	Relative weight
Being soft	3							3						9	3	3					3.0	8.0	8.0	8.0	1.0	1.2	3.6	7.1
Being crunchy	3			6										3	9	3					3.5	8.0	8.0	8.0	1.0	1.2	4.2	8.3
Being a dough that dissolves easily in the mouth								3						3	3	9					3.0	8.0	8.0	8.0	1.0	1.2	3.6	7.1
Have a color next to cream/beige			9							9											3.0	8.0	8.0	8.0	1.0	1.0	3.0	5.9
Have an uniform size		9																			4.0	8.0	8.0	8.0	1.0	1.0	4.0	7.9
Have an uniform thickness	9																				4.0	8.0	8.0	8.0	1.0	1.0	4.0	7.9
Have an appetizing aroma												9	6								3.0	8.0	8.0	8.0	1.0	1.0	3.0	5.9
Have a neutral flavor					3	3	3	6				3	9								4.0	8.0	8.0	8.0	1.0	1.2	4.8	9.5
Being fully				3	3		6	3			9					3					4.0	8.0	8.0	8.0	1.0	1.2	4.8	9.5
Being safe				6													9	9	9	9	4.0	9.0	9.0	9.0	1.0	1.0	4.0	7.9
Have a padromized caloric value					9	9	9	9													4.0	9.0	9.0	9.0	1.0	1.0	4.0	7.9
Being functional					9																5.0	9.0	9.0	9.0	1.0	1.5	7.5	14.9
Absolute weight	46.5	71.3	53.5	125.9	212.1	28.5	85.5	128.9	0.0	53.5	99.5	82.0	121.2	110.5	117.6	159.0	71.3	71.3	71.3	71.3	1605.4					Total:	50.5	100
Relative weight	2.9	4.4	3.3	7.8	13.2	1.8	5.3	8.0	0.0	3.3	5.3	5.1	7.6	6.9	7.3	9.9	4.4	4.4	4.4	4.4	120.0							100.0
Standards of prototype B	1 cm	15 cm	65%	19.7%	7.6%	52%	11%	16.51%		*5	*5	*5	*5	*5	*5	*5	$10^2/g$	aus/25g	$3×10^2/g$	$3×10^2/g$								

Fig. 2. Quality Matrix of a functional ready to bake dough for pies

4.2.6 Establishment of the designed quality

Now, it should be calculated the absolute weight of the quality characteristics at the bottom of the quality matrix. To do this, multiply the relative weight of each required quality by the numerical values of correlations, and add up these products vertically. The relative weight of each quality characteristic is the corresponding percentage of the absolute weight. For example, in Picutre 2, the absolute weight of 46.3 of the quality characteristic "dough thickness" was calculated by multiplying (7.1 x3) + (8.3 x 3) + (7.9 x9).

In the case of the functional dough for pies, were established physico-chemical, microbiological and sensory specifications for each quality characteristic of the finished product. The physico-chemical specifications were obtained by laboratory tests. The microbiological standards have been established according to the Brazilian law and the sensorial by a trained sensorial team.

5. Overview of the application of QFD in food products

QFD has been used in the food industry since 1987, i.e., its use is recent (Costa et al., 2001). But only after the 90's is that the articles were published showing the benefits of using the method in food products (Benner et al., 2003).

According to Souza Filho and Nantes (2004), the literature has dealt with much more organizational benefits and potential for improving the technical quality than the implementation process of QFD. Few articles describe how QFD has been used in real products and discuss their own experiences. Thus there isn't a deeper theoretical in the application of this method. To Charteris (1993) and Govers (1996), the strategic importance of QFD to contribute to the competitive advantage of firms may explain the reluctance of companies to share such important information on QFD.

Importantly, too, that one factor, perhaps one of the main, is the difficulty of finding scientific study of the application of QFD into the food industry, is due to the reliance on specific sensory attributes of each product. QFD was originally developed for the development of boats, automobiles and automobiles pieces. The technical characteristics related to this type of product are characteristics that have defined shapes and dimensions, specifications and parameters that can be measured accurately. However, the area of food has a different characteristic; the food may have different perceptions of consumer to consumer. Sensory perceptions are intrinsic to every human being and despite the food being technical specifications, mainly physico-chemical and microbiological parameters, the sensory perceptions are very difficult to specify. Allied to this, food ingredients have slightly larger deviations than pieces of heavy industry and those may change due to interactions between them or due to the process applied (Favaretto, 2007). Thus, the technicians of the food industry wishing to use the QFD tool must idealize the necessary changes in the method so that it becomes applicable in the development of a food product. It is important that simplifications are made to the product, its ingredients and their interactions so that the matrices can be used without any difficulty.

Cheng & Melo Filho (2010), in a survey on the contour and depth of application of QFD in the 500 largest Brazilian companies, came to the conclusion that is recent the introduction of the method in those companies and that is still a long way to go and a need for greater understanding and support of top management in the implementation of this method by companies.

Although the use of QFD is recent, some companies, however, begin to realize the advantages of their use and are already getting great results with the implementation of the method. Table 3 presents some applications of QFD in food product development in the last 12 years.

Authors	Applications of QFD	Results of the use of QFD
Antoni (1999)	Turkey dry fermented sausage	Easy understanding of the real needs to be addressed in the product. Interpretation of the first Quality Matrix as the voice of the customer divided into two parts: final consumer requirements and demands of the point of sale, unfolding them into the final product characteristics. Correlation between the raw material characteristics and the final product characteristics.
Viaene & Januszewska (1999)	Chocolate couverture	Approximation of the areas of Marketing and of Food Science and Technology, reduction of the final cost and increase of the success potential in launching the new product due to the participation of the consumers belonging to this target segment into the process.
Tumulero et al. (2000)	Salty crackers	Product improvement, sales increase and expansion of market share.
Marcos (2001); Marcos & Jorge (2002)	Table Tomato	Higher interaction of functional areas involved in the development process and establishment of sales strategies based on the analysis of the consumer market. Reduction of losses.
Chaves (2002)	Yogurt	Identification of the most important aspects from the standpoint of quality. Translation of the needs and desires of customers and structural changes to meet the expectations of a specific market.
Magalhães (2002)	Packaged pasteurized milk	Increase on the sales of the product
Stewart-Knox & Mitchell (2003)	Food products with reduced fat	Market identification, consumer awareness and participation of suppliers.

Table 3. Applications of QFD in food product development

Authors	Applications of QFD	Results of the use of QFD
Cortés & Da Silva (2005)	Yogurt	Translation of customer requirements and restructuring of the product according to market expectations. Commitment of all members of the company.
Gonçalves & Silva (2005)	Production of meals: integration of QFD x APPCC	Identification of the potential dangers of contamination in the generic process to produce meals. Verification of the phases of meals production that require greater attention from staff. The results of the items prioritized in the matrix served as an aid in the selection and training process of employees.
Waisarayutt & Tutiyapak (2006)	Instant rice noodles	Based on customer requirements, determination of the most important technical specifications of the product and the important parameters of the process. Development of control plans in accordance with the process parameters.
Delgado & Pedrozo (2007)	Micro food company in Peru, "Delicias del Sur".	Identification of customer needs and competitors' actions. Development of new food products based on differentiation through the incorporation of raw materials and inputs from local resources of functional character. Strengthening relationships in the areas of the organization.
Favaretto (2007)	Soft drink	Decrease in the time of product development, quality assurance of product and service offered to the customers. Culture change in the organization.
Ferreira & Miyaoka (2007)	Sweet Milk based on soy	Identification of real needs and desires of the audience. Serving an attractive niche market, who are lactose intolerant, vegetarians and people seeking a healthier diet. Improved quality and cost reduction.

Table 3. (continued): Applications of QFD in food product development

Authors	Applications of QFD	Results of the use of QFD
Matsunaga (2007)	Breaded chicken	Identification of the attributes most valued by consumers. Higher interaction between research and product development area and marketing area.
Miguel et al., (2007)	Consumer profile of pineapple "Pérola"	Identification of critical quality attributes at the purchase time, establishment of the degree of importance of each attribute required by the market and interpretation of sensory analysis. Identification of points to be improved within the supply chain, minimizing losses and maintaining and improving the quality of the final product.
Anzanello et al., (2009)	Christmas turkey	Identification of the process parameters, product and resources to be prioritized in the adoption of improvements. Identification of resources seen as bottlenecks to meet the demands prioritized in QFD. Global view while driving improvements in products.
Cheng & Melo Filho (2010)	Noble products with embedded (ready dishes like ham lasagna)	Anticipating the possible changes of habits and attitudes, adding value to the product.
Garcia (2010)	Fluid milk	Disclosure of the characteristics to be worked in order to improve product quality.
Kawai (2010)	Table tomato	Identification of the required characteristics by consumers and the establishment of a competitor that has the largest market share. Reduce losses and increase the income of the producer. Improved communication between sectors.
Pinto & Paiva (2010)	Functional dough ready for pies	Identification of the characteristics that consumers attach as of greater importance and prioritization those into the product development.

Table 3. (continued): Applications of QFD in food product development

Authors	Applications of QFD	Results of the use of QFD
Rodrigues (2010)	Improving food services at the Universidade Estadual de Campinas (Unicamp)	Identification of the quality characteristics most important and which should be tailored to the improvement of food service establishments of Unicamp. Improved workflow and execution of solutions more quickly and affordably.
Vatthanakul et al., (2010)	Product of gold kiwifruit leather	Defining the importance of the various sensory characteristics for the design of new products.
Zarei et al., (2011)	Canning industry	Identification of enablers of Lean Manufacturing viable to be implemented in practice in order to increase the lean production of food supply chain.

Table 3. (continued): Applications of QFD in food product development

Based on the work presented, it is emphasized that QFD enables the company to develop products that meet the diverse and growing demands of its customers, serving different market niches, in a short time, and is characterized by the efficiency in storage and transmission of information during the multifunctional product development activity. In addition, the use of the method QFD in several companies have pointed out others advantages as: resolving problems, reducing the development time and using of prototypes in a more objective way. The QFD is effective in order to direct, in a practice way, how to plan and conduct activities necessary to the process of the product development.

6. Sensory analysis and support tools to market research

The application of QFD in the food industry is complex and the literature does not allow establishing its full application in the development of these products. For this reason, adaptations that take into account the characteristics of the product "food" are necessary for the successful implementation of QFD in this industry. The adjustments should pursue a better integration of aspects of market research with the sensory evaluation of food, aiming to reveal the perception of consumers in a jointful way, in relation to the general market attributes (shape, size, ease of use, etc.) and to the sensory attributes that make them decide to accept and purchase of food. The integration of these aspects should also consider the current model of food consumption in the region and the market segment defined for the product.

In developing a new product is essential to optimize parameters such as shape, color, appearance, odor, flavor, texture, consistency and the interaction of different components, in order to achieve a full balance that translates into an excellent quality and have good acceptability. Thus, according to Monteleone and colleagues (1997), studying the relationships between the sensory attributes and the acceptance of a product can be very useful to formulate or improve a product, as well as to evaluate potential market opportunities. Understanding what drives the acceptance of a product is very important in light of a competitive market like nowadays. Meeting the most valued attributes makes more oriented the product development, benefiting not only the area of research and product development, as well as marketing, reducing distances between these two areas. In this sense, the methods of sensory evaluation of foods and evaluation of consumer desires can be an aid when applied the method QFD into the food product development.

In sensory analysis the differences between products (discriminatory tests), the intensity of a sensory attribute of quality (descriptive tests), or the degree of acceptance, preference or rejection for a product (affective tests) are measured by human senses. However, it is necessary to consider that the sensory perceptions can not be measured directly. Therefore to assess the individual stimulation received in the sensory evaluation is used scales that allow the quantification of them, as the specific objective of the evaluation (Bech et al. 1994; Chaves & Sproesser, 1996; Ferreira et al. 2000).

In acceptance tests, the hedonic scale is the most used and widespread. It is a structured scale, with nine sentences, balanced, considered easy to use and to understand (says the pleasant and unpleasant states in the body). The evaluation of the hedonic scale is converted into numerical scores, and statistically analyzed to determine the difference in the degree of preference between samples (ABNT, 1998, IFT, 1981; Land & Shepherd, 1988). Optionally you can use the hybrid hedonic scale, a continuous scale of 10cm, and with the advantages to meet the assumptions of the models of analysis of variance (ANOVA) as the normality of the residuals and homoscedasticity, and increase the discrimination between the samples.

The traditional methodologies for analyzing the affective tests data, like the ANOVA, used to compare more than two averages in the study, and the averages test, to determine the significance to a particular level of confidence, have shown limitations and shortcomings. For each product evaluated is obtained the average of the consumer´s group, assuming therefore that all respondents have the same behavior, ignoring their individuality. Thus, there may be occurring loss of important information about different market segments (Meilgaard et al. 2006; Polignano et al., 1999, Reis & Minim, 2006).

In order to analyze the affective data, taking into account the individual response of each consumer and not just the average of the consumer group that evaluated the products, it was developed a technique called Preference Mapping, which has been widely used by scientists of the area of sensory analysis (Behrens et al., 1999). This technique is often employed to identify groups of consumers who respond uniformly and that differ from other groups by age, sex, attitude, need, eating habits and/or responses to the product´s attributes. This gives the opportunity to interpretate the different areas of action of the market (Polignano, 2000, Westad et al., 2004).

As it can explain better the consumer preference for each attribute, the Preference Mapping is an interesting tool to collect the real needs of customers and turn them into project qualities. It can be very useful during the development of food products, particularly in the construction phase of the Designed Quality of the Quality Matrix (Polignano, 2000).

Cluster analysis is a set of statistical techniques whose objective is to seek a classification according to the natural relations that the sample shows, forming groups of objects (individuals, companies, cities or other experimental unit) according to the similarity in relation to some predetermined criteria, thus reducing the dimensionality of the data. According to Hair et al. (1998), the groupings (or clusters) resultants should have a high internal homogeneity (within groups) and high external heterogeneity (between groups).

Cluster analysis is useful in developing new products to meet consumer profiles. This can identify the profile of each group (age, marital status, psychological characteristics, etc.) that will define whether there are different demands (market segmentation). The profile is defined by the characteristics that make up the cluster, based on the concept of similarity.

In addition to these approaches, there is the publication of articles dealing with the use of Conjoint Analysis in the food industry. The Conjoint Analysis is a technique that allow to study the combined effect exerted by two or more independent variables on a dependent variable (Carneiro et al., 2003). Based on a decomposition analysis, the Conjoint Analysis determines the contribution of the studied factor levels expressed in samples or combinations of the consumer response (acceptance, preference or purchase intent). It is an analysis technique that can be used to identify the attributes/levels that most influence the selection, purchase and acceptance of products, after the sensory evaluation of them.

The Conjoint Analysis has been widely applied in the development of new products in all industrial sectors, in particular in the establishment of concepts, competitive analysis, selection of target market segment, the definition of price and advertising strategy (Drummond, 1998).

Another tool that has been used is the Repertory Grid. Kelly presented it in 1955, which advocates the theory that people act as scientists evaluating the world around them, creating hypotheses and establishing descriptors of what it is seen. The Repertory Grid method is a term used to describe a set of techniques related to the Kelly's theory (1955), which can be used to investigate the individual definitions in the perception of characteristics of his surroundings. This method is very flexible and allows it to be applied according to the researcher's interest (Mac Fie & Thomson, 1994).

A descriptor is defined as the way in which two items are similar and, in some way, different from a third party. In the sensory field, the samples are arranged in triads. In each set, two samples are kept together and one away from the taster and it describes how the samples are alike and how different. In addition to identifying the differences, the tasters should also describe the extremes of each descriptor raised, in order to build a scale so that samples can be quantified. The data collected are analyzed by GPA (Generalized Procrustes Analysis), which allow to be established a configuration with the main and common descriptors to the tasters. This is a multivariate analysis, that establish the consensus map of the data (Mac Fie & Thomson, 1994).

Another method that helps to identify the product perception by the consumer is called Focus Group. Focus Group can be defined as a planned session to obtain individual perceptions of such a product or service, in a peaceful environment, through the moderation of groups formed by six to nine people (MacFie & Thomson, 1994) or eight to twelve people (Fuller , 2011). It is a qualitative technique of group discussion, which allows interaction between people. The moderator leads the discussion to the topics of interest, listening to people, without interfering directly. This technique allows to raise the participants' perceptions about the subject matter. According to Fuller (2011), the main function of the Focus Group is to determine consumer reaction to the objects of study.

The Kano method has as it´s main goal to evaluate the influence of components of products in consumer satisfaction (Sauerwein et al., 1996). This method aims to rank the attributes in four characteristics groups according to the degree of care and satisfaction: indifferent (characteristics that do not affect consumer satisfaction), expected (mandatory characteristics), proportional (characteristics that customer satisfaction is proportional to the degree of care) and attractive (characteristics that customer satisfaction does not diminish if not offered, but increases if met) (Fonseca, 2002).

Using this model it is possible to improve the process of understanding the requirements from a classification that can help prioritize development resources. The identification of mandatory requirements for certain attributes and opportunities of future innovation, which customers have not even being classified as needs now, allowing the development and the use of criteria in a more efficient way for resource allocation. The process of translating requirements into product characteristics or services should also benefit from the information obtained using the Kano model (Guimarães, 2003).

Kano helps in the process of product development, prioritizing the characteristics that really have an impact on customer satisfaction. Beyond that, this may be very well combined with QFD, identifying the relative importance of the needs raised by consumers, contributing to the product development more focused and better targeted to the audience. The classification of characteristics can also be a guideline to define the attributes to be worked out for different market segments, creating product differentiation in that market.

7. Conclusion

The effective development of products has become the competitive advantage for many companies, especially in the food industry. For that, as shown in this chapter, projects must achieve the best levels of quality, of efficiency and speed in the elaboration of products. Certainly this requires an organizational effort of the entire company. What the firm plans, i.e., its development strategy, and how the company makes its planning – its development management - will determine the expressiveness of the product in the market. In this sense, it can be affirmed that the competitive advantage of these organizations is based on the capacity of its technical staff, in the procedures and organizational structure, in the strategies established to guide the process, in the methods used, in the way that the top management interacts with the process and yet, in the organization of the team according to the level of complexity of the project.

The QFD method as pointed out in this work is not a mechanism that addresses all of the aspects above, important for structuring the system of product development. However, by offering a way to treat the necessary information to the process and to plan the activities, the method has boosted the development system in several companies.

Ultimately, it is worth noting that, specifically for the food industry, some aspects can enhance the use of the method. Among them those can be mention:

Existence of a support infrastructure to the process, with sufficiently equipped laboratories and trained personnel to conduct the analysis.

The intrinsic quality of food products can be evaluated by physico-chemical, microbiological and sensory analyses. Thus, the values of the quality characteristics during construction of the matrices are determined by such methods. Various techniques of sensory analysis assist in analyzing and gathering market information and then building a quality matrix, as reported in this chapter.

Likewise, the pilot plants and experimental kitchens within companies, contribute greatly to the development of prototypes and conduction of activities in a more quickly way.

Knowledge and application of various statistical techniques in the process of product development.

Such techniques facilitate the identification of opportunities and strategic positioning of products, the identification of factors that affect customer preference and the market segmentation, supporting the use of the QFD method.

Assurance system of the consolidated quality

The consolidated quality management benefits directly the development process, as it aims to optimize the exchange of information, the commitment of the people involved in the activities, the specification of process parameters and the standardization of this procedures, as well, developing products with little variability. Certainly, these factors create a safer environment for the conduct of projects, facilitating, in particular, the specification of product technical parameters, of raw materials, packaging and of the manufacturing process.

8. References

ABNT (1998). Associação Brasileira de Normas Técnicas. *NBR 14141*: escalas utilizadas em análise sensorial de alimentos e bebidas. Rio de Janeiro, 1998.

Akao, Y. & Mazur, G. (2003).The leading edge in QFD: past, present and future. *International Journal of Quality & Reliability Management*. Vol. 20, No.1, pp. 20-35, ISSN 0265-671X.

Akao, Y. & Ohfuji, T. (1989). Recent aspects of Quality Function Deployment in service industries in Japan. *Proceedings of the International Conference on Quality Control*, ISBN 0-915299-41-0 3, Rio de Janeiro, 1989.

Antoni, I. (1999). *Desenvolvimento de um Embutido Fermentado de Carne de Peru pelo Método do Desdobramento da Qualidade*. 136 p. Dissertação (Mestrado em Tecnologia de Alimentos). Faculdade de Engenharia de Alimentos, Universidade Estadual de Campinas (Unicamp), Campinas.

Anzanello, M.; Lemos, F. & Encheveste, M. (2009). Aprimorando Produtos Orientados ao Consumidor Utilizando Desdobramento da Função Qualidade (QFD) e Previsão de Demanda. *Produto & Produção*, Vol. 10, No. 2, jun., 2009, pp. 01 - 27, ISSN 1516-3660.

Athayde, A. (1999). Indústrias agregam conveniências aos novos produtos. *Engenharia de Alimentos*, São Paulo, Vol. 24, 1999, pp. 39-41.

Bech, A.; Engelund, E.; Juhl, J.; Kristensen, K. & Poulsen, C. (1994). *QFood – optimal design of food products*. MAPP working Paper no.19. MAPP Centre, Aarhus. March, 1994; pp. 2-12; ISSN 09072101.

Behrens, J.; Silva, M. & Wakeling, I. (1999). Avaliação da aceitação de vinhos brancos varietais brasileiros através de testes sensoriais afetivos e técnica multivariada de mapa de preferência interno. *Revista da Sociedade Brasileira de Ciência e Tecnologia de Alimentos*, Campinas, Vol. 19, No. 2, May, 1999, ISSN 0101-2061.

Benner, M.; Linnemann, A.; Jongen, W. & Folstar, P. (2003). Quality function deployment (QFD) – can it be used to develop food products?, *Food Quality and Preference*, Netherlands, Vol. 14, pp. 327-339, ISSN 0950-3293.

Carneiro, J.; Silva, C.; Minim, V.; Regazzi, A.;Deliza, R. & Suda, I. (2003). Princípios básicos da Conjoint Analysis em estudos do consumidor. *Revista da Sociedade Brasileira de Ciência e Tecnologia de Alimentos*, Campinas, Vol. 37, No. supl, 2003, pp. 107-114, ISSN 0101-2061.

Chan, L. & Wu, M. (2002). Quality function deployment: a literature review. *European Journal of Operational Research, Vol.* 143, 2002, pp. 463-497, ISSN 0377-2217.

Charteris, W. (1993). Quality function deployment: a quality engineering technology for the food industry. *Journal of the Society of Dairy Technology*, Vol.46, No 1, 1993, February, 1993, pp. 12–21, ISSN 0037-9840.

Chaves, J. & Sproesser, R. (1996) *Práticas de laboratório de análise sensorial de alimentos e bebidas.* Publisher: UFV, Viçosa, Brazil.

Chaves, O. (2002). *Aplicação do método de desdobramento da função qualidade na industrialização do leite de consumo em Minas Gerais.* 86 p. Dissertação (Mestrado em Economia Rural). Departamento de Economia Rural, Universidade Federal de Viçosa, Viçosa.

Cheng, L. & Melo Filho, L. (2010). *QFD: desdobramento da função qualidade na gestão de desenvolvimento de produtos.* (2 ed). *Publisher:* Blucher, ISBN 9788521205418, São Paulo.

Clark, K. & Wheelwright. S. (1993). *Managing new product and process development.* Free Press, ISBN 0-02-905517-2, New York.

Cooper, R. (2001). *Winning at new products. Accelerating the process from idea to lanch.* (3ªed), Publisher: Basic Books. Cambridge, Massachusetts. ISBN 0738204633.

Cortés, D. & Da Silva, C. (2005). Revisão: Desdobramento da Função Qualidade – QFD – conceitos e aplicações na indústria de alimentos. *Brazilian Journal Food Technology*, Vol. 8, No. 3, 2005, pp. 200-209, ISSN 1517-7645.

Costa, A.; Dekker, M. & Jongen, W. (2001). Quality function deployment in the food industry: a review. *Trends in Food Science and Technology*, Vol.11, No. 9–10, 2001, pp. 306–314, ISSN 0924-2244.

Delgado, G. & Pedrozo, E. (2007). Inovação de Produtos Alimentícios: Alimentos funcionais a partir de produtos locais. *Proceedings of IV Convibra: Brazilian Administration Virtual Congress*, São Paulo/SP. Brazil, dez., 2007.

Drumond, F. (1998). *Ténicas Estastísticas para o Planejamento do Produto.* Fundação Christiano Ottoni, Belo Horizonte.

Earle, M. (1997). Changes in the food product development process. *Trends in Food Science & Technology*, Vol. 8, 1997, pp. 19-24, ISSN 0924-2244.

Favaretto, R. (2007). *Modelo de aplicação de QFD no desenvolvimento de Bebidas.* 96 p. Dissertação (Mestrado em Profissional em Engenharia Mecânica) – Faculdade de Engenharia Mecânica, Universidade Estadual de Campinas (Unicamp), Campinas.

Ferreira, G. & Miyaoka, A. (2007). *Estratégias de desenvolvimento do doce de leite à base de soja.* 30 p. Monografia (Graduação no Curso de Engenharia de Produção). Engenharia de Produção da Universidade Federal de Viçosa, Viçosa

Ferreira, V.; Almeida, T.; Pettinelli, M.; Silva, M.; Chaves, J. & Barbosa, E. (2000). *Análise sensorial: testes discriminativos e afetivos.* Sociedade Brasileira de Ciência e Tecnologia de Alimentos, Campinas.

Fonseca, M. (2002). *Uma abordagem para a redução de custos no desenvolvimento de produtos alimentícios.* 82 p. Tese (Mestrado em Engenharia de Produção). Faculdade de Engenharia de Produção, Universidade Federal do Rio de Janeiro, Rio de Janeiro.

Fuller, G. (2011). *New food product development: from concept to marketplace.* (3ed.), CRC Press, ISBN 13:978-1-4398-1865-7, Florida.

Garcia, A. (2010). *Uso do método DFQ (Desdobramento da Função Qualidade) para melhoria da qualidade do leite fluido.* 181 p. Tese (Doutorado em Tecnologia de Alimentos). Faculdade de Engenharia de Alimentos, Universidade Estadual de Campinas, Campinas.

Gonçalves, T. & Silva, C. (2005). Proposta de utilização do quality function deployment (QFD) no sistema de análise de pontos críticos de controle (APPCC) na produção de refeições. *Proceedings of XII SIMPEP*, Bauru, SP, Brazil, Nov., 2005.

Govers, C. (1996). What and how about quality function deployment (QFD). *International Journal of Production Economics*, New York, Vol.46-47, 1996, pp. 575-585, ISSN 0925-5273.

Guimarães, L. (2003). QFD - Analisando seus aspectos culturais organizacionais. *Revista Qualidade*, Vol., No. 128, Jan., 2003, pp.56-66.

Hair J.; Anderson, R.; Tatham, R. & Black, W (1998). Cluster analysis. In: *Multivariate data analysis*. Hair, J, Black, B, Babin, B & Anderson, R (5.ed.), pp. 469-518, Prentice Hall, Upper Saddle River.

IFT. (1981). Sensory Evaluation Division. Guidelines for the preparation and review of paper reporting sensory evaluation date. *Journal of Food Technology*, Vol.35, No.4, 1981, pp.16-17, ISSN 0022-1163.

Kawai, S. (2010). *Desenvolvimento de Tomate de Mesa, com o uso do método QFD (Quality Function Deployment), comercializado em um supermercado.* 217 p. Tese (Doutorado em Tecnologia Pós-Colheita). Faculdade de Engenharia Agrícola. Universidade Estadual de Campinas, Campinas.

Macfie, H. & Thomson, D. (1994). *Measurement of food preferences.* (1ed.), Springer, ISBN 9780834216792, New York.

Magalhães, G. (2002). *Incorporação da Qualidade Desejada pelos Consumidores ao Leite Pasteurizado Utilizando o Desdobramento da Função Qualidade.* 77 p. Dissertação (Mestrado em Ciência e Tecnologia de Alimentos). Programa de Pós-Graduação em Ciência e Tecnologia de Alimentos. Universidade Federal de Viçosa, Viçosa.

Marcos, S. & Jorge, J. (2002). Desenvolvimento de tomate de mesa, com o uso do método QFD (Desdobramento da Função Qualidade), comercializado em um supermercado. *Horticultura Brasileira*, Brasília, v. 20, n. 3, p. 490-496, setembro 2002.

Marcos, S. (2001) *Desenvolvimento de tomate de mesa, com o uso do método QFD (Quality Function Deployment), comercializado em um supermercado.* 199 f. Tese (Doutorado em Ciência de Alimentos). Faculdade de Ciência de Alimentos. Universidade Estadual de Campinas, Campinas.

Matsunaga, P. (2007). *Identificação de atributos sensoriais de pedaços empanados de frango mais valorizados pelo consumidor.* 121p. Dissertação (Mestre em Alimentos e Nutrição). Faculdade de Engenharia de Alimentos. Universidade Estadual de Campinas, Campinas.

Meilgaard, M.; Civille, G. & Carr, B. (2006). *Sensory Evaluation Techniques,* (4. ed.), FL: CRC Press, ISBN 0849338395, Boca Raton.

Melo Filho, L. & Cheng, L. (2007). QFD na garantia da qualidade do produto durante seu desenvolvimento: caso em uma empresa de materiais. *Produção,* Vol. 17, No.3,Dec., 2007, . p. 604-624, ISSN 0103-6513.

Miguel, A.; Spoto, M.; Abrahão, C. & Silva P. (2007) Aplicação do método QFD na avaliação do perfil do consumidor de abacaxi "Pérola". *Ciência e Agrotecnologia*, Vol. 31, No. 2, 2007, pp. 563-569, ISSN 1413-7054.

Mizuno, S. (1969). Company-wide quality control activities in Japan. *Reports of Statistical Application Research*, Vol.16, No.3, 1969, pp.68-77.

Monteleone, E.; Carlucci, A.; Caporale, G. & Pagliarini, E. (1997). Use of slope analysis to characterize preference for virgin olive oil. *Italian Journal of Food Service*, Vol. 9, No. 2, 1997, pp. 133-140, ISSN 1120-1770.

Pinto, A. & Paiva, C. (2010). Developing a functional ready to bake dough for pies using the Quality Function Deployment (QFD) method. *Revista da Sociedade Brasileira de*

Ciência e Tecnologia de Alimentos, Campinas, Vol.30, No.(Supl.1), mai., 2010, pp 36-43, ISSN 0101-2061.

Polignano, L. (2000). *Desenvolvimento de produtos alimentícios: implentação da ferramenta mapa de refrência e estudo da sua articulação com a matriz da qualidade.* 268 p. Dissertação (Mestre em Engenharia de Produção). Escola de Engenharia. Universidade Feredal de Minas Gerias, Minas Gerais.

Polignano, L.; Cheng, L. & Drumond, F. (1999). Utilização dos mapas de preferência como técnicas auxiliares do QFD durante o desenvolvimento de produtos alimentícios. *Proceedings of I Brazilian Conference on Management of Product Development*, Belo Horizonte, Brazil, 1999.

Reis, C. & Minim, V. (2006) *Testes de aceitação.* In: Análise sensorial: estudos com consumidores., Minim V., pp. 67-83, Publisher UFV, ISBN 85-7269-282-7, Viçosa, Brazil.

Rodrigues, N. (2010). *Aplicação da matriz da qualidade do QFD – Desdobramento Da Função Qualidade – para avaliar serviços de alimentação do campus da Unicamp.* 190 p. Tese (Doutorado em Tecnologia Pós-Colheita). Faculdade de Engenharia Agrícola. Universidade Estadual de Campinas, Campinas.

Sauerwein, E.; Bailom, F.; Matzler, K. & Hinterhuber, H. (1996). The Kano Model: how to delight your consumers. *Preprints of the IX International Working Seminar on Production Economics*, Innsbruck/Igls/Austria, February, 1996.

Siegel, S. & Castellan Jr, N. (2006). *Estatística não-paramétrica para ciências do comportamento.* (2 ed), Artmed, ISBN 85-363-0729-3, Porto Alegre.

Souza Filho, M. & Nantes, J. (2004). O QFD e a análise sensorial no desenvolvimento de produtos. *Proceedings of Production Engineering Symposium*, Bauru, Nov., 2004.

Stewart-Knox, B. & Mitchell, P. (2003). What separates the winners from the losers in new food product development? *Trends in Food Science and Technology*, Vol. 14, No.1-2, 2003, pp. 58-64, ISSN 0924-2244.

Tumelero, N.; Ribeiro, J. & Danilevicz, A. (2000). O QFD como ferramenta de priorização para o planejamento da qualidade. *Proceedings of Brazilian Conference on Management of Product Development*, São Carlos, Aug., 2000.

Vatthanakul, S.; Jangchud, A.; Jangchud, K.; Therdthai, N. & Wilkinson, B. (2010). Gold kiwifruit leather product development using quality function deployment approach. *Food Quality and Preference*, Vol.21, No.3, 2010, pp. 339-345, ISSN 0950-3293.

Viaene, J. & Januszewska, R. (1999). Quality function deployment in the chocolate industry. *Food Quality & Preference*, Vol. 10, pp. 377-385. ISSN, 0950-3293.

Waisarayutt, C. & Tutiyap, O. (2006). Application of Quality Function Deployment in Instant Rice Noodle Product Development. *The Kasetsart Journal: Natural Sciences.* Vol.40 (Suppl.) 162-171 p.

Westad, F.; Hersleth, M. & Lea, P. (2004) Strategies for consumer segmentation with applications on preference data. *Food Quality and Preference*, Vol. 15, No. 7-8 , 2004, pp.681-687, ISSN 0950-3293.

Zarei, M.; Fakhrzad M. & Paghaleh, M. J. (2011). Food supply chain leanness using a developed QFD model. *Journal of Food Engineering*, Vol. 102, pp. 25–33, ISSN 0260-8774.

23

Quality Preservation and Cost Effectiveness in the Extraction of Nutraceutically-Relevant Fractions from Microbial and Vegetal Matrices

Marco Bravi[1], Agnese Cicci[1] and Giuseppe Torzillo[2]
[1]Chimica Materiali Ambiente, Sapienza Università di Roma, Roma,
[2]CNR Istituto per lo Studio degli Ecosistemi, Sesto Fiorentino,
Italy

1. Introduction

Terrestrial and microbial vegetal matrices are a major source of nutraceutically and pharmaceutically relevant chemical compounds of different nature. In several cases the consumption of the raw vegetal or microbial matrix has been part of established diet regimes and has provided the consumers with a host of once unknown dietary benefits. Nowadays, while the consumption of the raw matrix still provides the original functional value, the separation of bioactive-enriched fractions has enabled the production of nutraceuticals, while the addition of nutraceutical fractions to food previously lacking or partially possessing them has lead to the industrial production of functionalised or fortified food respectively.

The separability of nutraceutically relevant fractions depends on the combination of several different features of the carrier matrix and of the fraction to be separated, namely: size, aggregation state and physical (hardness) and chemical (composition) features of the embodying matrix, chemical nature, bonding and degree of dispersion of the fraction of interest in the embodying matrix. Physical, chemical and electrostatic interactions between the embodying matrix, the fractions to be separated and exogenous agents (equipment and process auxiliary substances, and the environment) affect the desired separation; the chemical nature of the solvent (if any), the specific energy applied by the physical agent (if any), the frequency and intensity of the mechanical (e.g., ultrasound) and electromagnetic (e.g., microwave) field (if any), the processing time and temperature and the presence of specific case-by-case unwanted substances (e.g., water, oxygen, metal ions) play a role in the final outcome of the recovery process of the desired fraction.

2. Sub- and supercritical fluid extraction

Sub- and supercritical fluid extraction are promising separation processes aimed at replacing traditional lengthy, laborious, low selectivity and/or low extraction yield, toxic chemical-using separation processes (Herrero et al., 2006).

The low viscosity and (relatively) high diffusivity inherent in the supercritical state confers these solvents better transport properties than liquids solvents have. Furthermore, the "adjustability" of viscosity, density and solvent power (Del Valle and Aguilera, 1999) by acting on fluid density through the change of pressure and/or its temperature make them amply tunable to the specific separation needs. Last, but not least, using solvents generally recognized as safe (GRAS) helps meet the requirements of food and pharma processes. Carbon dioxide is the most commonly used because of its moderate critical temperature and pressure (31.1 °C and 7.39 MPa, respectively) and GRAS status. Supercritical CO_2 (Brunner, 2005): dissolves non-polar or slightly polar compounds; has a high solvent power for low molecular weight compounds, which decreases with increasing molecular weight; has high affinity with oxygenated organic compounds of medium molecular weight; has low solvent power for free fatty acids and their glycerides, even lower for pigments, and low solubility for water at temperatures below 100 °C; does not dissolve: proteins, polysaccharides, sugars and mineral salts; exhibits an increasing separating capability as pressure increases for compounds that are scarcely volatile, have a high molecular weight and/or are highly polar. Due to its low polarity, small amounts of "modifiers" (also called co-solvents), in the form of highly polar compounds such as ethanol or water, are often used to improve the separability of polar solutes.

Preparative systems for processing solid or liquid samples have different configurations; basically, they consist of solvent pump(s) delivering the main solvent and any required modifyer throughout the system, an extraction cell (separation from a liquid) or column (solid), and one or more separators in which the extract is collected when the solvent is expanded.

Extraction from solids is usually carried out discontinuously (batchwise) in a single stage because solids handling in pressurised vessels is troublesome and separation factors are high. Fluid mixtures often suffer from low separation factors and multi-stage contacting (countercurrently for highest effectiveness) becomes a necessity. When separation factors approach unity and many theoretical stages are required for the separation preparative chromatographic systems may be set up on a process scale (Brunner, 2005).

Pressurised liquids (e. g. water and ethanol at an intermediate temperature between their boiling points and their critical temperatures and under the appropriate pressure to maintain them in the liquid state) mitigate the drawbacks of supercritical fluids (e.g. supercritical carbon dioxide): their scarce affinity for polar solutes and extensive capital costs. Subcritical water has been successfully used for the extraction of essential oils, nutraceuticals and bioactives, among which polyphenols (King and Srinivas, 2009).

Water is a highly polar solvent at room temperature and atmospheric pressure due to hydrogen-bonding which reflects in its high dielectric constant. While at room temperature water is unsuitable as a solvent for non-polar compounds, at increased temperature a lowered dielectric constant (from 80 at 25 °C to 27 at 250 °C and 50 bar, intermediate between those of methanol, 33, and ethanol, 24, at 25 °C), viscosity and surface tension and an increased diffusivity confer an increased solvent power for non polar substances. Thermally labile compounds may be degraded at elevated temperatures (Teo et al., 2010) but newly formed bioactives (antioxidants) have also been observed (Plaza et al., 2010).

By adjusting the prevailing temperature and under the required pressure to remain in the liquid state, water may be tuned to the purpose, which may be that of an extraction solvent and/or reaction medium. The main parameters that influence the selectivity and extraction efficiency of pressurised water extraction include temperature, pressure, extraction time, flow rates and concentration of modifiers/additives. The residence time of the solute (reactant) in the aqueous medium becomes a critical parameter in conducting extractions above the boiling point of water and for optimizing reaction conditions (King and Srinivas, 2009).

2.1 Mathematical modelling of Pressurised Water Extraction (PWE)

The very core of PWE extraction may be divided in four sequential steps which take place in the extraction cell filled with sample materials: 1. desorption of solutes from the various active sites in the sample matrix; 2. diffusion of the extraction fluid into the matrix; 3. partitioning of solutes from the sample matrix into the extraction fluid and 4. chromatographic elution of the solutes out of the extraction cell to the collection vial. A two-step, single-site partition-based thermodynamic model is appropriate for PWE extraction (Kubátová et al., 2000 and 2002; Windal et al., 2000): 1. the compound is desorbed from its original binding sites in the sample matrix, under diffusion control but at a sufficiently fast rate to avoid being the overall limiting step; 2. the compound is eluted from the sample under thermodynamic partitioning control (K_D). The shape of an extraction curve would be defined by:

$$\frac{S_b}{S_o} = \frac{\left(1 - \dfrac{S_a}{S_o}\right)}{\dfrac{K_D \cdot m}{(V_b - V_a) \cdot \rho} + 1} + \frac{S_a}{S_o}$$

where S_a is the cumulative mass of the analyte extracted after volume V_a (ml), and S_b is the cumulative mass of the analyte extracted after volume V_b (data point b is next to data point a in experimental sequence). S_o is the initial total mass of analyte in the matrix. S_b/S_o and S_a/S_o are the cumulative fractions of the analyte extracted by the extraction fluid of the volume V_b and V_a, respectively. K_D is the distribution coefficient, ρ is the density of extraction fluid at given conditions (g/ml), and m is the mass of the extracted sample (g). The model depends on the extractant volume flowed, but not time.

The experimental device required for PWE is quite simple. Basically, the instrumentation consists of a water reservoir coupled to a high pressure pump to introduce the solvent into the system, a thermostating oven, where the extraction cell is placed and extraction takes place, and a restrictor or valve to maintain the pressure. Extracts are collected in a vial placed at the end of the extraction system. In addition, the system can be equipped with a coolant device for rapid cooling of the resultant extract (Herrero et al., 2006).

2.2 Predicting solubility via the hansen solubility parameter

Predicting solubility parameters capable of telling whether one substance can form a solution by dissolution in another can be done by several thermodynamical methods with a variable degree of theoretical base and reliability. Following the reasoning that dissolution is linked to likeness, i.e. similarity in bonding to itself, a solubility parameter can be defined for every substance as the square root of the cohesive energy density $\delta = (E/v)^{1/2}$ where v is the molar volume of the pure solvent, and E is its (measurable) energy of vaporization (Hildebrand and Scott, 1950). The total energy of vaporization of a liquid consists of multiple individual components, among which (atomic) dispersion forces, (molecular) permanent dipole–permanent dipole forces, and (molecular) hydrogen bonding (electron exchange) (Hansen, 1997). Dividing this by the molar volume gives the square of the total (or Hildebrand) solubility parameter as the sum of the squares of the Hansen D, P, and H components, that is:

$$\delta_T = E/V = E_D/V + E_P/V + E_H/V = \delta_D + \delta_P + \delta_H$$

Hansen parameters can be calculated from literature physical properties and solubility data, molecular structure, or group contribution methods (Hoy 1989; Hoftyzer and van Krevelen 1997; Stefanis and Panayitou 2008, Hansen 2007). These three parameters can be treated as coordinates for a point in the Hansen space; distance in the Hansen space predicts likeliness of reciprocal solubility of the substance pair. To calculate the distance (R_a) between Hansen parameters of substances S_1 and S_2 in Hansen space the following formula is used:

$$\left(R_a\right)^2 = 4\cdot\left(\delta_{D_2}-\delta_{D_1}\right)^2 + \left(\delta_{P_2}-\delta_{P_1}\right)^2 + \left(\delta_{H_2}-\delta_{H_1}\right)^2$$

which defines the radius of a sphere centered in substance S_1 and identifying the domain of substances which possess equal or better solubility than substance S_2.

An interaction radius (R_o), a characteristic quantity of the substance to be dissolved, can be also defined. The interaction radius effectively defines a metrics for the solubilisation of the substance under concern. Together, R_a and R_o determine whether the pair is within (solubility) range. The ratio R_a/R_o is called the relative energy difference (RED) of the system; if RED < 1 the molecules are alike and will dissolve, if RED = 1 the system will partially dissolve and if RED > 1 the system will not dissolve.

Correlation methods are available for the prediction of the temperature dependence of the solute solubility parameter, such as the Jayasri and Yaseen method (1980). The Hansen solubility prediction method lends itself to a computational identification of the optimal solvent, which corresponds (in principle) to the mixture exhibiting the lowest RED value. Using this approach, minimum diameter Hansen spheres can be obtained by changing the composition and the temperature of the (subcritical) solvent phase and solute-solvent interaction can be predicted and optimised.

Srinivas et al. (2009) studied the solvent characteristics of subcritical fluid solvents at different temperatures interacting with various bioactive compounds from their natural sources; they tested the applicability of this predictive method on valuable nutraceutical solutes extracted from natural sources. The use of this method will be presented with reference to the extraction of flavonoids from grape pomace.

Flavonoids consist of anthocyanins, flavonols, procyanidins with antimicrobial, antiviral and antioxidant as well as food colorant properties. Many investigations (e.g. Cacace and Mazza, 2002) have been carried out on extraction of anthocyanins from berry substrates using subcritical water and water–cosolvent systems, whose results have shown that a maximum extraction yield is obtained in the 120 to 160 °C temperature range. A lower solvent–feed ratio requirement compared to traditional solvent extraction techniques is reported (King et al., 2003). Studies on the effect of temperature and particle size on subcritical water extraction of anthocyanins from grape pomace pointed to an optimal yield at 120 °C using a 150 micron-sized substrate (King et al., 2007). These studies also indicated the possibility of thermal degradation of the target solutes occurring at higher temperatures.

Srinivas et al. (2009) applied the Hansen solubility method to the extraction of malvidin-3,5-diglucoside with water–ethanol mixtures by investigating: 1. solubility in the pure solvents as a function of temperature and 2. solubility in mixtures of the two solvents as a function of concentration. They predicted a slightly higher miscibility with ethanol (radius of 7.74 MPa$^{1/2}$ vs 9.73 Mpa$^{1/2}$) at different temperatures (25 to 75 °C for ethanol; 100 to 200 °C for water): although both ethanol and water can dissolve the target anthocyanins at different conditions, ethanol would be a preferred solvent for the extraction of anthocyanins. Mixture calculations show a significant decrease in the interaction radius of the Hansen sphere with addition of 10% ethanol. Although the maximum solvent power is predicted at 80% ethanol, the improvement over 10% ethanol is minimal. The optimal solvent temperature using hydroethanolic mixtures is in the range 25 to 75 °C. Results are broadly coherent with data reported by Monrad et al. (2010) and plotted in Figure 1.

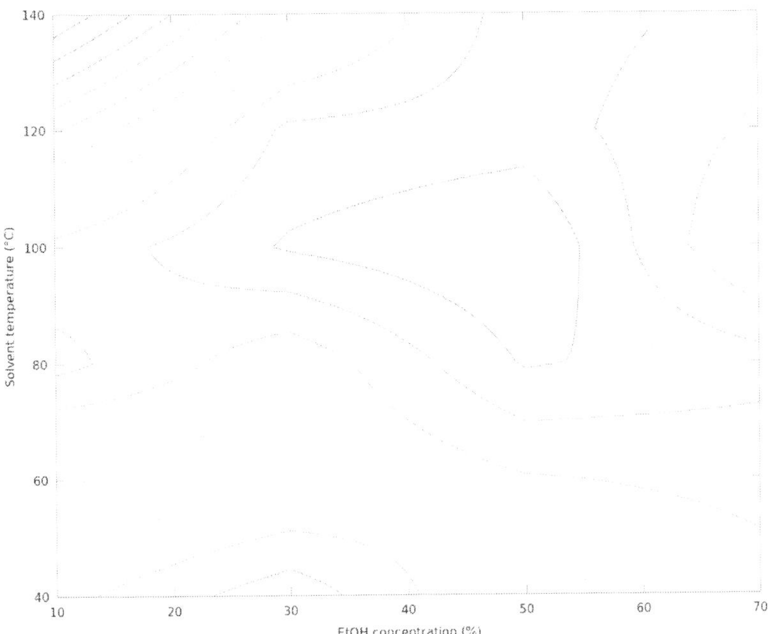

Fig. 1. Concentration of recovered Malvidin-3,5-O-diglucoside in hydroalcoholic extracts from subcritical extraction (Data from Monrad et al., 2010).

2.3 Multiple fluid process tuning for sub/supercritical fluid extraction

Multiple fluid processing involves the integration of two or more fluids held under pressure (above or below their critical temperature and pressure) applied as either mixtures or in a sequential manner for one or more unit processes (King and Srinivas, 2009). Their state may be variable (critical/subcritical/mixed). A large collection of solubility data is available in the literature (Gupta and Shim, 2007) and methods have been developed to predict it (such as the Hansen Solubility Parameter discussed above) to assist broad approaches in fraction separation and chemical transformation choices. Non polar gases can be tuned to the required solute capacity by adjusting their density (pressure) and to (ideally) match the required solubility parameter (polarity) by mixing them with polar modifier substances and possibly adjusting temperature and pressure. Polar substances, in their turn, may serve at one time as solvents and as reactants; water stands out in this respect because of its nil toxicity and cost and because it can potentially simplify process arrangement given that in most cases it is initially present in the vegetal matrix and its separation may be required for contact with non polar solvents (either organic or inorganic). Added substances may act as solubility enhancers (co-solvent: e.g. polar substances added to supercritical CO_2) or depressors (antisolvent: e.g. gases depending on their critical temperature); again, these added substances may also serve as reactants (e.g. hydrogen for hydrogenation reactions), which may be an atout despite the reduction in solvent power toward the other reactants. On the other hand, gas may be solubilised in subcritical liquid solvents, such as water. In addition to other properties (notably, for water, the dielectric constant) this may serve to tune pH by addition of CO_2, thereby obtaining a versatile medium with respect to acidic-based extraction chemistry and reactions (as in the extraction of flavonoids or in the pretreatment of lignocellulosic biomass) without the burden of a subsequent pH correction. The change in solvent power which can be obtained by the change of pressure (in the supercritical state) or temperature (in the subcritical state) permits the implementation of multiple unit operations (fractionation of multiple compounds, chemical reaction) with only one running fluid. If pressure change to carry out reciprocal solute/solvent separation and fractionations are deployed for solute separation and energy consumption must be reduced, membrane separations (expecially nanofiltration and reverse osmosis) may be implemented to obtain bulk molecule size-based separation and avoid recompressing the solvent across the full pressure range. Performing a sequential arrangement of the operations may make important savings in equipment capital cost possible. Membrane coupling may be equally useful for compressible fluids (e.g. CO_2) and scarcely compressible fluids (e.g. water or hydroalcoholic solutions); these latter tend to yield dilute solutions that need concentration and membrane processing can obtain that cheaply and neatly.

Antioxidant-containing matrices (e.g. grape and olive waste) are a fruitful area in which to apply combinations of mixed critical and subcritical fluid and unit processing steps. Residence time of the extracted solute in the hot pressurized water must be minimized to prevent degradation of the anthocyanin moieties or their possible reaction with sugars to other products. Some evidence that side reactions in pressurized water could be generating antioxidant moieties has been gathered (King and Srinivas, 2009). An example of pipelined processing is then offered by sunflower oil triglyceride conversion to free fatty acids (in subcritical water) followed by enzymatic esterification of the free fatty acids to FAMES in supercritical CO_2 using lipase catalysis (Baig et al., 2008).

2.4 Cost-effectiveness analysis in supercritical fluid extraction

The extraction of valuable materials from solid substrates by means of SCFs has been carried out on a commercial scale for more than two decades. Large-scale processes are related to the food industry like the decaffeination of coffee beans and black tea leaves and the extraction of bitter flavours (α-acids) from hops. Smaller scale processes comprise the extraction and concentration of essential oils, oleoresins and other high-value flavouring compounds from herbs and spices, and the removal of pesticides from plant material. The extraction of edible oils would be a large-scale process, but, as for all commodity products, the value-added is not high, so the economy of the process is the main problem and must be considered separately for each case.

Oil from oleaginous seeds is traditionally produced by hexane extraction from ground seeds, with a possible thermal degradation of the oil and an incomplete hexane elimination. SC-CO_2 extraction of oil from seeds has been proposed in many cases (Coriander, fennel, grape, hyprose, sunflower etc--see Reverchon and De Marco, 2006) given that oilseed triglycerides solubilise well in SC-CO_2 at 40 °C and pressure higher than 280 bar. Main parameters of this process are particle size, pressure and residence time. After extraction, the extract solution is sent to a separator working at subcritical conditions so that the reduced solubility enables recovery of oil and elimination of gaseous CO_2 from oil. Alternatively, temperature variations may be used to recover the oil extracted in a process operated at a fixed pressure; energy consumption can be reduced if thermal integration is properly deployed throughout the plant (Reverchon and De Marco, 2006).

While several seed oils have been extracted this way up to the pilot scale, here we take sunflower oil as an example.

The main models of SFE from vegetal matrices are derived from classical models of mass transfer and extraction kinetics.

Several models were already published; they mostly assume plug flow of the fluid through the fixed bed of the solid matrix and mass transfer control located in the fluid phase (as done by Lee et al., 1986), in the solid phase (e.g. Reverchon, 1986), or in both (e.g. Sovová et al., 1994).

Most of them consider only one pseudocomponent, "the solute", characterised by one (averaged) solubility value obtained experimentally. The solid matrix of the seeds is porous, homogeneous, and with constant physical properties during extraction. The physical properties of the fluid are constant as well. Usually, pressure and temperature gradients that appear in the fixed bed are neglected. Perrut et al. (1997) considered one solid phase and one fluid phase, where external mass transfer is the controlling step. Axial dispersion in the bed is neglected and all pores are pre-filled by the solvent during the initial increase of pressure in the extractor until equilibrium is reached between the solid and liquid phase (initial condition). Oil concentration y* at the solid – SCO2 interface depends on an equilibrium relation, function of pressure and temperature y* = f(x, P, T):

$$\rho_f \cdot \varepsilon \cdot \left(\frac{\partial x_f}{\partial t} + v \cdot \frac{\partial x_f}{\partial z} \right) = j_f$$

with

$$v = \frac{Q/\rho_f}{\text{extractor section}} \cdot \frac{1}{\varepsilon}$$

j is the flux of solute that is exchanged between the solid and the fluid phase based on global mass transfer coefficient considered at equilibrium.

$$j_f = a_p \, k_f \, \rho_f \, (y^* - x_f)$$

Perrut et al. (1997) introduced the notion of "transition concentration" x_t, meaning that oil concentration in SCO_2 is equal to the thermodynamic solubility value y_o regardless of its concentration in the solid concentration as long as this latter is sufficiently rich (above threshold value x_t which is normally in untreated seeds); when the concentration in the solid diminishes oil concentration is determined with a partition coefficient K.

$$x < x_t \quad f(x) = K \cdot x$$
$$x \geq x_t \quad f(x) = y_o$$

Three parameters are adjusted to experimental extraction data (solubility, partition coefficient, transition concentration).

More accurate modelling can be obtained by adding: other dispersion axes (e.g., axial dispersion in the reactor volume and radial dispersion in the seed volume as done by Cocero and García, 2001); volume compartments; a shrinking core model; a dual-zone (intact cells and broken cells; Figure 2) continuous modelling inside the particle (Sovová, 2005). This latter model acknowledges that many cells have been broken by the pre-treatment but, inside the seed, cells are intact so that oil from the broken cells is "free" and can be directly extracted while oil in the intact cells must diffuse through broken cells. Correspondingly one equation must be written for the broken cells and one for the intact cells, where r is the fraction of broken cells in a crushed seed.

$$r \cdot \rho_s \cdot (1 - \varepsilon) \cdot \frac{\partial x_1}{\partial t} = j_s - j_f$$

$$(1 - r) \cdot \rho_s \cdot (1 - \varepsilon) \cdot \frac{\partial x_2}{\partial t} = -j_s$$

The different fluxes between the broken cell and the fluid and between the intact and the broken cells are written (again resorting to the concept of transition concentration):

$$j_f = k_f\, a_p\, \rho_f\, (y^* - x_f) \; ; x_1 \neq x_t \text{ and } x_f < K\, x_t$$

$$j_s = k_s\, a_p\, \rho_s\, (x_2 - x_1)$$

The extraction of oil from oleaginous seeds might be one primary target of SFE if the added value of extracted vegetable oils were not very low. However, specialty oils or valuable components co-extracted with common oils could be reasonable target of SFE extraction (Brunner, 2005).

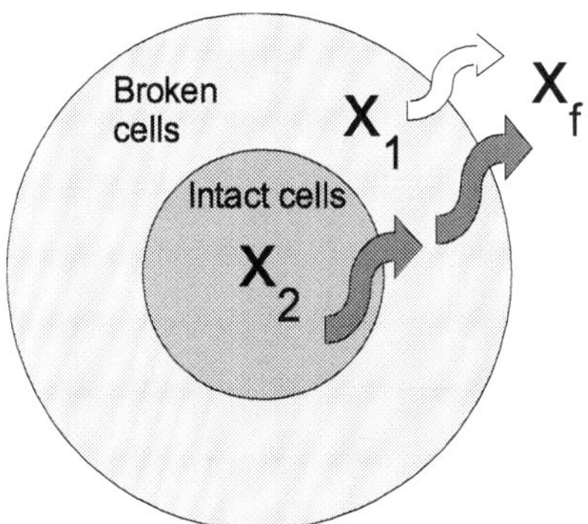

Fig. 2. Illustration of a dual-compartment modelling scheme for crushed seed extraction. (Reprinted from Boutin et al., 2011, with permission from Elsevier)

One drawback to widespread industrial scale adoption of supercritical fluid extraction use is the lack of realistic economical studies. In order to evaluate the feasibility of using SFE in the extraction of food products or components thereof, an estimate of the production cost must be provided, generally by costing an approximate plant design providing the separation. While the level of accuracy of the modelling (thermodynamics) and costing will affect the reliability of the estimate, a rigorous approach is generally out of the scientific investigator's reach and scope.

The assessment of the industrial economical feasibility of entire supercritical extraction processes has been carried out by a few investigators. Citrus peel oil deterpenation (Diaz et al., 2005), such as essential oil extraction from rosemary, fennel and anise (Pereira and Meireles, 2007) and sunflower oil (Bravi et al., 2002). Then, given that yield and quality of the product are simultaneous targets when investigating an extraction, an example of recovery yield and product quality bridging will be given along the lines of Bravi et al. (2003).

The solid substrate in most cases forms a fixed bed. The SCF flows through the fixed bed and extracts the product components until the substrate is depleted. This extraction from solids consists of two process steps, namely, the extraction, and the separation of the extract from the solvent. During the extraction the supercritical fluid is fed and evenly distributed at one end of the extractor where it flows, upward or downward, through a fixed bed of solid particles of the vegetal/microalgal matrix and dissolves the extractable components. The depletion boundary in the solid matrix will proceed in the direction of flow, as the concentration of the extracted components in the solvent. The shape of the concentration curve depends on the kinetic extraction properties of the solid material and the solvent power of the SCF which, in turn, depend on operating conditions. For the solid as well as for the solvent, the extraction is an unsteady process (Brunner, 1994).

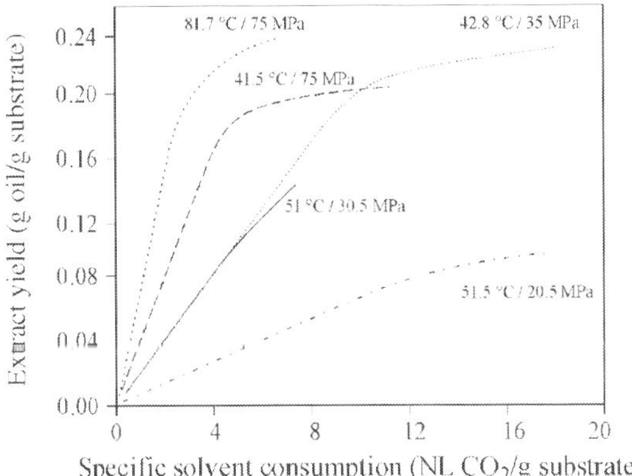

Fig. 3. Extraction yield vs mass of supercritical solvent flowed through the vegetal matrix bed (With kind permission from Springer Science+Business Media: Gas extraction: an introduction to fundamentals of supercritical fluids and the application to separation processes, 1994).

A time profile of the extraction yield will be typically shaped as any one of the curves in Figure 3. The initial part of the curve is a straight line (constant extraction rate). At later times the extraction rate decreases as the total amount of extractible substances in the substrate is approached. Any extractible component in the solid follows the extraction course pattern. If mass transfer is fluid phase side-limited the extraction profile is represented by a straight line (the slope equalling the extraction rate); however, if mass transfer is split among the contacted phases the extraction has an exponential course. If the solvent enters the extractor free from the extractible compounds and there are no irreversible reactions of the extractible compounds with the substrate, the matrix can be totally depleted; otherwise the extraction curve approaches a non null asymptote (Brunner et al., 2005).

High-pressure processes are particularly energy intensive and economic feasibility depends on energy integration; the solvent cycle scheme plays a central role in this respect (Brunner, 1994). In supercritical fluid processes the solvent can be recirculated either in supercritical or in liquid state; correspondingly, the piece of equipment of choice would be a compressor or a pump. Pumps have a lower capital cost than compressors, but pump-based solvent cycles also require several heat exchangers and condensers and additional heat energy at low extraction pressures. Compressors have a higher capital cost than pumps, but compressor-based solvent cycles require only one heat exchanger and a limited thermal energy supply; at extraction pressures lower than 300 bar, the compressor-based system has higher electrical energy consumption and lower energy consumption compared to the pump cycle (Diaz et al., 2009).

Objective functions have been annualised cost, utility cost (Cygnarowicz and Seider, 1989), energy consumption (Diaz et al., 2000), net profit (Diaz et al, 2003), product cost (Bravi et al, 2002). The approaches range from DAE model resolution (Bravi et al., 2002), nonlinear programming (Cygnarowicz and Seider, 1989), mixed integer nonlinear programming (Diaz et al., 2000, 2003, 2005; Espinosa et al., 2005). Mixed integer programming permits the association of binary variables to design options (e.g. potential process units). Diaz et al. (2005) and Espinosa et al. (2005) performed the optimal design of process and solvent cycle by formulating and solving two nonlinear programming problems.

Rigorous mass transfer models are rare in such models: a dynamic optimization model for an extraction column, also including energy and momentum balances in the packed bed was developed and coded in gPROMS by Fernandes et al. (2007).

Alternative approaches include experimental data-based process optimization, which typically end up in nonlinear correlations among process variables (Létisse et al., 2007 adopted extraction temperature, solvent flow rate and operating time).

As an example here we discuss the optimisation approach followed by Bravi et al. (2002 and 2003). The optimisation requires 1. devising an industrially-feasible process layout and 2. identify optimal operating conditions and assess SFE-extracted sunflower oil economic economic acceptability in the food market. This latter aim requires the setup of a comprehensive mathematical model of the whole process (extraction section and recovery section).

The adopted process included multiple batch extractors in parallel, each containing a fixed bed of seeds, and multiple CO_2 recovery stages operating at decreasing pressures (Figure 4). The optimisation problem size was reduced by adopting: 1. Perrut's (1997) non-continuous and piecewise-linear solid-fluid equilibrium mathematical model of the extraction phase and extraction conditions (40 °C and 280 bar); 2. oil approximation as a single component; 3. mass transfer resistance occurring only in the solvent phase; 4. negligibility of in-extraction enthalpy variations; 5. requirement of constant extract flow rate.

A comprehensive model for the entire process describing extractors, expansion valve, separator, compressors, solvent recovery vessels, heat exchangers, pipe union joints and branches was set up. The BWR equation of state was chosen as a trade-off between accuracy and computational weight.

The performance criterion (objective function) adopted for the optimisation was the unit oil production cost, estimated by rigorously accounting the operating costs directly referenced by the process design (e.g., compression costs and duty requirements) and applying short-cut techniques for all the remaining operating costs (e.g., manpower) and for the equipment cost estimates (by the cost index method). Aim was finding the minimum unit oil production cost as a function of the time allotted for the extraction phase on each seed batch and of the prevailing pressure in the oil separator at constant: size of each extraction vessel, number of simultaneously flowed extraction vessels and circulating solvent flow rate.

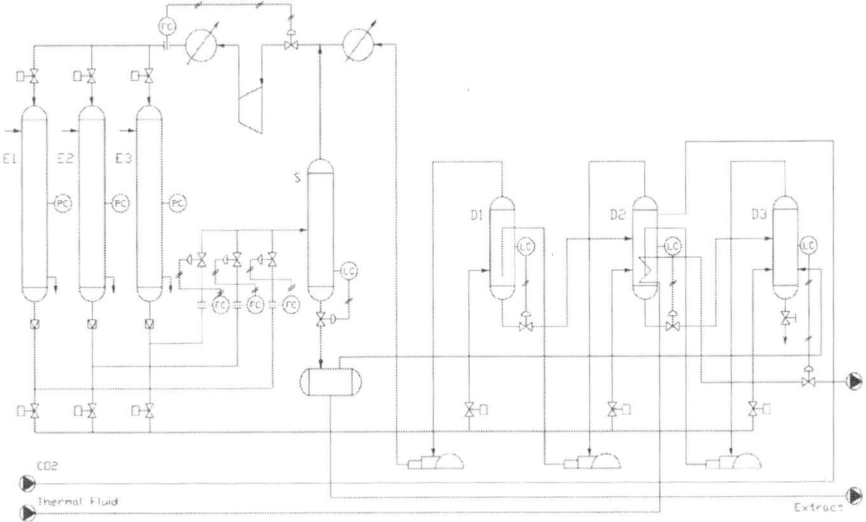

Fig. 4. Process scheme for contnuous sunflower oil extraction with supercritical CO_2 (Reprinted from Bravi et al., 2002, with permission from Elsevier).

A reduction of the required power expenditure for solvent recompression can be obtained by increasing the operating pressure of the oil separator but this reduces oil recovery. Cost minimisation led to optimal pressure in the separator of a single-extractor plant (100 bar). Furthermore, under the adopted equilibrium model, productivity collapses over time, when most of the seed bed has a very low oil concentration (Figure 5); from an economic standpoint product return decreases while the operating cost remains unchanged, thus leading to an increase of the average oil production cost. On the other hand, at the end of the extraction cycle the seed bed still contains an oil residue that cannot be exhausted by using re-circulated solvent but can be exhausted with hexane, yielding a lower-class product.

Reducing extraction time increases oil production rate at the expense of a lower recovery. In the example, extraction times of 20, 10, 5 minutes makes the best possible use of a plant with 3, 4, or 5 extractors. The authors found product cost to be minimum cost for a 4-extractor plant (0.67 Euro/kg)

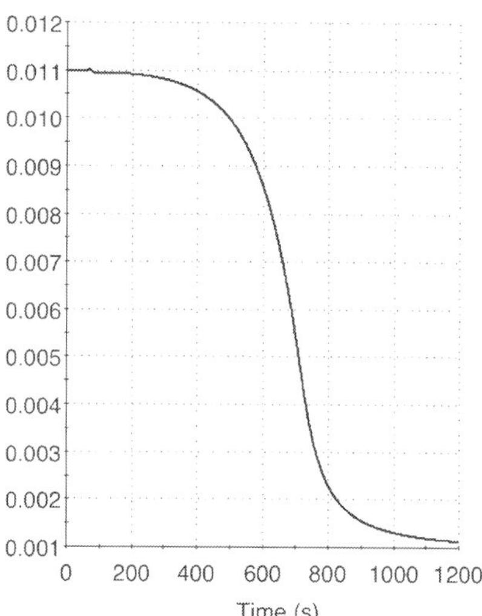

Fig. 5. Oil content in the supercritical phase leaving from a single extractor during a single extraction phase (Reprinted from Bravi et al., 2002, with permission from Elsevier)

Quality optimisation in extraction by SC-CO$_2$ can be obtained by suitably modelling the adopted quality parameters, which can be done relatively easily if these latter can be related to components whose concentration can be modelled by resorting to mass balances, mass transfer coefficients, and thermodynamic equilibria. Sunflower oil acidity modelling performed by Bravi et al. (2003) will be reported here as an example; another target may be the content of lipid-soluble alpha-tocopherol.

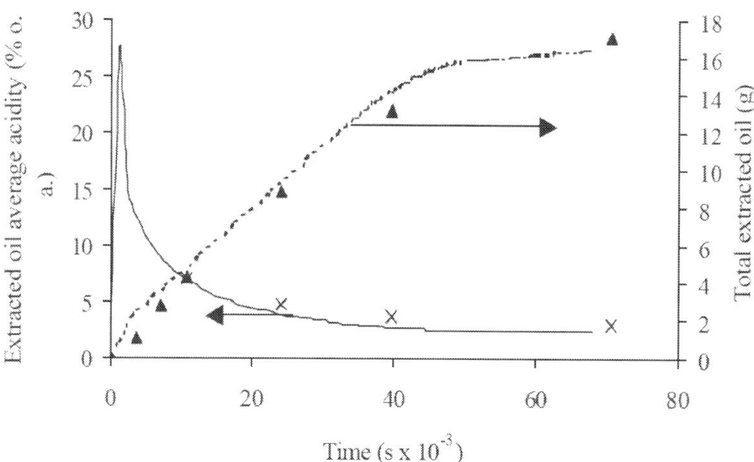

Fig. 6. Model prediction of instantaneous sunflower oil acidity and total extracted sunflower oil mass as a function of extraction time (From Bravi et al., 2003)

The mathematical model of this process takes into account the mass and enthalpy balances of carbon dioxide and oil in the extractor, the expansion valve and the separator. Oil is considered as being made of two components: free oleic acid (as acidity is customarily reported), and a triacyglyceridic pseudo-component. Perrut's et al. (1997) mathematical model of the oil extraction was rewritten to account for two simultaneously extracted components. The parameter estimation was carried out by optimising the model fit on the data obtained by means of a suitable oil quality-oriented experimentation.

Batch extraction runs were carried out on ground sunflower seeds in Perrut's conditions at different CO_2 flow rates and the composition was recorded on the collected oil in four well-mixed batches per extraction run. The results shows an initial constant-rate extraction phase after which productivity declines; the combined time profile of the collected oil acidity as a function of the amount of oil extracted clearly shows that the acid component tends to concentrate in the first oil fraction (Figure 6).

From the point of view of process optimisation an improvement of the quality of oil can be obtained by two means: 1. by eliminating the first portion of the extracted oil, which gives the largest contribution to acidity and/or 2. by extending the oil extraction degree; in order to keep some more oil extracted in the initial stages of the operation, much more must also be extracted in the final stages to ensure that the overall acidity is within the limits. Both of these measures increase the oil production cost because the oil extraction rate is maximum at the beginning of the operation and then decreases continuously and asymptotically to zero. As suggested by Bravi et al. (2002) the only partially exhausted oil matrix can be treated in a conventional extraction plant using hexane; we add here that the first extracted oil, which features an excessive acidity but is hexane-free, could be de-acidified and sold as a different product.

When optimising SFE yield in a stream concentrated in some desired component, identifying yield and extraction kinetics as a function of temperature, solids particle size distribution and CO_2-to-solids mass ratio are a key step before modelling can be applied. However, industrially-relevant practices may reduce yield compared to lab vales.

As an example, based on Molero Gómez's et al. (1996) set of oil yield-relevant set of optimal extraction conditions from grape seeds (pressure 200 bar, temperature 40°C, seed fragment size 0.35 mm, seed moisture content up to 6.5% and processing time 2 h), Bravi et al. (2007) carried out a study aimed at investigate the yield and extraction kinetics of α-tocopherol-enriched grape seed oil as a function of temperature, solids particle size distribution and CO_2 to solids mass ratio (hereinafter denoted as CSMR) and identifying the optimal extraction conditions defined as those ensuring a high yield in α-tocopherol-concentrated oil. Their experimental procedure did not include pre-soaking (elsewhere denoted as 'equilibration') of the seed matrix to better reflect the prospective process conditions in a forthcoming industrial use (From Bravi et al., 2003, with kind permission from AIDIC Servizi s.r.l.).

They confirmed that oil yield with CO_2 (14.4%) is slightly below that with hexane (15.4%) when the vegetal matrix is treated at a moderately low temperature and observed that temperature effects on yields must be analysed with care, as they may be the result of complex interactions between oil solubility in CO_2 and mass transfer coefficients, arguing that this may be due to a soaking degree which changes during the extraction itself. As Bravi et al. (2007) pointed out, this may lead to unexpected inverted yield vs temperature relationships.

The effect of fragment size is that of increasing the oil yield at any tested CSMR; this effect can be explained with the increase in the available surface area for mass transfer and with the reduction of the time required for the initial soaking of the vegetal matrix. From the practical point of view, the results of their work suggest that the grape seeds should be milled to a maximum size of 425 mm.

As far as α-tocopherol in the extract is concerned, its concentration in the oil extracted with SC-CO_2 increases with the extraction temperature; this result is coherent with the higher solubility of α-tocopherol in SC-CO_2 at 80°C than at 40°C measured by Chrastil (1982).

3. Optimisation of enzyme-assisted extraction

Whatever the solvent, purely solvent-based extraction (i.e., without the intervention of synergic agents) of bioactive compounds often suffers from low extraction yields, and requires long extraction times and leaves traces of the organic solvent used, generally toxic to some extent. When solvent extraction is carried out, resistance to solute migration to the bulk of the solvent may be controlled within the solid phase, within the liquid phase, or be split among the two. When the first, or even the last is true, non polar solvents may be able to overcome cellulosic barriers but may be unable to solubilise the desired compounds; in turn, polar solvents may be a suitable solvent for the desired compounds but may be unable to reach the solute location site inside the matrix. In both cases, reducing the mechanical hindrance to solute migration may make the operation significantly faster and reduce solvent requirements; as a side gain, it also reduces any degradation reaction that may affect the desired product during the extraction itself.

Enzymes, derived from bacteria, fungi, animal organs or vegetable/fruit extracts, have been used particularly for the treatment of plant material prior to conventional methods for extraction. Plant materials like vanilla, pepper, mace, mustard, fenugreek, rose, and citrus peel which have potential as a rich source of flavor have been studied for enzyme-assisted extraction of flavors. Similarly enzyme-assisted extraction of color has been studied in plant materials like marigold, safflower, grapes, paprika, tomato, alfalfa, and cherries (Sowbhagya and Chitra, 2010). Winemakers may make use of pectolytic enzymes to break down the middle lamella between the pulp cells and the pulp and skin cell walls releasing pigments, improving both juice yields and colour extraction (Ducruet et al., 1997). A list of some commercially relevant products recently obtained using enzyme-assisted extraction is reported by Puri et al. (2011).

Enzymes have been used to enhance extraction (e.g. flavonoids from plant material as by Kaur et al., 2010) while minimizing the use of solvents and heat and disrupt the pectin–cellulose complex (e.g. in citrus peel to enhance flavonoid production by Puri et al., 2011). A digestion step prior to extraction by solvents was found to be necessary to efficiently carry out the extraction from the raw material (Dheghan-Shoar et al., 2011).

Various enzymes such as cellulases, pectinases and hemicellulases are often required to disrupt the structural integrity of the plant cell wall; these enzymes hydrolyze cell wall components and degrade the pectin-cellulose complex in fruits, thereby increasing cell wall permeability, which results in higher extraction yields of bioactives.

The enzymatic reactions are usually conducted at low temperature (15 °C to 45 °C), the actual operating temperature being dictated by the trade-off between the two controlling phenomena (mass transfer enhancement by increased temperature which reduces required extraction time and thermal degradation of any thermolabile extracted compound). Above ~60 °C, heat alters the enzyme molecule irreversibly.

In order to carry perform cost-effective enzyme-assisted extractions the features and subtleties of enzyme catalysis must be considered and the appropriate enzyme or enzyme combination for the plant material selected must be identified.

Cost effectiveness and optimisation of enzyme-assisted extraction is obtained by identifying the optimal extraction conditions, aimed at maximising process profitability, which entails maximising the recovery rate of the target bioactive(s) while minimising their in-process degradation. Optimisation parameters, therefore, belong to two sets relevant to the enzymatic pretreatment and to the extraction phase respectively. The pretreatment should be optimised with regard to prevailing temperature and pH, pretreatment time, enzyme solution-to-solid ratio, solid particle size (distribution), enzyme load and enzyme composition; the extraction should be optimised with regard to prevailing temperature and pH, time, and solvent system deployed.

Synergism is due a special consideration when optimising enzyme-assisted extractions as it is for any enzyme-assisted hydrolysis. Considering the example of lycopene extraction from tomato waste, enzyme synergism may show up when they are acting simultaneously (e.g. Zuorro et al., 2011) or sequentially (Ruiz Teran et al., 2001). In the former case, enzyme preparations containing 50:50 pectinase and cellulase were found to have a significant synergistic effect (rate x18 w.r.t untreated material) with respect to simple enzymes (rate x3); in the latter, it was observed that cellulase (from *Trichoderma reesei*) and mixed enzyme cocktail (a mixture of arabinases, cellulases, hemicelullases, xylanases, and pectinases from *Aspergillus niger*) do not work efficiently when used together. Furthermore, results obtained using cellulase after the enzyme cocktail are similar to those observed when the cocktail is used alone. However, when cellulase was used first the extractive reaction proceeded with the highest efficiency (Table 1).

Enzymes used for pretreatment	Vanillin g/100 g
Water (control)	1.07 ± 0.08*
Cellulase + water	1.17 ± 0.06*
Cellulase + ethanol	2.70 ± 0.17*
Viscozyme + cellulase + water	1.17± 0.11
Viscozyme + cellulase + ethanol	2.66 ± 0.07
Cellulase + viscozyme + water	2.30 ± 0.10
Cellulase + viscozyme + ethanol	3.66 ± 0.04
Viscozyme + water	1.17 ± 0.05*
Viscozyme + ethanol	2.45 ± 0.21*

Table 1. Effect of enzyme treatment on vanillin extraction from vanilla beans. Viscozyme L. (Novozymes) is a mixture of arabinases, cellulases, hemicelullases, xylanases, and pectinases (Adapted from Ruiz Teran et al., 2001).

Enzyme-assisted extraction of bioactive compounds from plants has potential commercial and technical limitations: 1. the cost of enzymes (although this is going to decrease thanks to biofuel research); 2. inability of currently available enzyme preparations to hydrolyze some fractions of plant cell walls, limiting extraction yields of some compounds; 3. scale up to industrial scale may be troublesome because local process conditions in the equipment may be difficult to maintain at the larger scale.

Enzyme-assisted extraction optimization by traditional methods is time consuming and can ignore the interactions among various factors. The response surface method enables the evaluation of several process parameters simultaneously (along with their interactions up to the desired order) such as: time, temperature, pH, enzyme type and concentration. During cell wall degradation the polysaccharide–protein colloid can be degraded thereby creating an emulsion that interferes with the extraction. Therefore, non-aqueous systems are preferable for some materials because they minimize the formation of polysaccharide-protein colloid emulsions (Puri et al., 2011). Prior knowledge of the cell wall composition of the raw materials helps in the selection of an enzyme or enzymes useful for pretreatment; predictive methods would be useful to speed up the optimisation or keep up with product property changes, but nothing covering this gap has reached the open literature yet.

4. Optimisation of Microwave-Assisted Extraction (MAE)

Microwaves are electromagnetic waves in the frequency range 300 MHz to 300 GHz, that is between wavelengths of 1 cm and 1 m. Most small size microwave instruments operate at 2450 MHz and have an energy output between 600-700 W. At this frequency, the electric field changes the orientation of water molecules 2.45×10^9 times every second, while thermal agitation tends to restore the chaos inherent to the system. Thus creating an intense heat that can escalate as quickly as several degrees per second (depending on frequency and sample size, it has been estimated up to 100 °C/s at 4.9 GHz by Lew et al., 2002). The heating phenomenon is based on the interaction of the electrical field with the individual compounds of a material (possibly characterised by an inhomogeneous macrostructure). The transformation of electromagnetic energy in thermal energy occurs by two mechanisms: ionic conduction and dipole rotation. The ionic conduction generates heat due to the resistance of medium to ion flow. The migration of dissolved ions causes collisions between ions and molecules because the direction of the ions changes every time the electromagnetic field changes its orientation. Dipole rotation is related to the alternating movement of polar molecules, which try to line up with the electric field; thus only selective and targeted materials warm-up based on their dielectric constant.

The efficiency of the microwave heating depends on the dissipation factor of the material, tan δ, which measures the ability of the sample to absorb microwave energy and dissipate heat to the surrounding molecules as given by:

$$\varepsilon'' \tan \delta = \varepsilon'$$

where ε'' is the dielectric loss which indicates the efficiency of converting microwave energy into heat while ε' is the dielectric constant which measures the ability of the material to absorb microwave energy. The rate of conversion of electrical energy into thermal energy in the material is described by:

$$P = K\, f\, \varepsilon'\, E^2 \tan \delta$$

where P is the microwave power dissipation per unit volume, K is a constant, f is the applied frequency, ε' is the material's absolute dielectric constant, E is the electric field strength and tan ι is the dielectric loss tangent.

Multiple collisions from this agitation of molecules generate energy dissipation and therefore a temperature increase whose value depends on the local intensity of energy dissipation, on the local specific heat, and on the local conductivity by which heat diffuses to or from neighbouring areas of the material when these latter are cooler or hotter.

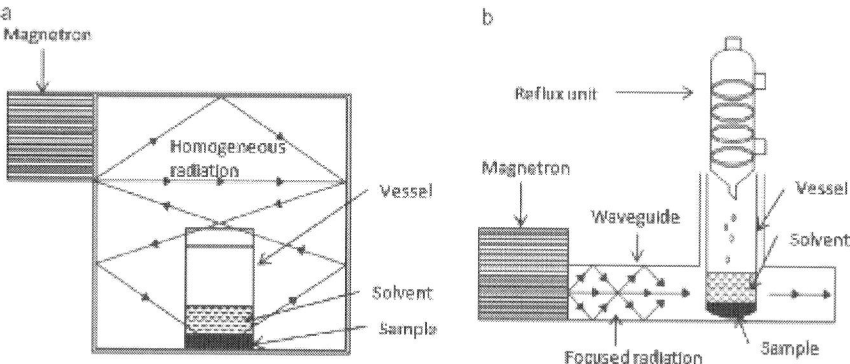

Fig. 7. (a) Closed type microwave system and (b) open type microwave system. (Source: Mandal et al., 2007)

Microwave-assisted extraction aims at supplying heating locally where the solvent or soaking medium is present, thus speeding up heating, with the aim of reducing bioactives degradation reactions and cause a discregation of the vegetal matrix structure. This latter aim requires that the basic structure of the extracted particles be stiff. Vegetal cells generally feature a stiff structure; microalgae lacking frustule or thick outer exopolysaccharide envelope may not benefit from microwave-assisted extraction (Pasquet et al., 2011), while wall-possessing microalgae benefit greatly from it (yields 3x-5x in a fraction of the time; Cravotto et al., 2008).

Microwave assisted extraction may be carried out in closed and open systems (Figure 7). Modifications of the basic scheme include operation under reduced pressure (for operation at reduced temperature and suction-enhanced migration of solutes); under nitrogen blanket (for increased protection against oxydation); without added solvent (resorting to constitutive/hydration water); with simultaneous microwave and ultrasound field (to facilitate the formation of cracks in the solid matrix, change the perceived polarity of the solvent and enhance the mass transfer) (Chan et al., in press).

Microwave-assisted extraction effectiveness strongly depends on: solvent choice, solvent to feed ratio, extraction time, microwave power, temperature, sample characteristic, effect of stirring.

Solvents which are suitable for conventional extraction techniques may not be suitable for microwave assisted extraction. Ethanol (used in the 40% to 100% concentration range) is a good microwave converter (good absorber as a dipole with a low specific heat) and is by itself suitable for extracting many active compounds from plants. However, modifiers can be added to solvents which are unsuitable per se to microwave capture in order to enhance their overall performance. Water was added as modifier to diethyl ether and ethanol or water can be added into poor microwave absorber such as hexane to enhance microwave heating efficiency (Alfaro et al., 2003; Ku et al., 2007). Room temperature ionic liquids interest lies in their negligible vapor pressure, wide thermal range in the liquid state, good thermal stability, tunable viscosity, miscibility with water and organic solvents, good solubility and extractability for various organic compounds (Du et al., 2007). Mixing carbonyl iron powders with the moist sample may increase absorption of microwave energy, particularly where solvent is limited (such as in SFMAE by Wang et al., 2006).

An optimum ratio of solvent to solid ratio ensures homogeneous and effective heating; above, heating may be insufficient and below mass transfer barrier may establish as the distribution of active compounds is concentrated in certain regions and their displacement out of cell matrix may be impaired (Mandal et al., 2010). The optimum ratio of solvent to solid ratio normally is in the interval 10--50 ml per g of solid matrix (Chan et al., in press).

Microwaves heat the sample locally and act as a driving force for damaging the plant matrix so that analytes can diffuse out and reach the solvent. Furthermore, the solvent viscosity and surface tension decrease improves mass transfer. Therefore, increasing the power will generally improve the extraction yield and result in shorter extraction time; however, the gain may be offset by thermal degradation of the solute, so that an optimal temperature (and microwave power) exist for each application (Chan et al, in press).

Process time (regardless of power) is from a few minutes to more than 1 h. Extending process time has been found to decrease the extraction yield due to thermal degradation, oxidation, or hydrolysis. The optimum microwave application mode should be adapted to the type of solid matrix and may be power-regulating (continuous medium power or pulsed high-power microwaves), for optimised degradation of the solid matrix; temperature-regulating (power is dependent variable), for optimised preservation of the bioactive extracts.

Prior to treatment the vegetal matrix is usually dried, powdered and sieved, avoiding too a small particle size which would be difficult to separate and would require equipment clean-up procedures later. Pretreating with water the vegetal matrix after drying may increase bioactives release thanks to the localised heating of the trapped soaking water which gives rise to cracks. The modified water content of the vegetal matrix also alters the balance between hydrolyzation (favoured) and oxidation of active compounds (Wang et al., 2006).

Stirring mitigates the negative effect of low solvent to feed ratio on extraction yield by improving mass transfer and thus avoiding the buildup of concentration gradients that impair the dissolution of bioactives bound to the sample matrix.

The determination of optimum MAE operating conditions is usually carried out through statistical optimization studies. A collection of optimised extraction conditions for a number of plant-sourced vegetal matrices is reported by Chan et al. (in press)..

A variation of MAE involves the deployment of microwaves without any added solvent. Historically, dry distillation was used by alchemists for sublimation and extraction. Solvent free microwave-assisted extraction (SFME) was conceived for laboratory scale applications in the extraction of essential oils from different kinds of aromatic plants. SFME is neither a modified microwave assisted extraction (MAE) which use organic solvents, nor a modified hydro-distillation which use a large quantity of water, both of which are more energy intensive due to the larger heat requirements to evaporate and condense the added solvents. Based on a relatively simple principle, this method involves placing plant material in a microwave reactor, without any added solvent or water. The heating undergone by constitutive water within the plant material expands the plant cells and leads to rupture of the glands and oleiferous receptacles, whereby essential oil is first freed and then evaporated together with the in situ water of the plant material. The vapours are continously condensed by a cooling system located outside the microwave oven. The water excess is refluxed to the extraction vessel in order to restore the in situ water to the plant material. A large number of different essential oils have been extracted by SFME (Lucchesi et al., 2004).

5. Optimisation of ultrasound-assisted extraction

Extraction enhancement by ultrasound is attributed to the propagation of ultrasound pressure waves, And to the resulting cavitation phenomena. The implosion of cavitation bubbles generates turbulence which accelerates diffusion and impingements and collisions which result in surface peeling, erosion, punching (Ugarte-Romero et al., 2006) and particle breakdown. This effect provides exposure of new surfaces further increasing mass transfer (Vilkhu et al., 2008).

Dry materials may swell, hydrate and increase their pore size under ultrasound treatment. Furthermore the particle size distribution of the vegetal matrix is shifted toward the smaller sizes, so that the cell surface directly exposed to extraction increases (Vinatoru, 2001).

Solvent selection is usually based on achieving high molecular affinity between the solvent and solute. When cavitation bubbles are generated in the bulk of the solvent phase by the ultrasound field, their hydrophobic surfaces increase the net hydrophobic character of the extraction medium (Vilkhu et al., 2008) so that its affinity toward non polar components is increased. Cavitation being initiated by drop of the total pressure below the saturation pressure of the solvent, ultrasound assistance in extraction with a given solvent will be affected by the physical properties of this latter (cavitation intensity decreases as vapour pressure and surface tension increase). When a supercritical solvent is used, cavitational events are impossible, since there is no liquid/gas phase boundary. However, other mechanisms are postulated, such as acoustic streaming and the presence of gas pockets in the solid causing cavitational collapse (Patist and Bates, 2008).

Ultrasound may also give rise to multiple simultaneous processes, such as extraction and (sono)chemical modification, whereby the food product may be modified by physical and chemical mechanisms. Cravotto et al. (2004), for instance, reported wax conversion to policosanol (common name for a mixture of C_{24}–C_{34} linear saturated fatty alcohols, a rich source of nutrients and pharmacologically active compounds) during rice bran extraction by using ultrasounds.

Although it is relatively easy to perform an ultrasound-assisted extraction at the laboratory scale, designing it for industrial scale is quite demanding. Some of the issues that need consideration when attempting to design an optimal ultrasound-assisted process at a significant scale have been reported by Vilkhu et al. (2008): 1. the nature of the tissue being extracted and the location of the components to be extracted with respect to tissue structures; 2. pretreatment of the tissue prior to extraction; 3. the nature of the component being extracted; 4. the effects of ultrasonics primarily involve superficial tissue disruption; 5. increasing surface mass transfer; 6. intra-particle diffusion; 7. loading of the extraction chamber with substrate; 8. increased yield of extracted components; 9. increased rate of extraction, particularly early in the extraction cycle enabling major reduction in extraction time and higher processing throughput.

Apparently, there is potential for ultrasonic cavitation to propagate free radicals (hydroxyl). Radical production should be quenched by the addition of small amounts of ethanol to cool cavitation bubbles and slow any radical-involving reactions (Vilkhu et al., 2008).

6. Optimisation for novel sources of bioactives: Microalgae production

The microalgae have in practice an interesting composition in regard to main components such as protein, polyunsaturated fatty acids (PUFA), pigments, and carbohydrates (Doucha 2009). The protein content is consistently high in micro algae. Some cyanobacteria (blue-green algae) are characterised by a high protein content (60-65%), not commonly found among higher plants. But, for full utilisation of the protein, special treatment of the microalgae is generally necessary. Moreover, microalgae are excellent producers of essential amino acids. However, until to date only three species are cultivated on industrial scale level: these are the cyanobacterium *Arthrospira*, the green algae *Chlorella* and *Dunaliella*. Their biomass is used for production of a rather limited range of products, most of them directed to the nutraceutical market. The success of these three species is due to the fact that they can be grown in a very selective medium (*Arthrospira* and *Dunaliella*), therefore contamination of parasites or competing organisms (microalgae, fungi, and others) is naturally prevented even in open reactors where it is possible to ensure a low cost of production for the biomass (Boussiba and Affalo, 2005), while *Chlorella* is endowed with a remarkably high growth rate sustained by organic acid addition in fermentor.

6.1 *Chlorella vulgaris*

Chlorella vulgaris cells contain β-1,3-glucan, polysaccharides and also a rich source of proteins, 8 essential amino acids, vitamins (B-complex, ascorbic acid), minerals (potassium, sodium, magnesium, iron, and calcium), β-carotene, chlorophyll, "CGF" (Chlorella growth factor), as well as other health-promoting substances (Hac´on-Lee et. al 2010). β-1,3-glucan is an active immunostimulator, a free- radical scavenger and a reducer of blood lipids (Ryll et al., 2003). However, various other health-promoting effects have been clarified (efficacy on gastric ulcers, wounds, and constipation; preventive action against atherosclerosis and hypercholesterolemia and antitumor action). *Chlorella vulgaris* biomass has colouring properties and has been tested with success as a pigment source for farmed products with functional activity (e.g. as antioxidants). The total annual production of *Chlorella* is estimated to be about 2000 t. The production process is based on the mixotrophic nature of the *Chlorella* strains and uses acetic acid as a carbon source. The production cost of biomass is not clear. However, based on claims that in those systems a high biomass concentration is achieved (more than 10 g/l) one may reach the conclusion that the production cost may be a figure close to those of open pond systems (10 -15 US$/Kg). However, the cost can raise to up to 30 US$/Kg, for example in the Central part of Europe where the adverse climatic conditions do not allow to grow the alga all the year around. Up to about 10 years ago most of the production took place in Taiwan and only 10-15% of the total production was carried out in green houses in Japan. The market for *Chlorella* products is limited to the Far East, mainly Japan. In some early works it was reported that *Chlorella* extracts may have affect the growth and production of lactic acid by lactic bacteria. The growth facilities are based on round concrete ponds mixed by a rotating arm which also provides CO_2 and acetic acid supply. Another development that should be mentioned is the attempt to set up new facilities for culturing *Chlorella* in Europe. One of them is located in Czech Republic (2 tons a year production capacity, Kopecky personal communication) which is based on inclined reactors that allow to maintain a fast flow rate of thin layer culture and as a result enables maintenance of high biomass concentrations and high volumetric productivity (Masojidek et al. 2010). The second facility is located in Germany, and uses tubular photobioreactors made with glass tubes arranged on a vertical fence and placed in a greenhouse. The total annual capacity is claimed to reach 150 tons per year (Pulz 2001). Although *Chlorella* market is still the largest one in term of gross revenue US$, one can expect that without developing new products and reducing the production cost *a sine qua non condition* for use of *Chlorella* biomass as feed additive in the animal feed market, it is difficult to expect an expansion of the production capacity.

6.2 *Arthrospira platensis*

At present *Arthrospira* (commercially indicated as *Spirulina*) represents the second most important commercial microalga in term of total market value US$ (after *Chlorella*), while in terms of total biomass produced, the *Arthrospira* market is twice or more of that occupied by *Chlorella* (Torzillo and Vonshak 2003). The major producers of *Arthrospira* are the DIC group of companies, Earthrise in California, USA, Hainan DIC Marketing in Hainan Island, China. On the whole these facilities produce about 1000 metric tons of *Arthrospira* annually (Belay et al. 2008; Sili et al. 2011). An other important *Arthrospira* producer is Cyanotech Corporation of Hawaii with an annual production of 300 tons. Other producers are located mainly in the Asia-Pacific region, particularly in China and India (Lee et. al 1997). The highest production capacity of *Arthrospira* biomass takes place in China. Recent estimates are that the total potential of the different sites of this country may exceed 2000 t. Production is carried out in raceway ponds of 2000-5000 m² in size and may contain between 400 and 1000 m³ of culture according to the dept adopted which can vary between 15 and 40 cm depending on season, desired algal density and, to a certain extent, the desired biochemical composition of the final product (Belay et. al 2008). The major share of the market for this organism is for health food involving crude biomass production. This has the advantages of simple processing (harvest and rudimentary handling) keeping production costs reasonably low, and of eluding the competition of the chemical industry which cannot match the wealth in nutritional bioactive components and attractiveness of natural products (Boussiba and Affalo 2005). *Arthrospira platensis* (Soletto et. al 2008, Harun et. al 2010) has commercialized as nutraceutical food, also a strong immune-stimulated molecule, Immulina®, can be extracted from it (Grzanna et. al 2006).

6.3 Dunaliella

This species is being grown as a source of beta-carotene. this carotenoid can accumulate in the cells grown under nutrient limitation and high sun light up to 12% of the dry weight (Ben-Amoz and Avron 1973). This pro-vitamin A product is widely used in the feed and food industry. Today, the product is available mainly in two forms: dried or extracted. Dried *Dunaliella*, in powder or pill form, is considered to be highest quality. The product is harvested by means of concentration and centrifugation, and thereafter dried by spray-drying. In this form, the product is mainly addressed to health food market for direct human consumption. The price is based on the beta-carotene content, and can reach about 2000 US$ per Kg of beta-carotene. The second kind of product is beta-carotene extracted into a vegetal oil. This product can be applied as a food colorant and a pro-vitamin additive for human consumption, for fish and poultry feed, or in the cosmetic industry as an additive to sunscreen products. From few reports it is estimated that the price, on the basis of beta-carotene content, varies in the range of US$ 500-600 per Kg of beta-carotene.

7. References

Alfaro M. J., Bélanger J. M. R., Padilla F. C., Jocelyn Paré J. R. (2003) Influence of solvent, matrix dielectric properties, and applied power on the liquid-phase microwave-assisted processes (MAP) extraction of ginger (*Zingiber officinale*). *Food Res Int* 36 (5): 499--504.

Baig M. N., Alenezi R., Leeke G.A., Santos R. C. D., Zetzl C., King J. W., Pioch D., Bowra S. (2008) Critical fluids as process environment for adding value and functionality to sunflower oil; a model system for biorefining, in: *Proceedings of the 11th Meeting on Supercritical Fluids*, Barcelona, Spain, May 4–7, 2008.

Belay A. (2008) Spirulina (*Arthrospira*): production and quality assurance. *In: Gershwin ME and Belay A (eds) Spirulina in human nutrition and health. pp1-25, CRC Press Taylor & Francis group, London, UK, pp.312.*

Ben-Amoz A., Avron M. (1983) On the factors which determine massive beta-carotene accumulation in the halotolerant alga Dunaliella bardawil. *Plant Physiol 72: 593-597.*

Boussiba S., Affalo C. (2005) An insight into the future of microalgal biotechnology. *Innovations in Food Tehnology (www.innovfoodtech.com).*

Boutin O., De Nadaïa A., Perez A. G., Ferrasse J.-H., Beltran M., Badens E. (in press) Experimental and modelling of supercritical oil extraction from rapeseeds and sunflower seeds, *Chem. Eng. Res. Des.*

Bravi M., Bubbico R., Manna F., Verdone N. (2002) Process optimisation in sunflower oil extraction by supercritical CO_2. *Chem. Eng. Sci.* 57: 2753 – 2764.

Brunner G. Gas extraction, in: An Introduction to Fundamentals of Supercritical Fluids and the Applications to Separation Processes, Springer, Berlin, 1994.

Cacace J. E., Mazza G. (2002) Extraction of anthocyanins and other phenolics from black currants with sulfured water. *J. Agric. Food Chem.* 50 (21): 5939--5946.

Carr A. G., Mammuccari R., Foster N. R. (2011) A review of subcritical water as a solvent and its utilisation for the processing of hydrophobic organic compounds. *Chem Eng J* 172: 1-17.

Chacón-Lee T. L. and González-Mariño G. E. (2010) Microalgae for "Healthy" Foods - Possibilities and Challenges. *Comp Rev Food Sci Food Saf, Vol. 9, (6): 655–675.*

Chana C.-H., Yusoffa R., Ngoha G.-C., Kung F. W.-L. (in press) Microwave-assisted extractions of active ingredients from plants – A review. *J, Chromatography A*

Cocero M. J., García J. (2001) Mathematical model of supercritical extraction applied to oil seed extraction by CO_2+ saturated alcohol--I. Desorption model. *J. supercrit. fluids* 20 (3): 229--243.

Cravotto G., Boffa L., Mantegna S., Perego P., Avogadro M., Cintas P. (2008) Improved extraction of vegetable oils under high-intensity ultrasound and/or microwaves. *Ultrason sonochem* 15 (5): 898--902.

Cygnarowicz M. L., Seider W. D. (1989) Effect of retrograde solubility on the design optimization of supercritical extraction processes. *Ind. Eng. Chem. Res.* 28 (10): 1497--1503.

Dehghan-Shoar Z., Hardacre A. K., Meerdink G., Brennan C. S. (2011) Lycopene extraction from extruded products containing tomato skin. Int. J. Food Sci. Technol. 46: 365–371.

Diaz M. S., Espinosa S., Brignole E. A. (2003) Optimal solvent cycle design in supercritical fluid processes. *Lat Am Appl Res* 33 (2): 161--165.

Diaz S., Brignole E. A. (2009) Modeling and optimization of supercritical fluid processes. *J. Supercrit Fluids* 47: 611–618.

Diaz S., Espinosa S., Brignole E. A. (2005) Citrus peel oil deterpenation with supercritical fluids Optimal process and solvent cycle design. *J. Supercrit Fluids* 35: 49–61.

Diaz S., Gros H., Brignole E. A. (2000) Thermodynamic modeling, synthesis and optimization of extraction--dehydration processes. *Comp. Chem. Eng.* 24 (9-10): 2069--2080.

Doucha J. and Lívanský K. (2009) Outdoor open thin-layer microalgal photobioreactor: potential productivity.*J Appl Phycol 21:111–117.*

Du F.Y., Xiao X. H., Li G. K. (2007) Application of ionic liquids in the microwave-assisted extraction of trans-resveratrol from *Rhizma Polygoni Cuspidati. J Chromatography A* 1140 (1-2): 56--62.

Ducruet J. A., Dong Canal-Llauberes R. M., Glories Y. (1997) Influence des enzymes pectolytiques séléctionées pour l'oenologie sur la qualité et la composition des vins rouges, *Rev Franc Oenol* 155: 16–19.

Espinosa S., Diaz M. S., Brignole E. A. (2005) Process optimization for supercritical concentration of orange peel oil. *Lat Am Appl Res* 35 (4) 321--326.

Fernandes J., Ruivo R., Mota J. P. B., Simoes P. (2007) Non-isothermal dynamic model of a supercritical fluid extraction packed column. *J supercrit. fluids* 41 (1) 20--30.

Garcia-Salas P., Morales-Soto A., Segura-Carretero A. and Fernández-Gutiérrez A. (2010) Phenolic-Compound-Extraction Systems for Fruit and Vegetable Samples. *Molecules, 15: 8813-8826.*

Grzanna R., Polotsky A., Phan P.V., Pugh N., Pasco D., and Frondoza C.G.(2006) Immolina, a High–Molecular-Weight Polysaccharide Fraction of Spirulina, Enhances Chemokine Expression in Human Monocytic THP-1 Cells. *J.l of Alternative and Complementary Medicine. June 2006, 12(5): 429-435.*

Gupta R. B., Shim J., Solubility in Supercritical Carbon Dioxide, CRC Press, Boca Raton, FL, USA, 2007.

Hansen CM. 2007. Hansen solubility parameters: a user's handbook. 2nd ed. Boca Raton, Fla.: CRC Press.

Harrod M., Macher M.-B., Hogberg J., Moller P. (1997) Hydrogenation of lipids at supercritical conditions, in: Proceedings of the Fourth Italian Conference on Supercritical Fluids and their Application, Capri, Italy, September 7–10, 1997, pp. 319–326.

Harun R., Singh M ., Forde G.M. and Danquah M. K. (2010) Bioprocess engineering of microalgae to produce a variety of consumer products. *Renewable and Sustainable Energy Reviews 14:1037–1047.*

Herrero M., Cifuentes A., Ibáñez E. (2006) Sub- and supercritical fluid extraction of functional ingredients from different natural sources: Plants, food-by-products, algae and microalgae A review. *Food Chem* 98: 136–148.

Herrero M., Plaza M., Cifuentes A., Ibáñez E. (2010) Green processes for the extraction of bioactives from Rosemary: Chemical and functional characterization via ultra-performance liquid chromatography-tandem mass spectrometry and in-vitro assays. *J. Chromat A*, 1217: 2512–2520.

Hildebrand, J., Scott, R. L., *The Solubility of Nonelectrolytes*, 3rd Ed., Reinhold, New York, 1950.

Jain T., Jain V., Pandey R., Vyas A., Shukla S. S. (2009) Microwave assisted extraction for phytoconstituents – An overview. *Asian J. Research Chem.* 2 (1): 19-25.

Kaur, A., Singh, S., Singh, R. S., Schwarz, W. H., Puri, M. (2010) Hydrolysis of citrus peel naringin by recombinant α-L-rhamnosidase from Clostridium stercorarium. J. Chem. Technol. Biotechnol. 85: 1419–1422.

King J. W., Gabriel R. D., Wightman J. D. (2003) Subcritical water extraction of anthocyanins from fruit berry substrates. *Proceedings of the 6th Intl. Symposium on Supercritical Fluids – Tome 1*; April 28–30, 2003; Versailles, France.

King J. W., Howard L. R., Srinivas K., Ju Z. Y., Monrad J., Rice L. Super Green 2007. Pressurized liquid extraction and processing of natural products. *Proceedings of the 5th Intl. Symposium on Supercritical Fluids*; November 28–December 1, 2007; Seoul, South Korea.

King J. W., Srinivas K. (2009) Multiple unit processing using sub- and supercritical fluids. *J. Supercrit Fluids* 47: 598–610.

Kubátová A., Jansen B., Vaudoisot J. F., Hawthorne S. B. (2002) Thermodynamic and kinetic models for the extraction of essential oil from savory and polycyclic aromatic hydrocarbons from soil with hot (subcritical) water and supercritical CO_2. *J. Chromat A* 975: 175–188.

Lavecchia R, Zuorro A. (2008) Improved lycopene extraction from tomato peels using cell-wall degrading enzymes. Eur Food Res Technol 228:153–8.

Lee Y. K. (1997) Commercial production of microalgae in the Asia-Pacific rim. *J Appl Phycol* 9: 403-4511.

Létisse M., Rozières M., Hiol A., Sergent M., Comeau L. (2006) Enrichment of EPA and DHA from sardine by supercritical fluid extraction without organic modifier: I. Optimization of extraction conditions. *J. supercrit. fluids* 38 (1): 27--36.

Lew A., Krutzik P. O., Hart M. E., Chamberlin A. R. (2002) Increasing Rates of Reaction: Microwave-Assisted Organic Synthesis for Combinatorial Chemistry. *J. Comb. Chem.*, 4 (2): 95–105.

Lu Y., Yue X.-F., Zhang Z.-Q., Li X.-X., Wang K. (2007) Analysis of *Rodgersia aesculifolia* Batal. Rhizomes by Microwave-Assisted Solvent Extraction and GC–MS. *Chromatographia* 66 (5-6): 443-446.

Lucchesi M. E., Chemat F., Smadja J. (2004) Solvent-free microwave extraction of essential oil from aromatic herbs: comparison with conventional hydro-distillation. *Journal of Chromatography A* 1043: 323–327.

Mandal V., Mandal S. C. (2010) Design and performance evaluation of a microwave based low carbon yielding extraction technique for naturally occurring bioactive triterpenoid: Oleanolic acid. *Biochem Eng J* 50 (1-2): 63--70.

Mandal V., Mohan Y., Hemalatha S. (2007) Microwave Assisted Extraction – An Innovative and Promising Extraction Tool for Medicinal Plant Research. Pharm Rev 1 (1, Jan-May): 7-18.

Masojidek J, Prasil O. (2010) The development of microalgal biotechnology in the Czech Republic. *J Ind Microbiol Biotechnol* 37:1307–1317.

Mercer P., Armenta E. E. (2011) Developments in oil extraction from microalgae. *Eur. J. Lipid Sci. Technol.* 24 (30-31): 539–547

Miller D. J., Hawthorne, S. B. (2000) Solubility of liquid organic flavor and fragance compounds in subcritical (hot/liquid) water from 298 to 473 K. *J Chem Eng Data* 45: 315–318.

Monrad J. K., Howard L. K., King J. W., Drinivas K., Mauromoustakos A. (2010) Subcritical Solvent Extraction of Anthocyanins from Dried Red Grape Pomace. *J. Agric. Food Chem.* 58: 2862–2868.

Pasquet V., Chérouvrier J. R., Farhat F., Thiéry V., Piot J. M., Bérard J. B., Kaas R., Serive B., Patrice T., Cadoret J. P., Picot L. (2011) Study on the microalgal pigments extraction process: Performance of microwave assisted extraction. *Proc Biochem* 46: 59–67.

Patist A., Bates D. Ultrasonic innovations in the food industry: From the laboratory to commercial production. *Inn Food Sci Emerg Technol* 9: 147 – 154.

Pereira C.G., Meireles M.A.A. (2007) Economic analysis of rosemary, fennel and anise essential oils obtained by supercritical fluid extraction. *Flavour Frag. J. 22 (5): 407-413.*

Plaza M., Amigo-Benavent M., del Castillo M. D., Ibáñez E., Herrero M. (2010) Facts about the formation of new antioxidants in natural samples after subcritical water extraction. *Food Res Int* 43: 2341–2348.

Plaza M., Herrero M., Cifuentes A., Ibáñez, E. (2009) Innovative Natural Functional Ingredients from Microalgae. *J. Agric. Food Chem.* 57: 7159–7170.

Pulz O. (2001) Photobioreactors: production systems for phototrophic microorganisms. *Appl Microbiol Biotechnol 57: 287–293.*

Puri M., Sharma D., Barrow C. J. (in press) Enzyme-assisted extraction of bioactives from plants. *Trends Biotech.*

Ryll J., Scheper T., Lotz M. (2003) Biotechnological production of β-1,3-Glucan in a technical pilot plant. Abstr. Eur. Workshop Microalgal Biotechnol., Germany, p. 56, 2003).Sili C., Torzillo G., Vonshak A. (2010) *Arthrospira.* In: Ecology of cyanobacteria. Whitton BA, Pott M (eds), Kluwer Academia Publishers. Dordrect/ London/ Boston (in press).

Sovová H. (2005) Mathematical model for supercritical fluid extraction of natural products and extraction curve evaluation. *J. supercrit. fluids* 33 (1): 35--52.

Sowbhagya H. B., Chitra V. N. (2010) Enzyme-Assisted Extraction of Flavorings and Colorants from Plant Materials. Crit Rev Food Sci Nutr 50 (2): 146-161.

Sowbhagya H. B., Chitra V. N. (2010) Enzyme-Assisted Extraction of Flavorings and Colorants from Plant Materials. Crit Rev Food Sci Nutr 50:146–161.

Srinivas K., King J. W., Monrad J. K., Howard L. R., Hansen C. M. (2009) Optimization of Subcritical Fluid Extraction of Bioactive Compounds Using Hansen Solubility Parameters. J Food Sci 74 (6): 342--354.

Teo C. C., Tan S. N., Yong J. W. H., Hew C. S., Ong E.S. (2010) Pressurized hot water extraction (PHWE). *J. Chromatography A.* 1217 (16): 2484--2494.

Torzillo G., Pushparaj B., Masojidek J. and A. Vonshak A. (2003) Biological constraints in algal biotechnology. *Biotechnol. Bioprocess Eng. 8: 339–348.*

Torzillo G., Vonshak A. (2003) Biotechnology for algal mass cultivation. In: Recent advances ion marine biotechnology (Fingerman M., Nagabhushanam R. eds) Volume 9 Biomaterials and Bioprocessing, Science Publishers, Inc. Enfield (NH), USA, pp. 45-77.

Ugarte-Romero E., Feng H., Martin E., Cadwallader R., Robinson J. (2006) Inactivation of *Echerichia coli* in apple cider with power ultrasound. *J Food Sci* 71: 102–109.

Vilkhu K., Mawson R., Simons L., Bates, D. (2008) Applications and opportunities for ultrasound assisted extraction in the food industry – A review. *Inn Food Sci Emerg Technol* 9: 161–169.

Vinatoru M. (2001) An overview of the ultrasonically assisted extraction of bioactive principles from herbs. *Ultrasonics Sonochemistry* 8: 303–313.

Wang Z., Ding L., Li T., Zhou X., Wang L., Zhang H., Liu L., Li Y., Liu Z., Wang, H., Zeng H., He H. (2006) Improved solvent-free microwave extraction of essential oil from dried *Cuminum cyminum* L. and *Zanthoxylum bungeanum* Maxim. *J. Chromatography A* 1102 (1-2): 11--17.

Windal I., Miller D. J., De Pauw E., Hawthorne S. B. (2000) Supercritical Fluid Extraction and Accelerated Solvent Extraction of Dioxins from High- and Low-Carbon Fly Ash. Anal. Chem. 72: 3916-3921.

Zuorro A., Fidaleo M., Lavecchia R. (2011) Enzyme-assisted extraction of lycopene from tomato processing waste. *Enz. Microb. Technol.* 49 (6-7): 567-573.